Embedded System Design with Arm Cortex-M Microcontrollers

Cem Ünsalan • Hüseyin Deniz Gürhan
Mehmet Erkin Yücel

Embedded System Design with Arm Cortex-M Microcontrollers

Applications with C, C++ and MicroPython

Cem Ünsalan
Marmara University
Istanbul, Turkey

Mehmet Erkin Yücel
Migros Ticaret AS
Istanbul, Turkey

Hüseyin Deniz Gürhan
Cizgi Teknoloji AS
Istanbul, Turkey

ISBN 978-3-030-88441-3   ISBN 978-3-030-88439-0   (eBook)
https://doi.org/10.1007/978-3-030-88439-0

This Springer imprint is published by the registered company Springer Nature Switzerland AG
The registered company address is: Gewerbestrasse 11, 6330 Cham, Switzerland

Contents

Introduction

1

1.1 Embedded Systems

The definition "embedded system" covers a broad range of devices used in almost all parts of our lives. Therefore, they have become indispensable tools in today's world. Moreover, as the devices we use become more intelligent, they will consist of one or many embedded systems in them. Since there are various embedded systems used in operation, it may be unfair to give a strict and limiting description for such a broad range of devices. However, we can make some general definitions as follows.

We can think of an embedded system as a computing device developed for solving a specific problem. To do so, it interacts with the environment as acquiring data, processes the acquired data, and produces the corresponding output accordingly. To perform all these operations, joint usage of hardware and software becomes mandatory. Hence, the designer developing the embedded system must know how the hardware works and how the dedicated software should be formed for it. More importantly, the designer should grasp the idea of jointly using hardware and software to get the best from the embedded system.

The embedded system will be working in stand-alone form most of the times. It may communicate with other nearby devices as with Internet of Things (IoT) applications. However, this does not mean that there is a server and all embedded systems connect to it to perform their operations. On the contrary, each embedded system can work in stand-alone form and share data to perform a complex operation.

Since the embedded system works in stand-alone form most of the times, it depends on battery or energy harvesting module to operate. Therefore, energy

---

Supplementary Information The online version contains supplementary material available at (https://doi.org/10.1007/978-3-030-88439-0_1).

C. Ünsalan et al., *Embedded System Design with Arm Cortex-M Microcontrollers*,
https://doi.org/10.1007/978-3-030-88439-0_1

dissipation becomes one of the main concerns in embedded system development. To note here, some embedded systems may be supplied by the main power line. However, the recent trend is forming a system such that it works independently. This leads to its deployment to remote locations where no main power line is available.

From our perspective, we can group embedded systems based on their hardware properties as follows. Field-programmable gate arrays (FPGA) form the first group. They provide the most flexible but hard to master hardware. Due to their properties, they have been extensively used in forming custom embedded systems at gate level. Therefore, we do not program an FPGA. Instead, we describe the system to be constructed by a hardware description language such as Verilog and VHDL.

The second hardware group in forming embedded systems consists of microcontrollers. A microcontroller has limited memory and computation power. However, it can be programmed in assembly or high-level language to perform the operations on it. The microcontroller also offers a cheap and energy efficient solution for embedded system formation. The most well-known microcontroller systems can be counted as Arduino and Arm® Cortex™-M based ones.

The microprocessors form the third embedded system group. In these, embedded Linux is used most of the times to control and organize operations. This also leads to graphical user interface (GUI) usage on them. Microprocessors also have fairly high memory and computation power compared to microcontrollers. However, energy dissipation and stand-alone usage are the main bottlenecks for them. The most well-known microprocessor-based embedded systems are the Raspberry Pi family and its clones.

The fourth hardware group for embedded systems consists of system on chip (SoC) devices. They have FPGA and microprocessor modules on them. Hence, they aim to benefit from both device properties. However, SoC programming and usage are still not as easy as a microcontroller (or microprocessor).

The mentioned four hardware groups have their dedicated development boards. Hence, they can be used easily in developing a prototype embedded system. They use cross compilers such that the code is written and debugged on PC. Then, it is embedded on the system. To note here, there is no GUI on an embedded system by default except embedded Linux-based boards.

Recently, new hardware options have emerged for embedded systems. One option is the joint usage of a microcontroller and microprocessor on the same dye. STMicroelectronics recently offered such devices containing both Arm® Cortex™-A and Cortex™-M CPUs. Hence, each can be used for specific applications. The next option is the development boards consisting of graphical processing units (GPU). These devices allow parallel processing via high-level programming languages. The recent advances in deep learning and neural networks also led to devices consisting of neural processing units (NPU or TPU) dedicated to neural network implementation. Such units are becoming part of almost all embedded systems nowadays.

Based on this brief summary, we expect two main trends for embedded system hardware in the future. The first trend is that we will see hardware components for embedded system implementation to become cheaper, be more powerful, consume less energy, and be easy to use. The second trend will be merging such devices. Hence, we expect to have an embedded system consisting of a microcontroller, microprocessor, GPU, and FPGA modules. They will work together to solve the problem at hand.

1.2 Microcontroller as Embedded System

As emphasized in the previous section, there are several hardware options which can be used in forming an embedded system. Although each group has its advantages, we picked the microcontroller as the embedded system in this book for the following reasons. Microcontrollers are simpler to program. They have a fairly wide usage area. More importantly, they can be handled by a wide range of audience. In fact, the most well-known example of this usage is the Arduino platform. The reader may ask the question "Then, why is the Arduino platform not being used in this book?" We deliberately used the STM32F429ZIT6 microcontroller based on the Arm® Cortex™-M architecture such that we can explain the working principles of microcontrollers in detail. Besides, we will introduce the Mbed platform which works similar to the Arduino platform. One final note, most of the concepts to be explained on microcontrollers in this book can be applied to other embedded system hardware groups with minor modification. Therefore, the reader can expand the usage area of these concepts. However, they should first be mastered on microcontrollers as explained in this book.

Each microcontroller family has its own programming language called assembly. The assembly language allows the programmer to reach and modify the micro-controller hardware at its lowest level. Although this is a fairly powerful tool, there are some shortcomings as well. First and most important of all, an assembly code written for a microcontroller family cannot be ported to a microcontroller from another family. Besides, writing and debugging code in assembly language is tedious. Hence, C became the de-facto programming language for microcontrollers. It offers flexibility in portability (to some degree). Besides, recent integrated development environments allow the user to control and debug the code running on the microcontroller fairly easily. As explained in the previous paragraph, the Arduino platform offers a unique option such that the code written there can be implemented on any microcontroller supporting the platform. Likewise, Arm® introduced the Mbed development environment for its microcontrollers. This platform depends on C++ language to operate. Hence, C++ has also become the preferred language for microcontrollers. Finally, MicroPython has emerged as the implementation of Python language on selected microcontrollers. It offers a unique property such that a software programmer fluent in Python programming on PC can start programming the microcontroller with the same syntax and rules. Besides, MicroPython also provides a unified approach among microcontrollers from different vendors. As an example, the microcontrollers in the Arduino and Mbed platforms can be programmed in the same way in MicroPython (as long as they support this option).

To note here, it is not always possible to use all microcontroller hardware directly in MicroPython. Therefore, the reader should take this into account while using it.

In this book, we will use C, C++, and MicroPython languages to program our microcontroller. While doing so, we will handle the hardware-based topics by the help of C language. Then, we will generalize these topics in C++ language. Finally, we will provide the MicroPython version of the code (if possible). Hence, a programmer expert in C language and a causal user can benefit from the book.

The STM32F429ZIT6 microcontroller to be used in the book cannot be used alone unless a dedicated development or specific board is constructed for it. The best option for us, as early learners, is using the available 32F429IDISCOVERY kit for this purpose. Therefore, we will introduce the embedded system properties using both the microcontroller and the kit.

We should also explain why the Arm® Cortex™-M architecture-based STM32F429ZIT6 microcontroller is picked in this book. Arm® Cortex™-M architecture has gained dominance among microcontrollers. The main reason is that Arm® forms the IP and companies use it with their custom peripheral units to form a physical microcontroller. Hence, a large ecosystem has been formed. Besides, Arm® Cortex™-M architecture offers advanced properties in low power usage. As explained before, this is extremely important for stand-alone embedded system implementation. Related to this, Arm® declared that its partners have shipped more than 180 billion Arm®-based chips [1]. Arm® also predicts that a trillion new IoT devices will be produced between now and 2035 [2]. Although this is a prediction, it shows the potential of growth in the embedded system market. Besides, it indicates that we will be surrounded by more and more embedded systems in the near future. Hence, today's students (from any discipline), new graduates, and experts should grasp the ideas in this book. Hence, they can get prepared to the opportunities in the job market as for today and the future.

1.3   About the Book

The aim of this book is teaching the embedded system concepts through micro-controller usage. Therefore, we follow the method of learning by doing as in all our previous books on this topic. We also picked the thematic application "robot vacuum cleaner" throughout the book. We explained components of this robot (both in hardware and software) as the end of chapter applications. Hence, the reader can see the overall layout of a complex embedded system formed step by step by its subparts.

We can list the topics to be covered in this book as follows. We will make an introduction by handling the microcontroller architecture in Chap. 2. Here, we will overview the CPU, memory, input and output ports, timer, ADC, DAC, and digital communication modules of the STM32F429ZIT6 microcontroller. We will briefly introduce the Arm® assembly language programming in this chapter as well. Finally, we will provide detailed information on the 32F429IDISCOVERY kit to be

used throughout the book. In other words, the second chapter summarizes properties
of the microcontroller to be used throughout the book.

We will evaluate the development environments to program the microcontroller
in Chap. 3. To do so, we will introduce the STM32CubeIDE, Mbed, Mbed Studio,
and MicroPython programming environments. STM32CubeIDE can be used to program the microcontroller via C and assembly languages. These are valuable options.
However, programming the microcontroller this way requires expertise. Instead, the
Mbed platform simplifies life for the programmer. Therefore, we will introduce
the web-based Mbed and its desktop version Mbed Studio next. The drawback
of these options is the limitation on reaching low-level hardware properties of
the microcontroller. Recently introduced MicroPython, as the implementation of
Python language for microcontrollers, also offers a good option to program the
microcontroller. Hence, we will cover it in Chap. 3.

We will introduce methods to interact the microcontroller with the outside world
in Chap. 4. Here, we will focus on digital input and output extensively. In Chap. 5,
we will focus on the interrupt concept which is extremely important in embedded
systems. Therefore, we will consider the occurrence of interrupts as well as ways to
handle them. We will also cover low power modes of the microcontroller in Chap. 5.
These will help us to run the microcontroller with minimum power dissipation
during its operation. Then, we will consider time-based operations in Chap. 6 in
connection with the timer, real-time clock, and watchdog timer modules. These
concepts are also extremely important in real-life applications when time-based
operations are needed. In Chap. 7, we will consider analog to digital conversion
(ADC) and digital to analog conversion (DAC). To do so, we will start with the
properties of analog and digital signals. Then, we will focus on the ADC and DAC
modules on the microcontroller. In Chap. 8, we will cover digital communication
concepts in embedded systems. To do so, we will consider the UART, SPI, I^2C,
CAN, USB, and other communication modules in the microcontroller. Next, we
will explore memory-based operations in Chap. 9. Through these we will evaluate
memory management in C, C++, and MicroPython languages. Besides, we will
introduce the direct memory access (DMA) and flexible memory controller (FMC)
modules in the microcontroller.

We will introduce the real-time operating system (RTOS) concepts to the reader
via FreeRTOS and Mbed OS in Chap. 10. They will lead to constructing complex
projects in an efficient manner. We will cover LCD usage and graphical user
interface (GUI) formation in Chap. 11. Hence, the reader can form a professional
GUI to interact with the microcontroller. We will introduce how basic digital signal
processing (DSP) applications can be done on the microcontroller in Chap. 12.
Likewise, we will consider the fundamental digital control (DC) and digital image
processing (DIP) concepts from the microcontroller perspective in Chaps. 13 and 14,
respectively. They can lead to more advanced embedded system applications such as
speech processing, computer vision, and machine learning. Finally, we will handle
advanced topics in Chap. 15.

Throughout the book, we target general audience both from engineering and
nonengineering disciplines and explain the topics accordingly. Hence, we tried to

avoid detailed hardware explanation from the electrical engineering perspective whenever possible. Therefore, a reader from nonengineering background can grasp the idea behind the embedded system topic. As for the software side, we assumed the reader has sufficient knowledge on at least one of C, C++, or Python languages. To note here, we will improve programming skills of the reader on these languages throughout the book. Here, our main advantage is implementing the code on the microcontroller and observing how it behaves step by step by the available debugging tools. Hence, we will make sure that the reader will gain in-depth knowledge on the general working principles of embedded systems (more specifically microcontrollers) both from hardware and software perspectives.

Sample codes in this book are available in the companion web site for the reader. Related to this, all STM32CubeIDE and Mbed Studio projects for the sample codes and end of chapter applications are available in the same web site. Course slides and figures used in the book are also available in the same web site for the reader and instructor. The solution manual is also available in the companion web site for the instructor only.

References

1. Arm. https://www.arm.com/company/news/2021/02/arm-ecosystem-ships-record-6-billion-arm-based-chips-in-a-single-quarter. Accessed 4 June 2021
2. Sparks, P.: The route to a trillion devices. The outlook for IoT investment to 2035. Arm (2017)

Microcontroller Architecture

2

2.1 The STM32F4 Microcontroller

The STM32F429ZIT6 microcontroller to be used throughout the book is based on the Arm® Cortex™-M4 architecture. For simplicity, we will call the microcontroller as STM32F4 from this point on. Functional block diagram of this microcontroller is as in Fig. 2.1. We will benefit from this representation in grouping hardware and peripheral units of the microcontroller as CPU, memory, general-purpose input and output ports, clock and timer modules, analog modules, digital communication modules, and other modules. We will introduce each unit next.

2.1.1 Central Processing Unit

Central processing unit (CPU) is responsible for organizing all operations within the microcontroller. This is done by the instructions provided to it by the programmer. To be more specific, the programmer constructs his or her algorithm and forms the corresponding C, C++, or assembly code. This code is debugged and embedded into flash memory of the microcontroller by the help of an integrated development environment (IDE, such as STM32CubeIDE) running on PC. We will explain the usage of STM32CubeIDE in detail in Chap. 3. The CPU executes the commands by using its resources (such as peripheral units). We will briefly explain this execution operation in Sect. 2.2.2. To note here, executing Python codes on the microcontroller is different. Therefore, we will handle it in detail in Sect. 3.3.

Supplementary Information The online version contains supplementary material available at (https://doi.org/10.1007/978-3-030-88439-0_2).

© Springer Nature Switzerland AG 2022
C. Ünsalan et al., *Embedded System Design with Arm Cortex-M Microcontrollers*,
https://doi.org/10.1007/978-3-030-88439-0_2

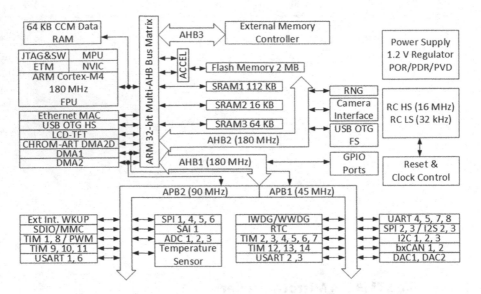

Fig. 2.1 Functional block diagram of the STM32F4 microcontroller

We provide block diagram of the CPU for the STM32F4 microcontroller in Fig. 2.2. This diagram gives insight about the inner working principles of the CPU as will be explained next. Before going further, we should mention that wire connections and their direction are not random in this figure. They indicate the direction of dataflow from and to the CPU.

We should take a close look at the blocks in Fig. 2.2. As mentioned before, the CPU of the STM32F4 microcontroller is based on the Arm® Cortex™-M4 architecture. Hence, the processor core (as indicated in the figure) has an accompanying floating-point unit (FPU). This unit is responsible for handling operations on floating-point numbers in hardware. Hence, they can be executed in a fast manner.

2.1.1.1 Nested Vectored Interrupt Controller
The nested vectored interrupt controller (NVIC), given in Fig. 2.2, is responsible for managing interrupt- and exception-based operations within the CPU. We will talk about what an interrupt is and how it is handled by NVIC in detail in Chap. 5.

2.1.1.2 Debug Access Port and Serial Wire Viewer
The debug access port (DAP) in Fig. 2.2 helps us to reach and monitor the CPU during operation. Hence, we can observe what is going on inside the CPU while executing code. In other words, DAP allows us to monitor CPU registers (to be explained in Sect. 2.1.1.6) without affecting the CPU operation.

The other option to perform the same operation is leading the CPU to the debug state. Here, the CPU operation is halted. Then, its registers can be monitored. We will talk about the debugging operation through STM32CubeIDE in Chap. 3. There

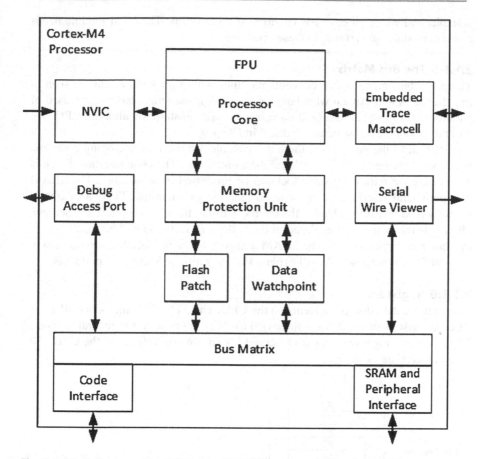

Fig. 2.2 Block diagram of the CPU for the STM32F4 microcontroller [5]

is also the serial wire viewer (SWV) performing a similar operation. Through it, the user can trace the program counter, variable and peripheral register values, exceptions, time stamp, and CPU cycles.

2.1.1.3 Memory Management Units
Data and codes are stored in memory as will be explained in Sect. 2.1.2. Memory protection unit, flash patch, and data watchpoints modules within the CPU are used to manage memory. We will explore usage of these modules in detail in Chap. 9.

2.1.1.4 Embedded Trace Macrocell Module
The embedded trace macrocell (ETM) module allows monitoring the code execution within the CPU (without stopping it). This is done by the GPIO pins. Therefore, the user can check when the CPU enters an interrupt, which task uses a common

variable, and where an error has occurred if there is any. The ETM module needs external modules to perform all these operations.

2.1.1.5 The Bus Matrix

There is a bus matrix which connects modules within the CPU as can be seen in Fig. 2.2. Let's first explain what bus means. In simplest terms, bus is composed of parallel wires in which data (and code blocks) are transferred within the CPU. We will talk about the bus structure in detail in Chap. 9.

Adjacent to the bus matrix, there are two blocks: the first one being code and the second one being SRAM and peripheral interface. The code interface block is responsible for retrieving code blocks to be processed from memory. The SRAM and peripheral interface block is important for two reasons. First, an external memory module can be added to the microcontroller through this interface. Second, all peripheral units (to be explored in the following sections) will be reached as if they are memory addresses. The SRAM and peripheral interface block module will be used for this purpose. We will explain this operation in Sect. 2.1.2 in detail.

2.1.1.6 Registers

Registers are data storage elements on the CPU. The STM32F4 microcontroller has 21 core registers in which 13 of them (R0 to R12) can be used for general purpose. The remaining eight registers are reserved for specific operations by the CPU. All core registers are as in Fig. 2.3.

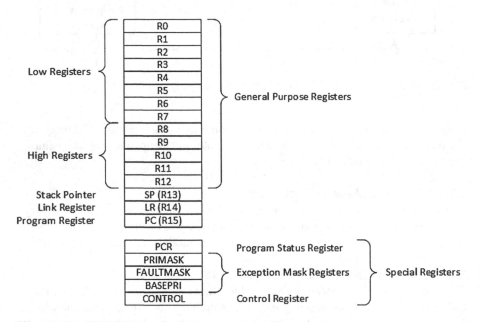

Fig. 2.3 The STM32F4 microcontroller core registers [5]

General-purpose registers (R0–R12) can be used for data processing operations. These are divided into two groups as low register (R0–R7) and high register (R8–R12) depending on their usage by 16- or 32-bit instructions [1].

The remaining eight registers are used for specific purposes. Among these, the program counter (R15, PC) is the register responsible for holding address of the instruction to be executed. Related to this, there is a link register (R14, LR) which holds return address during branching and exception operations such as interrupts. Stack pointer (R13, SP) is responsible for storing the location of local variables and is used when a function is called in code.

When an instruction is executed within the CPU, the output is summarized as being negative, zero, or an overflow, carry, or saturation has occurred. These results are stored as condition flags (each being one bit) within the program status register (PSR). Therefore, the programmer gets insight about the result of the operation. The PSR is one of the special registers as in Fig. 2.3.

We should take a closer look at the condition flags as negative (N), zero (Z), overflow (V), carry (C), and saturation (Q). The negative (N) flag is set to logic level 1 when the result of an operation is negative. It is set to logic level 0 otherwise. The zero (Z) flag is set to logic level 1 when the result of an operation is zero. It is set to logic level 0 otherwise. Overflow is a common problem in microcontrollers since numbers can be represented with finite number of bits in memory. The overflow (V) flag is set to logic level 1 when the result of an operation causes an overflow. It is set to logic level 0 otherwise. A carry bit is generated when the result of the addition operation is larger than the maximum number which can be represented at the destination register. Carry bit is also generated when subtracting a larger number from a smaller one. This is due to executing the subtraction operation as addition via two's complement representation in the microcontroller [9]. Finally, shift operations may generate carry when the shifted bit is logic level 1. The carry (C) flag is set to logic level 1 when the result of an operation generates a carry. It is set to logic level 0 otherwise. Sometimes numbers are represented between a specific interval. If they exceed this range, then they are saturated to maximum or minimum value of the interval. If the result of the saturated arithmetic is different from the unsaturated one, it means the operation causes a saturation. The saturation (Q) flag is set to logic level 1 when the result of an operation causes saturation. It is set to logic level 0 otherwise.

There are also four more special registers as can be seen in Fig. 2.3. These are PRIMASK, FAULTMASK, BASEPRI, and CONTROL. For more information on these registers, please see [1].

2.1.2 Memory

Data to be processed and code to be executed is stored in memory of the microcontroller. Although we will consider memory-based operations in Chap. 9 in detail, we provide introductory concepts here. Hence, the reader can grasp general characteristics of the memory and its usage.

2.1.2.1 Memory Types

There are two memory types in microcontrollers. There is random access memory (RAM) to store temporary data. Contents of RAM are lost when the power fed to it is turned off. There is also read-only memory (ROM) to store code or data permanently. Even if the power fed to it is turned off, contents of ROM are kept.

Before going further, we should explain roots of memory type naming convention. The reader may be confused with the name RAM. In fact, it tells us that any randomly picked region within RAM can be accessed (reached) at the same time duration. Likewise, ROM may be misleading such that we only read its content. So, when and how was code or data written to it? Well, sometime in the operation, we should have written data on ROM. In earlier times, there were dedicated devices just for this purpose. Nowadays, we have flash memory which works as ROM. Flash has characteristics such that its contents can easily be modified. At the same time, the content can be stored in flash for a sufficiently long time. Moreover, the distinction between RAM and flash is becoming fuzzy as technology is advancing. Hence, the same memory region can be used for both as RAM and flash as in the FRAM technology [3].

The STM32F4 microcontroller has two memory types as static RAM (SRAM) and flash. SRAM is a special memory type which does not need periodic refreshing as RAM. It also provides faster access to data. The STM32F4 microcontroller has 256 kB SRAM. The first 64 kB of SRAM is called core coupled memory (CCR) which provides fast code execution. For more information on this issue, please see [7]. There is also an extra 4 kB backup SRAM within the microcontroller which keeps data written to it even if the microcontroller goes into standby mode.

The STM32F4 microcontroller has 2 MB flash memory. This memory is divided into two banks and each being composed of 12 sectors. These sectors are formed by 16 kB, 64 kB, or 128 kB memory regions. This structure allows the programmer to erase a specific sector, bank, or the overall flash memory. We will provide more information on flash in Chap. 9.

2.1.2.2 Memory Map

Arm® partitioned the memory address range for its microcontroller cores based on usage. This partitioning is called memory map of the microcontroller and is given in Fig. 2.4. Since the STM32F4 microcontroller is based on the Arm® Cortex™-M4 architecture, it has the same memory map.

There is an extremely important concept to be mentioned here. The memory map of the STM32F4 microcontroller does not only represent SRAM and flash memory addresses. It also represents the interrupt and reset vector table, on-chip peripheral units, off-chip memory, off-chip peripherals, system components, and on- and off-chip debug components. Let's explain these concepts along with the memory map structure.

As can be seen in Fig. 2.4, the first 512 MB memory space is for code to be executed on the microcontroller. The next 512 MB memory space is used for on-chip SRAM. Code to be executed can also be stored here. However, this creates extra wait states and longer execution times. Therefore, running the code from this region is not recommended by Arm®. The next 512 MB memory space is used for

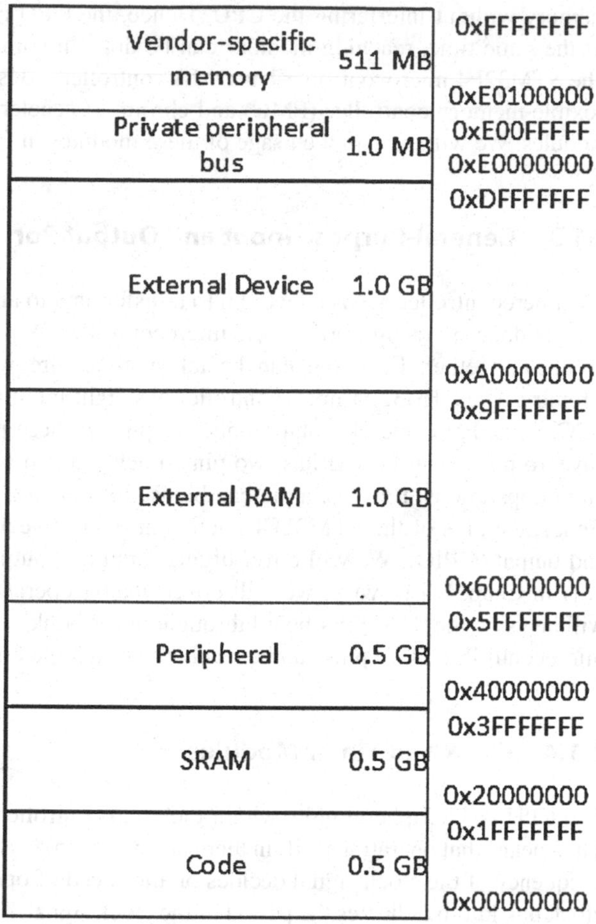

Fig. 2.4 The STM32F4 microcontroller memory map [5]

Vendor-specific memory	511 MB	0xFFFFFFFF — 0xE0100000
Private peripheral bus	1.0 MB	0xE00FFFFF — 0xE0000000
External Device	1.0 GB	0xDFFFFFFF — 0xA0000000
External RAM	1.0 GB	0x9FFFFFFF — 0x60000000
Peripheral	0.5 GB	0x5FFFFFFF — 0x40000000
SRAM	0.5 GB	0x3FFFFFFF — 0x20000000
Code	0.5 GB	0x1FFFFFFF — 0x00000000

all on-chip peripheral unit (GPIO, timers, ADC, and DAC) registers. All peripheral units are also treated as if they are separate memory locations. Hence, if the CPU tries to reach a peripheral unit, this is done in the same way as reaching a memory location. Although the user does not see this operation while writing code in C or C++ languages, it is important to keep this in mind. The next 2 GB memory space is used for external memory and peripheral units. The final 512 MB memory space is used for system peripheral units such as NVIC, memory protection unit (MPU), and SysTick (specific timer to be explored in Chap. 6). This memory space is also used for internal-external debug and vendor specific components.

2.1.2.3 Memory Related Modules
There are also memory related modules within the STM32F4 microcontroller. The first one is the direct memory access (DMA) module. We can explain the importance of this module as follows. The CPU is used in most memory reaching operations. However, it may be busy for some cases. The DMA module allows reaching

memory without interfering the CPU. Hence, the CPU can work on its own task. At the same time, reaching memory can be done in parallel by the DMA module. The STM32F4 microcontroller has DMA controllers. Besides these, STM32F4 has flexible memory controller (FMC) and chrom-art accelerator controller (DMA2D) modules. We will explain the usage of these modules in Chap. 9.

2.1.3 General-Purpose Input and Output Ports

The microcontroller needs a medium to transfer data to and from the outside world. This is done by using ports of the microcontroller. A port is a collection of pins grouped together. Each pin can be taken as a wire with its electronic control circuitry. The STM32F4 microcontroller has eight input and output ports called A, B, C, D, E, F, G, and H. Total number of pins in these ports is 114. Ports A to G have 16 pins each. Port H has two pins. Each pin can be used for both input and output operations. Besides, a pin can be used for analog and digital voltage levels. Hence, the pins of the STM32F4 microcontroller are called general-purpose input and output (GPIO). We will cover digital input and output operations using GPIO pins in Chap. 4. Likewise, we will cover analog operations in Chap. 7. Since we will be using the STM32F4 board throughout the book, we also summarize how the microcontroller GPIO pins can be reached through the board in Sect. 2.3.2.

2.1.4 Clock and Timer Modules

The CPU and peripheral units within the microcontroller are synchronous devices. This means that operations within them are synchronized by a common clock signal. Frequency of the clock signal decides on the speed of operation in actual time. The clock tick idiom indicates the period of the clock signal. The time needed to perform an operation in the microcontroller is indicated by clock pulses (or clock ticks). Therefore, the higher the clock frequency, the faster that operation is performed in actual time.

Clock signals are generated by oscillator circuitry which can be located inside or outside the microcontroller. The STM32F4 microcontroller has two internal oscillators formed by resistive capacitive (RC) circuitry. These are the 16 MHz high-speed internal (HSI) and 32 kHz low-speed internal (LSI) oscillator. Besides, the microcontroller has two oscillators formed by external crystal circuitry. These are the 4–26 MHz high-speed external (HSE) oscillator and 32768 Hz low-speed external (LSE) oscillator.

The maximum clock frequency that can be reached by the STM32F4 microcontroller is 180 MHz. Therefore, an operation requiring one clock cycle in the CPU actually requires $1/180 \times 10^{-6}$ s in actual time. Due to complexity of the microcontroller architecture and available peripheral units within the microcontroller, the clock source for different peripherals may not be the same. Therefore, the STM32F4 microcontroller has more than one internal and external clock source.

A clock controller module is used to manage all clocks and clock sources for this purpose.

The clock signal is used in timer modules as well. The timer module can be basically taken as a counter. It is responsible for keeping track of time elapsed between certain operations. The STM32F4 microcontroller has timer, system timer (SysTick), watchdog timer, and real-time clock (RTC) modules for time related operations. Clock and timer modules will be covered in detail in Chap. 6.

2.1.5 Analog Modules

The STM32F4 microcontroller has analog to digital converter (ADC) and digital to analog converter (DAC) as analog modules. Input and output from these modules are obtained from GPIO pins of the microcontroller. The ADC module generates digital representation of an analog voltage level fed to the microcontroller. The STM32F4 microcontroller has three such 12-bit ADC modules. The DAC module converts a digital representation to analog voltage level and feeds it to output. The STM32F4 microcontroller has a 12-bit DAC module with two output channels. The usage of ADC and DAC modules will be covered in detail in Chap. 7.

2.1.6 Digital Communication Modules

Digital communication modules are used to communicate with external devices using dedicated communication protocols. Although we will cover these modules in detail in Chap. 8, let's briefly summarize them. The STM32F4 microcontroller has three inter-integrated circuit (I^2C) modules; four universal synchronous/asynchronous receiver/transmitter (USART) modules; four universal asynchronous receiver/transmitter (UART) modules; six serial peripheral interface (SPI) modules; two controller area network (CAN) modules; one universal serial bus on-the-go full-speed (USB OTG_FS) module; one universal serial bus on-the-go high-speed (USB OTG_HS) module; and one Ethernet media access control (MAC) module with DMA controller. These modules will be extensively used when communicating with external devices or other microcontrollers.

2.1.7 Other Modules

The STM32F4 microcontroller has additional special purpose modules as well. These are the digital camera interface (DCMI) for receiving data from an external camera module; serial audio interface (SAI) for controlling audio peripheral units using different audio protocols; LCD-TFT controller (LTDC) for driving LCD and TFT panels; flexible memory controller (FMC) for interfacing with static memory, SDRAM memory, and 16-bit memory cards; secure digital input/output interface

(SDIO) for controlling multimedia, SD memory, and SDIO cards; and two inter-integrated sound (I2S) interfaces for audio data.

2.2   Assembly Language

When we write code in C or C++ language, in fact we are not writing in the format the microcontroller understands. The compiler (to be explored in Chap. 3) converts our code to machine language and embeds it to code region of the microcontroller memory. Hence, it can be executed by the CPU. Since the code in machine language is stored in memory, it is in binary form. Therefore, understanding it is neither trivial nor feasible for us.

Assembly language offers an option to write the code as close as possible to machine language but at the same time understandable by a human when inspected. In other words, we can write the code in assembly language in an editor (to be explored in Chap. 15) which can be understood when inspected. Besides, assembly language allows us to reach low-level hardware of the microcontroller. Therefore, using it provides a good insight in understanding inner working principles of the microcontroller. Hence, we cover assembly language programming concepts in this section. As a note, executing Python commands in the microcontroller is different from the topics introduced in this section. Therefore, we will cover them in Sect. 3.3.

2.2.1   The Arm® Cortex™-M4 Instruction Set

We have instructions defined by the language syntax when we consider C, C++, or Python languages. These instructions are independent of the hardware platform the code is executed on. Therefore, the code written for one platform can be ported to another one with minor modification. Since assembly language directly interacts with hardware of the microcontroller, this is not the case for it. In other words, there is no unique assembly language for all microcontrollers (or microprocessors in general). Instead, the assembly language for one microcontroller architecture (let's say Arm® Cortex™-M4) differs from another microcontroller architecture (such as Texas Instruments MSP430). This difference occurs by instruction set of the microcontroller.

The instruction set is composed of commands to reach and modify the lowest level hardware of the microcontroller. Therefore, the instruction set is also called software architecture of the microcontroller. The STM32F4 microcontroller is based on the Arm® Cortex™-M processor (more specifically the ARMv7E-M architecture). The total number of supported instructions in the Arm® Cortex™-M processor is 207. These can be divided into six subsets as general data moving and processing, memory access, arithmetic and logic, branch and control, floating-point, and other instructions. For more information on these instructions, please see [8].

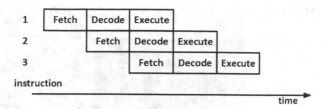

Fig. 2.5 The fetch, decode, and execute stages

2.2.2 Executing Machine Language Code in the Microcontroller

Although execution steps of a machine language code are an advanced topic beyond the concepts covered in this book, we feel they should be briefly summarized. This will give an insight on the operations performed by the CPU. As a reminder, the code written in C or C++, or assembly language, is converted to machine language by the compiler. Afterward, it is placed in the code region of the memory. Here, the memory address for the beginning of the code to be executed is predefined. As the reset signal is fed to the microcontroller, the program counter (PC) goes to this address and starts the fetch, decode, and execute stages as in Fig. 2.5. In the fetch stage, the instruction is fetched from the memory and placed in the pipeline. In the decode stage, the instruction is decoded to determine what operation to perform. In the execute stage, the instruction is executed.

The fetch, decode, and execute stages are handled independently. In other words, while one instruction is being executed, the next one is decoded, and the third one is fetched from code memory. Each stage requires one clock cycle for most instructions. However, some instructions, such as data transfer and branch, require more than one clock cycle for execution. Hence, they stall the pipeline. In order to decrease the delay caused by this operation, the three-stage pipeline has some additional properties. For more information on these, please see [1].

2.3 The STM32F4 Board

The STM32F4 microcontroller cannot be used alone unless a dedicated development or specific board is constructed for it. The best option for us, as early learners, is using the available 32F429IDISCOVERY kit for this purpose. For the sake of simplicity, we will call this board as STM32F4 from this point on. In this section, we will briefly introduce properties of the board. We will focus on these in detail in the following chapters whenever needed.

Fig. 2.6 The STM32F4
board

2.3.1 General Information

The STM32F4 board is as in Fig. 2.6. This board contains the following items:

- STM32F429ZIT6 microcontroller,
- On-board ST-LINK/V2-B embedded debugging tool interface,
- 2.4-inch QVGA TFT LCD,
- User and reset push buttons,
- Two user LEDs, LD3 being green and LD4 being red,
- I3G4250D three-axis MEMS gyroscope,
- 64 Mbit SDRAM,
- Micro-AB connector for USB OTG,
- 8 MHz external crystal.

As can be seen in the above list, the STM32F4 board offers a fairly wide range of options to use the STM32F4 microcontroller properties. They will help us to understand the microcontroller concepts better.

2.3.2 Pin Layout

The STM32F4 board has two headers (consisting of pins) such that a wire can be connected to them. Pins in these headers are directly connected to pins of the microcontroller. Hence, the user can reach the microcontroller pins through board headers. The pin layout of the STM32F4 board is as in Fig. 2.7.

As can be seen in Fig. 2.7, pins of the STM32F4 board are arranged in two headers, named P1 and P2. Usage areas of these pins are tabulated in Tables 2.1, 2.2, 2.3, and 2.4. Some of these pins are already connected to onboard modules and cannot be used for other purposes. We indicate these pins as "already used" in the mentioned tables. Unused pins can be used for more than one purpose. We will explore these usage areas in the following chapters. To note here, we only summarized the usage areas of the pins to be considered in this book.

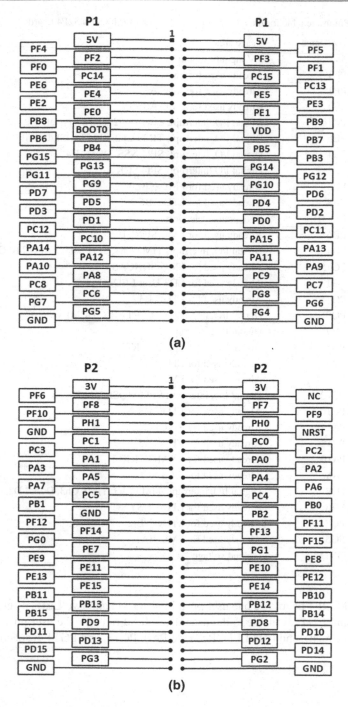

Fig. 2.7 Pin layout of the STM32F4 board [2]. (**a**) P1 header. (**b**) P2 header

Table 2.1 Pin usage table for the left column of P1 header of the STM32F4 board

Pin	Port name	Usage area
1	5V	5 V input or output
2	PF4	Already used for FMC A4
3	PF2	Already used for FMC A2
4	PF0	Already used for FMC A0
5	PC14	Digital I/O, interrupt, external 32768 Hz crystal
6	PE6	Digital I/O, interrupt, TIM9 Ch2, SPI4 MOSI, DCMI D7
7	PE4	Digital I/O, interrupt, SPI4 NSS, DCMI D4
8	PE2	Digital I/O, interrupt, SPI4 SCK
9	PE0	Already used for FMC NBL0
10	PB8	Already used for LTDC B6
11	BOOT0	Boot option selection
12	PB6	Already used for FMC SDNE1
13	PB4	Digital I/O, interrupt, TIM3 Ch1, SPI1 MISO, SPI3 MISO
14	PG15	Already used for FMC SDNCAS
15	PG13	Already used for LD3 User LED (Green)
16	PG11	Already used for LTDC B3
17	PG9	Digital I/O, interrupt, USART6 RX, DCMI VSYNC
18	PD7	Digital I/O, interrupt
19	PD5	Digital I/O, interrupt, USART2 RX
20	PD3	Already used for LTDC G7
21	PD1	Already used for FMC D3
22	PC12	Digital I/O, interrupt, SPI3 MOSI, UART5 TX
23	PC10	Already used for LTDC R2
24	PA14	Already used for SWCLK
25	PA12	Already used for LTDC R5
26	PA10	Already used for USART1 RX
27	PA8	Already used for I2C3 SCL
28	PC8	Digital I/O, interrupt, TIM3 Ch3, TIM8 Ch3, DCMI D2
29	PC6	Already used for LTDC HSYNC
30	PG7	Already used for LTDC CLK
31	PG5	Already used for FMC BA1
32	GND	Ground voltage

The STM32F4 board has one user push button (B1) and two LEDs as green and red (LD3 and LD4) available to the user. The GPIO pins for the user push button, LD3 LED (Green), and LD4 LED (Red) are PA0, PG13, and PG14, respectively.

Table 2.2 Pin usage table for the right column of P1 header of the STM32F4 board

Pin	Port name	Usage area
1	5V	5 V input or output
2	PF5	Already used for FMC A5
3	PF3	Already used for FMC A3
4	PF1	Already used for FMC A1
5	PC15	Digital I/O, interrupt, external 32768 Hz crystal
6	PC13	Digital I/O, interrupt
7	PE5	Digital I/O, interrupt, TIM9 Ch1, SPI4 MISO, DCMI D6
8	PE3	Digital I/O, interrupt
9	PE1	Already used for FMC NBL1
10	PB9	Already used for LTDC B7
11	VDD	VDD output
12	PB7	Digital I/O, interrupt, TIM4 Ch2, USART1 RX, I2C1 SDA, DCMI VSYNC
13	PB5	Already used for FMC SDCKE1
14	PB3	Digital I/O, interrupt, TIM2 Ch2, SPI1 SCK, SPI3 SCK
15	PG14	Already used for LD4 User LED (Red)
16	PG12	Already used for LTDC B4
17	PG10	Already used for LTDC G3
18	PD6	Already used for LTDC B2
19	PD4	Digital I/O, interrupt, USART2 RTS
20	PD2	Digital I/O, interrupt, UART5 TX, DCMI D11
21	PD0	Already used for FMC D2
22	PC11	Digital I/O, interrupt, SPI3 MISO, UART4 RX, USART3 RX, DCMI D4
23	PA15	Already used for Touch Panel TP_INT1
24	PA13	Already used for SWDIO
25	PA11	Already used for LTDC R4
26	PA9	Already used for USART1 TX
27	PC9	Already used for I2C3 SDA
28	PC7	Already used for LTDC G6
29	PG8	Already used for FMC SDCLK
30	PG6	Already used for LTDC R7
31	PG4	Already used for FMC BA0
32	GND	Ground voltage

2.3.3   Powering the Board and Programming the Microcontroller on It

The microcontroller on the STM32F4 board can be programmed easily by the onboard ST-LINK/V2-B debugger/programmer. To do so, we should connect the board to PC via mini USB cable through its USB connector CN1. Then, ST-LINK can be used for programming or debugging purposes. We will introduce methods to program the microcontroller this way in Chap. 3. The USB connection

Table 2.3 Pin usage table for the left column of P2 header of the STM32F4 board

Pin	Port name	Usage area
1	3V	3 V input or output
2	PF6	Digital I/O, interrupt, ADC3 IN4, TIM10 Ch1, SPI5 NSS, UART7 RX
3	PF8	Already used for L3GD20 MEMS SPI5 MISO
4	PF10	Already used for LTDC DE
5	PH1	Already used for external 8 MHz crystal
6	GND	Ground voltage
7	PC1	Already used for L3GD20 MEMS SPI5 NCS
8	PC3	Digital I/O, interrupt, ADC1 IN13, ADC2 IN13, ADC3 IN13, SPI2 MOSI
9	PA1	Already used for L3GD20 MEMS INT1
10	PA3	Already used for LTDC B5
11	PA5	Digital I/O, interrupt, ADC1 IN5, ADC2 IN5, DAC OUT2, TIM2 Ch1, TIM8 Ch1N, SPI1 SCK
12	PA7	Already used for ACP RST
13	PC5	Already used for OTG FS Over Current
14	PB1	Already used for LTDC R6
15	GND	Ground voltage
16	PF12	Already used for FMC A6
17	PF14	Already used for FMC A8
18	PG0	Already used for FMC A10
19	PE7	Already used for FMC D4
20	PE9	Already used for FMC D6
21	PE11	Already used for FMC D8
22	PE13	Already used for FMC D10
23	PE15	Already used for FMC D12
24	PB11	Already used for LTDC G5
25	PB13	Already used for USB OTG HS VBUS
26	PB15	Already used for USB OTG HS DP
27	PD9	Already used for FMC D14
28	PD11	Already used for LCD RGB TE
29	PD13	Already used for LCD RGB WRX DCX
30	PD15	Already used for FMC D1
31	PG3	Digital I/O, interrupt
32	GND	Ground voltage

for debugging/programming purposes can also be used to power the board. Hence, whenever the board is connected to PC, it runs by the provided power.

We can also use an external 5 V power supply to power the board. In this mode, 5V pin of the P1 header or 3V pin of the P2 header can be used to power the board externally. However, the STM32F4 microcontroller may not be programmed in this setting. Therefore, we suggest the reader to check [4] for further information on this topic.

Table 2.4 Pin usage table for the right column of P2 header of the STM32F4 board

Pin	Port name	Usage area
1	3V	3 V input or output
2	NC	–
3	PF7	Already used for L3GD20 MEMS SPI5 SCK
4	PF9	Already used for L3GD20 MEMS SPI5 MOSI
5	PH0	Already used for external 8 MHz crystal
6	NRST	External reset
7	PC0	Already used for FMC SDNWE
8	PC2	Already used for LCD RGB CSX
9	PA0	Already used for B1 user push button
10	PA2	Already used for L3GD20 MEMS INT2
11	PA4	Already used for L3GD20 LTDC VSYNC
12	PA6	Already used for L3GD20 LTDC G2
13	PC4	Already used for OTG FS power switch on
14	PB0	Already used for L3GD20 LTDC R3
15	PB2	Already used for BOOT1 boot option selection
16	PF11	Already used for FMC SDNRAS
17	PF13	Already used for FMC A7
18	PF15	Already used for FMC A9
19	PG1	Already used for FMC A11
20	PE8	Already used for FMC D5
21	PE10	Already used for FMC D7
22	PE12	Already used for FMC D9
23	PE14	Already used for FMC D11
24	PB10	Already used for LTDC G4
25	PB12	Already used for LTDC R5
26	PB14	Already used for USB OTG HS ID
27	PD8	Already used for FMC D13
28	PD10	Already used for FMC D15
29	PD12	Already used for LCD RGB RDX
30	PD14	Already used for FMC D0
31	PG2	Digital I/O, interrupt
32	GND	Ground voltage

The STM32F4 microcontroller operates within voltage levels 1.8 V to 3.6 V. We call the operation voltage value as the supply voltage (V_{DD}) throughout the book. Let's explain this voltage range in detail. The actual working voltage level for the microcontroller is 3.6 V. For some low-level operations, this supply voltage should be decreased. The microcontroller supply voltage may be decreased till 1.8 V to operate. However, some peripheral units will not work at this voltage level. For more information on this topic, please see [6].

2.4   Summary of the Chapter

We pick the STM32F4 microcontroller as the embedded system to be used in this book. Therefore, we explored its architecture in this chapter. To do so, we started with the hardware modules within the microcontroller as CPU, memory, GPIO ports, clock and timer modules, and digital communication modules. We will explore each module in detail in the following chapters. Observing all together with their interactions provides a better insight. Hence, we briefly introduced them here. The architecture of the microcontroller also consists of its software. Therefore, we covered the assembly language and how a code in assembly language is executed in the STM32F4 microcontroller. Although we will not extensively use assembly language programming throughout the book, we still believe that the reader should be aware of these topics, even in limited degree. We also introduced the STM32F4 board and its properties since we will use the STM32F4 microcontroller by the help of the board. Therefore, we started with the headers of the board. Then, we explained how the board can be powered and the STM32F4 microcontroller on it can be programmed.

   The reader should always remember that each microcontroller family and board developed by a vendor will have its specific properties. Hence, there is no unique recipe for all available boards and microcontrollers. However, we tried to explain the topics in this chapter as generic as possible. Hence, when another board and microcontroller is picked, the information provided here will still be useful. On the other hand, the information given in this chapter is crucial for the embedded system concepts to be covered in the following chapters. Hence, the reader should digest them before going further.

Problems

2.1. What are the other members of Arm® Cortex™-M architecture besides Cortex™-M4?

2.2. What are the main building blocks of the STM32F4 microcontroller?

2.3. How can code execution and data be observed in the STM32F4 microcontroller?

2.4. What does bus stand for in terms of the microcontroller architecture? How does the bus size affect microcontroller performance?

2.5. What is the fundamental difference between RAM and flash memory? What is the size of RAM and flash memory on the STM32F4 microcontroller?

2.6. Why do we need clock signals for microcontrollers?

2.7. Can an assembly code written for the MSP430 microcontroller be directly ported to the STM32F4 microcontroller?

2.8. What is the difference between the assembly and machine language?

2.9. Why do we need the STM32F4 board instead of the STM32F4 microcontroller alone?

2.10. Which methods can be used to power the STM32F4 board?

2.11. List the modules available on the STM32F4 board.

References

1. Arm: Cortex-M4 Devices Generic User Guide, arm dui 0553b edn. (2011)
2. Mbed: https://os.mbed.com/platforms/ST-Discovery-F429ZI/. Accessed 4 June 2021
3. Rzehak, V.: Low-Power FRAM Microcontrollers and Their Applications. Texas Instruments, slaa502 edn. (2019)
4. STMicroelectronics: Discovery kit with STM32F429ZI MCU, um1670 edn. (2020)
5. STMicroelectronics: STM32 Cortex-M4 MCUs and MPUs programming manual, pm0214 edn. (2020)
6. STMicroelectronics: STM32F405/415, STM32F407/417, STM32F427/437 and STM32F429/439 advanced Arm-based 32-bit MCUs, rm0090 rev 19 edn. (2021)
7. STMicroelectronics: Use STM32F3/STM32G4 CCM SRAM with IAR Embedded Workbench, Keil MDK-ARM, STMicroelectronics STM32CubeIDE and other GNU-based toolchains, an4296 rev 5 edn. (2021)
8. Ünsalan, C., Gürhan, H.D., Yücel, M.E.: Programmable Microcontrollers: Applications on the MSP432 LaunchPad. McGraw-Hill, New York (2018)
9. Ünsalan, C., Tar, B.: Digital System Design with FPGA: Implementation Using Verilog and VHDL. McGraw-Hill, New York (2017)

Software Development Platforms

3

3.1   The STM32CubeIDE Platform

C is the most effective language to program a microcontroller in terms of resource usage, speed of execution, and ease of programming. Therefore, it is taken as the de facto language for most microcontrollers. Here, we assume the reader has sufficient knowledge on fundamental C programming techniques. If this is not the case, we refer the reader to valuable books on this topic.

We will benefit from the STM32CubeIDE platform to program the STM32F4 microcontroller via C language. This platform has several advantages. First, it is free. Second, it is specifically developed for microcontrollers by STMicroelectronics. Third, it is possible to reach and modify all properties of the microcontroller including its peripheral units using STM32CubeIDE. Therefore, we picked it in this book. To note here, it is also possible to use C++ language to develop programs in STM32CubeIDE. However, we will focus on C language instead throughout the book.

To get familiar with STM32CubeIDE, we will start with its installation. Afterward, we will explain how to create and manage a project. Here, we will not adjust peripheral units of the microcontroller. Instead, we will focus on fundamental properties of creating and executing a project in STM32CubeIDE. As we master this step, we will see how the peripheral units can be modified in Sect. 3.1.5. To do so, we will benefit from STM32CubeMX which will be of great use in managing microcontroller hardware.

Supplementary Information The online version contains supplementary material available at (https://doi.org/10.1007/978-3-030-88439-0_3).

© Springer Nature Switzerland AG 2022

C. Ünsalan et al., *Embedded System Design with Arm Cortex-M Microcontrollers*,

https://doi.org/10.1007/978-3-030-88439-0_3

3.1.1 Downloading and Installing STM32CubeIDE

It may seem trivial; however, the first step in using the STM32CubeIDE platform is downloading and installing it. Therefore, we will briefly go over these steps first. Afterward, we will show the welcome screen of STM32CubeIDE and its perspectives. Hence, the reader will know what to expect at this step.

STM32CubeIDE can be downloaded from the web site https://www.st.com/en/development-tools/stm32cubeide.html after registering or logging in. When the download is complete, the reader should follow the steps given there for installation. When the installation is complete, the reader should set the workspace directory from the opening window. Afterward, STM32CubeIDE will launch.

3.1.2 Launching STM32CubeIDE

As the STM32CubeIDE launches, the welcome window will be as in Fig. 3.1. We will use this window for all our needs. Therefore, let's briefly summarize its basic sections.

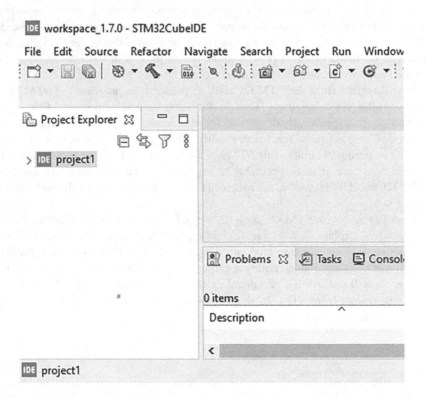

Fig. 3.1 STM32CubeIDE welcome window

As can be seen in Fig. 3.1, there exists the "Project Explorer" panel on the left side of the welcome window. The user can create a new project through it. As a new project is created, this panel summarizes all its properties. Code files will be opened as they are added to the project at the center of STM32CubeIDE. Hence, the reader will modify his or her code in this section. There will be other sections which summarize hardware properties of the microcontroller on the right side of the panel. The reader can choose available perspectives such as "C/C++," (default) "Debug," and "Device Configuration Tool" in STM32CubeIDE. There are also grouped buttons on the top of the panel. These will be of use in debugging and modifying the code. There are tabs which will summarize all build and debug operations at the bottom of the panel. We will explain all these properties in the following sections as we focus on parts of a project.

3.1.3 Creating a New Project

A project typically contains source, header, and includes files. STM32CubeIDE generates an executable file from these to be embedded to the microcontroller. This section is about creating a C project. We have two options here. The first one does not deal with peripheral and pin properties of the microcontroller. The second one deals with these. We will focus on the first option in this section to get familiar with the basics of STM32CubeIDE. We will focus on the second option in Sect. 3.1.5.

To create a new project, click "File," "New," and "STM32 Project." In the opening window, select the hardware platform to be used in the project. For our case, we should choose "STM32F429I-DISC1" from the "board selector" tab and press "Next." Afterward, the "Project Setup" window opens as in Fig. 3.2.

We should give a name to our project in this window. For our case, we set it as "project1." We should set the location of the project which may be done by the option "use default location." We should set the "Targeted Language" as C. We should set the "Targeted Binary Type" as "Executable." The final selection is titled "Targeted Project Type." Here, we have two options as "STM32Cube" and "Empty." The first option allows modifying the peripheral units. We will consider this option in Sect. 3.1.5. Hence, we select the "Empty" option to generate the project as for now. As we press the "Finish" button, a new project will be generated which can be observed under the "Project Explorer" tab as in Fig. 3.3.

As can be seen in Fig. 3.3, STM32CubeIDE added all necessary files to the project. The only step left is adding our C code to the "main.c" file generated under the "Src" folder. In this section, we will use the code given in Listing 3.1 as an example. To do so, open the "main.c" file by double-clicking on it. Add your code to this file. Then, save it by clicking the "Save" button on the upper left corner of the menu.

Project
Project Name: project1
☑ Use default location
Location: C:/Users/Cem/STM32CubeIDE/workspace_1.8.0   Browse...

Options

Targeted Language
◉ C ○ C++

Targeted Binary Type
◉ Executable ○ Static Library

Targeted Project Type
○ STM32Cube ◉ Empty

Fig. 3.2 Project setup window

Fig. 3.3 Project explorer tab

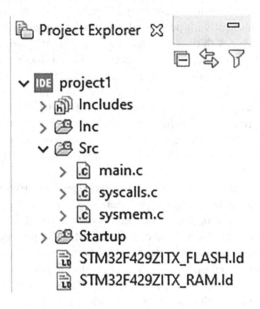

Listing 3.1 The first C code for the STM32F4 microcontroller

```c
int a = 1;
int b = 2;
int c;

int main(void)
{
int d;
```

```
c = a + b;
d = c;

while (1);
}
```

3.1.4   Building, Debugging, and Executing the Project

As we create our project, the next step is its execution. To do so, we will first consider the building and debugging steps. Afterward, we will focus on the execution step.

3.1.4.1 Building and Debugging the Project

The first step in executing the code on an embedded platform is building and debugging it. There is a button with hammer shape on STM32CubeIDE horizontal toolbar as "Build". As the user presses the right bottom arrow there, two options emerge as "Debug" and "Release". The first option builds the project with the necessary debug information. In this default option, no optimization is done on the built output. The second option is specific to deploying the built output to the microcontroller. Hence, the main focus here is optimizing output of the project and minimizing its size. Running steps in both options (such as warnings and errors) can be observed in the "CDT Build Console" window. If there is an error in a code line, it is indicated by a red cross.

The second step in executing the code is debugging and loading it to the target device. There is a button on STM32CubeIDE's horizontal toolbar as "Debug." It can also be reached from the "Run" tab. As we press this button, a new window appears asking for the "Edit Configuration" as in Fig. 3.4. Within this window, we can make all adjustments related to the debugging process. The default settings should be kept as they are in this window unless otherwise stated.

As we press "OK" in the "Edit Configuration" window, the code is debugged and downloaded to the microcontroller. Afterward, the STM32CubeIDE interface turns to the "Debug" perspective.

3.1.4.2 Executing the Project

As the debugged C code is downloaded to the microcontroller, the next step is its execution. At this stage, the code will be executed till the beginning of the main function. The microcontroller waits for further command. Buttons for program execution are located on the toolbar as in Fig. 3.5.

The name of each button can be observed by moving the cursor over it. These buttons and their function are explained briefly in the below list.

- **Skip all breakpoints**: Disable all previously set breakpoints. We will talk about what a breakpoint is in the following paragraphs.

Edit launch configuration properties

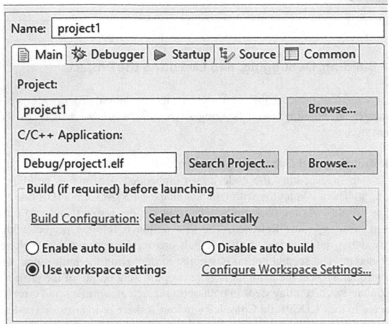

Fig. 3.4 The debug interface menu

Fig. 3.5 The execution session menu

- **Terminate and relaunch**: If a modification is done in the program while in debug mode, this button rebuilds and debugs it.
- **Resume**: Resumes execution of the code from the last executed location. Execution of the code continues until a breakpoint when this button is clicked.
- **Suspend**: Halts execution of the code. All windows used to observe software and hardware are updated with recent data.
- **Terminate**: Ends execution of the code. The debug session is also ended by this command.
- **Disconnect**: Disconnect from the target device.
- **Step into**: Executes the next line of code. If this line calls a function, the compiler executes the next line in it. Then, it stops.

Fig. 3.6 The Expressions
window

Expression	Type	Value
(x)= a	int	1
(x)= b	int	2
(x)= c	int	0
(x)= d	int	0

- **Step over**: Executes the next line of code. If this line calls a function, the compiler executes the function completely. Then, it stops.
- **Step return**: Completes execution of the function and exits the code.
- **Instruction stepping mode**: The C code is executed based on its generated assembly code within the step into, step over, and step return commands.

Observing variables, registers, or memory is important while developing the project. Code execution should halt in order to perform this operation. A breakpoint should be added to stop execution of the code at a specific code line. To do so, left-click on the desired code line and select "Add Breakpoint..." from the pop-up window. A new window appears asking for the breakpoint type. Here, we can select the "Regular" type. As the breakpoint is added, a blue circle will appear by the code line to indicate that there is a breakpoint there. The inserted breakpoint can be deleted by double-clicking on it.

The "Expressions" window (given in Fig. 3.6) can be used to observe selected variables. In order to add a variable to this window, select the variable to be observed and right-click on it. Select the "Add Watch Expression" option in the opening list. Then, click "OK." The reader can also double-click on the "Add new expression" button in the "Expressions" window and enter the name of the variable to the opened box.

We can define a variable either as local or global in C language. As the name implies, the global variable is available to all code sections. The local variable is only available to the function it is defined in. We will explore these variable types in detail in Chap. 9. Although local and global variables can be observed in the "Expressions" window, local variables can also be observed in the "Variables" window. Here, all local variables are automatically added to the mentioned window.

There is also "Live Expressions" window which can be enabled from the "Window/Show View" list besides other options. In this window, variables are updated in real time while executing the code. Hence, value of a variable can be observed without stopping the code.

3.1.5   Using STM32CubeMX to Modify Hardware of the Microcontroller

STM32CubeMX (available under STM32CubeIDE) can be used to control and modify hardware of the STM32F4 microcontroller. To note here, hardware properties can also be set via available libraries such as hardware abstraction layer (HAL). The advantage of STM32CubeMX is that operations are done visually on it. Moreover, it produces a template C code containing predefined functions to be used to control hardware properties. Therefore, we will be using this option throughout the book.

3.1.5.1 Creating a New Project Using STM32CubeMX
In order to modify hardware of the microcontroller, we can create a new project as explained in Sect. 3.1.3 with one difference. Since we will modify hardware of the microcontroller, we will set the "Targeted Project Type" as "STM32Cube" in Fig. 3.2. Afterward, a new window appears asking for the "Firmware Library Package Setup." At this step, leave all default settings as they are and press "Finish." The next pop-up window asks whether to initialize all peripheral units with their default mode. Click "No" since we will modify necessary pin properties as we need them. A window should open up as in Fig. 3.7.

We can set all system hardware properties through STM32CubeMX. At this step, let's assume that we want to turn on the green LED on the STM32F4 board whenever the user button on it is pressed. Although we will see how this can be done in detail in Chap. 4, let's create a simple project for this purpose. To do so, we should first modify the microcontroller pin properties. Hence, we should set the pin PA0 of the STM32F4 microcontroller as "GPIO_Input" and pin PG13 as "GPIO_Output." Fortunately, these modifications can be done in the STM32CubeMX interface easily. Within the "Pinout&Configuration" tab, locate the mentioned pins and left-click on them one by one. A pop-up window appears asking for which purpose the pin will be used for. Set the pin PA0 as "GPIO_Input" and pin PG13 as "GPIO_Output" and proceed to clock configuration by opening the "Clock Configuration" tab. We will benefit from the automatic clock setting property of STM32CubeMX here. To note here, this is the default setting and we suggest using this option whenever possible. Clock setup will be covered in detail in Chap. 6.

3.1.5.2 Generating and Modifying the Code
Now, we are ready to generate the code corresponding to the hardware setup. Before doing so, we can open the "Project Manager" tab. Here, we can set the properties of our project from the "Project," "Code Generator," and "Advanced Settings" tabs. We suggest keeping all settings as they are since these are generated automatically. The next step is generating the code related to the hardware setup. To do so, press the "Save" button. A pop-up window appears asking for generating the code. Press "OK" there.

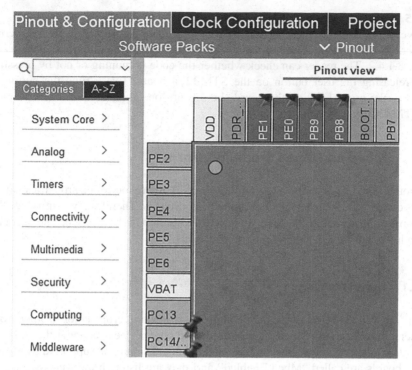

Fig. 3.7 Initial screen for STM32CubeMX

As the code is generated, we can open the "main.c" file under the "Src" folder. As this file is opened, the reader will observe that there are sections labeled "USER CODE BEGIN" and "USER CODE END" asking for the user to add his or her code snippets. This setup is done to ensure that other parts of the code related to hardware setup are not modified by mistake. Finally, we will test the project. To do so, add the following C code snippet in Listing 3.2 to the appropriate place in the opened "main.c" source file.

Listing 3.2 The C code snippet to be added

```
/* Infinite loop */
/* USER CODE BEGIN WHILE */
while (1)
{
if (HAL_GPIO_ReadPin(GPIOA, GPIO_PIN_0))
HAL_GPIO_WritePin(GPIOG, GPIO_PIN_13, GPIO_PIN_SET);
else
HAL_GPIO_WritePin(GPIOG, GPIO_PIN_13, GPIO_PIN_RESET);
/* USER CODE END WHILE */

/* USER CODE BEGIN 3 */
}
/* USER CODE END 3 */
```

3.1.5.3 Executing the Project

As we debug the project and embed the generated code on the STM32F4 microcontroller, it will be ready to be executed. Here, we will follow the same steps as in Sect. 3.1.4.2. The reader can check whether the code is running or not by pressing and releasing the user button on the STM32F4 board. Every time the button is pressed, the green LED on the board should turn on. This test ends our coverage of STM32CubeMX.

3.2 Mbed and Mbed Studio Platforms

Although STM32CubeIDE allows controlling all microcontroller properties, some users do not need such a detailed setup in their projects. Therefore, we introduce the web-based Mbed and its desktop version Mbed Studio in this section. The user can form his or her project faster via these platforms.

3.2.1 Mbed on Web

Arm® introduced Mbed as the online platform to develop projects for its microcontrollers. In order to use Mbed, the reader should open a free account at the web site https://os.mbed.com/ide/. Besides, the selected board should be supported by Mbed. Such boards are called "Mbed Enabled" and they are listed in the web site https://os.mbed.com/platforms/. Our STM32F4 board is Mbed enabled. Hence, we can use it with Mbed.

Mbed on web has three main advantages. First, it does not require downloading and installing any IDE on PC. It is structured such that the user develops his or her project on a web-based platform. As the project is built, Mbed creates a binary file (with extension ".bin") to be embedded on the microcontroller. Second, Mbed is formed such that a code written for one platform in the Mbed ecosystem can easily be ported to another platform as long as hardware requirements are satisfied. This is possible since Mbed simplifies and generalizes low-level hardware setup and usage. From this perspective, Mbed resembles the popular Arduino platform. Third, Mbed allows code sharing between community members since it is an online platform. This is done by posting and importing the generated code through the web-based interface. In fact, the reader can easily use a project posted by another community member by importing it to his or her workspace. Then, this project can be developed further by the reader for his or her own needs.

3.2.2 Managing a Project in Mbed

As we register to the Mbed web site, we will see the online compiler as in Fig. 3.8. Before creating our first project, we should add the development board to be used in operation. To do so, please click on the "No device selected" area on the top right

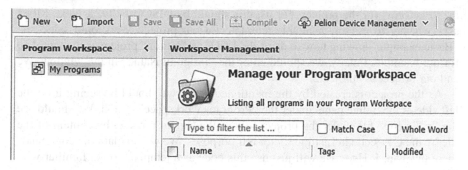

Fig. 3.8 Online Mbed compiler

Create new program for "DISCO-F429ZI"

This will create a new C++ program for "DISCO-F429ZI" in your workspace. You can always change the platform of this program once created.

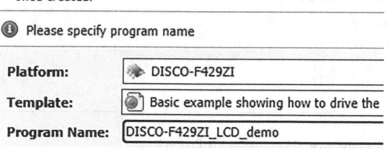

Fig. 3.9 Setting project properties under Mbed

section of the compiler. As the new window opens up, press the "Add Board" button on the bottom side of the window. Then, search for the STM32F4 board by typing "DISCO-F429ZI" in the search area and click on the board. Then, click on the "Add to your Mbed Compiler" button. Now, we should be seeing the board on the top right section of the online Mbed compiler.

3.2.2.1 Creating the Project

Let's consider a simple project and explain how to manage it step by step in Mbed. We will toggle the green LED on the STM32F4 board every 0.5 s in our project. To do so, we should benefit from the available template project. On the "New" selection section of the online compiler, create a new project. The "create new program" window will be as in Fig. 3.9. As can be seen in this figure, the user should select the "Platform" for the project. For us, this should be DISCO-F429ZI. Next, Mbed asks whether to use any available template for our project. We should select the "Basic example showing how to drive the LEDs and button" project as the template for our example. Finally, we should enter the "Program Name" in the window. We call our project as "project2."

As the project is created by the mentioned steps, we should be seeing it on the left side of the online panel under the "Program Workspace" region. We should see the "main.cpp" file under the project folder. We should replace the content of the file by the C++ code in Listing 3.3 for our purposes. We will explain the commands there in Chap. 4. Here, we will just use this code as a template to get familiar with Mbed.

Listing 3.3 The main file for our project under Mbed

```
#include "mbed.h"

#define WAIT_TIME 500000
DigitalOut greenLED(LED1);

int main()
{
while (true)
{
greenLED = !greenLED;
wait_us(WAIT_TIME);
}
}
```

3.2.2.2 Building and Executing the Project

As the project is created and its main file is modified, the next step is its execution. To do so, we should compile it. Hence, the user should press the "Compile" button at the top side of the online compiler. As the project is compiled, and assuming no problem has occurred, Mbed generates the file "project2.DISCO_F429ZI.bin" which can be found under the "Downloads" folder in the Windows operating system. In order to run it, the reader should connect the STM32F4 board to PC through its USB interface. As the board is connected to PC, it should be seen as a new USB device under the "Devices" folder. The reader should drag and drop the "project2.DISCO_F429ZI.bin" file to this folder to run it. Afterward, the result can be observed on the board.

3.2.3 Mbed Studio on Desktop

The procedure is straightforward in creating, building, and executing the project under Mbed. However, the user should have an active Internet connection to perform all these operations. This may not be possible for some cases. Besides, it is not possible to debug the code during its execution. We can benefit from Mbed Studio to overcome these shortcomings.

Fig. 3.10 Mbed Studio
launch window

Mbed Studio is the desktop version of Mbed. The reader can download it from the web site https://os.mbed.com/studio/. When the download is complete, the reader should follow the steps summarized there for installation. When the installation is complete, the reader should login to his or her Mbed account. Afterward, the program will launch and the IDE will be as in Fig. 3.10. As can be seen in this figure, Mbed Studio has a fairly compact interface. There exists a panel on the left side of the window. Through it, the user can create a new project, select the target, and build the project for the selected target. At the center side of the window, the main code file will be opened as it is added to the project. There are tabs which will summarize all build and debug operations at the bottom of the panel.

3.2.4   Managing a Project in Mbed Studio

We can create, build, debug, and execute a project under Mbed Studio. Before creating the project, we should connect our STM32F4 board to PC. Mbed Studio automatically detects the board and displays it under the "Target" section. Moreover, the reader can check the connection status of the board in the same section. If the board connection is active, then a green colored USB sign should be visible in front of the target board name.

Fig. 3.11 Creating the
project under Mbed Studio

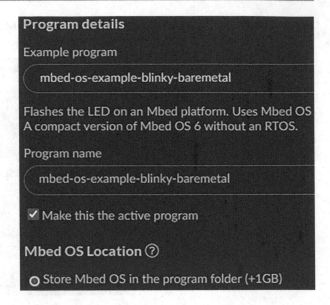

3.2.4.1 Creating the Project

We can create a new project under Mbed Studio with similar steps as in Mbed. To do
so, the reader should click on the "File" menu and select the "New program" there.
A new window opens up as in Fig. 3.11. Next, we should select a template program
as in Mbed. For our case, we can select "mbed-os-example-blinky-baremetal." We
should give a name to our project in the "Program name" window. Afterward, we
should press the "Add Program" button and let Mbed Studio import all necessary
libraries.

The code to be executed can be found in the "main.cpp" file under our project.
We will use the available code in our first project. We provide the content of the
main file in Listing 3.4 for completeness. This code performs the same operations
as in Listing 3.3.

Listing 3.4 The main file for our project in Mbed Studio

```
#include "mbed.h"

#define WAIT_TIME_MS 500
DigitalOut greenLED(LED1);

int main()
{
printf("This is the bare metal blinky example running on Mbed OS
    %d.%d.%d.\n", MBED_MAJOR_VERSION, MBED_MINOR_VERSION,
    MBED_PATCH_VERSION);

while (true)
{
greenLED = !greenLED;
thread_sleep_for(WAIT_TIME_MS);
}
}
```

3.2.4.2 Building, Debugging, and Executing the Project

There are three buttons on Mbed Studio's left panel with the hammer, triangle, and bug shape to build, run, and debug the project, respectively. Different from Mbed, the user does not have to drag and drop a file to execute the generated binary file. Instead, all these steps are handled by Mbed Studio. Hence, these operations are simplified further. As the program in Listing 3.4 is built and executed following the mentioned steps, the green LED on the STM32F4 board should be blinking every 0.5 s.

3.3 MicroPython

Python is a very popular programming language mainly targeting PC and single board computers like Raspberry Pi. It can also be used on microcontrollers. MicroPython is the required implementation of Python for this purpose. Therefore, we will introduce its setup, working principles, and usage in this section.

3.3.1 About Python

Python emerged as a high-level programming language and gained enormous success in data processing and scientific communities. There are several reasons for this popularity. First and most important of all, Python is a high-level programming language. This means that the programmer will not deal with low-level hardware modifications in it. Therefore, he or she can easily form a prototype code for a given problem. Second, Python can easily be installed and executed on Windows, Linux, or embedded Linux. Besides, the code written on Linux can be directly used in Windows or vice versa. Third, Python has several libraries covering most application areas. These further simplify life for the programmer. By the way, these libraries are provided by the Python user community free of charge. Moreover, almost all deep learning libraries are available in Python language. Hence, the programmer constructing such an application should use Python for this purpose.

We will use Python for two main reasons in this book. First, Python is the backbone of MicroPython to be introduced in the following section. Second, we will benefit from Python on PC during data transfer between the microcontroller and PC. Therefore, we assume that the reader has necessary background on Python. If this is not the case, we strongly suggest the reader to consult books on Python. We leave downloading and installation steps of Python to the reader. These can be handled by support from the Python web site https://www.python.org/.

3.3.2   Python for Microcontrollers: MicroPython

MicroPython is the implementation of Python language for microcontrollers. It allows executing high-level Python code on a selected microcontroller. To note here, MicroPython can only be used on a subset of microcontroller families supporting it. The STM32F4 microcontroller and STM32F4 board is supported by MicroPython. Hence, we benefit from it in this book.

MicroPython offers several advantages when used on microcontrollers. Its first advantage is allowing a Python programmer to use the microcontroller. Hence, such a programmer does not need to learn C or C++ language. The second advantage of MicroPython is its available libraries. Although this library collection is not as diverse as Python, it is expanding. Moreover, if the reader needs a specific library to be used under MicroPython, it can be constructed from scratch for the microcontroller at hand. We will provide a way to form such a library in Chap. 15. The third advantage of MicroPython is its being an interpreter-based language. This means that the programmer should compile, debug, and embed the code to the microcontroller in C or C++ languages. However, the code in MicroPython can be executed line by line by the read-evaluate-print loop (REPL, to be explained in detail in Sect. 3.3.5). As a final note, since C and C++ languages allow reaching low-level hardware properties of the microcontroller, they can be used to get the highest performance from it. When the performance is not the first priority, then MicroPython emerges as the high-level and fairly user-friendly option.

3.3.3   Setting up MicroPython on the STM32F4 Microcontroller

MicroPython has been compiled for selected microcontrollers and is available as a firmware file, with extension ".dfu," on its official web site https://micropython. org/download. The reader should download the most recent MicroPython version suitable for the microcontroller and development board at hand. For our case, we should check the "STM32 boards" section in the mentioned web site and select the latest version of the board file with the ".dfu" extension under the "Firmware for STM32F429DISC board" section for our STM32F4 microcontroller.

The dfu file extension stands for device firmware upgrade. Hence, we need to upgrade the firmware of our STM32F4 microcontroller to use MicroPython on it. We will benefit from the available sources to perform this operation in this book. The reader can check the provided references for more information on dfu file management on STM32 microcontrollers [2, 4].

As we download the appropriate dfu file from the MicroPython web site, the next task is loading it to the STM32F4 microcontroller as a firmware upgrade. To do so, we should download the STSW-STM32080 and STSW-LINK004 packs from the STMicroelectronics web site. The reader can use the "DFU file manager" program, within the STSW-STM32080 pack, and select "I want to EXTRACT S19, HEX or BIN from a DFU one." In the opening window, he or she should pick the downloaded dfu file for the STM32F4 microcontroller and select "Hex Extraction." The DFU file manager will create a "hex" file in the working directory when the reader clicks "extract." We can use the "STM32 ST-LINK Utility" program, within the STSW-LINK004 pack, to load the extracted hex file to the STM32F4 microcontroller. To do so, the reader should open the STM32 ST-LINK Utility program and select "Target, Program & Verify." In the opening window, he or she should browse the created hex file. Finally, the reader should click "Start" and wait until loading process is finished. Afterward, MicroPython will be ready to be used on the STM32F4 microcontroller.

The standard firmware mentioned in the previous paragraphs does not include some peripheral units or their subparts to be used throughout the book. Therefore, we formed a custom MicroPython firmware (with additions) and provided it in the companion book web site as a hex file. Therefore, we strongly suggest the reader to use it to benefit from the software provided in the book. To do so, necessary steps explained in the previous paragraph starting from loading the hex file to the STM32F4 microcontroller should be followed.

Since MicroPython has been loaded to the microcontroller as a firmware upgrade, it can be erased easily. Hence, the user can resume using the microcontroller by STM32CubeIDE or Mbed interfaces. To do so, the only step needed is programming the microcontroller via these interfaces. If the reader wants to use MicroPython again, the setup procedure should be applied again.

3.3.4   MicroPython Working Principles

While explaining the assembly language in Sect. 2.2, we mentioned that executing Python commands does not follow the same steps as in C or C++ language. In this section, we provide a summary of how a Python code is executed. Although we summarize the steps for Python on PC, they apply to MicroPython as well.

Let's start explaining the framework for executing the Python code from the beginning. Hence, we assume that the reader has formed a text file with extension ".py" by any editor. When this text file is called within the Python interpreter, it is converted to bytecode. We can call this step as converting the source file to bytecode. The bytecode is different from machine language and cannot be executed by the CPU. Instead, it is executed line by line by the Python virtual machine in terms of switch statements. For more information on these topics, please see [1].

3.3.5   Using MicroPython on the STM32F4 Microcontroller

We provide three different methods to use MicroPython on the STM32F4 microcontroller. The first method is based on read-evaluate-print loop (REPL). The second method is based on modifying the "main.py" file available under MicroPython. The third method is using an available Python IDE. We will explain these methods in detail next.

3.3.5.1 Read-Evaluate-Print Loop

The first method to use MicroPython on the STM32F4 microcontroller is based on the read-evaluate-print loop (REPL). REPL allows the programmer to execute the MicroPython code without any need for compiling or uploading it to the microcontroller. Hence, the programmer can easily test the code interacting with hardware. Moreover, we can execute a single line of code with this method such that we can observe its response on the microcontroller.

We can execute REPL by a terminal program (such as Tera Term) running on PC. In order to reach the STM32F4 board from PC, the terminal program should connect to it through the serial COM port associated with it. The port name can be found in "Device Manager" under the Windows operating system, with the name "STMicroelectronics STLink Virtual COM Port." For other operating systems, we kindly ask the reader to consult related resources. The baud rate within the terminal program should be set to 115200 bits/s.

As the connection between the terminal program and STM32F4 board is established, we should restart MicroPython (loaded onto the STM32F4 microcontroller) by pressing the reset button on the board. Afterward, we can execute the required MicroPython code line by writing it on the command prompt of the terminal program on PC. As an example, we can enter the `help()` command on the command prompt to list primary help topics in MicroPython. We can also use the tab button on the command prompt to list the content of a module in MicroPython. As an example, when we write `pyb.` on the command prompt and press the tab button, we can see all objects related to the `pyb` module.

3.3.5.2 Accessing and Modifying the Main File

When MicroPython is loaded to the flash memory of the STM32F4 microcontroller, there will be two files as "boot.py" and "main.py." The "boot.py" file contains initial setup commands. It is called first by the system. After setup, the "main.py" file is called. This file contains the user code to be executed. Hence, the second method to use MicroPython is by accessing and modifying content of the "main.py" file.

We can modify the content of the "main.py" file by using `rshell`. To do so, the user should install it by using the command `pip3 install rshell`. Afterward, it can be started by typing `rshell` on the command prompt on PC. Then, the user can reach the STM32F4 board by entering `connect serial COMx` to the command prompt. Here `x` stands for the COM port number the STM32F4 board is connected to.

As the connection to the board is established, content of the "main.py" file on the microcontroller can be modified using the command `cp Source /flash/main.py`. Here, the Python code written in the `Source` file is copied to the "main.py" file. The reader should also add path of the file in operation. Other Python files can be copied to the microcontroller flash memory using the command cp Source /flash. Again, path of the file must be provided here. The STM32F4 microcontroller must be reset by its reset button on the STM32F4 board after the files are copied.

The reader can also enter the MicroPython REPL over rshell using the command repl. We can test whether all the above steps have been implemented correctly by adding the command print("Hello world!") to the "main.py" file. As the code is executed, we expect to see the string Hello world! in the REPL window. The reader can exit from REPL using the command Ctrl-x.

The "main.py" file may become corrupt or the MicroPython code can enter an infinite loop during operation. For such cases, the reader should reset the STM32F4 microcontroller to its default settings. To do so, MicroPython's "factory reset the file system property" can be used. Flash memory of the microcontroller is returned to its initial state this way. To start MicroPython in this mode, press both the user and reset buttons on the STM32F4 board. Then, release the reset button. Green, red, and both green and red LEDs should turn on periodically. When both LEDs are turned on, release the user button. The green and red LEDs should flash quickly four times. Then, both LEDs should turn on. Wait until all LEDs are turned off and MicroPython starts in the factory reset mode.

3.3.5.3 Using an Available Python IDE

The third method in using MicroPython on the STM32F4 microcontroller is by an available Python IDE. Thonny is one such IDE developed by the Institute of Computer Science of the University of Tartu, Estonia. This IDE simplifies the operations introduced in the previous sections. Hence, the programmer can focus on the code. Let's explain how Thonny can be used for this purpose.

Thonny can be downloaded from its web site https://thonny.org/. As the program is set up and run, its main window will be as in Fig. 3.12. The upper part of the window is for writing and executing the Python or MicroPython code. The lower part of the window is the shell in which the reader can execute a Python code line by line and observe its output.

As can be seen in the lower part of the window in Fig. 3.12, the default interpreter under Thonny has been set to Python 3.7.7. Since we want to use Thonny for MicroPython, we should click "Run" and go to the "Select interpreter..." section. Then, we should select the "MicroPython (generic)" option as in Fig. 3.13. Then, we should press "OK."

As the generic MicroPython interpreter is selected in the previous step, Thonny automatically detects our STM32F4 board connected to PC. We should emphasize here that the ".hex" MicroPython file should have been loaded to the board beforehand as explained in Sect. 3.3.3. As the board is detected, the shell section of the IDE becomes as in Fig. 3.14. Hence, the reader can execute MicroPython commands there line by line as in the REPL window.

Thonny can also be used to execute the code, written in the upper part of the window in Fig. 3.12, on the STM32F4 microcontroller. To do so, we should press the green "Run current script" button with a white triangle on it. Although Thonny has a debug option, this is not available for the STM32F4 microcontroller. Hence, we

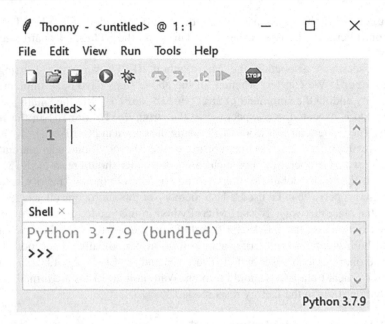

Fig. 3.12 Thonny, main window

Fig. 3.13 Thonny,
MicroPython interpreter
selection

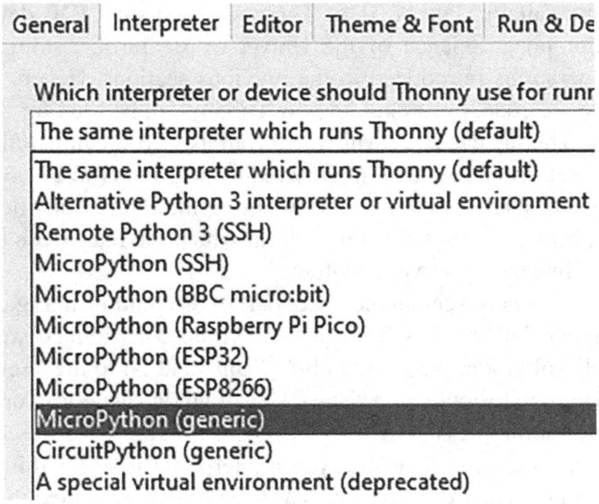

cannot benefit from it. However, the script can be easily executed this way. As a side note, the reader may come up with an error screen rarely while running Thonny. The code should be rerun in such cases. Besides, we expect to have more IDE platforms to support the STM32F4 board in the near future. They may also be used along with Thonny.

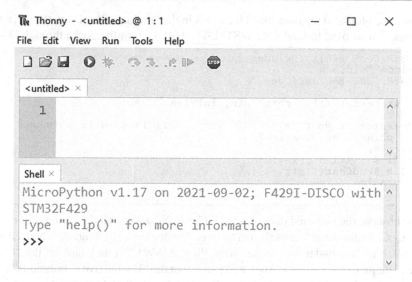

Fig. 3.14 Thonny, MicroPython command execution on the shell

3.4 Application: Tools for Analyzing the Generated Code

There are tools for analyzing the generated C, C++, and MicroPython codes. We will consider them in this section. Our main focus will be measuring the execution time and memory usage of the code or code block of interest. Specific for the STM32CubeIDE, we will also introduce the instrumentation trace macrocell (ITM) to analyze execution of the C code in more detail.

3.4.1 Analyzing the C Code in STM32CubeIDE

We will start with the ITM usage in this section. Then, we will show how the execution time can be measured under STM32CubeIDE. Next, we will provide a way to measure memory usage.

3.4.1.1 The Instrumentation Trace Macrocell Usage

ITM is used for observing variables and key code components during run time. We can start the ITM by opening the STM32CubeMX window and clicking "SYS" from the "System Core" menu. Then, we should select "Trace Asynchronous Sw" from the "Debug" dropdown list under the "SYS Mode and Configuration" window. Afterward, the pin PB3 of the STM32F4 board will be set as the "Serial Wire Output (SWO)" pin for the ITM module. Unfortunately, the SWO signal is physically disconnected from the PB3 pin via SB9 jumper on the STM32F4 board. Therefore, the jumper must be soldered to use the ITM. Afterward, the reader should add the below code blocks to the mentioned locations in the "main.c" file. Then, the `printf` function can be used to send data to STM32CubeIDE from the board through ITM.

```
/* USER CODE BEGIN Includes */
#include "stdio.h"
/* USER CODE END Includes */

int _write(int file, char *ptr, int len)
{
  /* Implement your write code here, this is used by puts and
      printf for example */
  int i=0;
  for(i=0 ; i<len ; i++)
    ITM_SendChar((*ptr++));
  return len;
}
```

To observe the obtained data in STM32CubeIDE debug window, click "Run" -> "Debug Configurations" and select "Debugger" in the opening window. Then, check the "Enable" box under the "Serial Wire Viewer (SWV)" block and set the "Core Clock" frequency accurately. Also, make sure that the "Enable live expressions" box is checked under the "Misc" block. Finally, click "Apply" and "Debug" buttons.

In the debug window, SWV related windows can be found under "Window" -> "Show View" -> "SWV." Here, we will only use the "SWV ITM Data Console," "SWV Data Trace Timeline Graph," and "SWV Data Trace" windows. The "SWV ITM Data Console" window is used to observe the `printf` function output in the code. The "SWV Data Trace Timeline Graph" window is used to graphically observe the selected variables in the code. Therefore, change of a variable over time can be observed in detail. The "SWV Data Trace" window is used to observe the history of selected variables in clock cycle resolution. Whenever a variable is changed in the code, its value and exact time in clock cycles are printed in this window. For further information on all these windows, please see [3].

In order to use the mentioned SWV windows, click the "Configure trace" button (represented by a screwdriver and wrench) in any of these windows. In the opening window, click the "Enable port 0" checkbox under the "ITM Stimulus Ports" block to observe the `printf` function output. In order to observe a variable, enable one of the four "Comparator" blocks and enter the name of the variable to the "Var/Addr" box. The configurations done under this window are applied to all SWV blocks. Finally, click "Start trace" button (represented as a red circle) in any of these windows before resuming the debug process.

3.4.1.2 Measuring the Execution Time

There is no predefined HAL function in STM32CubeIDE to count clock cycles and measure the execution time of a code block. Therefore, we will use the "data watchpoint and trace unit (DWT)" for this purpose. This unit has a specific DWT cycle count counter (DWT_CYCCNT) which holds the total number of clock cycles after enabling it. Therefore, we should set the "TRCENA" bit of "debug exception and monitor control register (DEMCR)" first. Then, we should enable the "CYCC-NTENA" bit of "DWT control register (DWT_CTRL)" to enable "DWT cycle count counter (DWT_CYCCNT)." Finally, we should clear "DWT_CYCCNT" by writing 0 to it before measuring the execution time. Afterward, cycle count can be obtained by reading the DWT_CYCCNT register on demand.

We provide an example on measuring the execution time of a code block below. The reader should first create a new STM32CubeIDE project and add the below code scripts to appropriate places in the "main.c" file. As can be seen in this example, the mentioned registers are not predefined. Hence, we should use their direct memory locations. We will explore how to do this in Chap. 9.

```c
/* USER CODE BEGIN PD */
volatile uint32_t *DWT_CTRL = (uint32_t *)0xE0001000;
volatile uint32_t *DWT_CYCCNT = (uint32_t *)0xE0001004;
volatile uint32_t *DEMCR = (uint32_t *)0xE000EDFC;
/* USER CODE END PD */

/* USER CODE BEGIN PV */
uint32_t coreClock;
uint32_t cycleCount;
float executionTime;

uint32_t i;
uint32_t data[1000];
/* USER CODE END PV */

/* USER CODE BEGIN 0 */
void startTiming(void)
{
*DEMCR = *DEMCR | 0x01000000;
*DWT_CTRL = *DWT_CTRL | 1;
*DWT_CYCCNT = 0;
}

uint32_t stopTiming(void)
{
return *DWT_CYCCNT;
}

float calculateTimeMs(uint32_t cycle, uint32_t frequency)
{
return (float)1000 * cycle / frequency;
}
/* USER CODE END 0 */

/* USER CODE BEGIN 2 */
coreClock = SystemCoreClock;
HAL_SuspendTick();

startTiming();
for (i = 0; i < 1000; i++)
{
data[i] = i;
}
cycleCount = stopTiming();
executionTime = calculateTimeMs(cycleCount, coreClock);
/* USER CODE END 2 */
```

3.4.1.3 Measuring Memory Usage

The reader can reach detailed memory usage of a code in STM32CubeIDE using its `Build Analyzer` window. Here, there are two sub-windows named "Memory Regions" and "Memory Details." In the first window, the reader can observe starting and ending addresses; free, used, and total memory sizes; and usage percentage of RAM, FLASH, and CCMRAM. In the second window, the reader can observe the detailed usage of each memory block separately.

3.4.2   Analyzing the C++ Code in Mbed Studio

We can analyze the generated C++ code under Mbed Studio. Therefore, we will consider measuring the execution time and memory usage for a given code next.

3.4.2.1 Measuring the Execution Time

There is the `Timer` module to perform time measuring operations under Mbed Studio. This module has the functions `start` and `stop` to start and stop the timer, respectively. There is also the function `elapsed_time` to calculate time difference between starting and stopping points. We provide a usage example for these functions in Listing 3.5.

3.4.2.2 Measuring Memory Usage

Mbed also provides a set of functions to monitor the stack and heap usage in run time. In order to use these functions, we should create an empty "mbed_app.json" file at the project root folder. Then, we should add the below code block to this file.

```
{
"target_overrides": {
"*": {
"platform.all-stats-enabled": true
} } }
```

As we perform all the mentioned operations, we will be ready to measure the memory usage for a given code. We provide a sample code to show how these operations can be done in Listing 3.5. Please note that the same code also provides an example of measuring the execution time for a code block.

Listing 3.5 Measuring the execution time and memory usage in Mbed Studio

```
#include "mbed.h"

mbed_stats_heap_t heapInfo;
mbed_stats_stack_t stackInfo;
uint32_t i;
uint32_t data[1000];

Timer myTimer;

int main()
{
myTimer.start();
```

```
for (i = 0; i < 1000; i++)
{
data[i] = i;
}
myTimer.stop();
printf("Execution time in microseconds: %llu\n", myTimer.
    elapsed_time());

mbed_stats_heap_get(&heapInfo);
printf("Heap size: %ld\n", heapInfo.reserved_size);
printf("Used heap: %ld\n", heapInfo.current_size);

mbed_stats_stack_get(&stackInfo);
printf("Main stack size: %ld\n", stackInfo.reserved_size);
printf("Used main stack: %ld\n", stackInfo.max_size);

while (true);
}
```

3.4.3   Analyzing the MicroPython Code

MicroPython also has functions to measure the execution time and memory usage for a given code. Therefore, we consider them next.

3.4.3.1 Measuring the Execution Time

There is the utime module to perform time measuring operations under MicroPython. This module has functions ticks_ms and ticks_us which can be used to obtain the current CPU time as milliseconds and microseconds, respectively. There is also another function ticks_diff to calculate the time difference between two time measurements. We provide a usage example for these functions in Listing 3.6.

3.4.3.2 Measuring Memory Usage

The reader can benefit from the function mem_info to observe the detailed memory usage under MicroPython. This function belongs to the module micropython. Hence, this module must be imported to use the function mem_info. Afterward, the current stack and heap usage in RAM can be printed in detail when this function is called. We provide a sample code to show how these operations can be done in Listing 3.5. Please note that the same code also provides an example of measuring the execution time for a code block.

Listing 3.6 Measuring the execution time and memory usage in MicroPython

```
import utime
import micropython

start = utime.ticks_us()
data = [i for i in range(1000)]
difference = utime.ticks_diff(utime.ticks_us(), start)
print(difference)
micropython.mem_info()
```

3.5   Summary of the Chapter

We introduced three software development platforms to manage the STM32F4 microcontroller in this chapter. We will be extensively using them in the following chapters for performing almost all operations on the microcontroller. To be more specific, we will be using STM32CubeIDE as the coding environment for C and assembly language-based programming. We also introduced the STM32CubeMX along with STM32CubeIDE. Both platforms offer the highest level of control while programming the microcontroller. Some projects do not need such a detailed control. Therefore, the reader can benefit from more user-friendly environments as Mbed on web and Mbed Studio on PC. Hence, we introduced them next as possible development platforms for C++ based projects. These development platforms offer a smooth transition from Arduino IDE. We introduced MicroPython as the implementation of the Python language on microcontrollers. This option opens up a new way of microcontroller programming. To emphasize again, information provided in this chapter will be extensively used throughout the book. Therefore, we strongly suggest the reader to master them.

Problems

3.1. Download and install STM32CubeIDE on your computer.

(a) Create an empty project to execute the C code in Listing 3.1 on your STM32F4 board.
(b) Repeat part (a) for the C code in Listing 3.2.

3.2. Open an account on the Mbed web site. Create a project to execute the C++ code in Listing 3.3 on your STM32F4 board.

3.3. Download and install Mbed Studio on your computer. Form a project to execute the C++ code in Listing 3.4 on your STM32F4 board.

3.4. Set up MicroPython on your microcontroller. Write a simple code to print "Hello World!" on the PC terminal. Use the methods introduced in Sect. 3.3.5 to execute your code.

References

1. Kamthane, A.N., Kamthane, A.A.: Programming and Problem Solving with Python, 2nd edn. McGraw-Hill, New York (2019)
2. STMicroelectronics: Getting started with DfuSe USB device firmware upgrade STMicroelectronics extension, um0412 edn. (2009)
3. STMicroelectronics: STM32CubeIDE User Guide, um2609 rev 3 edn. (2021)
4. STMicroelectronics: USB DFU protocol used in the STM32 bootloader, an3156 edn. (2021)

Digital Input and Output

4

4.1 Bit Values as Voltage Levels

We do not need to know the inner working principles of the microcontroller, such as voltage levels representing bits, when it is not interacting with the outside world. Instead, we can track all operations within the microcontroller by binary logic levels 0 and 1 (as bits). However, if we start interacting the microcontroller with the outside world, then abstract representations such as logic level 0 or 1 do not mean much. Therefore, we should associate these logic levels with actual voltage values used in the microcontroller. We will deal with this next.

There are several voltage standards to represent logic levels 0 and 1 in digital systems [1, 7]. The STM32F4 microcontroller is compatible with the JEDEC standards JESD36 and JESD52. To be more specific, the microcontroller supports 3.3 V standard low voltage complementary metal-oxide semiconductor (LVCMOS) and transfer logic (LVTTL) levels. We provide these in Fig. 4.1. In this figure, V_{DD} represents the supply voltage of the microcontroller.

Let's summarize voltage values in Fig. 4.1. When logic level 1 is fed to output from a pin, the voltage value there will be between V_{DD} and V_{OH} for both LVTTL and LVCMOS. Likewise, when logic level 0 is fed to output, the voltage value at the pin will be between 0 V and V_{OL}. The voltage value to logic level correspondence also applies to input pin of the microcontroller. When we apply a voltage value between V_{DD} and V_{IH} to a pin, it will be recognized as logic level 1 in input. Likewise, applying a voltage value between V_{IL} and 0 V to an input pin will be recognized as logic level 0. Hence, there are tolerance values for both input

Supplementary Information The online version contains supplementary material available at (https://doi.org/10.1007/978-3-030-88439-0_4).

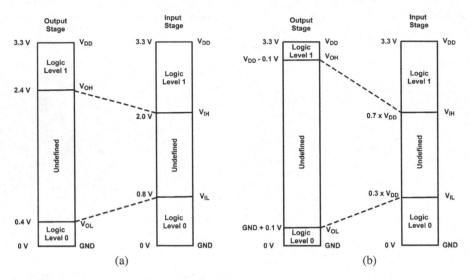

Fig. 4.1 3.3 V logic levels. (**a**) LVTTL. (**b**) LVCMOS

and output voltage values. This should be taken into account during digital I/O operations. Voltage values outside the tolerances are taken as uncertain within the microcontroller.

4.2 Interfacing Voltage Levels with the Microcontroller

While using a GPIO pin of the microcontroller, some applications may require special consideration. These can originate from input with contact closures (such as switch or button) or input with voltage values outside the microcontroller's tolerance levels. Likewise, when the microcontroller drives output loads with high current requirements, the user should be cautious. We will explain methods to be used in such scenarios next.

4.2.1 Digital Input from a Switch or Button

We can get digital input from a switch or button. In this setup, when the button is pressed (or switch is closed), a voltage value is applied to the microcontroller pin. The user should pay attention to two issues here. The first one is the setup (and related hardware) required for interfacing the switch or button to the microcontroller. The second one is avoiding switch bouncing.

4.2.1.1 Setting Up the Switch or Button

There are two input setup options when a switch or button is connected to the pin of a microcontroller. In the first setup, the microcontroller gets V_{DD} V (logic level 1) on its pin when the button (or switch) is interacted. Likewise, it gets 0 V (logic level 0) when the button or switch is left to its natural state. This is called active high input. In the second setup, the microcontroller gets 0 V (logic level 0) on its pin when the button (or switch) is interacted. It gets V_{DD} V (logic level 1) when the button or switch is left to its natural state. This is called active low input. Active high or low inputs are not related to the microcontroller. Depending on the application, the input can be adjusted as active high or low using a pull-up or pull-down resistor. If input pin is left floating, then the input state cannot be determined. The pull-up and pull-down resistors are used to pull the input to V_{DD} or 0 V when the normally closed switch or button is interacted. The pull-up and pull-down resistors are used to pull the input to V_{DD} or 0 V when the normally open switch or button is not interacted. These resistors can be externally connected or internal pull-up/pull-down resistors of the microcontroller can be used. Sample circuit diagrams for the active high/low input setup using normally opened and closed push buttons are as in Fig. 4.2.

The preferred setup for button interfacing is active high in the STM32F4 board. Therefore, when the user button on the board (connected to pin PA0) is pressed, it will generate logic level 1. When the button is released, it will generate logic level 0. Also, the user button is normally opened.

4.2.1.2 Avoiding Switch Bouncing

Switch bouncing is a serious problem when a button (or switch) is used in a digital system. Bouncing may occur due to the following reason. If the button (or switch) is interacted or activated, two metal parts connect and disconnect several times before an actual stable connection. Hence, the device may generate output more than once, typically 10 to 100 times over a period of about 1 ms [2]. GPIO pins of the microcontroller can capture these input values which result in false or multiple triggering while reading the input in the code.

Either a software- or hardware-based solution can be used to eliminate the switch bouncing problem. The simplest software solution is adding a delay while reading the input in the code. Hence, once an input is captured, successive inputs are ignored during the delay. Although this method is easy to implement, actual inputs will also be eliminated at the same time. Moreover, this solution may block other operations and can cause unpredictable problems in the code. Therefore, it should be used with precaution. We provide a sample code on this issue in the following section.

We can also use a shift register-based software solution to avoid switch bouncing. In this setup, the microcontroller captures the input values in 1 ms or a smaller time interval. These values are shifted through a shift register. During switch bouncing, the shift register content will be formed by logic level 0 and 1 values. However, the content will be formed by the same value after the bouncing effect. When such a pattern is observed, the stable input can be captured. This method may also block other software routines. However, the blocking period is usually much shorter

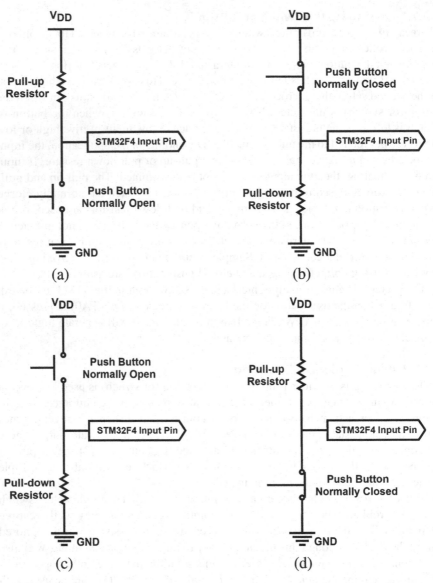

Fig. 4.2 Active high and low input circuit diagrams for the push button. (**a**) Active low input, normally open push button. (**b**) Active low input, normally closed push button. (**c**) Active high input, normally open push button. (**d**) Active high input, normally closed push button

compared to the plain delay method. We provide a usage example of this method in the following sections.

There are also hardware-based solutions to avoid the switch bouncing problem. To do so, the most feasible circuitry is adding a low-pass RC filter (composed of a

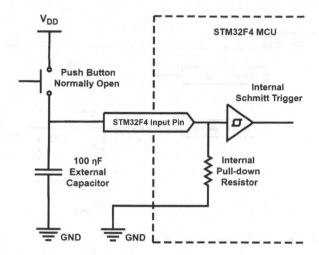

Fig. 4.3 Hardware solution for the switch bouncing problem

resistor and capacitor) followed by Schmitt trigger to the pin. Each digital I/O pin of the STM32F4 microcontroller has a Schmitt trigger. The internal pull-down resistor available at the pin hardware can be used in forming the RC filter. Resistance values for these are typically 40 KΩ. An external capacitor should be connected to the pin as well. Hence, the actual setup to eliminate switch bouncing via RC circuitry will be as in Fig. 4.3 based on the available hardware.

4.2.2 Digital I/O with High Voltage Values

Digital input to the microcontroller can originate from a source, such as sensor or external device, with high voltage values corresponding to logic level 0 and 1. Directly connecting such a source to the microcontroller can damage or burn out the GPIO pin circuitry. It may even damage other parts of the microcontroller. Likewise, some loads may require high voltage values to operate. Hence, the microcontroller should feed the necessary voltage levels to them. There are two methods to handle such situations. The first one is using a level shifter between the high voltage source or load and microcontroller. This method can be used for both high voltage input and output operations. The second method is using high voltage tolerant pins of the microcontroller, if there are any. To note here, even if there are such pins on the microcontroller, they can only be used for input.

Let's first consider level shifters with two options as unidirectional and bidirectional setup. Within the unidirectional setup, we can only convert a high voltage to a low value. In the bidirectional setup, we can convert both high and low voltage values to each other. The basic unidirectional high to low voltage level shifter is in fact a voltage divider formed by resistors. This shifter contains two resistors, as R1 and R2, which divide the high voltage by the ratio R1/(R1 + R2). We provide this setup in Fig. 4.4.

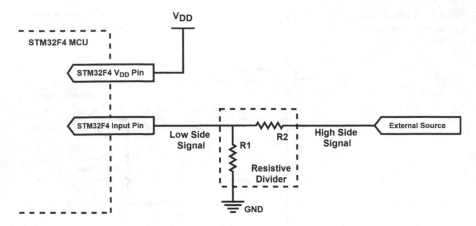

Fig. 4.4 Basic unidirectional voltage level shifter

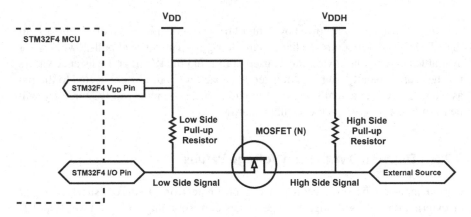

Fig. 4.5 Basic bidirectional voltage level shifter

The basic bidirectional voltage level shifter contains a single N-channel MOS-FET and a pair of pull-up resistors [3]. This setup is as in Fig. 4.5. We can divide the circuit in this figure into two parts as low and high side. Let's consider the scenario in which the low side is set as output and high side as input. Here, when the low side signal is set to V_{DD}, the MOSFET turns off, and high side signal is pulled up to V_{DDH} (high voltage level) through the pull-up resistor. When the low side signal is set to 0 V, the MOSFET source pin is set to ground voltage. Hence, the MOSFET turns on and the high side is pulled down to 0 V. Let's next consider the scenario in which the high side is set as output and low side as input. Here, the MOSFET substrate diode conducts current and pulls the low side signal down to 0.7 V when the high side signal is set to 0 V. When the high side signal is set to V_{DDH}, the substrate diode does not conduct current. Hence, the low side signal is pulled up to V_{DD} through the pull-up resistor.

The reader does not have to form the mentioned voltage level shifter circuitry from scratch. There are ready to use breakout circuits (boards) in the market which contain multiple voltage level shifters. The reader can benefit from them.

The next option in handling (relatively) high voltage inputs is using tolerant pins. The STM32F4 microcontroller has 5 V input pins tolerant to voltage levels representing binary values in TTL form. We refer the reader to check the microcontroller datasheet to see all available pins [5]. Since we benefit from the STM32F4 board, all pins except PA4 and PA5 are tolerant to 5 V input. To use them, we should disable their pull-up/pull-down resistor. The reader can use these pins when a high voltage input is to be connected to the GPIO pin of the microcontroller.

4.2.3 Digital Output to a Load Requiring High Current and Voltage Values

Most microcontrollers can feed 3.3–5 V and 25–40 mA through their GPIO pins. These values are 3.3 V and 25 mA for separate pins of the STM32F4 microcontroller [5]. Besides, the total output current that can be drawn from all pins of the STM32F4 microcontroller at once cannot exceed 120 mA. We may want to drive heavy loads, such as a motor or power LEDs, by the microcontroller. Voltage and current output from the microcontroller may not be sufficient for them. There are three methods to handle these cases using transistors, optocouplers, and relays. We will consider them next.

The first method in handling a load requiring high voltage and current is using a transistor circuit between the load and microcontroller pin. This setup is only suitable for DC voltages. The reader can use either a BJT or MOSFET for this purpose [4]. BJTs can handle up to 100 mA current and 20 V voltage. Multiple transistors connected to each other, such as Darlington bridge configuration, can also be used for higher current gain. There are boards and ICs available for this purpose. The standard example is the ULN2003 chip in the market. The MOSFET can handle very high currents and be easily connected to the microcontroller. However, the reader should pay attention to the gate-source threshold of MOSFET since it may exceed the power supply range of the microcontroller. We provide typical connection options between the BJT, MOSFET, and microcontroller in Fig. 4.6.

As can be seen in Fig. 4.6, BJT or MOSFET turns off when the microcontroller pin is set to logic level 0 (in other words 0 V). Hence, the current cannot flow through the load. When the microcontroller pin is set to logic level 1 (in other words 3.3 V), small current flows through the bias resistor and base-emitter junction of the BJT. Hence, BJT turns on and current flows from the collector (connected to load) to the emitter pin (connected to ground). MOSFET turns on and current flows from drain to source (connected to ground). To note here, V_{DD2} is the separate DC power supply voltage where load is connected to. The V_{DD2} voltage level can be same or higher than V_{DD}.

The second method in handling a load requiring high voltage and current is using an optocoupler between the load and microcontroller pin. This setup is only suitable

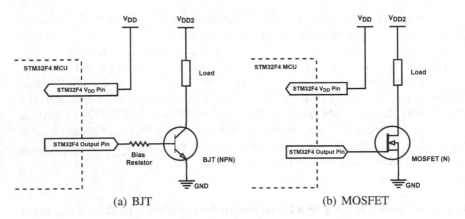

Fig. 4.6 Driving a heavy load with transistor. (**a**) BJT. (**b**) MOSFET

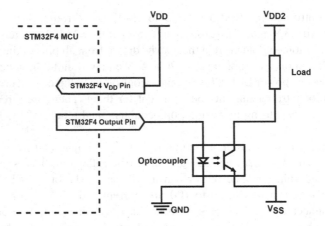

Fig. 4.7 Driving a heavy load with optocoupler

for DC voltages. Optocoupler is an electronic component made up by a simple phototransistor and LED that separates two circuits from each other using light. This setup allows handling loads with current and voltage values up to 3 A and 20 V, respectively. Moreover, the optocoupler completely separates two sides of the circuit (load and microcontroller) including their ground. This means different supply voltages and grounds can be used in the microcontroller and load connected to it. A typical connection between an optocoupler and microcontroller is as in Fig. 4.7.

As can be seen in Fig. 4.7, when the microcontroller pin is set to logic level 0 (in other words 0 V), the LED turns off. Hence, the phototransistor turns off and current does not flow from V_{DD2} to V_{SS} through load. When the microcontroller pin is set to logic level 1 (in other words 3.3 V), the LED illuminates the phototransistor. Upon receiving light, the phototransistor turns on and current flows from its collector to emitter.

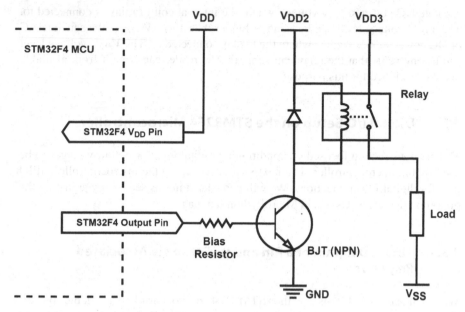

Fig. 4.8 Driving a heavy load with relay

The third method in handling a load requiring high voltage and current is using a relay between the load and microcontroller pin. Unlike the previous two options, a relay can control both AC and DC voltages. Relay is an electromechanical or solid-state element that can open and close. The electromechanical relay is composed of a contact switch that opens or closes when current flows in its internal coil or coils. It can handle very high loads having current and voltage values such as 10–15 A or 110–220 V, respectively. The solid-state relay (SSR) usually consists of an optocoupler at its input stage and triac or MOSFET at its output stage for AC or DC output, respectively. SSRs can handle very high loads such as the ones requiring 500 A current.

Some small power, electromechanical or solid state, relays can be directly connected to the GPIO pin of the microcontroller. However, a transistor or another supplementary circuit is required to drive the relay itself most of the times. For such a case, typical connection between the relay and microcontroller will be as in Fig. 4.8.

As can be seen in Fig. 4.8, when the microcontroller pin is set to logic level 0 (in other words 0 V), BJT turns off and current cannot flow through internal coil of the relay. Hence, internal contact switch does not close and load current does not flow. When the microcontroller pin is set to logic level 1 (in other words 3.3 V), BJT turns on and current flows through internal coil of the relay. This closes the internal contact switch of relay and high current can flow through the load. The diode connected against the coil is of type fly back. It protects the circuit from any back electromotive force (EMF) by the coil of the relay. To note here, V_{DD2} is the separate DC power supply voltage where the internal coil of relay is connected to. The V_{DD2} voltage level can be same or higher than V_{DD}. V_{DD3} is the separate AC or DC power supply voltage where the load is connected to. The V_{DD3} voltage level can be any value that the relay can support. The reader can benefit from available boards developed for this purpose.

4.3   Digital I/O Setup on the STM32F4 Microcontroller

The previous section focused on conditioning digital input or output voltages to be used by the microcontroller. The next step is setting up the microcontroller GPIO pins for digital I/O applications. We will consider it in this section by adjusting the pin properties via C, C++, and MicroPython languages.

4.3.1   Circuit Diagram of a Pin and Its Setup via Associated Registers

We tabulated GPIO pins of the STM32F4 microcontroller available on the STM32F4 board in Sect. 2.3.2. Here, we will take a close look at the circuit diagram of a generic pin of the STM32F4 microcontroller as in Fig. 4.9. As can be seen in this figure, there are two blocks as input and output driver. There are data registers, multiplexers, and transistors to control these blocks and read/write data to the pin. Besides, diodes are placed to protect the pin from external voltage spikes. There are also controllable pull-up and pull-down resistors in the figure.

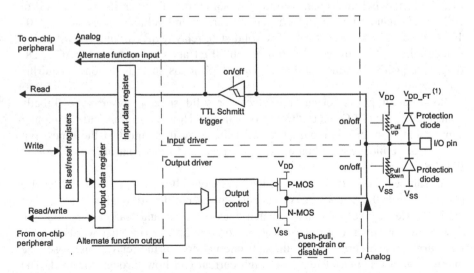

Fig. 4.9 Circuit diagram of a pin [6]

All components in Fig. 4.9 are controlled by their dedicated registers. Let's explain how a pin can be set up this way. When the pin is configured as input, the output buffer (driver transistors) is disabled. When the pin is configured as output, the output buffer is enabled. Whether the pin is configured as input or output, the input block is always active. The GPIO mode register, MODER, controls these blocks. In case the pin is configured as input, pull-up and pull-down resistors can be enabled or disabled if required. These resistors are configured by the pull-up/pull-down register, PUPDR. The pin value can be read through input data register, IDR. In case the pin is configured as output, output type and speed can be adjusted with output type register, OTYPER, and output speed register, OSPEEDR, respectively. The pin value can be set to logic level 1 or 0 through the bit set/reset register, BSRR, and bit reset register, BRR, or directly accessing the output data register, ODR. If the digital alternate functions (such as I2C and UART) will be used, then the input or output will be directed to those peripheral units. This is controlled by two alternate function registers, AFR [2]. Similarly, if the analog input or output will be used, then the I/O pin is directed to these peripheral units using the analog switch control register, ASCR. The Schmitt trigger is used to prevent noise and switch bouncing at the input pin. It is always turned on when the pin is used as GPIO. When an analog operation is done via the pin, the Schmitt trigger is turned off to save power.

4.3.2 GPIO Registers in Memory Map of the STM32F4 Microcontroller

The previous section focused on the usage of GPIO registers for controlling hardware components. In this section, we will deal with the structure of these registers. In fact, these registers are reached and processed as if they are memory addresses in Arm® architecture. This structure is called memory mapped IO. The memory address-based operation is not specific to GPIO registers. Other peripheral unit registers are also treated in the same way in Arm® architecture. However, we will only provide the GPIO memory map in this chapter to emphasize working principles of this setup.

All peripheral unit registers are mapped to memory addresses between 0x40000000 and 0x60000000 in the STM32F4 microcontroller (or in general the Arm® Cortex™-M4 architecture). Each peripheral unit has its dedicated range within this address. For example, the GPIO registers are represented between memory addresses 0x40020000 and 0x40022BFF. Below, we provide how memory addressing is done for them.

```
PERIPH_BASE = 0x40000000

AHB1PERIPH_BASE = PERIPH_BASE + 0x00020000

GPIOA_BASE = AHB1PERIPH_BASE + 0x00000000
GPIOB_BASE = AHB1PERIPH_BASE + 0x00000400
GPIOC_BASE = AHB1PERIPH_BASE + 0x00000800
GPIOD_BASE = AHB1PERIPH_BASE + 0x00000C00
GPIOE_BASE = AHB1PERIPH_BASE + 0x00001000
```

```
GPIOF_BASE = AHB1PERIPH_BASE + 0x00001400
GPIOG_BASE = AHB1PERIPH_BASE + 0x00001800
GPIOH_BASE = AHB1PERIPH_BASE + 0x00001C00
GPIOI_BASE = AHB1PERIPH_BASE + 0x00002000
GPIOJ_BASE = AHB1PERIPH_BASE + 0x00002400
GPIOK_BASE = AHB1PERIPH_BASE + 0x00002800

GPIOx_MODER = GPIOx_BASE + 0x00000000
GPIOx_OTYPER = GPIOx_BASE + 0x00000004
GPIOx_OSPEEDR  = GPIOx_BASE + 0x00000008
GPIOx_PUPDR = GPIOx_BASE + 0x0000000C
GPIOx_IDR = GPIOx_BASE + 0x00000010
GPIOx_ODR = GPIOx_BASE + 0x00000014
GPIOx_BSRR = GPIOx_BASE + 0x00000018
GPIOx_LCKR = GPIOx_BASE + 0x0000001C
GPIOx_AFRL = GPIOx_BASE + 0x00000020
GPIOx_AFRH = GPIOx_BASE + 0x00000024
GPIOx_BRR = GPIOx_BASE + 0x00000028
```

As can be seen here, there is a base address for all peripheral units. Moreover, there is an extra offset value for each GPIO port. Registers for each port are also specified by additional address offset values on top of the previous base address and offset values. By summing a specific GPIO base and register offset values, content of the register can be accessed. For example, to read the content of the IDR register (GPIOx_IDR), the memory address can be obtained by summing (GPIOx_BASE + 0x00000010).

4.3.3   Setting Up GPIO Registers

We can adjust GPIO port properties through its dedicated registers. The direct way to perform this is using assembly language and reaching the required memory address. Although this is a valid option, it is not user-friendly. Instead, we can benefit from the available functions in C, C++, and MicroPython languages for this purpose. Hence, the user will not deal with low-level operations.

4.3.3.1 C Language
STM32CubeIDE allows a simpler method to set up port registers when the C language is used in programming. To be more specific, HAL library functions are used for this purpose. HAL is the hardware abstraction layer which allows programmers to form device-independent code while accessing the hardware. Hence, the same code may be used in other devices with minor modifications.

A GPIO pin within a port can be configured in several modes as input, output, analog, external interrupt, or alternate function. In this chapter, we will only explain configuring the GPIO pin as input or output using STM32CubeMX. When the GPIO pin is configured as input, it can be left floating or its weak internal pull-up or pull-down resistor can be enabled by software. When the GPIO pin is configured as output, it can be configured as open-drain or push-pull mode. Its drive speed can also be selected depending on the application.

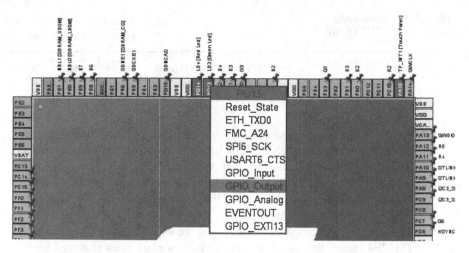

Fig. 4.10 GPIO configuration via STM32CubeMX

A sample procedure while configuring and using a GPIO pin is as follows. We will assume that the project has already been created in STM32CubeIDE as explained in Sect. 3.1.5. To modify the pin usage, first open the "Device Configuration Tool" window by clicking ".ioc" file in the project explorer window. Then, click the GPIO pin on the STM32F4 microcontroller and select the GPIO type as in Fig. 4.10 on the "Device Configuration Tool" window. If the GPIO is selected as input, then its pull-up/pull-down resistor can be activated using "GPIO Pull-up/Pull-down" option under the "System Core—GPIO" section. If the mode is selected as output, then speed can be configured through "Maximum output speed" and mode can be configured through "GPIO mode" under the "System Core—GPIO" section. The main file will be generated when the ".ioc" file is saved with changes. Hence, STM32CubeIDE will form necessary functions to set up GPIO registers via HAL library functions.

4.3.3.2 C++ Language

We can benefit from Mbed and available functions there to adjust GPIO pin properties. To do so, we should first explain how a pin is represented in Mbed. A pin is defined by its port name and number. For example, the zeroth pin in port A is defined by the constant **PA_0**. Mbed also represents the onboard LEDs and buttons by generic names. For the STM32F4 board, the green LED, red LED, and user buttons are called **LED1**, **LED2**, and **BUTTON**, respectively. These definitions make the generated code more readable and general. Hence, when another Mbed enabled board is used, the same definitions will hold. We next tabulate the pin names and constants corresponding to them for the STM32F4 board for Mbed in Table 4.1

GPIO pin setup is done by class definitions under Mbed. Let's start with the output port. There are two classes as **DigitalOut** and **PortOut** for this purpose. The user can manage a single pin via **DigitalOut**. The usage of this class is as

Table 4.1 Pin names under Mbed for the STM32F4 board

Pin	Device	Mbed	
		Pin name	Generic name
PG13	LD3 LED (green)	PG_13	LED1
PG14	LD4 LED (red)	PG_14	LED2
PA0	User button	PA_0	BUTTON

follows. We should give a name to the pin we would like to use. Assume that we want to call the green LED on the STM32F4 board by the name greenLED. We can define this LED under Mbed as DigitalOut greenLED(PG_13). From this point on, the class member greenLED can be used to reach and modify pin PG13 which is defined as output. We can manage more than one pin at a time via PortOut. To do so, the class usage is slightly different such that a mask should be formed beforehand to indicate that pin. As an example, if we want to define the green LED on the STM32F4 board, first of all we should form a mask such that its 13th bit (starting from the least significant bit) should be set to logic level 1. All other entries should be set to logic level 0. Hence, we should have #define GRN_LED_MASK 0x2000 or #define GRN_LED_MASK 0b10000000000000. In the first and second definitions, the hexadecimal and binary representations are used, respectively. Then, we can define our green LED as PortOut greenLED(PortG, GRN_LED_MASK).

We can modify and read a pin by the classes DigitalIn and PortIn. The DigitalIn class allows reading the pin entry and modifying its connected resistor by PullUp, PullDown, PullNone, or OpenDrain properties. Let's consider the user button on the STM32F4 board. We can modify and read contents of this button as follows. Assume that we would like to call the button as button. We can do so by DigitalIn button(PA_0, PullDown). Here, we define the pin PA0 as input and set the resistor connected to it in pull-down mode. Afterward, we can use the class definition button to reach and read the button value.

We can perform the same operation by the class PortIn as well. To do so, we should define a mask as in PortOut usage. For our case, this mask will be as #define BUTTON_MASK 0x0001 or #define BUTTON_MASK 0b00000001 for the hexadecimal and binary representations, respectively. Then, we can define the button class as PortIn button(PortA, BUTTON_MASK). We will consider the usage of the button class in Sect. 4.4.2.

4.3.3.3 MicroPython

We can also set up and reach GPIO pins of the STM32F4 microcontroller via MicroPython. To do so, we should first initialize the related pin using the function pyb.Pin(pinName, mode=pinMode, pull=pinPushPull, af=pinAlternateFunction). Here, pinName stands for the pin name. There are different formats for entering this input. However, we will use the format 'Pxy' in this book. Here, x and y represent the port name and pin number, respectively. As an example, we should enter 'PG13' to represent the 13th pin of port G. pinMode is used for selecting one of the predefined six modes as Pin.IN, Pin.OUT_PP,

PIN.OUT_OD, Pin.AF_PP, Pin.AF_OD, and Pin.ANALOG. Here, Pin.AF_PP and Pin.AF_OD are used for alternate functions and Pin.ANALOG is used for analog purposes. Hence, we do not consider them in this chapter. Pin.IN mode is used for setting the desired pin as digital input. Pin.OUT_PP mode is used for setting the desired pin as digital output with push-pull resistor control. Pin.OUT_OD mode is used for setting the desired pin as digital output with open-drain control. If a pull-up resistor is used in this mode, output is connected to ground with one transistor and to V_{DD} with the pull-up resistor. pinPushPull is used for enabling or disabling pull-up/pull-down resistors. Pin.PULL_NONE disables both pull-up and pull-down resistors for the desired pin. It is also the default selection. Pin.PULL_UP enables the pull-up resistor and Pin.PULL_DOWN enables the pull-down resistor. pinAlternateFunction is used for selecting the alternate function for the desired pin. We will explain this in the following chapters when other peripheral units are introduced.

4.4   Digital I/O Usage on the STM32F4 Microcontroller

We can use a GPIO pin after setting up its properties. We provide ways to do this via C, C++, and MicroPython languages. We also provide usage examples in each language.

4.4.1   C Language Usage

We can use GPIO pins through HAL library functions in C language. After configuring the pin, the level of the pin can be read using the function HAL_GPIO_ReadPin() in the code if it is configured as input. If the GPIO pin is configured as output, then it can be set to logic level 1 or 0 by the function HAL_GPIO_WritePin() in the code. Alternatively, the function HAL_GPIO_TogglePin() can be used to toggle the output level. We provide detailed explanation of the mentioned functions next.

```
GPIO_PinState HAL_GPIO_ReadPin(GPIO_TypeDef *GPIOx,uint16_t
    GPIO_Pin)
/*
GPIOx: The port name where x can be (A...H) to select the
    specific port for the device.
GPIO_Pin: The pin to be read. This parameter can be one of
    GPIO_PIN_0 ...  GPIO_PIN_15
*/

void HAL_GPIO_WritePin(GPIO_TypeDef *GPIOx, uint16_t GPIO_Pin,
    GPIO_PinState PinState)
/*
GPIO_Pin: The pin to be written to. This parameter can be one of
    GPIO_PIN_0 ... GPIO_PIN_15
PinState: Specifies the value to be written to the selected bit.
    This parameter can be one of the GPIO_PinState value
GPIO_BIT_RESET: To clear the port pin
```

```
GPIO_BIT_SET: To set the port pin
*/
```

void HAL_GPIO_TogglePin(GPIO_TypeDef *GPIOx, uint16_t GPIO_Pin)
```
/*
GPIO_Pin: The pin to be toggled. This parameter can be one of
    GPIO_PIN_0 ... GPIO_PIN_15
*/
```

Next, we will provide example codes to show how these functions can be used to control GPIO pins. Our first code, given in Listing 4.1, uses the HAL library functions to toggle the green LED (connected to pin PG13) on the STM32F4 board in an infinite loop using delays. In order to observe how this code runs, create an empty project under STM32CubeIDE and configure pin PG13 as output. Then, use the code given in Listing 4.1 to toggle the green LED on the STM32F4 board. The code also consists of an alternative way of performing the same operation. This section is labeled as "Alternatively" within the code.

Listing 4.1 Toggle the onboard green LED by delay functions, the C code

```
/* Infinite loop */
/* USER CODE BEGIN WHILE */
while (1)
{
HAL_GPIO_WritePin(GPIOG, GPIO_PIN_13, GPIO_PIN_SET);
HAL_Delay(1000);

HAL_GPIO_WritePin(GPIOG, GPIO_PIN_13, GPIO_PIN_RESET);
HAL_Delay(1000);

// Alternatively
/*
HAL_GPIO_TogglePin(GPIOG,GPIO_PIN_13);
HAL_Delay(1000);
*/

/* USER CODE END WHILE */
/* USER CODE BEGIN 3 */
}
/* USER CODE END 3 */
```

The second code, given in Listing 4.2, toggles the red and green LEDs on the STM32F4 board via user button presses. Here, pins PG13 and PG14 (connected to the green and red LEDs) are set as output. The pin PA0 (connected to the user button) is set as input. In order to observe how this code runs, create an empty project and configure the input and output pins via STM32CubeMX interface. Then, use the code given in Listing 4.2 to toggle the green and red LEDs via the user button. Within the code, there is an additional delay function to prevent switch bouncing due to button presses.

Listing 4.2 Toggle the onboard red and green LEDs by the user button presses, the C code

```
/* USER CODE BEGIN PV */
uint32_t counter = 0;
/* USER CODE END PV */

/* USER CODE BEGIN 2 */
HAL_GPIO_WritePin(GPIOG, GPIO_PIN_13, GPIO_PIN_SET);
HAL_GPIO_WritePin(GPIOG, GPIO_PIN_14, GPIO_PIN_RESET);

/* USER CODE END 2 */

/* Infinite loop */
/* USER CODE BEGIN WHILE */
while (1)
{
if (HAL_GPIO_ReadPin(GPIOA, GPIO_PIN_0) == GPIO_PIN_SET)
{
HAL_Delay(100); // Avoid Switch Bouncing
HAL_GPIO_WritePin(GPIOG, GPIO_PIN_13, GPIO_PIN_RESET);
HAL_GPIO_WritePin(GPIOG, GPIO_PIN_14, GPIO_PIN_SET);
}
else
{
HAL_Delay(100); // Avoid Switch Bouncing
HAL_GPIO_WritePin(GPIOG, GPIO_PIN_13, GPIO_PIN_SET);
HAL_GPIO_WritePin(GPIOG, GPIO_PIN_14, GPIO_PIN_RESET);
}
/* USER CODE END WHILE */
/* USER CODE BEGIN 3 */
}
/* USER CODE END 3 */
```

The third code, given in Listing 4.3, has two counters. The first counter can be used to turn on and off the green and red LEDs on the STM32F4 board via increasing its value at each button press. When mod four of the first counter has value zero, both LEDs turn off. When mod four of the first counter value has value one, only the red LED turns on. When mod four of the first counter value has value two, only the green LED turns on. When mod four of the first counter value has value three, both LEDs turn on. The second counter increases in a continuous manner in an infinite loop.

Listing 4.3 Turn on the green and red LEDs by specific button press count values, the C code

```
/* USER CODE BEGIN PV */
uint32_t counter1 = 0, counter2 = 0;
/* USER CODE END PV */

/* Infinite loop */
/* USER CODE BEGIN WHILE */
while (1)
{
if (HAL_GPIO_ReadPin(GPIOA, GPIO_PIN_0) == GPIO_PIN_SET)
{
while (HAL_GPIO_ReadPin(GPIOA, GPIO_PIN_0) == GPIO_PIN_SET);
counter1++;

if (counter1 % 4 == 0)
{
```

```
HAL_GPIO_WritePin(GPIOG, GPIO_PIN_13, GPIO_PIN_RESET);
HAL_GPIO_WritePin(GPIOG, GPIO_PIN_14, GPIO_PIN_RESET);
}
else if (counter1 % 4 == 1)
{
HAL_GPIO_WritePin(GPIOG, GPIO_PIN_13, GPIO_PIN_RESET);
HAL_GPIO_WritePin(GPIOG, GPIO_PIN_14, GPIO_PIN_SET);
}
else if (counter1 % 4 == 2)
{
HAL_GPIO_WritePin(GPIOG, GPIO_PIN_13, GPIO_PIN_SET);
HAL_GPIO_WritePin(GPIOG, GPIO_PIN_14, GPIO_PIN_RESET);
}
else if (counter1 % 4 == 3)
{
HAL_GPIO_WritePin(GPIOG, GPIO_PIN_13, GPIO_PIN_SET);
HAL_GPIO_WritePin(GPIOG, GPIO_PIN_14, GPIO_PIN_SET);
}
}
counter2++;
/* USER CODE END WHILE */

/* USER CODE BEGIN 3 */
}
/* USER CODE END 3 */
```

In Listing 4.3, there is no function added to the code to avoid switch bouncing. Instead, a hardware circuit is formed by the internal pull-down resistor and 100 ηF external capacitor (available on the STM32F4 board). Still some glitches may be observed when the button is pressed. There is a while loop placed in the code to prevent the multiple toggling operations at each button press. However, these loops block other routines, such as counter increment. The reader can observe this by placing a breakpoint to the counter increment line and pressing the button.

The fourth code, given in Listing 4.4, is the modified form of the code given in Listing 4.3. It uses an additional shift register to eliminate switch bouncing.

Listing 4.4 Turn on the green and red LEDs by specific button press count values with switch bouncing handled, the C code

```
/* USER CODE BEGIN PD */
#define DEBOUNCE_PATTERN 0x7FFF
/* USER CODE END PD */

/* USER CODE BEGIN PV */
uint32_t shiftRegButton;
uint32_t counter1 = 0, counter2 = 0;
/* USER CODE END PV */

/* Infinite loop */
/* USER CODE BEGIN WHILE */
while (1)
{
shiftRegButton = (shiftRegButton << 1) + HAL_GPIO_ReadPin(GPIOA,
    GPIO_PIN_0);
HAL_Delay(1);

if (shiftRegButton == DEBOUNCE_PATTERN)
{
counter1++;
```

```
if (counter1 % 4 == 0)
{
HAL_GPIO_WritePin(GPIOG, GPIO_PIN_13, GPIO_PIN_RESET);
HAL_GPIO_WritePin(GPIOG, GPIO_PIN_14, GPIO_PIN_RESET);
}
else if (counter1 % 4 == 1)
{
HAL_GPIO_WritePin(GPIOG, GPIO_PIN_13, GPIO_PIN_RESET);
HAL_GPIO_WritePin(GPIOG, GPIO_PIN_14, GPIO_PIN_SET);
}
else if (counter1 % 4 == 2)
{
HAL_GPIO_WritePin(GPIOG, GPIO_PIN_13, GPIO_PIN_SET);
HAL_GPIO_WritePin(GPIOG, GPIO_PIN_14, GPIO_PIN_RESET);
}
else if (counter1 % 4 == 3)
{
HAL_GPIO_WritePin(GPIOG, GPIO_PIN_13, GPIO_PIN_SET);
HAL_GPIO_WritePin(GPIOG, GPIO_PIN_14, GPIO_PIN_SET);
}
}

counter2++;
/* USER CODE END WHILE */

/* USER CODE BEGIN 3 */
}
/* USER CODE END 3 */
```

In Listing 4.4, there is a 16-bit shift register for each pin input and the read value is shifted through it. When the 0x7FFF pattern is observed for the user button, then LEDs turn on and off accordingly. The reader can observe that this code does not block the counter increment by placing a breakpoint to the counter increment line and pressing the button.

4.4.2 C++ Language Usage

Class definitions in Sect. 4.3.3.2 allow us to use them in digital output and input. Let's start with digital output first. We can change the voltage level at a certain pin by the function write() under class DigitalOut. Hence, if we want to feed logic level 1 to the green LED, after defining it as in Sect. 4.3.3.2, then we can use the command greenLED.write(1). The same function also allows directly reaching the pin by the command greenLED=1. The reader can pick one of these options as desired. If we want to use the PortOut class, then we should use the mask definition as greenLED = GRN_LED_MASK.

We can read a pin value by the function read() under the class DigitalIn. This function allows reading the pin content directly. Assume that the user button has been defined beforehand as in Sect. 4.3.3.2. Then, we can read its value by the code button.read(). We can also directly use the definition. Assume that we would like to check whether the pin value is equal to logic level 1. We can do so in two different ways as if (button.read()==1) or if (button==1). We can also use the PortIn class for this purpose.

We next provide examples on the usage of digital I/O under Mbed. The first example, blinking the green LED on the STM32F4 board, is the C++ version of the code given in Listing 4.1. In fact, this is the default code for generating a project under Mbed as given in Sect. 3.2. We modify it to show available setups for this operation as in Listing 4.5. The reader should enable and disable related parts of the code to see how the same operation can be done in different ways. To execute the code in Listing 4.5 in Mbed Studio, we should copy and paste it to the "main.cpp" file in our active project.

Listing 4.5 Toggle the onboard green LED by delay functions, the C++ code

```
#include "mbed.h"

//DigitalOut greenLED(LED1);
//DigitalOut greenLED(PG_13);

#define GRN_LED_MASK 0x2000
//#define GRN_LED_MASK 0b10000000000000
PortOut greenLED(PortG, GRN_LED_MASK);

int main()
{
while (true)
{
greenLED = 0;
//greenLED.write(0);
//greenLED = 0;

thread_sleep_for(1000);

//greenLED = 1;
//greenLED.write(1);
greenLED = GRN_LED_MASK;

thread_sleep_for(1000);

//Alternatively
/*
greenLED = !greenLED;
thread_sleep_for(1000);
*/
}
}
```

Our second Mbed example on the usage of GPIO functions is the modified form of the code in Listing 4.2. As a reminder, this code toggles the green and red LEDs based on the user button presses. We provide the C++ code for this operation in Listing 4.6.

Listing 4.6 Toggle the onboard red and green LEDs by the user button presses, the C++ code

```
#include "mbed.h"

DigitalOut greenLED(LED1);
```

```
DigitalOut redLED(LED2);
//DigitalIn button(BUTTON1);

#define BUTTON_MASK 0x0001
//#define BUTTON_MASK 0b00000001
PortIn button(PortA, BUTTON_MASK);

int main()
{
greenLED = 1;
redLED = 0;
while (true)
{
if (button == 1)
{
thread_sleep_for(100);
greenLED = 0;
redLED = 1;
}
else
{
thread_sleep_for(100);
greenLED = 1;
redLED = 0;
}
}
}
```

Our third Mbed example on the usage of GPIO functions is the modified form
of the code in Listing 4.3. As a reminder, this code turns on and off the green and
red LEDs based on the user button count values. We provide the C++ code for this
operation in Listing 4.7.

Listing 4.7 Turn on the green and red LEDs by specific button press count values, the C++ code

```
#include "mbed.h"

DigitalOut greenLED(LED1);
DigitalOut redLED(LED2);
DigitalIn button(BUTTON1);

int counter1 = 0, counter2 = 0;

int main()
{
greenLED = 0;
redLED = 0;

while (true)
{
if (button)
{
while (button);
counter1++;

if (counter1 % 4 == 0)
{
greenLED = 0;
redLED = 0;
}
else if (counter1 % 4 == 1)
```

```
{
greenLED = 0;
redLED = 1;
}
else if (counter1 % 4 == 2)
{
greenLED = 1;
redLED = 0;
}
else if (counter1 % 4 == 3)
{
greenLED = 1;
redLED = 1;
}
}

counter2++;
}
}
```

Our fourth Mbed example on the usage of GPIO functions is the modified form of the code in Listing 4.7. As in Listing 4.4, we use an additional shift register to eliminate switch bouncing in this code. We provide the C++ code for this operation in Listing 4.8.

Listing 4.8 Turn on the green and red LEDs by specific button press count values with switch bouncing handled, the C++ code

```
#include "mbed.h"
#define DEBOUNCE_PATTERN 0x7FFF

DigitalOut greenLED(LED1);
DigitalOut redLED(LED2);
DigitalIn button(BUTTON1);

int counter1 = 0, counter2 = 0;
uint32_t shiftRegButton;

int main()
{
greenLED = 0;
redLED = 0;

while (true)
{
shiftRegButton = (shiftRegButton << 1) + button;
thread_sleep_for(1);

if (shiftRegButton == DEBOUNCE_PATTERN)
{
counter1++;

if (counter1 % 4 == 0)
{
greenLED = 0;
redLED = 0;
}
else if (counter1 % 4 == 1)
{
greenLED = 0;
redLED = 1;
```

```
}
else if (counter1 % 4 == 2)
{
greenLED = 1;
redLED = 0;
}
else if (counter1 % 4 == 3)
{
greenLED = 1;
redLED = 1;
}
}

counter2++;
}
}
```

4.4.3   MicroPython Usage

We can also benefit from MicroPython to use GPIO pins. To do so, we should apply the following procedure. After the desired pin is initialized as introduced in Sect. 4.3.3.3, it can be controlled using the function Pin.value(value). If the pin is configured as input, then we should use this function as value = Pin.value() to get the pin value. If the pin is configured as output, it is used as Pin.value(1) or Pin.value(0) for setting or resetting the output, respectively. Here, functions Pin.high() and Pin.on() can also be used instead of Pin.value(1). Likewise, functions Pin.low() and Pin.off() can be used instead of Pin.value(0).

Now, we will redo the examples given in the previous sections using MicroPython. First, we toggle the green LED (connected to pin PG13) on the STM32F4 board in an infinite loop using delays as in Listings 4.1 and 4.5. The MicroPython code for this example is given in Listing 4.9.

Listing 4.9   Toggle the onboard green LED by delay functions, the MicroPython code

```
import pyb

def main():
    greenLED = pyb.Pin('PG13', mode=pyb.Pin.OUT_PP)
    while True:
        greenLED.value(1)
        pyb.delay(1000)
        greenLED.value(0)
        pyb.delay(1000)

main()
```

Second, we repeat the examples in Listings 4.2 and 4.6. As a reminder, we toggle the red and green LEDs on the STM32F4 board via user button presses here. The MicroPython code for this example is given in Listing 4.10.

Listing 4.10 Toggle the onboard red and green LEDs by the user button presses, the MicroPython code

```
import pyb

def main():
    greenLED = pyb.Pin('PG13', mode=pyb.Pin.OUT_PP)
    redLED = pyb.Pin('PG14', mode=pyb.Pin.OUT_PP)
    userButton = pyb.Pin('PA0', mode=pyb.Pin.IN, pull=pyb.Pin.
        PULL_DOWN)
    while True:
        if userButton.value() == 1:
            pyb.delay(100)
            greenLED.value(0)
            redLED.value(1)
        elif userButton.value() == 0:
            pyb.delay(100)
            greenLED.value(1)
            redLED.value(0)

main()
```

Third, we repeat the examples in Listings 4.3 and 4.7. The MicroPython code for this example is given in Listing 4.11.

Listing 4.11 Turn on the green and red LEDs by specific button press count values, the MicroPython code

```
import pyb

def main():
    counter1 = 0
    counter2 = 0
    greenLED = pyb.Pin('PG13', mode=pyb.Pin.OUT_PP)
    redLED = pyb.Pin('PG14', mode=pyb.Pin.OUT_PP)
    userButton = pyb.Pin('PA0', mode=pyb.Pin.IN, pull=pyb.Pin.
        PULL_DOWN)
    while True:
        if userButton.value() == 1:
            while userButton.value() == 1:
                pass
            counter1 = counter1 + 1
            if counter1 % 4 == 0:
                greenLED.value(0)
                redLED.value(0)
            elif counter1 % 4 == 1:
                greenLED.value(0)
                redLED.value(1)
            elif counter1 % 4 == 2:
                greenLED.value(1)
                redLED.value(0)
            elif counter1 % 4 == 3:
                greenLED.value(1)
                redLED.value(1)
            counter2 = counter2 + 1

main()
```

Fourth, we repeat the examples in Listings 4.4 and 4.8. As a reminder, these are the modified codes such that a shift register is added to eliminate switch bouncing.

Therefore, we also modify the MicroPython code in Listing 4.11 in the same way and provide the modified code in Listing 4.12.

Listing 4.12 Turn on the green and red LEDs by specific button press count values with switch bouncing handled, the MicroPython code

```python
import pyb

def main():
    DEBOUNCE_PATTERN = const(0x7FFF)
    shiftRegButton = 0
    counter1 = 0
    counter2 = 0
    greenLED = pyb.Pin('PG13', mode=pyb.Pin.OUT_PP)
    redLED = pyb.Pin('PG14', mode=pyb.Pin.OUT_PP)
    userButton = pyb.Pin('PA0', mode=pyb.Pin.IN, pull=pyb.Pin.
        PULL_DOWN)
    while True:
        shiftRegButton = ((shiftRegButton << 1) + userButton.
            value()) & 0xFFFF
        pyb.delay(1)
        if shiftRegButton == DEBOUNCE_PATTERN:
            counter1 = counter1 + 1
            if counter1 % 4 == 0:
                greenLED.value(0)
                redLED.value(0)
            elif counter1 % 4 == 1:
                greenLED.value(0)
                redLED.value(1)
            elif counter1 % 4 == 2:
                greenLED.value(1)
                redLED.value(0)
            elif counter1 % 4 == 3:
                greenLED.value(1)
                redLED.value(1)
        counter2 = counter2 + 1

main()
```

4.5   Application: Digital Input and Output Operations in the Robot Vacuum Cleaner

In this chapter, we implement the three features of our robot vacuum cleaner using digital IO functions. Therefore, we start with controlling the LEDs and buttons on the robot. We next measure the dust chamber on the robot. If it is full, then we inform the reader with the LED. Finally, we detect obstacles on the right and left side of the robot with digital distance sensors. We provided the equipment list, circuit diagram, detailed explanation of design specifications, and peripheral unit settings in a separate document in the folder "Chapter4\EoC_applications" of the accompanying supplementary material. In the same folder, we also provided the complete C, C++, and MicroPython codes for the application.

4.6   Summary of the Chapter

Through digital input and output, the STM32F4 microcontroller can feed voltage levels (corresponding to bit values) to a connected device or electronic component (such as LED). Likewise, the microcontroller can receive bit values (converted from voltage levels) from the connected device or electronic component (such as button and voltage source). These let the microcontroller interact with the outside world. Therefore, we introduced digital input and output, with all its components, in this chapter. Hence, we started with the standards on representing a bit value as voltage level. The aim here was to emphasize that the bit value is converted to voltage level when fed to output from the microcontroller. Likewise, received voltage level is converted to the corresponding bit value when read from input of the microcontroller. We next considered different structures such as interfacing the microcontroller with inputs having a high voltage value or output to loads requiring high current or voltage. Through these, the microcontroller can interact with a wide variety of external devices and electronic components. Finally, we focused on how digital input and output can be set up and used in the STM32F4 microcontroller. We also applied these methods on a real-world system as the end of chapter application. This will be the backbone of applications to be considered in the following chapters.

Problems

4.1. Summarize voltage standards to represent logic level 0 and 1 in a digital system.

4.2. What should we understand from the undefined region in Fig. 4.1?

4.3. What does active high and low setup mean in button interfacing? Which setup has been used in the STM32F4 board?

4.4. Form a circuit for bidirectional communication between the STM32F4 micro-controller and a sensor. The sensor has 5 V input-output voltage level. The sensor cannot use the same pin for both input and output. Therefore, two different pins on the STM32F4 board should be used.

4.5. The STM32F4 microcontroller will be used for a smart home system. It is necessary to control two bulbs connected to the lighting system. Each bulb needs 220 V AC voltage to operate. Hence, design a circuit between the bulbs and STM32F4 board.

4.6. Design a hardware circuit to eliminate switch bouncing in a push button. Setup for the button is active low input, normally open.

4.7. Can we detect switch bouncing by code only?

4.8. Parity check is an important method used in digital communication and data storage. We will simulate this method by forming a character array with nine entries. Each array entry can be 0 or 1. In actual implementation, eight bits represent the transmitted data. The ninth bit is added to satisfy the parity condition (whether being even or odd parity). Form a project such that when the (selected) parity condition is satisfied in the character array, the green LED on the STM32F4 board is turned on. Otherwise, the red LED on the board is turned on. Form the project using:

(a) C language under STM32CubeIDE.
(b) C++ language under Mbed.
(c) MicroPython.

4.9. We have an array composed of 16 entries each being either 0 or 1. Form a project to check the array entries periodically, with 1 s period. To do so, we can benefit from an appropriate delay function in C, C++, or MicroPython languages. If the checked array value is 0, then the red LED on the STM32F4 board should turn on. Otherwise, the green LED on the board should turn on. Form the project using:

(a) C language under STM32CubeIDE.
(b) C++ language under Mbed.
(c) MicroPython.

4.10. Form a project such that red and green LEDs on the STM32F4 board start toggling at 1 s interval. To do so, we can benefit from an appropriate delay function in C, C++, or MicroPython languages. There should be a 0.5 s offset between the red and green LEDs. The toggling should continue ten times. Then, the toggle period should increase to 2 s. After ten more toggles, the period should decrease to 1 s again. This operation should continue periodically with this setup indefinitely. Form the project using:

(a) C language under STM32CubeIDE.
(b) C++ language under Mbed.
(c) MicroPython.

4.11. Restructure the project in Problem 4.10 such that period changes occur by button presses. Hence, when the user button on the STM32F4 board is pressed once, toggling of the red and green LEDs should start with 1 s interval. When the user button is pressed twice, the toggling period should become 2 s. When the button is pressed thrice, the toggling period should decrease to 1 s again. This operation should continue periodically with this setup indefinitely.

4.12. We would like to form a digital dice. We throw the dice by pressing the user button on the STM32F4 board. Output of the digital dice is the red and green LEDs on the board. Since we have two LEDs, the dice can produce four values as 00 (both LEDs off), 01 (only the red LED is on), 10 (only the green LED is on), and 11 (both LEDs are on). Form a project to satisfy these constraints using:

(a) C language under STM32CubeIDE. Please use the functions srand() and rand() to generate random numbers.
(b) C++ language under Mbed. Please use the functions srand() and rand() to generate random numbers.
(c) MicroPython. Please use the function pyb.rng() to generate random numbers.

References

1. Analog Devices: Low Voltage Logic Interfacing, mt-098 rev.0, 01/09, wk edn. (2009)
2. Horowitz, P., Hill, W.: The Art of Electronics, Cambridge University Press, 3rd edn. Newness (2015)
3. NXP: Level shifting techniques in I2C-Bus design, an10441 edn. (2007)
4. Sedra, A.S., Smith, K.C.: Microelectronic Circuits, 5th edn. Oxford University Press, Oxford (2004)
5. STMicroelectronics: STM32F427xx STM32F429xx, docid024030 rev 10 edn. (2018)
6. STMicroelectronics: STM32F405/415, STM32F407/417, STM32F427/437 and STM32F429/439 advanced Arm-based 32-bit MCUs, rm0090 rev 19 edn. (2021)
7. Texas Instruments: Logic guide, SDYU001AB, (2017)

Interrupts and Power Management

5

5.1 The Interrupt Concept in Embedded Systems

The interrupt usage is almost inevitable in embedded systems. On the other hand, interrupts depend on a signal generated either by hardware or another dedicated software module. Hence, handling and using the interrupt is slightly different from standard programming practices. Therefore, we will start with explaining the general interrupt concepts in this section. Then, we will explore how interrupts are handled in the STM32F4 microcontroller.

5.1.1 Interrupts in General

An embedded system should be interacting with the outside world by its nature. Moreover, it should respond to a request coming from the user. One option to handle such requests is forming an infinite loop in the code and checking requests in a continuous manner. Although this method is useful in its own rights, it blocks the CPU. To overcome this problem and freeing up the CPU from waiting in an infinite loop, a different programming approach is needed. The interrupt concept comes into play for this purpose. Within this approach, asynchronous or periodic requests can be handled without letting the CPU to wait for them indefinitely. One can think of interrupt as a response mechanism to an input signal such that when it occurs, a predefined function is executed by the CPU. The difference of the interrupt is that

Supplementary Information The online version contains supplementary material available at (https://doi.org/10.1007/978-3-030-88439-0_5).

when the required input is not fed to the microcontroller, the CPU performs its usual synchronous operation. Hence, it does not wait for an event to occur in the infinite loop.

Let us summarize the overall interrupt operation on an actual example. Assume that the microcontroller controls a DC motor by sending a signal to it periodically. We will explain how this can be done in Chap. 13. The user can press a button to start or stop the motor. Another button can be used to change rotation direction of the motor. These button presses can happen any time. Hence, the CPU should handle these asynchronous inputs besides feeding the signal to the motor. Now, let us formalize this setup by using interrupts.

In our example, the interrupts are generated when two separate buttons connected to the microcontroller pins are pressed. Hence, we have GPIO interrupts. To handle these, we should set up the microcontroller such that related GPIO pins should be able to generate interrupt requests. In other words, the related pins should be allowed to generate interrupts. Hence, the programmer should adjust properties of each pin based on its registers. The user should also set priority to each interrupt request. In other words, the user should rank the interrupt requests coming from the start-stop and rotation direction buttons. The usual practice here is that the request coming from the start-stop button should be handled first. Hence, it should have precedence over the rotation direction button interrupt request. As these settings are done, the pins will be ready to form interrupt requests. Besides, the CPU should be set to accept and handle these requests. This is done by setting a special register within the CPU.

As the microcontroller is set up to handle the interrupt requests, let us observe what happens when they occur. When a button (either controlling the start-stop or rotation direction) connected to the pin is pressed, the voltage difference there is sensed by an edge detector circuit connected to the pin of the microcontroller. As a result, a flag is set indicating that the pin generated an interrupt request.

A special module handles this interrupt request. To do so, it first checks whether the request is allowed or not. Then, the module checks priority of the incoming interrupt such that it should precede the existing operation. We have three scenarios based on the current state of the CPU. We will explain them in detail in Sect. 5.1.2.3.

What we mean by responding to the interrupt request is as follows. The CPU is diverted to the function to be executed as a response to the interrupt request. This function is called interrupt service routine (ISR) or callback function. For our example, the motor start-stop and rotation direction change operations should be done within their specific callback functions. As a note, these functions are defined within the code. However, they are not called by their name in the code. Instead, the hardware signal generated by the associated button press calls each function. As the CPU starts executing the callback function, the related flag should be reset. Hence, the CPU knows that the interrupt is handled and there is no need to pay attention to it again. As the operation within the callback function is done, the CPU turns back to its previous operation.

As can be seen in our example, the asynchronous button press is processed by the interrupt operation such that it is handled by a different mechanism. The reader should pay attention to some key issues here. The first and most important one is that the time spent in the callback function should be short. Hence, the usual synchronous operations will not be affected much. The second issue is that, since callback is yet another function, data exchange between it and the main function should be handled appropriately. One option here is using global variables. Third, the interrupt request flag should be reset when the callback function is executed. Hence, the CPU will not be confused by assuming that the interrupt request is not handled and try to serve it again.

The GPIO interrupt example we have provided falls into the maskable interrupts case. This means that the user can disable an interrupt request within this group. Hence, it may not be served at all. There is another group called non-maskable interrupts. As the name implies, an interrupt request within this group must be handled. These interrupts represent key operations such that when not served, they cause the microcontroller to malfunction. Reset is one such interrupt. As a matter of fact, when we debug our code and embed it on the microcontroller, a reset signal is generated such that the CPU goes to the beginning of the code to be executed. We may also generate the reset signal by pressing the reset button on the board. When this signal is generated, the CPU stops what it is doing and turns back to the beginning of the code.

5.1.2   Interrupts in the STM32F4 Microcontroller

Interrupts fall into the general exceptions category in Arm® architecture. An exception can be taken as an operation changing program execution flow of the CPU. It can either originate from an internal source, such as faulty operations within the CPU, or from hardware peripheral unit, such as GPIO pins. We will not deal with the former exception concepts in this book. Instead, we direct the reader to Yiu's [3] excellent book on Arm® Cortex™-M microcontrollers for exceptions originating from faulty operations. We will only focus on interrupt handling and usage instead. To note here, there is also event under the exception category. Different from the interrupt, the event does not direct the CPU to the callback function. Instead, it is used to trigger the peripheral units most of the times. For more information on events, please see [2].

We should also mention operation modes of the CPU based on interrupts. When executing a sequential code, the CPU is in thread mode. When an interrupt request comes and the CPU starts executing the related callback function, then it will be in handler mode. The thread and handler modes have different ways of reaching registers. However, this topic is beyond the scope of this book. Therefore, we again direct the reader to Yiu's [3] book on this topic. We will briefly mention the thread and handler modes in the following sections whenever needed.

We provide block diagram of the STM32F4 interrupt-based modules in Fig. 5.1. As can be seen in this figure, there are four main blocks related to interrupt handling in the STM32F4 microcontroller. The first block contains peripheral units (such as timers, ADC, and UART modules) which can generate interrupt requests. Within

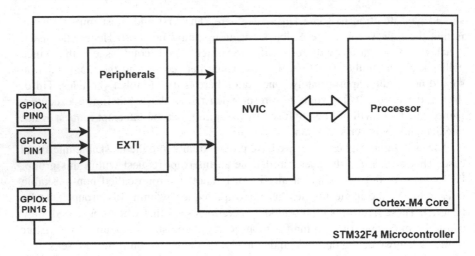

Fig. 5.1 Block diagram of the STM32F4 interrupt-based modules

the second block, GPIO pin interrupts are handled by the extended interrupts and events controller (EXTI). In the third block, all interrupt requests (whether generated by a peripheral unit or EXTI) are handled by the nested vector interrupt controller (NVIC). In the fourth block, accepted interrupt requests are processed by the CPU. Next, we will focus on each module separately to emphasize its role in interrupt handling and execution stages.

5.1.2.1 Interrupt Operations in Peripheral Units

As explained in the previous section, peripheral units can generate interrupt requests. This is done in several ways. For example, a GPIO pin can generate an interrupt request when it observes a low to high or high to low voltage transition. The timer can generate periodic interrupt requests. The ADC module can generate an interrupt request when the analog to digital conversion operation is finished. The UART module can generate an interrupt request after receiving or submitting a packet. We will focus on interrupt-based operations for each peripheral unit in its dedicated chapter except GPIO interrupts. The EXTI module handles GPIO interrupt operations in the STM32F4 microcontroller. Therefore, we will introduce this module next.

There is a general structure for interrupt operation via peripheral units. Each peripheral unit has a specific register to enable or disable its interrupts. Besides, the unit has a flag indicating the interrupt request. In other words, if an interrupt request is generated by the peripheral unit, a flag is raised to indicate that. Then, the request is handled by the NVIC module.

5.1.2.2 Extended Interrupts and Events Controller

The STM32F4 microcontroller has a dedicated module, called extended interrupts and events controller (EXTI), to handle external interrupts from some peripheral

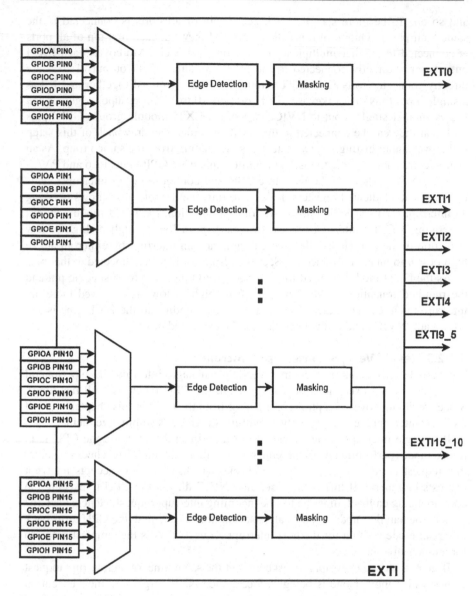

Fig. 5.2 Detailed block diagram of the EXTI module

units. In this chapter, we will focus on this module while handling GPIO interrupts. As can be seen in Fig. 5.1, the EXTI module stands between the GPIO pins and NVIC module. We provide a more detailed block diagram of the EXTI module in Fig. 5.2.

As can be seen in Fig. 5.2, GPIO pins are grouped within themselves and fed to separate multiplexers. Each multiplexer has an output pin called EXTI0, EXTI1, and so on. To be more specific, each zeroth pin of all ports is connected to the same multiplexer. Output of it is called EXTI0. Likewise, each first pin of all ports is connected to another multiplexer having output EXTI1. Moreover, each EXTI output is not directly connected to the NVIC module. EXTI outputs 1 to 4 are directly connected to NVIC. EXTI outputs 5 to 9 are grouped together and fed as a single input to NVIC. Likewise, EXTI outputs 10 to 15 are grouped together and fed as another single input to NVIC. Hence, the EXTI module groups GPIO pins such that they can be connected to the NVIC module. The drawback of this setup is that we cannot distinguish two interrupts originating from the same group. As an example, it is not possible to distinguish interrupts from GPIO pins PA6 and PA7.

The EXTI module can be used to enable and configure GPIO interrupts. This is done by its dedicated registers. Related to this, the module also allows masking an interrupt for its specific output. As for configuration, the EXTI module allows generating a GPIO interrupt request by observing a low to high or high to low transition in the pin. Hence, the user can generate an interrupt by either an active high or low input (as introduced in Sect. 4.2) from the button connected to that pin.

As the EXTI module is used to enable the GPIO interrupt for a specific pin and the required transition (either low to high or high to low) is observed there, an interrupt request is generated. Hence, a bit corresponding to the EXTI line is set. Afterward, NVIC handles the operations to be explained next.

5.1.2.3 Nested Vectored Interrupt Controller

The nested vectored interrupt controller (NVIC) manages interrupts from peripheral units and internal exceptions originating from the CPU (such as fault handling). Since we do not cover exceptions originating from the CPU in this chapter, we will focus on interrupt handling operations within NVIC. Let's summarize them.

If a single interrupt request comes from a peripheral unit while the CPU is in thread mode (performing its usual sequential operation), then NVIC checks whether that request is allowed or not. It also checks whether the CPU accepts interrupt requests in general. If this is the case, then NVIC directs the CPU to execute the corresponding callback function for the incoming interrupt request. Hence, the CPU goes to the handler mode. As the callback function is executed, the CPU turns back to thread mode waiting for the next interrupt request. This is the simplest scenario for an interrupt operation.

If more than one interrupt request comes at the same time, or an interrupt request comes while another one is being serviced, then NVIC organizes their execution based on their priority level. To do so, interrupt requests coming from different sources should be ordered. This is done within NVIC. Although there is a standard order for interrupt priority levels, the user can adjust them by his or her needs while forming the project. The important point here is that a priority level cannot exceed three important (and non-maskable) interrupts.

As the priority level is set and a new interrupt request comes while another one is being serviced, the following scenarios can occur. First, if the CPU is in thread mode, then the priority of the incoming interrupt precedes it. Second, the CPU may be in handler mode responding to another interrupt with higher priority. In this case, response to the incoming low priority interrupt should wait till the end of this operation. Then, the low priority interrupt is handled. Third, the CPU may be in handler mode responding to an interrupt request with lower priority. Since the new incoming interrupt request has higher priority, the low priority one is suspended and the incoming new interrupt is handled. Afterward, the CPU turns back to handle the low priority interrupt request.

5.1.2.4 Interrupt Operations in the CPU

As the NVIC module handles the interrupt request and directs the program flow, then the request is processed by the CPU. The first operation here is checking whether the interrupt requests are generally allowed by the CPU. If this condition is satisfied and CPU allows processing interrupt requests, then the corresponding callback function is executed by the CPU. Hence, CPU goes to handler mode. To note here, it is the user's responsibility to form the callback function. Besides, the corresponding interrupt flag should be reset within this function. Hence, the CPU will not serve it again. This reset operation is done automatically for most peripheral units. As the interrupt is serviced, then the CPU turns back to the thread mode again.

5.2 Interrupt Setup in the STM32F4 Microcontroller

We should prepare the STM32F4 microcontroller to respond to interrupts. We show how this can be done in C, C++, and MicroPython languages in this section. To note here, while setting up interrupt properties in C language, we will use the STM32CubeMX interface.

5.2.1 Interrupt Setup via C Language

STM32CubeIDE adjusts all interrupt related hardware properties, such as interrupt priority and NVIC module setup, automatically as the project is compiled. Let's look at an example while configuring a GPIO pin interrupt.

Assume that the project has already been created in STM32CubeIDE following the steps in Sect. 3.1.3. First, open the "Device Configuration Tool" window by clicking the ".ioc" file in the project explorer window. Hence, STM32CubeMX starts. Then, click on the GPIO pin on the STM32F4 microcontroller and select the GPIO type as "GPIO_EXTI" for pin interrupt source as in Fig. 5.3. Under the "System Core—GPIO" section, GPIO mode can be selected among the following options as "external interrupt mode with rising edge trigger detection," "external interrupt mode with falling edge trigger detection," "external interrupt mode with rising/falling edge trigger detection," "external event mode with rising edge trigger detection," "external event mode with falling edge trigger detection," and "external event mode with rising/falling edge trigger detection." The difference between the event and interrupt at this stage is as follows. If an event-based option is selected, then the CPU goes to the main loop in the code after waking up from a sleep mode

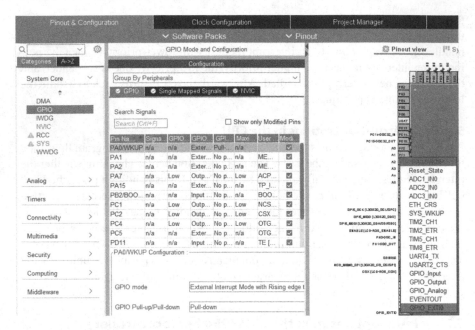

Fig. 5.3 Interrupt configuration in STM32CubeMX

to be explained in Sect. 5.4. In the interrupt selection, the CPU starts executing the callback function after waking up from the sleep mode. Select appropriate type such as "external interrupt mode with falling edge trigger detection" here. In the same section, the pull-up/pull-down resistor can be activated using the "GPIO Pull-down" option. Select pull-up/pull-down resistor here. Afterward, enable the EXTI line interrupt under NVIC tab located in the same window or under "System Core—NVIC" section. The reader can also change the interrupt priority under this section.

5.2.2   Interrupt Setup via C++ Language

In order to set up GPIO interrupts in Mbed, we should first adjust GPIO pins. This can be done by the class `InterruptIn`. The usage of this class is similar to the class definition `DigitalIn` introduced in Sect. 4.4.2. Different from there, this class definition also enables pin interrupts. We should also enable the general interrupts in the main function. This can be done by the function `__enable_irq()`. Likewise, we can disable maskable interrupts by the function `__disable_irq()` available under Mbed. Hence, we can free the CPU to respond maskable interrupts if needed.

While setting up the interrupt, we should also form the callback function under Mbed. Assume that we want to toggle the green LED on the STM32F4 board at each user button press. We can define a function performing this operation as a response to the GPIO interrupt originating from the user button. We will form our interrupt class as `InterruptIn button(BUTTON, PullDown)`. Then, we can form our callback function.

5.2.3 Interrupt Setup via MicroPython

GPIO interrupts can also be controlled in MicroPython. The function `pyb.ExtInt(pin, mode, pull, callback)` is used to configure interrupt properties for the desired pin. Here, `pin` is the pin name the interrupt source is connected to. `mode` is the interrupt mode and can be `pyb.ExtInt.IRQ_RISING` for triggering on rising edge, `pyb.ExtInt.IRQ_FALLING` for triggering on falling edge, or `pyb.ExtInt.IRQ_RISING_FALLING` for triggering on both edges. `pull` is the internal pull-up/pull-down resistor. It can be `pyb.Pin.PULL_NONE` for disabling the pull-up/pull-down resistor, `pyb.Pin.PULL_UP` for enabling the pull-up resistor or `pyb.Pin.PULL_DOWN` for enabling the pull-down resistor. `callback` is the name of the interrupt callback function in MicroPython. It has one input as the line that triggers the interrupt.

The pyb module in MicroPython has no predefined function for changing interrupt priorities. Therefore, we created a custom function `set_interrupt_priority(irq_number, priority_level)`. Here, `irq_number` is the predefined number for each interrupt source. `priority_level` is the desired priority level which can be between 0 and 15. We provide this function in Listing 5.1. Here, `Priority_base_address` is the base address for interrupt priority level registers. Each 8-bit priority level register can be accessed using this base register address and related interrupt source number. A list of all interrupt sources and related interrupt source numbers are also provided in this file. Finally, it can be seen that priority level is shifted by 4 bytes before it is saved to the related register. The reason for this is that the priority level is kept in highest 4 bits of the related register.

Listing 5.1 Custom function for changing interrupt priorities in MicroPython

```
# Import stm module to project
import stm

# Constant base address for interrupt priority level register
Priority_base_address = const(0xE000E400)

# Constant variables for interrupt number definitions
WWDG_IRQn = const(0) # Window WatchDog Interrupt
PVD_IRQn = const(1) # PVD through EXTI Line detection Interrupt
TAMP_STAMP_IRQn = const(2) # Tamper and TimeStamp interrupts
    through the EXTI line
RTC_WKUP_IRQn = const(3) # RTC Wakeup interrupt through the EXTI
    line
FLASH_IRQn = const(4) # FLASH global Interrupt
RCC_IRQn = const(5) # RCC global Interrupt
EXTI0_IRQn = const(6) # EXTI Line0 Interrupt
EXTI1_IRQn = const(7) # EXTI Line1 Interrupt
EXTI2_IRQn = const(8) # EXTI Line2 Interrupt
EXTI3_IRQn = const(9) # EXTI Line3 Interrupt
EXTI4_IRQn = const(10) # EXTI Line4 Interrupt
```

```
DMA1_Stream0_IRQn = const(11) # DMA1 Stream 0 global Interrupt
DMA1_Stream1_IRQn = const(12) # DMA1 Stream 1 global Interrupt
DMA1_Stream2_IRQn = const(13) # DMA1 Stream 2 global Interrupt
DMA1_Stream3_IRQn = const(14) # DMA1 Stream 3 global Interrupt
DMA1_Stream4_IRQn = const(15) # DMA1 Stream 4 global Interrupt
DMA1_Stream5_IRQn = const(16) # DMA1 Stream 5 global Interrupt
DMA1_Stream6_IRQn = const(17) # DMA1 Stream 6 global Interrupt
ADC_IRQn = const(18) # ADC1, ADC2 and ADC3 global Interrupts
EXTI9_5_IRQn = const(23) # External Line[9:5] Interrupts
TIM1_BRK_TIM9_IRQn = const(24) # TIM1 Break interrupt and TIM9
    global interrupt
TIM1_UP_TIM10_IRQn = const(25) # TIM1 Update Interrupt and TIM10
    global interrupt
TIM1_TRG_COM_TIM11_IRQn = const(26) # TIM1 Trigger and
    Commutation Interrupt and TIM11 global interrupt
TIM1_CC_IRQn = const(27) # TIM1 Capture Compare Interrupt
TIM2_IRQn = const(28) # TIM2 global Interrupt
TIM3_IRQn = const(29) # TIM3 global Interrupt
TIM4_IRQn = const(30) # TIM4 global Interrupt
I2C1_EV_IRQn = const(31) # I2C1 Event Interrupt
I2C1_ER_IRQn = const(32) # I2C1 Error Interrupt
I2C2_EV_IRQn = const(33) # I2C2 Event Interrupt
I2C2_ER_IRQn = const(34) # I2C2 Error Interrupt
SPI1_IRQn = const(35) # SPI1 global Interrupt
SPI2_IRQn = const(36) # SPI2 global Interrupt
USART1_IRQn = const(37) # USART1 global Interrupt
USART2_IRQn = const(38) # USART2 global Interrupt
EXTI15_10_IRQn = const(40) # External Line[15:10] Interrupts
RTC_Alarm_IRQn = const(41) # RTC Alarm (A and B) through EXTI
    Line Interrupt
OTG_FS_WKUP_IRQn = const(42) # USB OTG FS Wakeup through EXTI
    line interrupt
DMA1_Stream7_IRQn = const(47) # DMA1 Stream7 Interrupt
SDIO_IRQn = const(49) # SDIO global Interrupt
TIM5_IRQn = const(50) # TIM5 global Interrupt
SPI3_IRQn = const(51) # SPI3 global Interrupt
DMA2_Stream0_IRQn = const(56) # DMA2 Stream 0 global Interrupt
DMA2_Stream1_IRQn = const(57) # DMA2 Stream 1 global Interrupt
DMA2_Stream2_IRQn = const(60) # DMA2 Stream 2 global Interrupt
DMA2_Stream3_IRQn = const(59) # DMA2 Stream 3 global Interrupt
DMA2_Stream4_IRQn = const(60) # DMA2 Stream 4 global Interrupt
OTG_FS_IRQn = const(67) # USB OTG FS global Interrupt
DMA2_Stream5_IRQn = const(68) # DMA2 Stream 5 global Interrupt
DMA2_Stream6_IRQn = const(69) # DMA2 Stream 6 global Interrupt
DMA2_Stream7_IRQn = const(70) # DMA2 Stream 7 global Interrupt
USART6_IRQn = const(71) # USART6 global interrupt
I2C3_EV_IRQn = const(72) # I2C3 event interrupt
I2C3_ER_IRQn = const(73) # I2C3 error interrupt
FPU_IRQn = const(81) # FPU global interrupt
SPI4_IRQn = const(84) # SPI4 global Interrupt

# Function to set ptiority level of selected interrupt source
def set_interrupt_priority(irq_number, priority_level):
        stm.mem8[Priority_base_address + irq_number] =
            priority_level << 4
```

5.3 Interrupt Usage in the STM32F4 Microcontroller

As the setup is done, we can use interrupts for our specific purposes. Here, we provide C, C++, and MicroPython language-based examples. To be more readable, we handled the same examples in three languages. Hence, the reader can grasp the interrupt usage in any language. To note here, we only considered GPIO interrupts in this section since we have not covered the remaining peripheral units yet. We will cover further interrupt examples in the following chapters as other peripheral units are introduced.

5.3.1 Interrupt Usage via C Language

HAL library has special functions to handle interrupts. These will be of use when an interrupt-based code is written in C language under STM32CubeIDE. Next, we only provide the one related to GPIO interrupts in detail.

```
void HAL_GPIO_EXTI_Callback (uint16_t GPIO_Pin)
/*
GPIO_Pin: The pin to be connected to the corresponding EXTI line.
    This parameter can be one of GPIO_PIN_0 ... GPIO_PIN_15
*/
```

When a GPIO interrupt occurs, the NVIC module directs the CPU to its corresponding ISR. The HAL library handles the ISR such that within it the `HAL_GPIO_EXTI_Callback` is called. Without complicating the concepts, we can safely say that `HAL_GPIO_EXTI_Callback` is the callback function called when a GPIO interrupt occurs. We should define this function within the "main.c" file under STM32CubeIDE. To note here, this function is generated in weak form (without any content) automatically in the file "stm32l4xx_hal_gpio.c" when a project is generated. Hence, when the reader forgets defining the actual `HAL_GPIO_EXTI_Callback` function, the code still runs. When the user defines his or her function, then this weak function is ignored.

We next provide two examples on the usage of GPIO interrupt functions. The first example, with the C code given in Listing 5.2, aims to familiarize the reader with the setup and usage of interrupts. Here, when the onboard button (connected to PA0 pin of the STM32F4 board) is pressed and released, the green and red LEDs (connected to PG13 and PG14 pins of the STM32F4 board) toggle. The toggling operation is done in the callback function. As a side note, this example performs the same operation as in Listing 4.2 but with interrupt usage.

Listing 5.2 Toggle the onboard red and green LEDs by GPIO interrupt, the C code

```
/* USER CODE BEGIN PV */
int counter1 = 0, counter2 = 0;
/* USER CODE END PV */

/* USER CODE BEGIN 0 */
void HAL_GPIO_EXTI_Callback(uint16_t GPIO_Pin)
```

```
{
if (GPIO_Pin == GPIO_PIN_0)
{
HAL_GPIO_TogglePin(GPIOG, GPIO_PIN_13);
HAL_GPIO_TogglePin(GPIOG, GPIO_PIN_14);
}
counter2++;
}
/* USER CODE END 0 */

/* USER CODE BEGIN 2 */
HAL_GPIO_WritePin(GPIOG, GPIO_PIN_13, GPIO_PIN_SET);
HAL_GPIO_WritePin(GPIOG, GPIO_PIN_14, GPIO_PIN_RESET);
/* USER CODE END 2 */

/* Infinite loop */
/* USER CODE BEGIN WHILE */
while (1)
{
counter1++;
/* USER CODE END WHILE */
/* USER CODE BEGIN 3 */
}
/* USER CODE END 3 */
```

In order to execute the code in Listing 5.2, create an empty project under STM32CubeIDE and configure the LED pins of the STM32F4 microcontroller as output. Then, configure the button pin as GPIO_EXTI and select the mode as "external interrupt mode with falling edge trigger detection." As the project is compiled and run, the CPU executes the callback function when the button is pressed and then released. In the callback function, the green and red LEDs toggle according to the interrupt source pin. By adding a breakpoint, the reader can observe that the counter in the infinite loop increases while the CPU is not in handler mode. The counter in the callback function increases only once the button is pressed and then released.

The second example aims to familiarize the reader with the nested interrupt and priority mechanisms. To do so, we should connect an external normally open button in active high configuration and a capacitor to the pin PA5 as in Fig. 5.4. Afterward, we should configure the onboard and external button pins as GPIO_EXTI and select their mode as "external interrupt mode with falling edge trigger detection."

We should modify the callback function in Listing 5.2 with the new callback functions as in Listing 5.3. We should also change the interrupt priority of EXTI line 0 interrupt to 1 and EXTI line[9:5] interrupt to 2 while configuring interrupt sources. The reader can execute the new code with the modified setup to see the effect of these changes.

Listing 5.3 The effect of priority on GPIO interrupts, the C code

```
/* USER CODE BEGIN PV */
int counter1 = 0, counter2 = 0;
/* USER CODE END PV */

/* USER CODE BEGIN 0 */
void HAL_GPIO_EXTI_Callback(uint16_t GPIO_Pin)
```

Fig. 5.4 External button connection, second interrupt example with C code

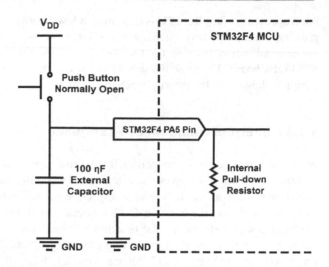

```
{
if (GPIO_Pin == GPIO_PIN_0)
{
HAL_GPIO_TogglePin(GPIOG, GPIO_PIN_13);
HAL_Delay(5000);
}
if (GPIO_Pin == GPIO_PIN_5)
{
HAL_GPIO_TogglePin(GPIOG, GPIO_PIN_14);
HAL_Delay(5000);
}
counter2++;
}
/* USER CODE END 0 */

/* USER CODE BEGIN 2 */
HAL_GPIO_WritePin(GPIOG, GPIO_PIN_13, GPIO_PIN_SET);
HAL_GPIO_WritePin(GPIOG, GPIO_PIN_14, GPIO_PIN_RESET);
/* USER CODE END 2 */

/* Infinite loop */
/* USER CODE BEGIN WHILE */
while (1)
{
counter1++;
/* USER CODE END WHILE */

/* USER CODE BEGIN 3 */
}
/* USER CODE END 3 */
```

The second example is formed in two separate parts. In the first part, a low priority interrupt comes when the CPU is handling a high priority interrupt. Hence, the low priority interrupt is served after the high priority one is served. First, press the onboard button on the STM32F4 board. Release the onboard button, wait for 1 s, and press and release the external button. You can observe that the second press will not be recognized until four more seconds pass. In the second part, a high priority interrupt comes when the CPU is executing a low priority interrupt. Due to interrupt priority, the high priority interrupt is served immediately. Now, press and release the external button and wait for 1 s. Then, press and release the user button on the STM32F4 board. This time the second press will be recognized immediately. This example shows how the priority mechanism works.

5.3.2   Interrupt Usage via C++ Language

As we adjust all necessary properties for the interrupt setup, we can form our main function in Mbed. Here, we should decide on when the interrupt occurs. We have two options here as the falling or rising edge of the button press. In the former case, we should use the function `fall` associated with the `InterruptIn` class. In the latter case, we should use the function `rise` associated with the `InterruptIn` class. Assume that we select the falling edge to generate the interrupt. Hence, our setup will be `center.fall(&ISR_callback)`. Here, the user-defined function `&ISR_callback` is called when the button interrupt occurs and all conditions are satisfied. We provide an example on the usage of all settings in Listing 5.4. To execute this code in Mbed Studio, we should copy and paste it to the "main.cpp" file in our active project.

Listing 5.4   Toggle the onboard green LED using an interrupt, the C++ code

```
#include "mbed.h"

InterruptIn userButton(BUTTON1);
DigitalOut greenLED(LED1);

void getButtons()
{
greenLED = !greenLED;
}

int main()
{
greenLED = 0;
userButton.fall(&getButtons);
__enable_irq();

while (true);
}
```

As can be seen in Listing 5.4, the CPU waits in an infinite loop formed by the code line `while(true)`. When the user presses the button, this loop is interrupted. The CPU executes the function `ISR_callback`. Afterward, it turns back and waits in the infinite loop for another interrupt.

5.3.3 Interrupt Usage via MicroPython

We can redo the examples given in Sect. 5.3.1 using MicroPython. First, we will toggle the onboard red and green LEDs using onboard push button. Initially, green LED is turned on and red LED is turned off. When the user push button is pressed, both LEDs are toggled and the second counter is increased by one to keep the number of interrupts. The MicroPython code for this example is given in Listing 5.5.

Listing 5.5 Toggle the onboard LEDs using an interrupt, the MicroPython code

```
import pyb
counter2 = 0
greenLED = pyb.Pin('PG13', mode=pyb.Pin.OUT_PP)
redLED = pyb.Pin('PG14', mode=pyb.Pin.OUT_PP)

def getButtons(id):
    global counter2
    if id == 0:
        greenLED.value(not greenLED.value())
        redLED.value(not redLED.value())
        counter2 = counter2 + 1

def main():
    global counter2
    counter1 = 0
    userButtonInt = pyb.ExtInt('PA0', pyb.ExtInt.IRQ_RISING, pyb.
        Pin.PULL_DOWN, getButtons)
    greenLED.value(1)
    redLED.value(0)
    while True:
        counter1 = counter1 + 1

main()
```

Next, we can add 5 s delay after toggling LEDs in the callback function for our example in Listing 5.5 to show interrupt priorities in MicroPython. Also, we will change the priority level of EXTI line 0 interrupt to 1 and the priority level of EXTI line[9:5] interrupt to 2. Here, user push button is connected to EXTI line 0 interrupt source, and an external push button is connected to EXTI line[9:5] interrupt source. Here, we should copy the "mympfunctions.py" file in Listing 5.1 to the microcontroller flash and import the module mympfunctions in the "main.py" file in order to use the function set_interrupt_priority. MicroPython code for this example is given in Listing 5.6.

Listing 5.6 Toggle the onboard LEDs using an interrupt with priority settings, the MicroPython code

```
import pyb
from mympfunctions import *

counter2 = 0
greenLED = pyb.Pin('PG13', mode=pyb.Pin.OUT_PP)
redLED = pyb.Pin('PG14', mode=pyb.Pin.OUT_PP)

def getButtons(id):
    global counter2
```

```
    if id == 0:
        greenLED.value(not greenLED.value())
        pyb.delay(5000)
    if id == 5:
        redLED.value(not redLED.value())
        pyb.delay(5000)
    counter2 = counter2 + 1

def main():
    counter1 = 0
    userButtonInt = pyb.ExtInt('PA0', pyb.ExtInt.IRQ_RISING, pyb.
        Pin.PULL_DOWN, getButtons)
    extButtonInt = pyb.ExtInt('PA5', pyb.ExtInt.IRQ_RISING, pyb.
        Pin.PULL_DOWN, getButtons)
    greenLED.value(1)
    redLED.value(0)
    set_interrupt_priority(EXTI0_IRQn, 1)
    set_interrupt_priority(EXTI9_5_IRQn, 2)
    while True:
        counter1 = counter1 + 1

main()
```

The code in Listing 5.6 shows that interrupt priorities can be set in MicroPython as well. When the onboard push button is pressed, the green LED toggles and 5 s delay starts. During this time, pressing external push button connected to pin PA5 does not affect the operation and the red LED cannot be toggled until the 5 s delay is over. If the onboard push button is pressed after the external push button, the green LED toggles immediately without waiting for the 5 s delay. The reason is that the interrupt priority of EXTI line 0 is higher than EXTI line[9:5].

5.4 Power Management in Embedded Systems

Power management is the second topic to be considered in this chapter. We will start with explaining its importance in embedded applications. Power management is also closely related to the interrupt usage. Therefore, we will next form the link between power management and interrupt usage. When talking about power management, we should also understand how a power supply works. Hence, we will consider battery and briefly summarize its properties in this section.

5.4.1 Importance of Power Management in Embedded Applications

Embedded systems can be used in stand-alone applications in which they depend on battery. In order to extend working time of the overall system, we have three options. First, we can select a battery with larger capacity. Second, we can design a system with battery recharging capability. These two options are related to the power supply side. The third option is decreasing power consumption of the embedded system. Due to importance of the problem, recent microcontrollers provide power management options to the user. Arm® Cortex™-M architecture is well-known in this respect. To note here, improving the battery capacity and using microcontroller power management options together provide the best possible combination.

Power management in microcontrollers starts by decreasing the power consumption of the device. To do so, all peripheral units should be turned off when not used so that they do not dissipate power. This is done by inhibiting the clock signal sent to them. As a note, all peripheral units are turned off after resetting the microcontroller. It is the responsibility of the programmer to turn on the peripheral unit to be used when needed. Besides, the user should also select the appropriate power mode for the microcontroller.

5.4.2 The Link Between Power Management and Interrupt Usage

The microcontroller performs its operations in active (run) mode. However, this mode has the highest power consumption. Therefore, we should let the microcontroller run in an appropriate low power mode whenever possible.

One of the most important usage areas of interrupts is letting the CPU in low power modes. This is done as follows. If the operation to be performed by the CPU is infrequent, one option is letting the CPU to run continuously in an infinite loop to wait for the event to be responded. Since the event is rare, most of the power will be dissipated while waiting for the event to occur. The next option is setting the system such that it stays in an appropriate low power or sleep mode. When the external event occurs, it triggers an interrupt. Afterward, the CPU wakes up, performs the operation related to the event in active mode, and goes back to its previous low power mode again. Hence, the system stays in active mode for a short period of time or when only needed. For the remaining times, the system stays in low power mode and does not dissipate power as it does in the active mode.

Let's consider an actual example on the usage of interrupts to decrease the overall power consumption in operation. In our example, we press a button to open a gate. We can safely assume that the button press is rare. If the interrupt-based setup is not used, the system waits in active mode all the times. Meanwhile, the power consumption will be maximum. On the other hand, if the button press generates an interrupt, then we can let the system to sleep till the interrupt occurs. Hence, it dissipates the lowest possible power during this time. The system will be in active mode only while executing the callback function. This will lead to the overall power consumption to be minimal. This scenario indicates the importance of interrupt usage in power management.

5.4.3   Battery as Power Supply

The embedded system is powered by battery most of the times when used in a stand-alone application. Hence, the reader should know fundamental properties of the battery such that the stand-alone application can be realized efficiently.

There are rechargeable and disposable batteries. As the name implies, the rechargeable battery can be used more than once. Although this may seem reasonable, the charging circuit increases the overall cost. Besides, rechargeable batteries are expensive. As a final note, lifetime of a rechargeable battery is around 1000–2000 charging. On the other hand, the disposable battery can be used without any charging circuitry. It is cheap. However, the disposable battery can only be used once and should be replaced when depleted.

Capacity of a battery is provided in milliampere hour (mAh) by the manufacturer. This indicates the current value the battery can provide within 1 h in predefined voltage level. The other explanation is as follows. Capacity indicates how many hours can the battery be used when the mentioned current is drawn from it. Let's consider an example to explain this concept. Assume that we have a battery with 2000 mAh capacity. Assume further that we draw 150 mA current during 15 s and 2 mA current during 45 s from the battery within 1 min. We can calculate the time needed for the battery to deplete as follows. Every minute, the average current needed will be $150 \times 1/4 + 2 \times 3/4 = 39$ mA. Therefore, the time needed for the battery to deplete can be found as 2000 mAh / 39 mA = 51.28 h in this usage scenario.

As a final note, the battery may work in different characteristics when excessive or very low current is drawn from it. However, the average current value is taken for granted for most calculations. Besides, each battery from the same manufacturer may have slightly different characteristics. Hence, nominal values are used in most calculations. We will follow the same reasoning in this book as well. The user should take these deviations into account as safety margins.

5.5   Power Management in the STM32F4 Microcontroller

We will focus on power management issues in the STM32F4 microcontroller in this section. Therefore, we will first summarize the power management features followed by power supply options for the microcontroller. Then, we will analyze the power modes available in the STM32F4 microcontroller. Finally, we will introduce STM32CubeMX properties for power usage analysis.

5.5.1   Power Management Features

The STM32F4 microcontroller has several key features related to power management. These can be summarized as follows. The microcontroller allows multiple independent power supply usage. This leads to reducing power consumption while some peripherals are supplied at higher voltages. The microcontroller has several power modes. These allow the programmer to benefit from the microcontroller resources while saving power. Multiple interrupt sources can be used to wake up the CPU from a low power mode. Hence, different pin and peripheral configurations may be applied. For example, the STM32F4 microcontroller can be woken up with user interaction by a button connected to a GPIO pin or periodically using timers. The microcontroller has a battery backup domain. This can be used to keep some peripheral units active (including RTC to be introduced in Chap. 6) and certain backup registers (to be introduced in Chap. 9) without the main power supply. This allows saving the last state of the microcontroller when the main power is lost.

5.5.2   Power Supply Options

We introduced powering options for the STM32F4 board in Sect. 2.3.3. Here, we will focus on feeding power to the STM32F4 microcontroller. We provide the power supply options for our microcontroller in Table 5.1. To note here, other members of the STM32F4 microcontroller family may have more power supply options. Since we will not deal with them in this book, we did not consider them. The reader can see these options at [1, 2].

Table 5.1 deserves special consideration. Here, V_{DD} supplies all GPIO ports, the reset circuitry, temperature sensor, all internal clock sources, and the standby circuitry. V_{DD} also supplies an embedded voltage regulator which supplies all digital circuitry, such as CPU, digital peripheral units, SRAM, and flash. Regulator is an electronic component or circuitry (depending on the type) which generates a fixed voltage output regardless of changing input voltage level. The voltage regulator in the STM32F4 microcontroller operates in different modes depending on the power mode the microcontroller is in. Since we will only focus on the power modes in this book, we lead the reader to references for further information on regulator operating modes [1, 2].

V_{DDA} is the power supply for analog peripheral units since analog and digital voltage levels may be different in operation. For such cases, a different analog power supply is needed. To improve conversion accuracy, the ADC module has an independent power supply which can be separately filtered and shielded from noise on the PCB. The ADC voltage supply input is available on a separate

Table 5.1 The STM32F4 microcontroller power supply options

Power supply	Description	Voltage level (V)
V_{DD}	Main power supply	1.7–3.6
V_{DDA}	Analog power supply	1.7–3.6
V_{BAT}	Backup power supply	1.65–3.6
V_{REF}	ADC/DAC reference voltage	1.7–3.6

V_{DDA} pin. An isolated supply ground connection is provided by the pin V_{SSA} which is connected to ground externally on the STM32F4 board. A backup battery can be connected to the V_{BAT} pin to supply the backup domain. However, this pin is connected to V_{DD} in the STM32F4 board. Hence, we cannot benefit from it. V_{REF} pins provide reference voltage to the ADC and DAC modules. We will explore how these reference voltage values are used in Chap. 7.

The STM32F4 microcontroller also has a power supply supervisor circuitry. It can be used to track, monitor, and manage power supplies mentioned in Table 5.1. We refer the reader to the references for further information on the usage of this circuitry [1, 2].

5.5.3   Power Modes

The STM32F4 microcontroller supports four power modes as run (active), sleep, stop, and standby. These are tabulated in Table 5.2.

Let's take a closer look at Table 5.2. As can be seen in this table, our first option for operation is the run mode. This mode does not have any power saving property. Hence, all peripherals and clock sources can be activated within it. The user has five options for setting the main CPU clock between 180 and 120 MHz in the run mode based on the regulator used. Our second option is the sleep mode. The CPU is stopped in this mode. However, SRAM and all peripherals are active. These can wake up the CPU by an interrupt. Our third option is the stop mode. This mode is called deep sleep mode in Arm® literature. The stop mode achieves the lowest power

Table 5.2 The STM32F4 microcontroller power modes

Power mode	Regulator mode	CPU/SRAM/peripherals active	Max. clock speed (MHz)
Run	Main-scale 1 (OverDrive)	All active	180
	Main-scale 1	All active	168
	Main-scale 2 (OverDrive)	All active	168
	Main-scale 2	All active	144
	Main-scale 3	All active	120
Sleep	Main-scale 1 (OverDrive)	SRAM/peripherals	180
	Main-scale 1	SRAM/peripherals	168
	Main-scale 2 (OverDrive)	SRAM/peripherals	168
	Main-scale 2	SRAM/peripherals	144
	Main-scale 3	SRAM/peripherals	120
Stop	Main-scale 3	SRAM	–
	Main-scale 3 (UnderDrive)	SRAM	–
	Low power-scale 3	SRAM	–
	Low power-scale 3 (UnderDrive)	SRAM	–
Standby	Power-down mode	–	–

consumption while retaining the contents of SRAM and registers. The CPU can be woken up from the stop mode by any of the EXTI line (the EXTI line source can be one of the 16 external lines, the PVD output, the RTC alarm/wake-up/tamper/time stamp events, the USB OTG FS/HS wakeup, or the Ethernet wakeup). Our fourth option is the standby mode. Only the RTC, 4 kB of backup SRAM, and 20 backup registers (80 bytes) are active in this mode. The CPU can exit from the standby mode either by an external reset pin, a watchdog reset, a rising edge on one of the enabled WKUPx pins or when a RTC event occurs. After waking up from standby mode, code execution restarts in the same way as after a reset operation.

5.5.4   STM32CubeMX for Power Usage Analysis

STM32CubeMX has a power consumption calculator (PCC) tool which can be used to estimate active, inactive, and average current consumption values for a given scenario. The PCC tool also estimates the battery life when the one to be used in operation is selected. The PCC tool can be used by setting up multiple steps such that different power modes, clock source and frequency, and peripheral units can be selected at each step. Hence, the user can check complex usage scenarios with the tool.

In order to use the PCC tool, the reader should create a project as explained in Sect. 3.1.5. Then, the reader should open the "Device Configuration Tool" window by clicking the ".ioc" file in the project explorer window. Afterward, he or she should click on the "Tools" tab as in Fig. 5.5. On the left side menu, a V_{DD} value and battery can be selected. On the right-hand side of the view, a new step can be added by clicking the "New Step" button. While adding or editing a step, power and memory settings, clocks, peripheral units, and other additional settings, such as duration of a step, can be adjusted. When a new step is added and there is a transition error between power modes, then the step turns to red on the list. Results are shown as graphs in the display section of the tool.

We next provide an example on the usage of the PCC tool. Within this example, we will calculate the current consumption when the CPU runs for 10 s when the code in Listing 5.2 is executed. To do so, open the ".ioc" file in the project explorer window, click on "Tools," tab and add a new step. Here, choose the power mode as "RUN," power scale as "Scale-3-Low," memory fetch type as "Flash," and CPU frequency as "25 MHz." Enter the step duration as 10 s or 10000 ms and click the "Enable IPs from Pinout" button. You may also enable IPs (peripherals) from peripheral selection menu in the middle. Once the configuration is complete, click the "Add" button. You can observe the current consumption profile graph. As can be seen there, approximately 11.41 mA average current is consumed in operation. Now, edit the added step and change the step duration as 0.1 s or 100 ms. Add another step, choose power mode as "STOP_NM (Normal Mode)" and clock configuration as "Regulator_LP," and set the step duration as 9.9 s or 9900 ms. This time, the

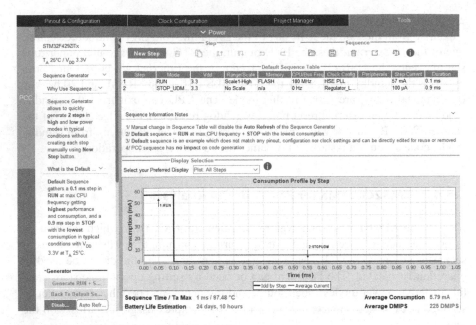

Fig. 5.5 The PCC tool under STM32CubeMX

average current consumption drops down to approximately $401.2~\mu A$. We will use the PCC tool in the following sections to emphasize the usage of power modes in general.

5.6   Usage of Power Modes in Code

We provide C, C++, and MicroPython codes on the usage of low power modes on actual examples in this section. To note here, C and MicroPython programming allows using all power modes. For the C++-based codes, we will use whichever option is available.

5.6.1   Power Modes in C Language

Power modes can be entered and exited by executing HAL library functions in C language. We will provide these next.

5.6.1.1 Sleep Mode

HAL library functions related to the sleep mode are as follows. The function `HAL_PWR_EnterSLEEPMode` can be used to enter the sleep mode. The function `HAL_PWR_EnableSleepOnExit` enables sleep-on-exit when returning from handler to thread mode. Likewise, the function `HAL_PWR_DisableSleepOnExit` disables sleep-on-exit when returning from the handler to thread mode. We provide detailed information on these functions next.

```
void HAL_PWR_EnterSLEEPMode (uint32_t Regulator, uint8_t
    SLEEPEntry)
/*
Regulator: Specifies the regulator state. This parameter is not
    used for the STM32F4 family. Hence, we should set it as
    PWR_MAINREGULATOR_ON.
SLEEPEntry: Specifies if the sleep mode is entered with WFI or
    WFE instruction. This parameter can be one of
    PWR_SLEEPENTRY_WFI or PWR_SLEEPENTRY_WFE
*/

void HAL_PWR_EnableSleepOnExit (void)

void HAL_PWR_DisableSleepOnExit (void)
```

We next provide an example on the usage of sleep mode in C language. To do so, we repeat the example given in Listing 5.2 to toggle onboard LEDs using GPIO interrupts. However, instead of actively waiting in an infinite loop when the button is not pressed, this time we let the CPU wait in sleep mode. To do so, the reader should create an STM32CubeIDE project first. Then, the infinite loop in Listing 5.2 should be replaced by the code block in Listing 5.7. As the project is executed, the reader can observe that both counters in the infinite loop and callback function do not increase when the button is not pressed.

Listing 5.7 A sample code for the usage of sleep mode functions, in C language

```
/* Infinite loop */
/* USER CODE BEGIN WHILE */
while (1)
{
HAL_SuspendTick();
HAL_PWR_EnterSLEEPMode(PWR_MAINREGULATOR_ON, PWR_SLEEPENTRY_WFI);
HAL_ResumeTick();

counter1++;
/* USER CODE END WHILE */

/* USER CODE BEGIN 3 */
}
/* USER CODE END 3 */
```

The reader can observe that there are two extra functions used in Listing 5.7 as HAL_SuspendTick() and HAL_ResumeTick(). These are related to the SysTick and clock setup to be explained in detail in Chap. 6. However, we had to cover them here since these two modules work in sleep mode. SysTick automatically generates interrupts at every millisecond and wakes up the CPU. To avoid this, we had to disable it before the sleep operation and enable it afterward as in Listing 5.7.

We can analyze the power consumption of the modified project by the PCC tool. Here, the GPIOA and GPIOG ports are enabled. The power scale is set to "Scale-3-Low," memory fetch type to "Flash," and CPU frequency to "25 MHz" for both run and sleep modes. Under these conditions, the average current consumption will be 11.1 mA and 3.1 mA for the run and sleep modes, respectively.

5.6.1.2 Stop Mode

HAL library functions related to the stop mode are as follows. The function HAL_PWR_EnterSTOPMode can be used to enter the stop mode 0 or 1 by specifying the regulator state. Likewise, the function HAL_PWREx_EnterUnder DriveSTOPMode leads the microcontroller to the underdrive mode with low power consumption. We provide detailed information on these functions next.

```
void HAL_PWR_EnterSTOPMode (uint32_t Regulator, uint8_t
    STOPEntry)
/*
Regulator: Specifies the regulator state. This parameter is not
    used for the STM32F4 family. Hence, we should set it as
    PWR_MAINREGULATOR_ON.
STOPEntry: Specifies if stop mode is entered with WFI or WFE
    instruction. This parameter can be one of PWR_STOPENTRY_WFI
    or PWR_STOPENTRY_WFE.
*/

HAL_StatusTypeDef HAL_PWREx_EnterUnderDriveSTOPMode (uint32_t
    Regulator, uint8_t STOPEntry)
```

We will next consider usage of the stop mode by repeating the example in Listing 5.2. This time we let the CPU wait in stop mode instead of actively waiting in an infinite loop when the user button is not pressed. As in the previous example, the reader should first create a project. Then, he or she should use the code block in Listing 5.8 to replace the infinite loop in Listing 5.2 and run the project. The reader can observe that the counter in the infinite loop does not increase when the button is not pressed. To note here, the functions HAL_SuspendTick() and HAL_ResumeTick() are used to stop and start the SysTick, respectively. They are added to avoid starting the SysTick during the debug session. They may not be needed if the project is compiled and run without debugging it.

Listing 5.8 A sample code for the usage of stop mode functions, in C language

```
/* Infinite loop */
/* USER CODE BEGIN WHILE */
while (1)
{
HAL_SuspendTick();
HAL_PWR_EnterSTOPMode(PWR_LOWPOWERREGULATOR_ON, PWR_STOPENTRY_WFI
    );
HAL_ResumeTick();

counter1++;
/* USER CODE END WHILE */

/* USER CODE BEGIN 3 */
}
/* USER CODE END 3 */
```

We can analyze power consumption of the modified project by the PCC tool. Here, the GPIOA and GPIOG ports are enabled. The power scale is set to "Scale-3-Low," memory fetch type to "Flash," and CPU frequency to "25 MHz" for both run mode. Also the clock configuration is set to "Regulator_LP" for stop mode. Under these conditions, the average current consumption will be 11.1 mA and 290 μA for the run and stop modes, respectively.

5.6.1.3 Standby Mode

HAL library functions related to the standby mode are as follows. The function HAL_PWR_EnterSTANDBYMode can be used to enter the standby mode. The function HAL_PWR_EnableWakeUpPin can be used to enable the wake-up PINx functionality. The macro __HAL_PWR_CLEAR_FLAG should be used to clear the flag if it will be used in the wake-up process for the standby mode. We provide detailed information on these functions next.

```
void HAL_PWR_EnterSTANDBYMode (void)

void HAL_PWR_EnableWakeUpPin (uint32_t WakeUpPinx)
/*
WakeUpPinx: Specifies which Wake-Up pin to enable. This
    parameter can only be PWR_WAKEUP_PIN1.
*/

__HAL_PWR_CLEAR_FLAG (_FLAG_)
/*
__FLAG__: specifies the flag to clear. This parameter can be one
    of the PWR_FLAG_WU or PWR_FLAG_SB
*/
```

We next provide an example on the usage of standby mode. Here we wake up the CPU with the onboard button connected to PA0, toggle the onboard LEDs connected to pins PG13 and PG14, and then put CPU back to standby mode. To do so, the reader should first create an empty project and configure the LED pins of the STM32F4 microcontroller as output. Then, he or she should configure the button pin as SYS_WKUP and select the mode as "external interrupt mode with falling edge trigger detection." Then, the reader should use the code block in Listing 5.9 and run the project.

Listing 5.9 A sample code for the usage of standby mode functions, in C language

```
/* USER CODE BEGIN 2 */
HAL_GPIO_TogglePin(GPIOG,GPIO_PIN_13);
HAL_Delay(1000);
HAL_GPIO_TogglePin(GPIOG,GPIO_PIN_14);
HAL_Delay(1000);

__HAL_PWR_CLEAR_FLAG(PWR_FLAG_WU);
HAL_PWR_EnableWakeUpPin(PWR_WAKEUP_PIN1);
HAL_PWR_EnterSTANDBYMode();
/* USER CODE END 2 */
```

As the project is executed, the green and red LEDs turn on one after the other with 1 s delay. Then, they turn off together. However, we do not turn off the LEDs in code. Instead, they turn off since the GPIO ports connected to them are closed. As the CPU turns on after the standby mode, it resets itself. Hence, the LEDs turn on and off again.

We can analyze the power consumption of the modified project by the PCC tool. Here, the GPIOA and GPIOG ports are enabled. The power scale is set to "Scale-3-Low," memory fetch type to "Flash," and CPU frequency to "25 MHz" for the run mode. Also the clock configuration is set to ALL CLOCKS OFF for the standby mode. Under these conditions, the average current consumption will be 11.1 mA and 2.2 μA for the run and standby modes, respectively.

5.6.2   Power Modes in C++ Language

The recent Mbed release only has one function sleep() related to sleep modes and power management operations. Therefore, there is not much we can do to adjust the power modes manually. Instead, we should depend on the sleep() function for this operation. We direct the reader to the Mbed web site for further information on this topic.

We provide a sample code on the usage of the sleep() function in Listing 5.10. This code works as follows. Device settings for the red and green LEDs on the STM32F4 board are done first. Then, interrupt settings for the user button are done. The green LED toggles within the callback function. The red LED toggles within the infinite loop under the main function. The sleep() function is also in the infinite loop. Therefore, in the first run, the red LED toggles and CPU goes to sleep waiting for a GPIO interrupt. As the user presses the button on the STM32F4 board, the CPU wakes up and toggles the green LED. Afterward, the CPU goes to thread mode and toggles the red LED and sleeps again. This example shows how the sleep mode can be used under Mbed in its basic sense.

Listing 5.10 A sample code for the usage of sleep mode, in C++ language

```
#include "mbed.h"

DigitalOut greenLED(LED1);
DigitalOut redLED(LED2);
InterruptIn userButton(BUTTON1);

void getButtons(){
greenLED = !greenLED;
}

int main()
{
greenLED=0;
redLED = 0;
userButton.fall(&getButtons);
__enable_irq();

while (true)
{
redLED = !redLED;
sleep();
}

}
```

5.6.3 Power Modes in MicroPython

MicroPython supports active, sleep, stop, and standby power modes. These can be used by their dedicated functions as follows. The function pyb.wfi() puts the CPU in sleep mode. The CPU turns back to run mode again when any internal or external interrupt occurs. The SysTick interrupt occurs every 1 ms in MicroPython. Hence, the CPU can be kept in this mode with the maximum duration of 1 ms. The function pyb.stop() puts the CPU in stop mode. The CPU returns to run mode again when any external interrupt or RTC event occurs. The function pyb.standby() puts the CPU in standby mode. The CPU resets itself when any external wake-up or tamper interrupt, or RTC event, occurs.

Now, we will repeat the examples given in Sect. 5.6.1 using MicroPython. First, we will redo the example in Listing 5.7 for observing the sleep mode. MicroPython uses SysTick interrupt in background. Hence, it must be disabled with code line stm.mem32[0xE000E010] &= ~1. The code for this example is given in Listing 5.11.

Listing 5.11 A sample code for the usage of MicroPython sleep mode function

```
import pyb
greenLED = pyb.Pin('PG13', mode=pyb.Pin.OUT_PP)
redLED = pyb.Pin('PG14', mode=pyb.Pin.OUT_PP)
counter2 = 0

def getButtons(id):
    global counter2
    if id == 0:
        greenLED.value(not greenLED.value())
        redLED.value(not redLED.value())
        counter2 = counter2 + 1

def main():
    counter1 = 0
    userButtonInt = pyb.ExtInt('PA0', pyb.ExtInt.IRQ_RISING, pyb.
        Pin.PULL_DOWN, getButtons)
    greenLED.value(1)
    redLED.value(0)
    stm.mem32[0xE000E010] &= ~1
    while True:
        counter1 = counter1 + 1
        print(counter1)
        pyb.wfi()

main()
```

Next, we will repeat the example given in Listing 5.8 for observing the stop mode. The MicroPython code for this example is given in Listing 5.12.

Listing 5.12 A sample code for the usage of MicroPython stop mode function

```
import pyb
import stm
counter2 = 0
greenLED = pyb.Pin('PG13', mode=pyb.Pin.OUT_PP)
redLED = pyb.Pin('PG14', mode=pyb.Pin.OUT_PP)
```

```
def getButtons(id):
    global counter2
    if id == 0:
        greenLED.value(not greenLED.value())
        redLED.value(not redLED.value())
        counter2 = counter2 + 1

def main():
    counter1 = 0
    userButtonInt = pyb.ExtInt('PA0', pyb.ExtInt.IRQ_RISING, pyb.
        Pin.PULL_DOWN, getButtons)
    greenLED.value(1)
    redLED.value(0)
    stm.mem32[0xE000E010] &= ~1
    while True:
        counter1 = counter1 + 1
        print(counter1)
        pyb.stop()

main()
```

Finally, we will repeat the example in Listing 5.9 for observing the standby mode. Here, we need to add two code lines stm.mem32[stm.PWR + stm.PWR_CR] |= (1«2) and stm.mem32[stm.PWR + stm.PWR_CSR] |= (1«8) to clear the wake-up flag and enable wake-up property for pin PA0, respectively. The code for this example is given in Listing 5.13.

Listing 5.13 A sample code for the usage of MicroPython standby mode function

```
import pyb
import stm
greenLED = pyb.Pin('PG13', mode=pyb.Pin.OUT_PP)
redLED = pyb.Pin('PG14', mode=pyb.Pin.OUT_PP)

def main():
    userButton = pyb.Pin('PA0', mode=pyb.Pin.IN, pull=pyb.Pin.
        PULL_DOWN)
    greenLED.value(1)
    pyb.delay(2000)
    redLED.value(1)
    pyb.delay(2000)
    stm.mem32[stm.PWR + stm.PWR_CR] |= (1 << 2)
    stm.mem32[stm.PWR + stm.PWR_CSR] |= (1 << 8)
    while True:
        pyb.standby()

main()
```

5.7   Application: Interrupt-Based Operations and Power Management for the Robot Vacuum Cleaner

In this chapter, we update the digital IO operations introduced in Chap. 4 using interrupts and power modes. Hence, we benefit from the advanced features for our robot vacuum cleaner. We provided the equipment list, circuit diagram, detailed explanation of design specifications, and peripheral unit settings in a separate document in the folder "Chapter5\EoC_applications" of the accompanying supplementary material. In the same folder, we also provided the complete C, C++, and MicroPython codes for the application.

5.8   Summary of the Chapter

This chapter focused on two important embedded system concepts as interrupts and low power modes. We handled them in the same chapter since the latter becomes more useful with the usage of the former. The interrupt concept may seem unusual to a programmer at first sight. However, it is a must for an embedded system designer. Therefore, we considered the interrupt concept in detail starting from its conceptual explanation to its usage in the STM32F4 microcontroller. Here, we only used GPIO interrupts since digital input and output has been introduced in the previous chapter. We will handle the remaining interrupts as we introduce other peripheral units in the following chapters. As for low power modes, we started explaining them by a general introduction to the battery as power supply. Then, we explored the power management and low power mode operations in embedded systems. As the end of chapter application, we extended our application introduced in the previous chapter by adding interrupt and low power mode usage to it. We will be extensively using interrupt-based programming and low power modes from this point on. Therefore, the concepts introduced in this chapter are extremely important. Hence, we kindly ask the reader to master them.

Problems

5.1. What is the difference between ISR and callback function?

5.2. What is the difference between the interrupt and event in the STM32F4 microcontroller?

5.3. Explain the usage area of the edge detection module in Fig. 5.2.

5.4. Can we eliminate switch bouncing problem by an interrupt-based setup?

5.5. There are two push buttons connected to pins PB0 and PB1 of the STM32F4 board. Set up the interrupt and associated callback function to count the number of button presses. If the first button is pressed more than the second button, then only the red LED on the STM32F4 board turns on. If the second button is pressed more than the first button, then only the green LED on the STM32F4 board turns on. If both buttons are pressed equally, then both LEDs turn on. Form the project using:

(a) C language under STM32CubeIDE.
(b) C++ language under Mbed.
(c) MicroPython.

5.6. Repeat Problem 4.8 such that the parity check is done when the user button on the STM32F4 board is pressed and a port interrupt is generated. Hence, form a callback function for the operation.

5.7. Repeat Problem 4.9 such that each array entry will be processed by a button press on the STM32F4 board. Hence, each array entry will be handled in a callback function associated with the port interrupt.

5.8. Repeat Problem 4.11 by using port interrupts and associated callback functions.

5.9. Repeat Problem 4.12 by using port interrupts and associated callback functions.

5.10. We will implement an IoT system that uses the STM32F4 microcontroller and analog temperature sensor. The sensor is connected to the ADC1 module of the STM32F4 microcontroller. The system will wake up from the standby mode and measure ambient temperature every minute. Assume that the system needs 20 ms to initialize; measure the temperature in 100 ms; and save the result to SRAM in 5 ms. Moreover, assume that the STM32F4 microcontroller draws 10 mA and 3 μA current in active (run) and standby modes, respectively. The sensor draws 1 mA and 1 μA current in active (run) and standby modes, respectively. The system will be supplied by two AAA alkaline batteries connected in series. Each battery has capacity 1250 mAh. Calculate how long can the system work in this setup by:

(a) hand.
(b) the STM32CubeIDE PCC tool.
(c) Compare the results in part (a) and (b).

5.11. What are the power supply options for the STM32F4 microcontroller?

5.12. Add power management option to Problem 5.5 such that the STM32F4 microcontroller waits in an appropriate low power mode when not used.

5.13. Add power management option to Problem 5.6 such that the STM32F4 microcontroller waits in an appropriate low power mode when not used.

5.14. Add power management option to Problem 5.7 such that the STM32F4 microcontroller waits in an appropriate low power mode when not used.

References

1. STMicroelectronics: STM32F427xx STM32F429xx, docid024030 rev 10th edn. (2018)
2. STMicroelectronics: STM32F405/415, STM32F407/417, STM32F427/437 and STM32F429/439 advanced Arm-based 32-bit MCUs, rm0090 rev 19th edn. (2021)
3. Yiu, J.: The Definitive Guide to Arm Cortex-M3 and Cortex-M4 Processors, 3rd edn. Newnes, London (2014)

References

1. ...

Timing Operations

6

6.1 Clock Signals in Embedded Systems

In order to understand time-based operations, we should first understand the clock signal and oscillator used to generate it. Therefore, we will start with these in this section. Afterward, we will focus on clock signals in the STM32F4 microcontroller.

6.1.1 What Is a Clock Signal?

Clock is basically a periodic signal oscillating between high and low states with predefined frequency (specifically called clock frequency). A sample clock signal is given in Fig. 6.1. Voltage levels for the high and low clock states can differ between different embedded systems. These voltage levels are not the main focus of this book since we do not deal with the low-level hardware design of an embedded system. However, the clock frequency has utmost importance since it defines the speed of operations within the embedded system.

We should answer one fundamental question before going further. Why does an embedded system (such as microcontroller) need a clock signal? In order to answer this question properly, we should understand the basic structure of an embedded system. Let's focus on a microcontroller consisting of large number of logic gates. Through these, data transfer and basic operations are performed. A logic gate is formed by transistors with limited operation speed. Hence, there are times in which the logic gate is in undefined state such that it neither represents logic level 0 nor

Supplementary Information The online version contains supplementary material available at (https://doi.org/10.1007/978-3-030-88439-0_6).

C. Ünsalan et al., *Embedded System Design with Arm Cortex-M Microcontrollers*, https://doi.org/10.1007/978-3-030-88439-0_6

Fig. 6.1 A sample clock
signal

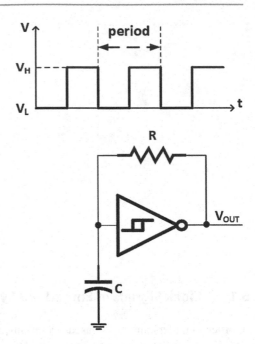

Fig. 6.2 Basic Schmitt
trigger RC oscillator

1. A synchronization signal is needed to avoid these undefined states. Clock signal
is used for this purpose such that at its each rising (or falling) edge, operations
are performed. No operation is done between rising (or falling) edges of the clock
signal. Hence, undefined states in logic gates are avoided. As a result, all operations
within the microcontroller are done synchronously.

6.1.2   Oscillator as the Clock Signal Source

Oscillator is an electronic circuit producing a periodic square wave which is the
clock signal for our case. Therefore, an embedded system that needs a clock signal
should have an oscillator or oscillators. An oscillator can be described by its two
properties as accuracy and response time. Accuracy indicates how precise is the
clock signal frequency. Response time indicates the time needed for the oscillator
to stabilize such that it can generate the clock signal with the desired frequency.

There are mainly two oscillator types as resistive-capacitive (RC) and crystal.
A basic Schmitt trigger RC oscillator is given in Fig. 6.2. Here, frequency of the
clock signal is determined by the R and C component values. Working principle of
the RC oscillator is briefly as follows. The Schmitt trigger output goes to high level
when the capacitor is charged. It goes to low level when the capacitor is discharged.
Hence, the desired periodic clock signal can be produced. RC oscillators are mainly
used as internal oscillators in microcontrollers since their components can be added
to the hardware production process. Therefore, there is no additional component

Fig. 6.3 A crystal oscillator

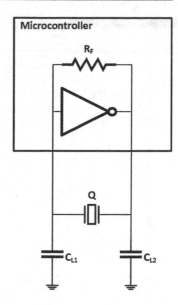

needed while using them. Besides, RC oscillators are cheap and have fast response time. However, their accuracy is low because of the passive components used in constructing the oscillator. Moreover, the accuracy can be affected from changing temperature and supply voltage values.

A crystal oscillator circuit is given in Fig. 6.3. Here, frequency of the clock signal is determined by the crystal which is a filter passing only signals with specific frequency. When the crystal oscillator is used, components other than the crystal and capacitors are placed within the microcontroller. However, the crystal cannot be placed inside the microcontroller. Therefore, it should be added to the system as an external unit. Crystal oscillators have relatively low response time. On the other hand, accuracy of crystal oscillators is superior compared to RC oscillators.

Let's focus on Fig. 6.3. In this figure, Q is the quartz crystal which is the main component of the oscillator. In simplest terms, the quartz crystal is represented by an RLC circuit with a parallel capacitor added to it. This setup leads to a frequency range that can be generated. Afterward, the exact clock signal frequency can be determined by the load capacitance value obtained by external capacitors C_{L1}, C_{L2}, and the stray capacitance C_s (not shown in the figure). The inverter circuit represented by the NOT gate in the figure aims to generate a phase shift. Hence, oscillation within the circuit can be generated. The feedback resistor R_F, connected in parallel to the inverter, ensures the oscillator to work in linear region.

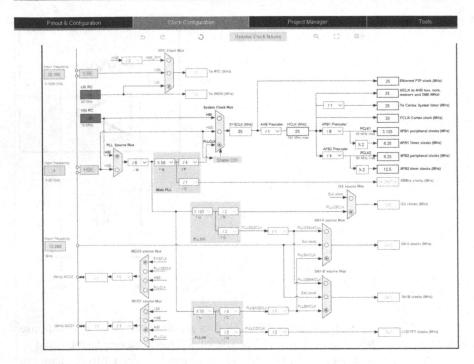

Fig. 6.4 The STM32F4 microcontroller clock tree

6.1.3 Managing Clocks in the STM32F4 Microcontroller

The STM32F4 microcontroller has several clocks and clock signal sources. The reason for this is as follows. When more than one time-based operation is performed within the microcontroller, clocks working with different frequencies may be needed. In such operations, when one desired frequency is low, the other may be high. Therefore, different clocks should be used.

Block diagram of the clock tree for the STM32F4 microcontroller, obtained from STM32CubeMX, is given in Fig. 6.4. As can be seen in this figure, all clocks used in the microcontroller can be derived from LSI, LSE, HSI, and HSE clocks. We explain them in detail next.

The STM32F4 microcontroller has two low speed clocks as LSI and LSE. LSI is the low speed internal clock generated from an internal 32 kHz RC oscillator. LSE is the low-speed external clock. There are two ways to generate it. The first one is connecting an external 32768 Hz crystal to the OSC32_IN and OSC32_OUT pins of the STM32 microcontroller. The second way is connecting an external clock to the OSC32_IN pin of the STM32F4 microcontroller. Here, the OSC32_OUT pin should be left at high impedance. In this setup, the clock frequency can be selected between 0 and 1000 kHz. Unfortunately, the STM32F4 board does not have an LSE crystal. Therefore, the second way cannot be used in this book.

Fig. 6.5 "RCC mode and configuration" window

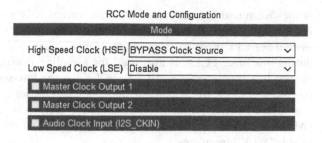

The STM32F4 microcontroller has two high speed clocks as HSI and HSE. HSI is the high-speed internal clock generated from an internal 16 MHz RC oscillator. HSE is the high-speed external clock. There are two ways to generate it. The first one is connecting an external crystal to the OSC_IN and OSC_OUT pins of the STM32F4 microcontroller. This way, the clock frequency can be set between 4 and 26 MHz. The second way is connecting an external clock to the OSC_IN pin of the STM32F4 microcontroller. Here, the OSC_OUT pin should be left at high impedance. In this setup, the clock frequency can be set between 1 and 50 MHz. The STM32F4 board has an onboard 8 MHz HSE crystal.

Low- and high-speed clocks introduced in the previous paragraphs can be selected from STM32CubeMX. To do so, the reader should select the RCC module from the "System Core" drop-down menu. The opened "RCC Mode and Configuration" window will be as in Fig. 6.5. Here, HSE and LSE clock sources can be selected as "Disable," "BYPASS clock source," or "Crystal/Ceramic Resonator" from the related drop-down list. Also, two clock signals can be directed to output pins by selecting "Master Clock Output 1" and "Master Clock Output 2" boxes. Master clock output 1 signal can be fed by the LSE, HSE, HSI, or main PLL clock. Master clock output 2 signal can be fed by the SYSCLK, PLLI2S clock, HSE, or main PLL clock. Selected clock source frequencies can be divided by their prescalers.

Other clocks used inside the STM32F4 microcontroller are as follows. The IWDG clock is used by independent watchdog timer and fed by the LSI clock. RTC clock is used by real-time clock module and fed by either the LSI, LSE, or scaled HSE clock. I^2S clock is used by inter-IC sound peripheral unit and fed by either the PLLI2S or external I2S_CKIN pin. Here, PLLI2S can be fed by the HSI or HSE clock. SAI clocks are used by the synchronous audio interface peripheral unit and fed by either the PLLSAI, PLLI2S, or I2S_CKIN pin. Here, PLLSAI can be fed by the HSI or HSE clock. LCD-TFT clock is used by the LCD-TFT controller module and fed by PLLSAI. SYSCLK is the system clock. Hence, all clocks used inside the microcontroller, except the ones mentioned before, are derived from it. SYSCLK can be fed by the HSI, HSE, or main PLL. Its maximum frequency can be set to 180 MHz. Here, the main PLL can be fed by the HSI or HSE clock. HCLK is the processor clock used by CPU core, advanced high-performance bus (AHB), memory, and DMA. Its maximum frequency can be set to 180 MHz. HCLK is a gated clock which means it can be enabled or disabled by a control bit. FCLK is the free-running processor clock. This means it is not gated. In other words, it cannot be stopped by a control bit. FCLK continues to run in the background even if the CPU goes to sleep mode. Hence, it can control interrupts and events during this time. FCLK is also used for sampling interrupts and clocking debug blocks. It is synchronous to HCLK. PCLK1 is the low-speed advanced peripheral bus (APB1) clock. Its maximum frequency can be set to 45 MHz. APB1 timers use double PCLK1 frequency. PCLK2 is the high-speed advanced peripheral bus (APB2) clock. Its maximum frequency can be set to 90 MHz. APB2 timers use double PCLK2 frequency.

All prescalers used in the clock tree are actually frequency dividers. These are constructed using cascaded flip-flops. In operation, the input frequency is divided by two after each flip-flop. Hence, we can obtain f/2, f/4, f/8, $\cdots$, frequency valued clock signals with this setup. For more information on frequency dividers by flip-flops, please see [2].

Although we are free to choose the mentioned clock sources, the best option is leaving them to be set by STM32CubeMX. Hence, we can benefit from the automatic clock configuration option there. When one of the HCLK, FCLK, PCLK1, PCLK2, or system timer clock frequencies is entered to the related box in STM32CubeMX, all other clocks and PLL blocks are configured automatically if possible. Otherwise, STM32CubeMX warns the user as "No solution found using the current selected sources. Do you want to use other sources?" The reader can change the clock source and repeat the process. If no clock source can provide the desired frequency, then a different value should be selected.

6.2   Timers in Embedded Systems

The clock signals are fed to timer modules for time-based operations. Therefore, we will introduce timers in this section. Different embedded systems may have timer modules with different operating characteristics. Hence, we will only focus on timers in the STM32F4 microcontroller in this section. The reader should also pay attention to the plural usage of the timers. This indicates that there is more than one timer module in the microcontroller. We will explain why this is the case.

6.2.1   What Is a Timer?

Timer is basically a counter such that the count value changes (going up, down, or up and down) by a trigger. If we pick the trigger to be the rising or falling edge of a periodic clock signal, then the count value changes based on the clock. This way, we can determine the actual time within our application running on the microcontroller. Let's provide an example for this operation.

Assume that period of the clock signal fed to the timer module is $1/f_{Timer}$ s. We set the timer module to count up starting from zero. Assume further that the count reaches the value $(N - 1)$ at a specific time instant. Hence, we can deduce

Fig. 6.6 A basic timer module

that the time passed between the start of the operation and specific time instant is N/f_{Timer} s. As can be seen in this example, we can deduce the actual time value by the help of the timer module. This allows performing time-based operations such as creating specific time intervals, periodically obtaining data, measuring the time duration between two events, or creating time related output by the microcontroller.

We can also use the prescaler, basically a frequency divider, to change period of the clock signal fed to the timer module. This allows us to obtain different time values by the same clock signal. We provide a basic timer block composed of the prescaler and counter in Fig. 6.6. In this figure, f_{CLK} represents the frequency of the clock signal fed to the timer module. This signal is modified by the prescaler such that the new clock signal has frequency f_{Timer}.

We next provide the actual usage example of the timer module. We know that the counter within the timer changes its value at every $1/f_{Timer}$ s. Let's create a 10 s time interval using the clock signal with 180 MHz f_{CLK} frequency. We should count 1800 million clock signal edges to obtain the 10 s time interval unless we use the prescaler. Instead, we can divide f_{CLK} by 18000. Hence, we can obtain the 10 s time interval by counting 100000 clock signal edges with frequency f_{Timer}.

The timer module works independent of the CPU. Therefore, the CPU does not allocate its resources to determine the required time instant. We can benefit from timer interrupts such that separately working CPU and timer module can achieve a time-based operation. Let's provide a simple example for this case. Assume that we want to toggle an LED every second. The timer module can perform the counting process to reach 1 s in background without interfering the CPU. As the toggling time is reached, the timer module generates an interrupt and the related callback function is executed by the CPU to toggle the LED. The CPU can perform other tasks assigned to it except the callback function execution period. Therefore, the CPU does not dissipate its valuable resources to the time calculation process.

6.2.2 Introducing Timers in the STM32F4 Microcontroller

The STM32F4 microcontroller has four different timer types as base, watchdog, system timer (SysTick), and real-time clock (RTC). These are used for specific purposes. The base and system timer modules can be used for general time related operations. The watchdog timer can be used for resetting the CPU when an error halts code execution. The RTC can be used for determining the exact date and time. Hence, specific actions can be taken based on this information. We will explain each timer type in detail next.

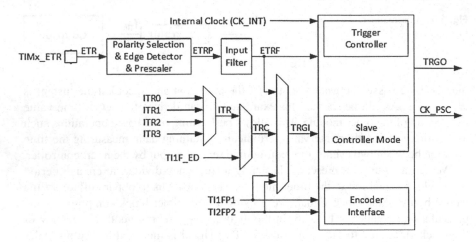

Fig. 6.7 Base timer trigger and clock controller block

6.2.3   Base Timers in the STM32F4 Microcontroller

There are 14 base timers in the STM32F4 microcontroller. These are called TIM1, TIM2, $\cdots$, TIM14. All of these timers share similar properties. Hence, we will explain how they work based on one such timer. Then, we will summarize their differences in Sect. 6.2.3.6. The STM32F4 microcontroller base timer consists of four blocks as trigger and clock controller, counter, input capture, and output compare. Let's look at these blocks in detail.

6.2.3.1 Trigger and Clock Controller Block
The trigger and clock controller block in the base timer is used to start/stop, reset, or update the timer counter using one of the trigger inputs. This block is given in Fig. 6.7. The trigger source can be the internal clock signal CK_INT, external clock signal ETR, timer signals TI1FP1 or TI2FP2, or internal trigger signals ITR0-3. The internal clock, CK_INT, is actually the PCLKx signal obtained from the STM32F4 microcontroller clock tree. The external clock, ETR, is the signal obtained from the pin TIMx_ETR. Other inputs and properties of the trigger and clock controller block will be explained in Sect. 6.2.7.

6.2.3.2 The Counter Block
The counter block of the base timer is given in Fig. 6.8. Here, the internal clock signal, CK_PSC, is divided by the 16-bit prescaler first to obtain the counter clock, CK_CNT. In every clock edge, content of the counter register, CNT counter, is updated according to the selected counting mode to be explained next.

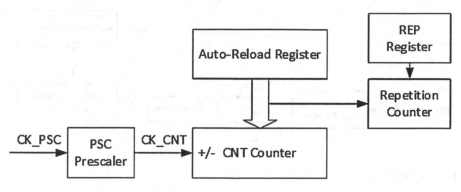

Fig. 6.8 Base timer counter block

6.2.3.3 Counting Modes

There are five counting modes in the STM32F4 microcontroller base timers. These are up mode, down mode, center aligned mode 1, center aligned mode 2, and center aligned mode 3.

In the up mode, the CNT counter is increased by one at every CK_CNT edge. When the CNT counter reaches the value in the auto-reload register ARR, the CNT counter is reset. The counter overflow event is generated. Then, up counting restarts from zero again.

In the down mode, the CNT counter is decreased by one at every CK_CNT edge. When the CNT counter reaches zero, the CNT counter is reset. The counter underflow event is generated. Then, down counting restarts from the ARR value again.

In the center aligned modes, the CNT counter is increased from zero to the ARR value. As this value is reached, the counter overflow event is generated. Then, the CNT counter is decreased from the ARR value to zero. The counter overflow event is generated again. The difference between the three center aligned modes occurs while generating output compare interrupt flags. In center aligned mode 1, the flag is generated when the counter is counting down. In center aligned mode 2, the flag is generated when the counter is counting up. In center aligned mode 3, the flag is generated when the counter is counting up or down.

The base timer has a repetition counter which can be used in all mentioned counting modes. If the repetition counter value is set to zero (as the default case), the counter generates an update event at every overflow or underflow. If the repetition counter is set to a value different from zero, then the overflow or underflow decreases the set value by one. When the repetition counter reaches zero, an update event is generated. The repetition counter is reset to its initial value and the operation repeats again. This allows increasing the period of update event generation.

The counter block also has an auto-reload preload mechanism related to counting modes. When this property is enabled and the ARR value is changed on the run, the

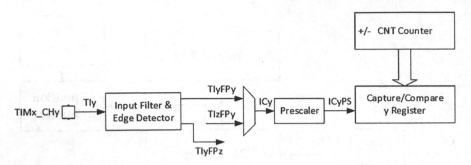

Fig. 6.9 Base timer input capture block

counter block continues to run according to the old ARR value. The new value is
written to ARR when an update event is generated.

6.2.3.4 The Input Capture Block

The input capture block in the base timer is used to copy the value inside the counter
register to the capture/compare register when the selected input signal transition is
detected. This transition can be low to high (rising edge), high to low (falling edge),
or both (rising and falling edges). The input signal should be connected to the related
timer channel (tabulated in Table 6.1) as well. Hence, we can measure period of a
signal by using the capture mode with rising or falling edges, pulse duration of a
signal by using capture mode with both edges, or duration of an operation (e.g.,
time between pressing and releasing a button) using different capture modes.

The input capture block is given in Fig. 6.9. Here, the signal coming from the pin
TIx can be filtered first to remove unwanted noise. Then, polarity of the selected
input signal can be set using the edge detection and polarity selector blocks. The
signal edge to perform the capture operation can be selected as rising, falling, or
both edges. Then, the divider block is used to decide the number of signal edges
to perform the capture event. This allows making more robust measurements under
noise.

The input capture mode has three options as input capture direct, input capture
indirect, and input capture triggered by TRC. The input capture direct mode is used
to perform input capture from a signal (more precisely, the input signal) connected
to its own channel. The input capture indirect mode is used to perform input capture
from a signal connected to its complementary channel. For channel1, this is the
signal coming from channel2. Hence, channel2 must be set to input capture direct
mode to use this mode. Finally, input capture triggered by the TRC mode is used to
perform input capture from a signal connected to its selected ITRx signal if slave
mode is selected. The combined channels option is used to perform special input
capture modes to read encoders, hall sensors, or pulse width modulation (PWM)
inputs. We will not consider these modes in this book. Please see [1] for further
information on them.

Table 6.1 Input signals for
each channel

Channel	Input 1	Input 2	Input 3
Channel 1	MC pin 1	MC pin 2	TRC signal
Channel 2	MC pin 2	MC pin 1	TRC signal
Channel 3	MC pin 3	MC pin 4	TRC signal
Channel 4	MC pin 4	MC pin 3	TRC signal

We should also mention the valid input signals for each channel. Each channel of a base timer has three inputs. We tabulate them in Table 6.1. In this table, MC pins 1 to 4 stand for the microcontroller pins connected to channels 1 to 4, respectively. TRCs signal stands for one of the internal trigger signals (ITRx).

When the direct mode is selected in the input capture mode, each channel performs the capture operation from its own pin. As an example, when channel 1 is selected, the capture operation is done from the input signal coming from MC pin 1 as in Table 6.1. This is the standard input capture operation. Assume that we want to measure frequency of a PWM signal (to be introduced in Chap. 7). We connect it as an input signal to channel 1 MC pin 1 and perform the capture operation to make measurements and calculate the frequency value accordingly.

When the indirect mode is selected in the input capture mode, each channel performs the capture operation from its other pin. As an example, when channel 1 is selected, the capture operation is done from the input signal coming from MC pin 2 as in Table 6.1. This allows measuring different time values for the same signal. The best example for this scenario is measuring the frequency and duty cycle of a PWM signal. One way of doing this is feeding this signal to the two pins connected to two different channels and measuring frequency and duty cycle separately. The input capture indirect mode allows using one pin and performs calculations. For our example, we can feed the PWM signal to MC pin 1 for channel 1 and select the input capture mode. We can measure frequency of the signal this way. At the same time, we can select the input capture indirect mode for channel 2 and use MC pin 1 again. We can measure duty cycle of the signal this way. Hence, we use one pin for two different measurements.

The input capture block can also generate an interrupt as follows. When the selected edge type is detected, the value inside the counter register is copied to the capture/compare register. Capture/compare interrupt flag (CCxIF) is set and an interrupt is generated. As a side note, the capture/compare interrupt should have been enabled beforehand.

6.2.3.5 The Output Compare Block

The output compare block in the base timer is used for comparing the value inside the counter register with the capture/compare register. When both values are equal, transition of the output signal can be changed or an update event can be generated. Hence, we can generate the desired output signal with specific frequency and pulse duration. We can also use this block to generate update events at specific time instants without generating an output signal.

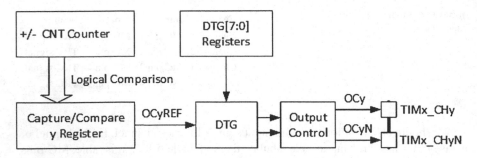

Fig. 6.10 Base timer output compare block

The output compare block of the base timer is given in Fig. 6.10. The output signal OCy is set, reset, or toggled when the value in the counter register equals the value written to the capture/compare register. Then, the capture/compare interrupt flag (CCxIF) is set and an interrupt is generated if the capture/compare interrupt has been enabled beforehand. OCyN, as the complementary signal of OCy, can also be generated. We will use this block while generating the PWM signal in Chap. 7.

The user can select one of the four output compare modes as output compare no output, output compare CHx, output compare CHxN, and output compare CHx CHxN. The output compare no output mode allows using the output compare mode for internal timing purposes without creating any output signal. The output compare CHx mode creates output signal at the related pin. The output compare CHxN mode creates a complementary output signal from the related pin. The output compare CHx CHxN mode creates two output signals as CHx and CHxN. PWM or forced output signals can be generated from related pins using suitable output compare selections. We will explore these options further in Chap. 7.

6.2.3.6 Summary of the STM32F4 Microcontroller Base Timers

The STM32F4 microcontroller has 14 base timers as mentioned before. We tabulate their properties in Table 6.2. In this table, resolution indicates how many bits does the timer counter has. The higher the number of bits, the better the resolution of the timer.

As can be seen in Table 6.2, TIM6 and TIM7 have no capture/compare channels. Therefore, they are mainly used for triggering the DAC module or basic time operations with 16-bit counter. TIM1 and TIM8 are advanced control timers with complementary output. Hence, they can be used for generating three phase PWM signals with programmable inserted dead times. They also have all features of complete general-purpose timers. TIM2 and TIM5 have 32-bit counters. TIM3 and TIM4 have 16-bit counters. These four timers have four independent channels with full capture/compare modes. TIM9, TIM10, TIM11, TIM12, TIM13, and TIM14 have 16-bit counters. However, they can be only used in up mode. TIM9 and TIM12 have two capture/compare channels. The remaining timers have only one such channel.

Table 6.2 Summary of the STM32F4 microcontroller base timers

Timer	Resolution (bits)	Mode	Prescaler (bits)	Capture/compare channels	Max. timer clock (MHz)
TIM1, TIM8	16	Up, down, center aligned	16	4	180
TIM2, TIM5	32	Up, down, center aligned	16	4	90
TIM3, TIM4	16	Up, down, center aligned	16	4	90
TIM6, TIM7	16	Up	16	0	90
TIM9	16	Up	16	2	180
TIM10, TIM11	16	Up	16	1	180
TIM12	16	Up	16	2	90
TIM13, TIM14	16	Up	16	0	90

6.2.4 System Timer in the STM32F4 Microcontroller

System timer, also called SysTick, is mainly used by the real-time operating system (RTOS). However, it can also be used as a dedicated timer. Since SysTick is extensively used by RTOS, it is fed by the main system clock. This clock can be divided by 1 or 8 according to the prescaler selection. SysTick has a 24-bit down-counter to create 1 ms time intervals.

6.2.5 Watchdog Timers in the STM32F4 Microcontroller

A software failure can create an infinite loop and halt code execution within the microcontroller. Resetting the CPU is the best option to recover from such situations. Watchdog is a special timer which resets the CPU if its counter counts to zero from its initial value (called timeout). The counter must be reloaded by a code to prevent the reset operation. If the code cannot be executed until the reload operation, the watchdog timer resets the CPU.

There are two different watchdog timers in the STM32F4 microcontroller. These are the independent watchdog (IWDG) and window watchdog (WWDG). IWDG is a classic watchdog timer as given in Fig. 6.11. It can be seen that it is clocked from the 32 kHz LSI RC oscillator. This means that IWDG does not depend on the main system clock and can operate in standby and stop modes. IWDG also has an 8-bit prescaler to divide this clock further.

All operations within the IWDG are controlled by its key register. We can start the IWDG by writing 0xCCCC to it. Initially, the watchdog counter is set to its maximum value, 0x0FFF. As the operation starts, this counter value decreases by one at every clock edge. If 0xAAAA is written to the key register before the count value reaches zero, then we kick the dog. Hence, the watchdog counter can be set to a new value and it restarts counting down from it. If we do not kick the dog, then the IWDG resets the CPU as the watchdog counter reaches zero.

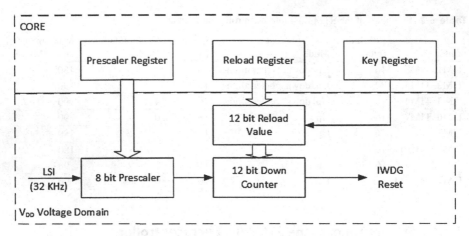

Fig. 6.11 IWDG block diagram

The WWDG does not have a key register as in IWDG. Instead, kicking the dog is done by writing a new value to its count register before timeout. In other words, the WWDG has a 7-bit counter which counts down starting from 127 at every clock edge. When this counter reaches 63, the CPU is reset by WWDG. Hence, this is the timeout value. We should kick the dog before the counter reaches this value. Therefore, we should write a new value to the counter before it reaches 63. In this case, the CPU is not reset and the counter restarts counting down from the new entered value. The WWDG is fed by the main system clock which is divided by 4096 initially. The clock frequency can be divided further by its prescaler.

The WWDG also has a window property. Hence, there is a window register which keeps a value between 127 and 63. During down counting of the WWDG counter, kicking the dog should be performed between the window register value and 63. In this setting, the CPU is not reset. If kicking the dog is not done within the mentioned time duration, then the CPU is reset by the WWDG. As a side note, IWDG resets the CPU after a predefined time. On the other hand, the WWDG resets the CPU if it is not kicked before predefined time is elapsed or it is kicked before the window value is reached. This has the following advantage. While using the IWDG, a bug may happen in a loop consisting of the watchdog reset line. As a result, the CPU stays within the loop but not resetting the CPU. When we use WWDG for such a scenario, the reset request will be done early, and the CPU understands that there is a problem in the code.

Both IWDG and WWDG can also be used as regular down counting timers. Since the STM32F4 microcontroller has several timers, we will not explore this property further. For more information on this property, please see [1].

6.2.6   Real-Time Clock in the STM32F4 Microcontroller

Real-time clock (RTC) is a unique binary coded decimal (BCD) timer which can run independently as long as it is powered properly. It works even during reset operation or under low power modes. It keeps track of the exact time as seconds, minutes, hours (in 12- or 24-h format), day (day of the week), date (day of the month), month, and year in two 32-bit registers in BCD format. The subseconds is also tracked within the RTC and kept in another register. Daylight saving compensation can also be configured in RTC through software.

RTC has 20 backup registers each being able to store 4 bytes. These keep program-specific data after reset or shut down when the system is powered by external battery. These registers are reset automatically if an unintended access to the device is detected using tamper detection property.

RTC has two programmable alarms with interrupt generation property. These are called Alarm A and B. These alarms can be used by enabling and setting a date to them. The date of the alarm is kept in separate registers. Hence, these two alarms work independent of each other. RTC also has an automatic wake-up property. It can be used to wake the microcontroller up from a low power mode periodically and perform desired actions. RTC has also an additional digital calibration feature to compensate for any inaccuracy at the clock signal generated by the crystal oscillator. For more information on this property, please see [1].

6.2.7   Advanced Base Timer Operations in the STM32F4 Microcontroller

We explained the trigger and clock controller block under base timers in Sect. 6.2.3. As a reminder, this block can be used to start/stop, reset, or update the timer counter using one of the trigger inputs. If the trigger input is selected as CK_INT or ETR, the clock is fed from internal PCLKx signal or external TIMx_ETR. In this setting, the base timer can be used alone.

The STM32F4 microcontroller allows using more than one base timer in master and slave setup. To use the first base timer as slave, one of its internal trigger signals, ITR0-3, should be selected as trigger input. The master base timer should create the ITR signal as output. Hence, a link is created between the master and slave timers through the ITR signal. As a result, the master timer controls operations in the slave timer.

When a timer is selected as slave, it can be used in one of the four modes as reset, gated, trigger, and external clock mode 1. In the reset mode, rising edge of the selected trigger input (TRGI) resets the counter and content of the timer registers is updated. In the gated mode, the counter clock is enabled when TRGI is high and disabled when it is low. This means the counter is started or halted by the master timer. In the trigger mode, the counter is started at the rising edge of the TRGI signal. In the external clock mode 1, rising edges of the TRGI signal update the counter according to the counter mode.

Table 6.3 The STM32F4 microcontroller ITRx connections

Slave timer	ITR0	ITR1	ITR2	ITR3
TIM1	TIM5_TRGO	TIM2_TRGO	TIM3_TRGO	TIM4_TRGO
TIM2	TIM1_TRGO	TIM8_TRGO	TIM3_TRGO	TIM4_TRGO
TIM3	TIM1_TRGO	TIM2_TRGO	TIM5_TRGO	TIM4_TRGO
TIM4	TIM1_TRGO	TIM2_TRGO	TIM3_TRGO	TIM8_TRGO
TIM5	TIM2_TRGO	TIM3_TRGO	TIM4_TRGO	TIM8_TRGO
TIM8	TIM1_TRGO	TIM2_TRGO	TIM4_TRGO	TIM5_TRGO
TIM9	TIM2_TRGO	TIM3_TRGO	TIM10_TRGO	TIM11_TRGO
TIM12	TIM4_TRGO	TIM5_TRGO	TIM13_TRGO	TIM14_TRGO

There are eight different modes to create a trigger output for the master timer. These modes are reset, enable, update, compare pulse, OCREF1, OCREF2, OCREF3, and OCREF4. In the reset mode, master and slave timers are reset at the same time. In the enable mode, the counter enable signal is used as the trigger output (TRGO) to start several timers at the same time. Hence, they can be synchronized. In the update mode, the update event is selected as TRGO. In the compare pulse mode, a positive pulse is fed to TRGO when the CC1IF flag is set. In OCREFx mode, the OCxREF signal generated by the output compare mode is used as TRGO. These trigger outputs can also be used to trigger other peripheral units such as ADC and DAC.

When the master timer is used in the update mode and slave master is used in external clock mode 1, slave counter is updated whenever an update event occurs in the master timer. This way, we connect these two timers having a-bit and b-bit counters in series and obtain one timer with (a+b)-bit counter. More than two timers can also be connected this way and counters with higher bit numbers and larger resolution can be obtained.

All ITRx signal connections for the STM32F4 microcontroller master and slave timers are given in Table 6.3. This table indicates which of the two timer modules can be used in master/slave form. As an example, if TIM1 is selected as slave, only TIM5, TIM2, TIM3, or TIM4 can be used as master. To note here, we cannot form a master/slave connection between any two timers. Table 6.3 also indicates which signal should be fed to a selected slave timer. As a side note, each slave timer has four trigger inputs as ITR0, ITR1, ITR2, and ITR3. For each timer, these are connected to different master timers. Therefore, as we select a master timer, we should select the connected trigger input for the slave timer accordingly.

6.3   Timer Setup in the STM32F4 Microcontroller

We will explain how to initialize the clock and timer modules in this section. To do so, we will consider C, C++, and MicroPython based operations. These will lead to the usage of the clock and timer modules in Sect. 6.4.

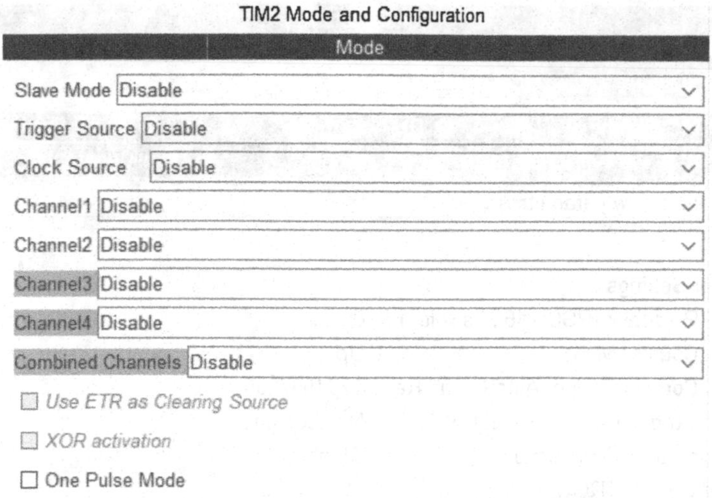

TIM2 Mode and Configuration

Mode

Slave Mode [Disable   ⌄]

Trigger Source [Disable   ⌄]

Clock Source   [Disable   ⌄]

Channel1 [Disable   ⌄]

Channel2 [Disable   ⌄]

Channel3 [Disable   ⌄]

Channel4 [Disable   ⌄]

Combined Channels [Disable   ⌄]

☐ Use ETR as Clearing Source

☐ XOR activation

☐ One Pulse Mode

Fig. 6.12 TIMx mode and configuration window

6.3.1   Timer Setup via C Language

STM32CubeMX offers a detailed and visually instructive environment to set up and initialize timer modules. Therefore, we will use it to set up the base timers, SysTick, watchdog timers, and RTC.

6.3.1.1 Setting Up Base Timers

To set up base timers, go to the "Timers" drop-down menu in "Pinout and Configuration" window and select the desired TIMx peripheral under STM32CubeMX. Opening TIMx mode and configuration window will be as in Fig. 6.12. As mentioned before, different base timers may have different properties and capture/compare channels. Hence, this window may differ for the selected TIM peripheral. In the opened TIMx mode and configuration window, as in Fig. 6.12, the user can select the internal clock or ETR2 external clock from the "Clock Source" drop-down list to use it for basic counter operations.

If we use the selected TIMx module in slave mode, we can choose external clock mode 1, reset, gated, or triggered mode from the associated drop-down list. If the slave mode is selected, external trigger source can be selected from the trigger source drop-down list. Here, trigger outputs coming from other timers (ITRx), channel 1 edge detection signal (TI1_ED), or filtered and synchronized external timer input signals (TI1FP1 and TI2FP2) are valid options.

We can select the desired input capture or output compare properties from each Channelx drop-down list. Within STM32CubeMX, the input capture mode has three options as input capture direct, input capture indirect, or input capture triggered by TRC. The user can select one of the four output compare modes as output compare no output, output compare CHx, output compare CHxN, and output compare CHx CHxN.

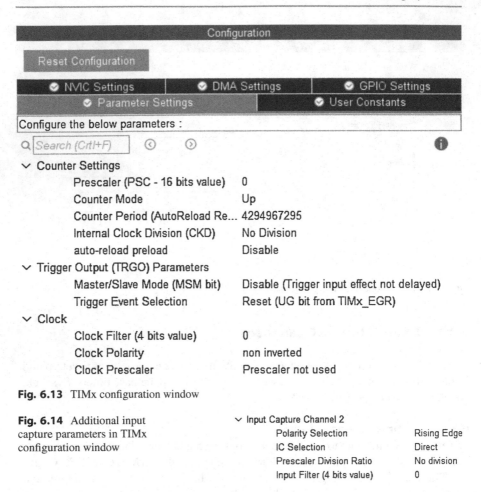

Fig. 6.13 TIMx configuration window

Fig. 6.14 Additional input
capture parameters in TIMx
configuration window

When the basic counter or input capture mode is selected, the configuration window will be as in Fig. 6.13. The user can set the Prescaler value for the clock signal, Counter Period value for ARR register, Counter Mode, Internal Clock Division for timer input filters, repetition counter (available for other timer modules), and auto-reload preload property from the counter settings drop-down list in the parameter settings subwindow. The user can also set master/slave mode and trigger event selection for creating trigger output from the trigger output (TRGO) parameters drop-down list.

When the input capture mode is selected, additional Input Capture Channel x drop-down list will be opened as in Fig. 6.14. Here, the user can set the polarity to select input signal edge type (whether from low to high or high to low), prescaler division ratio to set number of edges to perform input capture, and input filter.

The base timer interrupts can be enabled from the NVIC settings window. Here, all timer interrupts except for TIM1 and TIM8 can be enabled by selecting the

TIMx global interrupt box. Different timer interrupts can be enabled separately for TIM1 and TIM8. Update event interrupts are enabled by selecting the TIMx update interrupt box. Capture/compare interrupts are enabled by selecting the TIMx capture/compare interrupt box. Slave mode trigger interrupt is enabled by the TIMx trigger and commutation interrupts.

6.3.1.2 Setting Up the System Timer

The system timer (SysTick) does not have a setup operation under STM32CubeMX. The main reason is that it runs in background independently. The user can modify only its related interrupt settings if it will be used. We will explain this option in detail in Sect. 6.4.

6.3.1.3 Setting Up Watchdog Timers

STM32CubeMX can be used to set up watchdog timers in the STM32F4 micro-controller. To do so, the user should go to the system core drop-down menu in the "Pinout and Configuration" window under STM32CubeMX and select the IWDG or WWDG peripheral. In the opening "Mode and Configuration" window, we should tick the "activated" checkbox. The opening configuration window will be as in Fig. 6.15 for IWDG. Here, IWDG counter clock prescaler is used to divide the 32 kHz LSI clock. The IWDG down-counter reload value is used to select the timeout value. As an example, if we want 1 s timeout value, we can set the prescaler to 16 and counter to 2000.

The opening configuration window will be as in Fig. 6.16 for WWDG. Here, the WWDG counter clock prescaler is used to further divide PCLK1 which was divided by 4096 beforehand. WWDG down-counter value is used to select the timeout value. This value can be entered between 64 and 127 since the WWDG resets the system when this counter goes below 64. The WWDG window value is used to set the window value.

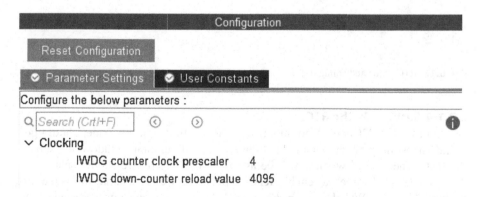

Fig. 6.15 IWDG configuration window

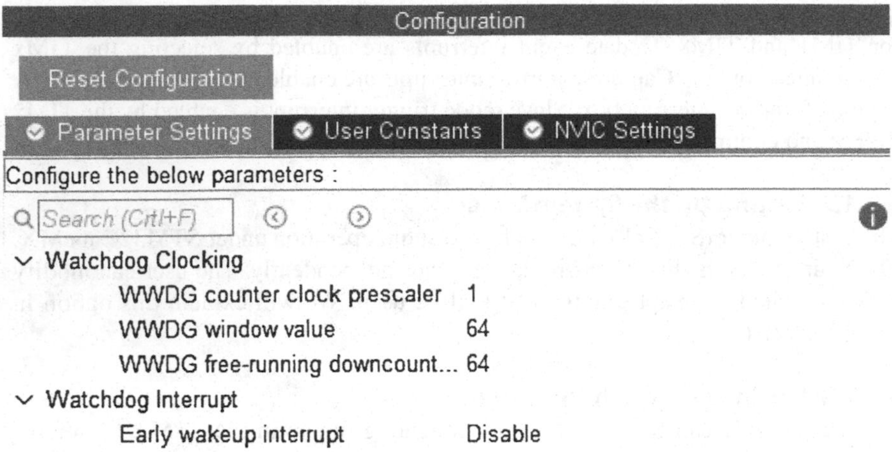

Configuration

Reset Configuration

✓ Parameter Settings   ✓ User Constants   ✓ NVIC Settings

Configure the below parameters :

🔍 Search (Crtl+F)   ⊲   ⊳   ℹ

∨ Watchdog Clocking
          WWDG counter clock prescaler   1
          WWDG window value   64
          WWDG free-running downcount... 64
∨ Watchdog Interrupt
          Early wakeup interrupt   Disable

Fig. 6.16 WWDG configuration window

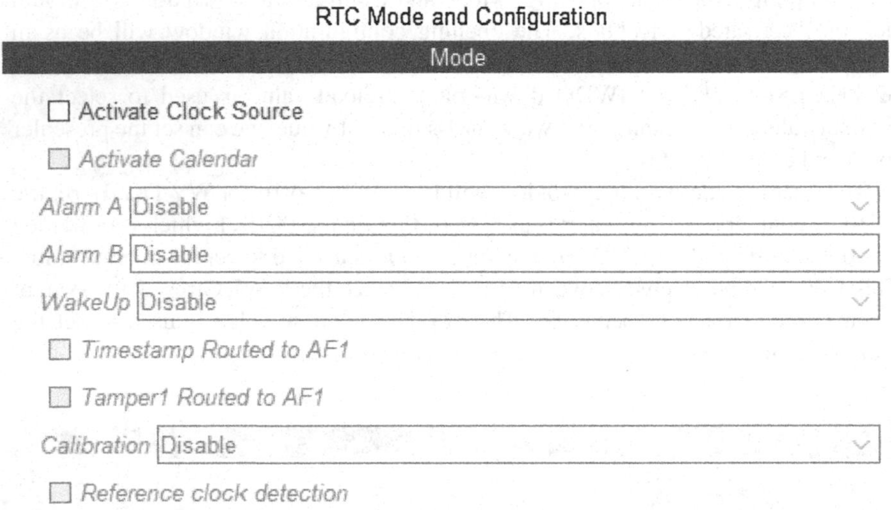

RTC Mode and Configuration

Mode

☐ Activate Clock Source

☐ Activate Calendar

Alarm A | Disable   ∨

Alarm B | Disable   ∨

WakeUp | Disable   ∨

☐ Timestamp Routed to AF1

☐ Tamper1 Routed to AF1

Calibration | Disable   ∨

☐ Reference clock detection

Fig. 6.17 RTC mode and configuration window

6.3.1.4 Setting Up the RTC

We can use STM32CubeMX to initialize the RTC. To do so, the reader should go
to the Timers drop-down menu in the "Pinout & Configuration" window and select
the RTC. The opened window will be as in Fig. 6.17. Here, we should select the
"Activate Clock Source" to enable RTC. Afterward, we can enable the wake-up
property from the "WakeUp" drop-down list. Likewise, we can activate the calendar
by ticking the "Activate Calendar" checkbox. Finally, we can activate alarms from
"Alarm A" and "Alarm B" drop-down lists.

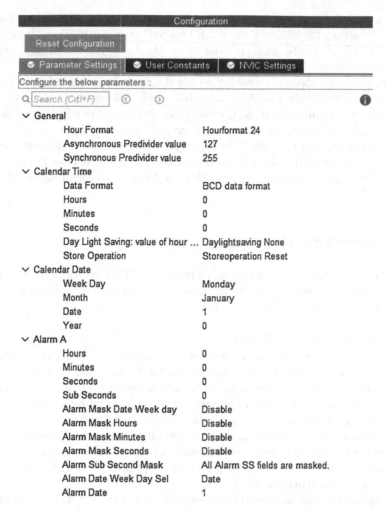

Fig. 6.18 RTC configuration window

The configuration window for RTC will be as in Fig. 6.18. Within this window, the reader can select the hour format as "Hourformat12" or "Hourformat24" within the "General" drop-down list. We can use the asynchronous and synchronous predivider values together to obtain 1 Hz clock from RTC. As a side note, the asynchronous predivider value can be selected between 0 and 127. The synchronous predivider value can be selected between 0 and 32767.

The reader can set the desired RTC time from the "Calendar Time" and "Calendar Date" drop-down lists. Daylight saving adjustment can also be configured within the "RTC configuration" window. Here, the options are "Daylightsaving None," "Daylightsaving Sub1h," or "Daylightsaving Add1h". The reader can set the desired alarm time in the "Alarm" drop-down list. To do so, we should first enter the time of the alarm as hour, minute, second, and subsecond. Then, day of the alarm can be selected by setting the "Alarm Date Week Day Sel" property. If "Date" is selected here, "Alarm Date" can be between 1 and 31. If "Weekday" is selected here, "Alarm Date" can be between "Monday" and "Sunday."

If the wake-up property is enabled within the configuration window, it can be modified from the "Wake Up" drop-down list. Here, wake-up clock is used which can be obtained by dividing the RTC or it can be directly set to 1 Hz. The wake-up counter can be used to set the desired wake-up interval.

RTC interrupts can be enabled from NVIC settings. Alarm interrupts can be enabled by selecting RTC Alarm A and B interrupts through EXTI line 17 checkbox. Wake-up interrupt is enabled by selecting RTC wake-up interrupt through EXTI line 22 checkbox.

6.3.2   Timer Setup via C++ Language

We can use Mbed to set up and use timers in the STM32F4 microcontroller in C++ language. Here, we have options for the base timers, watchdog timer, and RTC. We will evaluate each module in detail next.

6.3.2.1 Setting Up Base Timers

The base timer can be used under Mbed by its three classes `Timer`, `Timeout`, and `Ticker`. The timer class can be created by writing `Timer TimerX`. Here, `TimerX` is the name of the created `Timer` class and can be named as desired. The timer class is used to perform high-resolution counting operation by its 64-bit counter. It can be started, stopped, reset, or read for measuring precise time durations.

The timeout class can be created by writing `Timeout TimeoutX`. Here, `TimeoutX` is the name of the created `Timeout` class and can be named as desired. The timeout class is used to generate an interrupt after a set delay time. This interrupt is generated only once.

The ticker class can be created by writing `Ticker TickerX`. Here, `TickerX` is the name of the created `Ticker` class and can be named as desired. The ticker class is used to generate periodic interrupts at desired time intervals.

6.3.2.2 Setting Up the Watchdog Timer

Mbed has a `Watchdog` class to control the IWDG watchdog timer. There is only one instance for the watchdog timer and can be used directly by `Watchdog::get_instance()`. Or, this instance can be referenced as `Watchdog &WatchdogX = Watchdog:: get_instance()`. Here, `WatchdogX` is the name of the reference and can be named as desired. Then, this reference can be used throughout the code.

6.3.2.3 Setting Up the RTC

Mbed has no specific class to control RTC. Instead, RTC can be set to the desired date using the function `set_time(time_t t)`. Here, `t` is the number of seconds since `January 1, 1970`, which is the universal UNIX timestamp. Instead of calculating this value to set the desired time, the function `time_t mktime(tm *timeptr)` can be used. Through it, we can calculate the desired time value via the standard date format. Here, the `tm struct` has members `tm_year`, `tm_mon`, `tm_mday`, `tm_hour`, `tm_min`, and `tm_sec`. `tm_year` is the desired year and can be entered as the desired year minus 1900. `tm_mon` is the desired month of the year and can be entered between 0 and 11 to represent months January to December, respectively. `tm_mday` is the desired day of the month and can be entered between 0 and 31. `tm_hour`, `tm_min`, and `tm_sec` represent hour, minute, and second of the desired date. They can be entered between 0 and 23, 0 and 59, and 0 and 60, respectively.

6.3.3   Timer Setup via MicroPython

We can use MicroPython to set up and use timers in the STM32F4 microcontroller as well. Here, we have options for base timers, watchdog timer, and the RTC. We will evaluate each module in detail next.

6.3.3.1 Setting Up Base Timers

MicroPython has the `Timer` class to perform base timer operations. The base timer is created using the function `pyb.Timer(id)`. Here, `id` represents the timer number. As the base timer is created, it can be initialized by the function `init(freq, prescaler, period, mode, div)`. Here, `freq` is used to set the desired timer frequency. The `prescaler` is used to divide the timer clock. `period` is used to enter the period value. `mode` is used to select the counting mode as `Timer.UP`, `Timer.DOWN`, or `Timer.CENTER`.

As the base timer is initialized, its capture/compare channel can be initialized using the function `channel(mode, pin)`. Here, `mode` is used to select the capture/compare mode and can be one of `Timer.PWM`, `Timer.PWM_INVERTED`, `Timer.OC_TIMING`,   `Timer.OC_ACTIVE`,   `Timer.OC_INACTIVE`, `Timer.OC_TOGGLE`, `Timer.OC_FORCED_ACTIVE`, `Timer.OC_FORCED_INACTIVE`, `Timer.IC`, `Timer.ENC_A`, `Timer.ENC_B`, or `Timer.ENC_AB`. Here, `Timer.IC` is the input capture mode and `Timer.OC_TIMING` is the output compare mode without output and used for internal timing purposes. `Timer.ENC_A`, `Timer.ENC_B`, and `Timer.ENC_AB` are special input capture modes. They are used for encoder operations. We will not explain them further in this book. All other output compare modes will be explained in Chap. 7 related to the PWM-based DAC operations. When the input capture mode is selected, additional `polarity` input is used to select the signal edge. This value can be `Timer.RISING`, `Timer.FALLING`, or `Timer.BOTH`. When the output compare modes are selected, additional `compare` and `polarity` inputs can be used to set the compare register value and output polarity. The polarity can be `Timer.HIGH` or `Timer.LOW`. `pin` is the pin object which shows the pin connected to the selected channel, and alternate function for this channel is configured automatically.

6.3.3.2 Setting Up the Watchdog Timer

MicroPython has a WDT class to control the IWDG watchdog timer. This timer is created using the function machine.WDT(timeout). When the function is called, watchdog timer is started using the timeout value in milliseconds. This value cannot be changed or stopped after initialization. The timeout resolution must be in seconds and minimum timeout value must be 1 s.

6.3.3.3 Setting Up the RTC

MicroPython has the RTC class to control RTC. Hence, the RTC object can be created using the function pyb.RTC. Its time can be set using the function datetime(year, month, day, weekday, hours, minutes, seconds, subseconds). Here, weekday represents the day of the week and can be entered between 1 and 7 for Monday through Sunday. Also, subseconds can be entered between 0 and 255 and it counts down from the entered number. The function calibration(cal) can be used to calibrate the RTC by adding or subtracting number of cal ticks from the RTC clock over a 32 s period. cal value can be selected between -511 and 512.

6.4   Timer Usage in the STM32F4 Microcontroller

As the desired timer module is set in the STM32F4 microcontroller, it can be used to serve its purpose. We will introduce how this can be done in this section. To do so, we will evaluate the timer usage in C, C++, and MicroPython languages. While doing this, we will provide usage examples as well.

6.4.1   Timer Usage in C Language

We can use the timer module in C language after setting its properties in STM32CubeMX as explained in Sect. 6.3. Next, we will provide the usage options for base timers, SysTick, watchdog timers, and RTC via related HAL library functions. As a side note, C language provides the most detailed options to use all available timer modules in the STM32F4 microcontroller.

6.4.1.1 Usage of Base Timers

After setting the desired base timer properties in Sect. 6.3, related timer initialization code is created under the function MX_TIMx_Init. This function is called automatically in the "main.c" file. Then, specific HAL library functions can be used to control the base timer. The functions HAL_TIM_Base_Start and HAL_TIM_Base_Stop can be used to start or stop the base timer when it is not used with interrupts. This mode is used when capture/compare channels are used or base timer is used to trigger other peripherals. The functions HAL_TIM_Base_Start_IT and HAL_TIM_Base_Stop_IT can be used to start or stop the base timer when it is used with interrupts. This mode is used when specific time or time intervals are created to perform the desired time-based tasks without polling CPU. HAL_TIM_PeriodElapsedCallback is the callback function which is executed when triggered by update event. More information on these functions are given next.

```
HAL_TIM_Base_Start(TIM_HandleTypeDef *htim)
/*
htim: pointer to the TIM_HandleTypeDef struct.
*/

HAL_TIM_Base_Stop(TIM_HandleTypeDef *htim)

HAL_TIM_Base_Start_IT(TIM_HandleTypeDef *htim)

HAL_TIM_Base_Stop_IT(TIM_HandleTypeDef *htim)

HAL_TIM_PeriodElapsedCallback(TIM_HandleTypeDef *htim)
```

The functions HAL_TIM_IC_Start_IT and HAL_TIM_IC_Stop_IT can be used to start and stop input capture with interrupts on the selected channel. When the input capture is done, the callback function HAL_TIM_IC_CaptureCallback is executed. Inside it, captured values can be read using the function HAL_TIM_ReadCapturedValue. More information on these functions are given next.

```
HAL_TIM_IC_Start_IT(TIM_HandleTypeDef *htim, uint32_t Channel)
/*
htim: pointer to the TIM_HandleTypeDef struct.
Channel: variable used to select the channel for input capture
    operation. It can be entered as TIM_CHANNEL_x where x is the
    channel number.
*/

HAL_TIM_IC_Stop_IT(TIM_HandleTypeDef *htim, uint32_t Channel)

HAL_TIM_IC_CaptureCallback(TIM_HandleTypeDef *htim)

HAL_TIM_ReadCapturedValue(TIM_HandleTypeDef *htim, uint32_t
    Channel)
```

The functions HAL_TIM_y_Start and HAL_TIM_y_Stop can be used to start and stop the desired output compare modes. Functions HAL_TIM_y_Start_IT and HAL_TIM_y_Stop_IT can be used for the same purpose with additional interrupt capability. Here, y can be OC for output compare, PWM for PWM mode, and OnePulse for one pulse mode. The callback function HAL_TIM_OC_DelayElapsedCallback is executed when the counter equals to the value inside the capture/compare register. More information on these functions are given next.

```
HAL_TIM_OC_Start(TIM_HandleTypeDef *htim, uint32_t Channel)
/*
htim: pointer to the TIM_HandleTypeDef struct.
Channel: variable used to select the channel for output compare
    operation. It can be entered as TIM_CHANNEL_x where x is the
    channel number.
*/

HAL_TIM_OC_Stop(TIM_HandleTypeDef *htim, uint32_t Channel)
```

```
HAL_TIM_OC_Start_IT(TIM_HandleTypeDef *htim, uint32_t Channel)

HAL_TIM_OC_Stop_IT(TIM_HandleTypeDef *htim, uint32_t Channel)

HAL_TIM_OC_DelayElapsedCallback(TIM_HandleTypeDef *htim)

HAL_TIM_PWM_Start(TIM_HandleTypeDef *htim, uint32_t Channel)

HAL_TIM_PWM_Stop(TIM_HandleTypeDef *htim, uint32_t Channel)

HAL_TIM_PWM_Start_IT(TIM_HandleTypeDef *htim, uint32_t Channel)

HAL_TIM_PWM_Stop_IT(TIM_HandleTypeDef *htim, uint32_t Channel)

HAL_TIM_OnePulse_Start(TIM_HandleTypeDef *htim, uint32_t Channel)
/*
Channel: variable used to select the channel for output compare
    operation. It can be entered as TIM_CHANNEL_x where x is the
    channel number. Only channel 1 and 2 can be used here.
*/

HAL_TIM_OnePulse_Stop(TIM_HandleTypeDef *htim, uint32_t Channel)

HAL_TIM_OnePulse_Start_IT(TIM_HandleTypeDef *htim, uint32_t
    Channel)

HAL_TIM_OnePulse_Stop_IT(TIM_HandleTypeDef *htim, uint32_t
    Channel)
```

6.4.1.2 Usage of the System Timer
There is just one callback function named HAL_IncTick(void) for SysTick. This function is called every millisecond. We can use it to run time-based operations as such.

6.4.1.3 Usage of Watchdog Timers
After setting the desired watchdog timer properties in Sect. 6.3, related timer initialization code is created under the functions MX_IWDG_Init or MX_WWDG_Init based on the selected watchdog timer. The functions are called automatically in the "main.c" file. Afterward, specific HAL library functions can be used to control the watchdog timers. Functions HAL_IWDG_Refresh and HAL_WWDG_Refresh can be used to refresh IWDG and WWDG peripherals, respectively. More information on these functions are given next.

```
HAL_IWDG_Refresh(IWDG_HandleTypeDef *hiwdg)
/*
hiwdg: pointer to the IWDG_HandleTypeDef struct.
*/

HAL_WWDG_Refresh(WWDG_HandleTypeDef *hwwdg)
/*
hwwdg: pointer to the WWDG_HandleTypeDef struct.
*/
```

6.4.1.4 Usage of the RTC

After setting the desired RTC properties in Sect. 6.3, related timer initialization code is created under the function MX_RTC_Init. This function is called automatically in the "main.c" file. Then, specific HAL library functions can be used to control the RTC. Functions HAL_RTC_SetTime and HAL_RTC_SetDate can be used to set time of the day or exact calendar time. Functions HAL_RTC_GetTime and HAL_RTC_GetDate can be used to obtain the current day time or calendar time. More information on these functions are given next.

```
HAL_RTC_SetTime(RTC_HandleTypeDef *hrtc, RTC_TimeTypeDef *sTime,
    uint32_t Format)
/*
hrtc: pointer to the RTC_HandleTypeDef struct.
sTime: pointer to the Time struct.
Format: Time format is selected as binary with RTC_FORMAT_BIN or
    BCD with RTC_FORMAT_BCD.
*/

HAL_RTC_SetDate(RTC_HandleTypeDef *hrtc, RTC_DateTypeDef *sDate,
    uint32_t Format)

HAL_RTC_GetTime(RTC_HandleTypeDef *hrtc, RTC_TimeTypeDef *sTime,
    uint32_t Format)

HAL_RTC_GetDate(RTC_HandleTypeDef *hrtc, RTC_DateTypeDef *sDate,
    uint32_t Format)
```

The function HAL_RTC_SetAlarm_IT is used to set the desired alarm with interrupt. The function HAL_RTC_AlarmAEventCallback is executed when an alarm is generated. The set alarm can be deactivated using the function HAL_RTC_DeactivateAlarm. More information on these functions are given next.

```
HAL_RTC_SetAlarm_IT(RTC_HandleTypeDef *hrtc, RTC_AlarmTypeDef *
    sAlarm, uint32_t Format)

HAL_RTC_AlarmAEventCallback(RTC_HandleTypeDef *hrtc)

HAL_RTC_DeactivateAlarm(RTC_HandleTypeDef *hrtc, uint32_t Alarm)
```

6.4.1.5 Timer Usage Examples via C Language

Let's provide examples on the usage of timer modules via C language. Our first example is toggling the red and green LEDs on the STM32F4 board. More precisely, the green LED, connected to pin PG13 of the board, and red LED, connected to pin PG14 of the board, will be toggled at 1 s intervals. There will be a 0.5 s delay between the toggling operations in the red and green LEDs. To do so, we use the external HSE clock. Then, the SYSCLK frequency is set to maximum 180 MHz in the clock configuration window. Afterward, we enable the base timer TIM1 by selecting the internal clock as the clock source. We set the prescaler to 3599, counter period to 49999, and counter mode to up mode to obtain $((49999+1) \times (3599+1))/180000000 = 1$ s update event period. The green LED is toggled using this update event. Also, Channel 1 is configured as "Output Compare No Output."

We picked this mode to use the output compare mode to create the desired delay between toggling red and green LEDs. Here, we do not create an output signal. After the "Output Compare No Output" is selected from the related drop-down list, related parameters are added to the "Parameter Settings" window. Among these, "Pulse" value should be set to 25000 to obtain (25000 × (3599+1)) / 180000000 = 0.5 s delay time. Finally, we enable the TIM1 update interrupt and TIM1 capture/compare interrupt from the "NVIC Settings" window.

We provide the C code for our first example in Listing 6.1. Please add the given code parts to the related locations in your "main.c" file. Here, we use the code line if(__HAL_TIM_GET_FLAG(&htim1, TIM_FLAG_UPDATE) != RESET) __HAL_TIM_CLEAR_FLAG(&htim1, TIM_FLAG_UPDATE) to clear the update event flag in case it is set before the timer has started. When the timer init function is executed, this flag is set beforehand and must be cleared to prevent erroneous run.

Listing 6.1 Toggling the red and green LEDs using the TIM1 update and compare interrupts, in C language

```
/* USER CODE BEGIN 0 */
void HAL_TIM_PeriodElapsedCallback(TIM_HandleTypeDef *htim)
{
HAL_GPIO_TogglePin(GPIOG, GPIO_PIN_13);
}

void HAL_TIM_OC_DelayElapsedCallback(TIM_HandleTypeDef *htim)
{
HAL_GPIO_TogglePin(GPIOG, GPIO_PIN_14);
}
/* USER CODE END 0 */

/* USER CODE BEGIN 2 */
if (__HAL_TIM_GET_FLAG(&htim1, TIM_FLAG_UPDATE) != RESET)
__HAL_TIM_CLEAR_FLAG(&htim1, TIM_FLAG_UPDATE);
HAL_TIM_Base_Start_IT(&htim1);
HAL_TIM_OC_Start_IT(&htim1, TIM_CHANNEL_1);
/* USER CODE END 2 */
```

When the code in Listing 6.1 is executed, the red LED toggles first after 0.5 s using the compare interrupt. Then, the green LED toggles after 1 s using the update interrupt. Then, they keep toggling with 1 s intervals and 0.5 s between them.

Our second example is on turning on and off the green LED, connected to pin PG13 of the STM32F4 board. More precisely, the green LED will be turned on if pressing time of the user button, connected to pin PA0 of the STM32F4 board, is more than 3 s. If this time is less than 3 s, then the green LED turns off. To do so, we use the internal HSI oscillator to generate the 180 MHz SYSCLK. Afterward, we enable the base timer TIM2 by selecting Channel 1 in the "Input Capture" direct mode. We use Channel 1 of TIM2 since the user button on the STM32F4 board is connected to this channel. We select "Polarity Selection" as "Both Edges" from "Input Capture Channel 1" drop-down list. We also set the "Prescaler" to 0 and "Counter" period to 0xFFFFFFFF to obtain (0xFFFFFFFF + 1)/(PCLK1 × 2) = 47.72 s capture time. Here, we use PCLK1 × 2 since TIM2 is fed by twice the

PCLK1 frequency. Finally, we enable the capture/compare interrupt by selecting TIM2 global interrupt box in the "NVIC Settings" window.

We provide the C code for our second example in Listing 6.2. Please add the given code parts to the related places in your "main.c" file. When the user button on the STM32F4 board is pressed, its output goes from low to high and captured value is written to the variable `icRisingEdgeValue`. When the user button is released, its output goes from high to low and captured value is written to the variable `icFallingEdgeValue`. Then, we calculate the difference between these two values. If the second value is smaller than the first one, it means that counter is reset after first capture. Hence, the constant `0xFFFFFFFF + 1` is added to the difference. As the time duration of the user button press is calculated this way, the green LED is turned on or off accordingly.

Listing 6.2 Measuring the button pressing time using the input capture mode, in C language

```c
/* USER CODE BEGIN PV */
uint32_t icRisingEdgeValue = 0;
uint32_t icFallingEdgeValue = 0;
uint32_t icDifferenceValue = 0;
float pressingTime = 0;
uint8_t icFlag = 0;
/* USER CODE END PV */

/* USER CODE BEGIN 0 */
void HAL_TIM_IC_CaptureCallback(TIM_HandleTypeDef *htim)
{
if (htim->Channel == HAL_TIM_ACTIVE_CHANNEL_1)
{
if (icFlag == 0)
{
icRisingEdgeValue = HAL_TIM_ReadCapturedValue(&htim2,
    TIM_CHANNEL_1);
icFlag = 1;
}
else
{
icFallingEdgeValue = HAL_TIM_ReadCapturedValue(&htim2,
    TIM_CHANNEL_1);
if (icFallingEdgeValue > icRisingEdgeValue)
{
icDifferenceValue = icFallingEdgeValue - icRisingEdgeValue;
}

else
{
icDifferenceValue = 0xFFFFFFFF + 1 + icFallingEdgeValue -
    icRisingEdgeValue;
}
pressingTime = ((float)icDifferenceValue) / (2 *
    HAL_RCC_GetPCLK1Freq());
if (pressingTime > 3.0)
HAL_GPIO_WritePin(GPIOG, GPIO_PIN_13, GPIO_PIN_SET);
else
HAL_GPIO_WritePin(GPIOG, GPIO_PIN_13, GPIO_PIN_RESET);
icFlag = 0;
}
}
}
```

```
/* USER CODE END 0 */

/* USER CODE BEGIN 2 */
HAL_TIM_IC_Start_IT(&htim2, TIM_CHANNEL_1);
/* USER CODE END 2 */
```

In our third example, the green LED connected to pin PG13 of the STM32F4
board is toggled at 1 s intervals using SysTick. To do so, we use the external HSE
clock. Then, the SYSCLK frequency is set to 180 MHz and "Cortex System Timer"
is set to the same value by setting its prescaler to 1. We provide the C code for our
third example in Listing 6.3. Please add the given code parts to the related places in
your "main.c" file.

Listing 6.3   Toggling the green LED using SysTick, in C language

```
/* USER CODE BEGIN PV */
uint16_t sysTickCnt = 0;
/* USER CODE END PV */

/* USER CODE BEGIN 0 */
void HAL_IncTick(void)
{
sysTickCnt++;
if (sysTickCnt >= 1000)
{
HAL_GPIO_TogglePin(GPIOG, GPIO_PIN_13);
sysTickCnt = 0;
}
}
/* USER CODE END 0 */
```

In our fourth example, the green LED connected to pin PG13 of the STM32F4
board is toggled at 1 s intervals using the TIM1 update event. To do so, we use
the same clock and timer configurations given in the first example. We also start the
IWDG module with 4 s timeout value. To do so, we activate IWDG from the "IWDG
Mode and Configuration" window. Then, its prescaler is set to 32 and down-counter
reload value is set to 4000. Hence, we obtain $4000 \times 32 / 32000 = 4$ s timeout value.
We provide the C code for our fourth example in Listing 6.4. Please add the given
code parts to the related places in your "main.c" file.

Listing 6.4   Using IWDG to reset the system, in C language

```
/* USER CODE BEGIN 0 */
void HAL_TIM_PeriodElapsedCallback(TIM_HandleTypeDef *htim)
{
if (htim->Instance == TIM1)
{
HAL_IWDG_Refresh(&hiwdg);
HAL_GPIO_TogglePin(GPIOG, GPIO_PIN_13);
}
}

void HAL_GPIO_EXTI_Callback(uint16_t GPIO_Pin)
{
if (GPIO_Pin == GPIO_PIN_0)
{
```

```
while (1);
}
}
/* USER CODE END 0 */

/* USER CODE BEGIN 2 */
if (__HAL_TIM_GET_FLAG(&htim1, TIM_FLAG_UPDATE) != RESET)
__HAL_TIM_CLEAR_FLAG(&htim1, TIM_FLAG_UPDATE);
HAL_TIM_Base_Start_IT(&htim1);
/* USER CODE END 2 */
```

In Listing 6.4, we refresh the IWDG module in TIM1 update interrupt callback function every second. Hence, if there is no problem, then the green LED keeps toggling. However, when the user button connected to pin PA0 of the STM32F4 board is pressed, the system is stuck in an infinite loop. As a result, the green LED stops toggling. After 4 s, system is reset by IWDG and the green LED starts toggling again. This example shows how effective the watchdog timer is in an unexpected infinite loop trap.

Our fifth example deals with setting the RTC alarm and generating an interrupt accordingly. The green LED, connected to pin PG13 of the STM32F4 board, is turned on within the callback function. To do so, we first select the RTC clock. We use the external HSE clock and divide it by 25 to obtain 320 kHz RTC clock since the STM32F4 board does not have an LSE crystal. Then, we activate RTC and enable its "Alarm A." We set the asynchronous predivider value to 124 and synchronous predivider value to 2559 in the "Parameter Settings" window. Hence, we obtain 1 Hz from the 320 kHz RTC. Afterward, we should enter the current date and time as "Calendar Time" and set alarm time to 1 min after the calendar time. Finally, we enable the alarm interrupt from the "NVIC Settings" window. We provide the C code for the fifth example in Listing 6.5. Please add the given code parts to the related places in your "main.c" file. As the code is executed, it can be seen that the green LED toggles in the designated time.

Listing 6.5   Using the alarm property of RTC, in C language

```
/* USER CODE BEGIN 0 */
void HAL_RTC_AlarmAEventCallback(RTC_HandleTypeDef *hrtc)
{
HAL_GPIO_TogglePin(GPIOG, GPIO_PIN_13);
}
/* USER CODE END 0 */
```

6.4.2   Timer Usage in C++ Language

We will next consider the timer usage in C++ language under Mbed. To do so, we will consider base timers, watchdog timer, and RTC. We will provide usage examples for these timers as well.

6.4.2.1 Usage of Base Timers

Mbed uses the `std::chrono` API to measure the elapsed time. If a timer is created as `Timerx`, `Timerx.elapsed_time().count()` is used to measure the elapsed time from start in microseconds. Casting can be done to the count value to measure the elapsed time in milliseconds, seconds, or seconds as float. To do so, the pattern is `chrono::duration_cast<chrono::milliseconds>` `(Timerx.elapsed_time()).count()`,   `chrono::duration_cast` `<chrono::seconds>(Timerx.elapsed_time()).count()`, and `chrono::` `duration<float>(Timerx.elapsed_time()).count()`   for   milliseconds, seconds, and seconds as float, respectively.

After the `Timeout` class is created, the function `attach(callback, time)` can be used to execute the callback function `callback` after desired seconds set by the `time` value. When microsecond time delay is needed, the function `attach_us(callback, time)` should be used. Here, the `time` value represents the time in microseconds. The `detach` function is used to detach the callback function from the timer.

6.4.2.2 Usage of the Watchdog Timer

The function `start(uint32_t timeout)` is used to start the watchdog timer with the given `timeout` value. If the input of the function is empty, it is started with the maximum timeout value. Watchdog timer can be stopped using the function `stop`. Or, it can be refreshed using the function `kick`. Status of the watchdog timer can be observed using the function `is_running`. Entered timeout or maximum available timeout values can be obtained using the functions `get_timeout` and `get_max_timeout`, respectively. These functions can be used as `Watchdog::get_instance().kick()` or as `WatchdogX.kick()` if referenced as `Watchdog &WatchdogX = Watchdog::get_instance()`.

6.4.2.3 Usage of the RTC

After the RTC date is set, current time can be observed using the function `ctime(time_t *timer)`. This function calculates the current time from the `timer` value and returns it in the format `Www Mmm dd hh:mm:ss yyyy`. Here, `Www` is the weekday. `Mmm` is the month (in string). `dd` is the day of the month. `hh:mm:ss` is the time, and `yyyy` is the year. The `timer` value in the function can be obtained from RTC using `time(NULL)`.

6.4.2.4 Timer Usage Examples via C++ Language

Next, we provide Mbed-based timer usage examples. We start with repeating the first example given in Sect. 6.4.1. Mbed does not have a function to be used for the output compare mode for timing purposes. Hence, we used two ticker classes to toggle the red and green LEDs in 1 s intervals in our example. We also used another timeout class to create 0.5 s delay between the red and green LEDs. We provide the C++ code for our first example in Listing 6.6.

Listing 6.6 Toggling the green and red LEDs using timer interrupts, in C++ language

```
#include "mbed.h"

DigitalOut greenLED(LED1);
DigitalOut redLED(LED2);
Timeout timeoutDelay;
Ticker ticker1;
Ticker ticker2;

void toggleGreenLED()
{
greenLED = !greenLED;
}

void toggleRedLED()
{
redLED = !redLED;
}

void delay05()
{
toggleRedLED();
ticker1.attach(&toggleRedLED, 1s);
}

int main()
{
timeoutDelay.attach(&delay05, 500ms);
ticker2.attach(&toggleGreenLED, 1s);

while (1);
}
```

In Listing 6.6, `timeoutDelay` is the created `Timeout` class, and it calls the function `delay05()` once to create 0.5 s delay between LEDs. In the function `delay05()`, the red LED connected to pin PG14 of the board is toggled using the function `toggleRedLED()`. `ticker1`, the created `Ticker` class, is also attached to the function `toggleRedLED()` to toggle the red LED with 1 s intervals. `ticker2`, the second created `Ticker` class, is attached to the function `toggleGreenLED()` at the beginning of the `main` function to toggle the green LED with 1 s interval.

We next repeat the second example given in Sect. 6.4.1 now in C++ language. Mbed does not have a function for the input capture mode. Hence, we use the GPIO interrupts and `Timer` class to measure the duration between successive interrupt requests. We provide the corresponding C++ code for this example in Listing 6.7.

Listing 6.7 Measuring button pressing time using GPIO interrupts and timer, in C++ language

```
#include "mbed.h"

InterruptIn userButton(BUTTON1);
DigitalOut greenLED(LED1);
Timer timer1;

uint32_t elapsedMs;

void triggerRise()
{
```

```
timer1.start();
}

void triggerFall()
{
timer1.stop();
elapsedMs = chrono::duration_cast<chrono::milliseconds>(timer1.
    elapsed_time()).count();
if (elapsedMs > 3000)
greenLED = 1;
else
greenLED = 0;
timer1.reset();
}

int main()
{
userButton.rise(&triggerRise);
userButton.fall(&triggerFall);

while (1);
}
```

In Listing 6.7, the timer1 is started at the rising edge of the GPIO interrupt generated by the user button press. At the falling edge of the user button press, the timer1 is stopped and elapsed time is measured in milliseconds. If the elapsed time is greater than 3000 ms, then the green LED on the STM32F4 board turns on. Otherwise, the green LED turns off. Finally, the timer1 is reset to measure the next pressing time.

The fourth example in Sect. 6.4.1 can also be done via Mbed. We provide the corresponding C++ code in Listing 6.8. Here, the watchdog timer is started with 4000 ms timeout value. The watchdog is reset and the green LED is toggled every second. When the user presses the onboard button, the system is stuck in an infinite loop. The green LED stops toggling. After 4 s, the system is reset by the watchdog timer. Hence, the green LED starts toggling again.

Listing 6.8 Using the watchdog timer to reset stuck system, in C++ language

```
#include "mbed.h"

DigitalOut greenLED(LED1);
InterruptIn userButton(BUTTON1);
Ticker ticker1;

void toggleGreenLED()
{
greenLED = !greenLED;
Watchdog::get_instance().kick();
}

void trigger()
{
while (1);
}

int main()
{
Watchdog &wdt = Watchdog::get_instance();
ticker1.attach(&toggleGreenLED, 1s);
wdt.start(4000);
```

```
userButton.rise(&trigger);

while (1);
}
```

In our final Mbed-based timer example, we use the RTC to observe the current time when the user button is pressed. We provide the C++ code for our example in Listing 6.9. Within the code, we first set the date as "31 July 2020 18:18:18." When the user button is pressed, the current time based on set date is displayed. To do so, we set a flag in the callback function `trigger` when the user button is pressed. Then, we use the `printf` function in the `main` function to perform the printing operation. We followed such a procedure since it is recommended not to use the `printf` function inside callback functions since it is slow. When `printf` is being executed within the callback function, another interrupt may emerge. This either causes a problem or other interrupts may be missed. Therefore, we control the `printf` function by a flag outside the callback function.

Listing 6.9 Using RTC to display the current time, in C++ language

```
#include "mbed.h"

InterruptIn userButton(BUTTON1);

tm date;
uint8_t flag = 0;

void trigger()
{
flag = 1;
}

int main()
{
userButton.rise(&trigger);
date.tm_year = 120;
date.tm_mon = 6;
date.tm_mday = 31;
date.tm_hour = 18;
date.tm_min = 18;
date.tm_sec = 18;
set_time(mktime(&date));

while (true)
{
if (flag == 1)
{
time_t seconds = time(NULL);
printf("Current Time = %s\r\n", ctime(&seconds));
flag = 0;
}
}
}
```

6.4.3   Timer Usage in MicroPython

We finally consider the timer usage via MicroPython. To do so, we will consider base timers, watchdog timer, and RTC. We will also provide usage examples for these timers as well.

6.4.3.1 Usage of Base Timers

The function callback(fun) can be used to set the callback function fun after the base timer is created and initialized. After the timer is created as Timerx, it can be used as Timerx.callback(fun).

The function callback(fun) can be used to set the capture/compare callback function fun after the capture/compare channel is created and initialized. After the timer channel is created as TimerChannelx, it can be used as TimerChannelx.callback(fun). The functions capture() and compare() of the created capture/compare channel can be used to get the captured time or set the desired compare value.

6.4.3.2 Usage of the Watchdog Timer

There is only one function to control IWDG in MicroPython. This function is feed. It is used to refresh the watchdog timer to prevent system reset.

6.4.3.3 Usage of the RTC

The function datetime() can be used to obtain the current time after the RTC is set. Please remember that the same function was used for setting the desired time. However, it returns the current time when used with no input argument. The function wakeup(timeout, callback) can be used to execute the callback function in every timeout interval given in milliseconds. This function can be used to wake up from stop or standby modes.

6.4.3.4 Timer Usage Examples via MicroPython

We can use the timer modules in the STM32F4 microcontroller via MicroPython. We provide such examples in this section. We will start with repeating the first example given in Sect. 6.4.1 for this purpose. We provide the MicroPython code for this example in Listing 6.10.

Listing 6.10 Toggling the red and green LEDs using the TIM1 update and compare interrupts, in MicroPython

```
import pyb

greenLED = pyb.Pin('PG13', mode=pyb.Pin.OUT_PP)
redLED = pyb.Pin('PG14', mode=pyb.Pin.OUT_PP)

def toggleGreenLED(timer):
    greenLED.value(not greenLED.value())

def toggleRedLED(timer):
    redLED.value(not redLED.value())

def main():
```

```
timer1 = pyb.Timer(1, prescaler=3359, period=49999, mode=pyb.
    Timer.UP)
timer1CH1 = timer1.channel(1, pyb.Timer.OC_TIMING, polarity=
    pyb.Timer.HIGH, compare=24999)
timer1.callback(toggleGreenLED)
timer1CH1.callback(toggleRedLED)

main()
```

In Listing 6.10, SYSCLK is set to 168 MHz in MicroPython. Hence, we set the prescaler to 3359, counter period to 49999, and counter mode to "Up mode" to obtain $((49999+1) \times (3359+1))/168000000 = 1$ s update event period within the code. As a result, the red LED toggles first after 0.5 s using the compare interrupt. Then, the green LED toggles after 1 s using the overflow interrupt. Then, both LEDs keep toggling with 1 s interval and 0.5 s between them.

We next repeat the second example given in Sect. 6.4.1 using MicroPython. We provide the corresponding MicroPython code used for this purpose in Listing 6.11. As mentioned in the previous example, SYSCLK is set to 168 MHz in MicroPython. Hence, PCLK1 is set to 42 Mhz. Therefore, we set the Timer 2 prescaler to 7 and period to 0x0FFFFFFF to obtain $(0x0FFFFFFF + 1)/(PCLK1 \times 2/(7 + 1)) = 25.57$ s capture time in the code. We use Channel 1 of Timer 2 since the user button on the STM32F4 board is connected to this channel. When the user button is pressed, its output goes from low to high and the captured value is written to the variable icRisingEdgeValue. When the user button is released, its output goes from high to low and the captured value is written to the variable icFallingEdgeValue. Then, the difference between these two values is calculated. If this difference is greater than 3 s, the green LED is turned on. Otherwise, the green LED is turned off.

Listing 6.11 Measuring button pressing times using input capture mode, in MicroPython

```
import pyb

timer2 = pyb.Timer(2, prescaler=7, period=0x0FFFFFFF, mode=pyb.
    Timer.UP)
timer2CH1 = timer2.channel(1, pyb.Timer.IC, pin=pyb.Pin.board.PA0
    , polarity=pyb.Timer.BOTH)
greenLED = pyb.Pin('PG13', mode=pyb.Pin.OUT_PP)
calcFlag = 0
icRisingEdgeValue = 0
icFallingEdgeValue = 0
icDifferenceValue = 0
pressingTime = 0

def ic_callback(tim):
    global calcFlag
    global icRisingEdgeValue
    global icFallingEdgeValue
    if pyb.Pin.board.PA0.value():
        icRisingEdgeValue = timer2CH1.capture()
    else:
        icFallingEdgeValue = timer2CH1.capture()
        calcFlag = 1

def main():
```

```
    global calcFlag
    global pressingTime
    global icDifferenceValue
    timer2CH1.callback(ic_callback)
    while True:
        if calcFlag == 1:
            if icFallingEdgeValue > icRisingEdgeValue:
                icDifferenceValue = icFallingEdgeValue -
                    icRisingEdgeValue
            else:
                icDifferenceValue = 0x0FFFFFFF + 1 +
                    icFallingEdgeValue - icRisingEdgeValue
            pressingTime = (icDifferenceValue)/(10500000)
            print(pressingTime)
            if pressingTime > 3.0:
                greenLED.value(1)
            else:
                greenLED.value(0)
            calcFlag = 0

main()
```

We can repeat the fourth example given in Sect. 6.4.1 using MicroPython as well. We provide the corresponding MicroPython code for this case in Listing 6.12. Within the code, the watchdog timer is started with 4000 ms timeout value. The callback function `toggleGreenLED` is called every second. Within it, the green LED toggles and the WDT is reset. However, the `flag` variable should have value 0 for these operations. When the user presses the button, the `flag` variable is set to 1. Hence, toggling the LED and resetting the WDT is blocked. Therefore, the CPU resets after 4 s and the green LED restarts toggling.

Listing 6.12 Using the watchdog timer to reset the system, in MicroPython

```
import pyb
import machine

greenLED = pyb.Pin('PG13', mode=pyb.Pin.OUT_PP)
wdt = machine.WDT(timeout=4000)
flag = 0

def toggleGreenLED(timer):
    global flag
    if flag == 0:
        greenLED.value(not greenLED.value())
        wdt.feed()

def trigger(pin):
    global flag
    flag = 1

def main():
    pyb.ExtInt('PA0', mode=pyb.ExtInt.IRQ_RISING, pull=pyb.Pin.
        PULL_NONE, callback=trigger)
    timer1 = pyb.Timer(1, prescaler=3359, period=49999, mode=pyb.
        Timer.UP)
    timer1.callback(toggleGreenLED)

main()
```

We finally repeat the fifth example given in Sect. 6.4.2 using MicroPython. We provide the corresponding MicroPython code in Listing 6.13. Here, we set the current date to "Friday July 31 2020 18:18:18:000" first. When the user button is pressed, the current time based on the set date is displayed. Here, we set a flag when the button is pressed and perform the printing operation in the main function using this flag. This procedure is followed since it is recommended not to use print function inside the callback function.

Listing 6.13 Using RTC to display the current time, in MicroPython

```
import pyb

flag = 0

def trigger(pin):
    global flag
    flag = 1

def main():
    global flag
    pyb.ExtInt('PA0', mode=pyb.ExtInt.IRQ_RISING, pull=pyb.Pin.
        PULL_NONE, callback=trigger)
    rtc = pyb.RTC()
    rtc.datetime((2020, 7, 31, 5, 18, 18, 18, 0))
    while True:
        if flag == 1:
            print(rtc.datetime())
            flag = 0

main()
```

6.5 Application: Timing Operations in the Robot Vacuum Cleaner

In this chapter, we perform the time-based operations in our robot vacuum cleaner. Therefore, we measure the speed of our robot with encoders and photosensors connected to its wheels. We also set the RTC. Hence, the robot automatically wakes up at the same time every day, sweeps the room, and returns to standby mode. We provided the equipment list, circuit diagram, detailed explanation of design specifications, and peripheral unit settings in a separate document in the folder "Chapter6\EoC_applications" of the accompanying supplementary material. In the same folder, we also provided the complete C, C++, and MicroPython codes for the application.

6.6   Summary of the Chapter

Time is one of the most valuable assets in today's world. Hence, time-based operations are crucial in embedded applications. Therefore, we explored how time-based operations are done in the STM32F4 microcontroller (as an embedded system) in this chapter. To do so, we started with the oscillators in the STM32F4 microcontroller and STM32F4 board as the main building block of clock modules in the microcontroller. Then, we explored different timer modules available in the STM32F4 microcontroller. These timers are based on clock signals to operate. Besides, they have different usage areas. We covered these by providing usage examples to each in this chapter. Finally, we added time-based properties to our application considered in the previous chapters as the end of chapter application.

Problems

6.1. Why should there be more than one timer module in an embedded system?

6.2. What types of oscillators are available in the STM32F4 microcontroller and on the STM32F4 board? Compare properties of these oscillators.

6.3. How can we solve the switch bouncing problem by using timer modules?

6.4. How does a prescaler (more specifically frequency divider) work?

6.5. Form a project to count the number of user button (on the STM32F4 board) presses within 1 min. Each button press should evoke an interrupt. Use appropriate timer in operation. Form the project using:

(a) C language under STM32CubeIDE.
(b) C++ language under Mbed.
(c) MicroPython.

6.6. Add power management option to Problem 6.5 such that the STM32F4 microcontroller waits in an appropriate low power mode when not used.

6.7. Form a project with the following specifications. When the user presses the button on the STM32F4 board, the green LED on the board turns on. The LED should turn off after 5 s. Use an appropriate timer module in operation. Form the project using:

(a) C language under STM32CubeIDE.
(b) C++ language under Mbed.
(c) MicroPython.

6.8. Add power management option to Problem 6.7 such that the STM32F4 microcontroller waits in an appropriate low power mode when not used.

6.9. A stepper motor is a special type of brushless DC motor, and its rotation is divided into number of equal steps. Stepper motors have different number of coils and by energizing each coil in sequence, stepper motor rotates for one step at each time. The speed of the stepper motor can be adjusted by changing delay between each phase. Form a project with the following specifications. There is a stepper motor with four phases and 200 steps for one rotation. The phases are connected to pins PE2, PE3, PE4, and PE5 of the STM32F4 board. The stepper motor has three different speeds as fast, normal, and slow which can be changed using the onboard push button. Delay between phases is set to 5 ms, 20 ms, and 50 ms for each mode, respectively. The stepper motor stops operating automatically after 20 turns. Delay between phases and total run time of stepper motor are controlled using TIM2 module. Add power management option to your solution such that the STM32F4 microcontroller waits in an appropriate low power mode when not used. Form the project using:

(a) C language under STM32CubeIDE.
(b) C++ language under Mbed.
(c) MicroPython.

6.10. Form a project with the following specifications. The green LED on the STM32F4 board toggles every hour, every day at 9:00 AM, and every week Mondays at 9:00 AM. Use an appropriate timer module in operation. Add power management option to your solution such that the STM32F4 microcontroller waits in an appropriate low power mode when not used. Form the project using:

(a) C language under STM32CubeIDE.
(b) C++ language under Mbed.
(c) MicroPython.

6.11. Repeat Problem 6.10 such that the green LED will turn on at the designated time. Then, it will turn off after 10 s.

6.12. Repeat Problem 4.9 with an appropriate timer module. Add power management option to your solution such that the STM32F4 microcontroller waits in an appropriate low power mode when not used.

6.13. Repeat Problem 4.10 with an appropriate timer module. Add power management option to your solution such that the STM32F4 microcontroller waits in an appropriate low power mode when not used.

6.14. Repeat Problem 4.11 with an appropriate timer module. Add power management option to your solution such that the STM32F4 microcontroller waits in an appropriate low power mode when not used.

6.15. Form a project with the following specifications. The IWDG module is started to reset the STM32F4 microcontroller after 4 s. The push button is used to kick the watchdog timer when pressed. Finally, the Timer 13 module is configured to generate interrupts with 1 s interval. In the first 4 s, the Timer 13 is used to reset the IWDG counter. After 4 s, the timer toggles the onboard green LED. Add power management option to your solution such that the STM32F4 microcontroller waits in an appropriate low power mode when not used. Form the project using:

(a) C language under STM32CubeIDE.
(b) C++ language under Mbed.
(c) MicroPython.

6.16. Form a project with the following specifications. The Timer 3 module is configured as master to toggle its output with 0.1 Hz frequency. The Timer 2 module is configured as slave to toggle the onboard green LED with 1 s interval. The Timer 2 module is enabled/disabled by the Timer 3 module output signal. Add power management option to your solution such that the STM32F4 microcontroller waits in an appropriate low power mode when not used. Form the project using C language under STM32CubeIDE.

References

1. STMicroelectronics: STM32F405/415,   STM32F407/417,   STM32F427/437   and STM32F429/439 advanced Arm-based 32-bit MCUs, rm0090 rev 19th edn. (2021)
2. Ünsalan, C., Tar, B.: Digital System Design with FPGA: Implementation Using Verilog and VHDL. McGraw-Hill, New York (2017)

Conversion Between Analog and Digital Values

7

7.1 Analog and Digital Values

From an embedded system perspective, there are two types of values as analog and digital. We will consider these in detail in this section. We will also consider how digital values can be represented in code form.

7.1.1 Analog Values in Physical Systems

In order to explain analog values better, let's take temperature readings of a room. We can measure the temperature value at any time instant. Hence, we can have infinite number of observations at hand. Each observation can be represented by infinite precision depending on the measuring device. In short, we can obtain infinite number of observations, each having infinite precision. These are the two defining properties of analog values.

Temperature is not the only analog value in real world. In fact, measurements taken from actual physical systems will be in analog form most of the times. We will need a sensor to make such measurements. Although sensors differ based on their application areas, they share a common property. A sensor converts one value to another form. For our purposes, we will pick sensors converting a physical measurement to electrical form either as voltage or current. Hence, it can be processed in the embedded system. We will introduce specific sensors for the application at hand whenever needed throughout the book.

Supplementary Information The online version contains supplementary material available at (https://doi.org/10.1007/978-3-030-88439-0_7).

C. Ünsalan et al., *Embedded System Design with Arm Cortex-M Microcontrollers*, https://doi.org/10.1007/978-3-030-88439-0_7

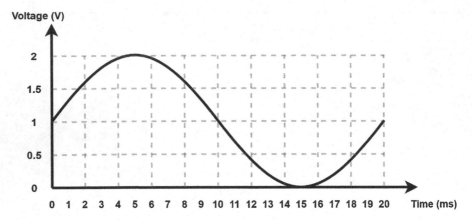

Fig. 7.1 A sample analog signal

We can represent analog values in time. Such a representation will be called analog signal. We provide a sample analog signal in Fig. 7.1. The reader should always remember that this signal is composed of infinite number of values. Amplitude of the signal at any time instant will have infinite precision as well.

7.1.2　Digital Values in Embedded Systems

An embedded system has limited memory space and processing power by its nature. In other words, we cannot store or process infinite numbered and ranged analog values in the embedded system. The suitable form for such systems is digital. The basic digital representation is the bit which can take two values as logic level 0 or 1. In order to have a more general digital representation, we will need type definitions. Since these depend on the programming language used in the embedded system, we will cover them in Sect. 7.1.3.

Before going further, we should briefly explain how a digital value is represented in an embedded system. To do so, let's focus on the simplest digital form as one bit which can have a value as logic level 0 or 1. As explained in Chap. 4, each logic level is represented by a voltage value. However, instead of using the analog voltage there, its level is used. Therefore, all processes based on the bit are based on predefined voltage levels, not the exact analog voltage value. This is the main difference of digital representation.

As in analog signals, we can have digital signals. Here, we assume that we are keeping successive digital values in successive memory locations. We provide a sample digital signal in Fig. 7.2. We will consider properties of digital signals further in Chap. 12.

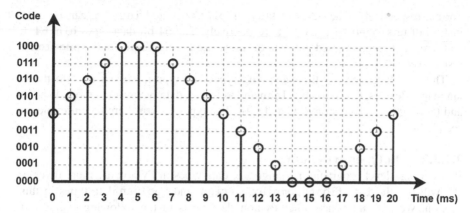

Fig. 7.2 A sample digital signal

7.1.3 Digital Values in Code

A single bit representing logic level 0 or 1 is not sufficient to cover all digital values. Each programming language has its own format, as data types, to represent them. Hence, we allocate fixed number of bits to a digital value in a specific format. Let's consider them in C, C++, and Python languages.

7.1.3.1 Data Types in C and C++ Languages

Data types in C and C++ languages have slight variations based on the embedded platform used. Therefore, we pick the representation used by Arm® Cortex™-M-based microcontrollers. In this representation, the C and C++ data types share the same properties. The only difference is that C++ has extra data types to be explained next.

The C language has five fundamental data types for data storage [1]. These are char, short, int, float, and double. The former three keep only the integer part of the number. The latter two data types can keep the integer and fractional parts of the number. For more information on the integer and fractional number representations, please see [6]. The mentioned data types also have their type modifiers as signed, unsigned, and long. These define the number of bits assigned to the number. As an example, the char data type has 8 bits assigned to it. If its unsigned version is used, then all 8 bits are assigned to represent the value of the number. If the signed bit representation of char data type is used, then the most significant bit is assigned to sign representation. The remaining seven bits are assigned to represent the value of the number.

The reader can also benefit from the integer data types defined under the "stdint.h" header file for microcontrollers. We can summarize them based on the number of bits assigned as follows. The 8-bit data types uint8_t and int8_t can be used instead of char and signed char, respectively. The 16-bit data types uint16_t and int16_t can be used instead of unsigned short and short, respectively. The 32-bit data types `uint32_t` and `int32_t` can be used instead of `unsigned int` and `int`, respectively. The 64-bit data types `uint64_t` and `int64_t` can be used instead of `unsigned long long` and `long long`, respectively.

The C++ language has the same data types and type modifiers as with the C language. Besides, it also has `boolean` and `wchar_t` data types. For a more detailed and thorough representation of C and C++ data types in Arm® architecture, please see [1, 3].

7.1.3.2 Data Types in Python

Being a high-level programming language, data types in Python are more diverse. We will use the `int`, `float`, `complex`, `string`, and `boolean` data types in this section. We may introduce other Python data types in the following chapters if needed. As the name implies, the `int` data type can store integer numbers. There is no limit in terms of the number of assigned bits to an integer representation as long as the memory requirements are satisfied. The `float` data type can be used to store numbers having integer and fractional parts. The maximum float number can be `1.8e308`. A value higher this limit is taken as infinite. The minimum number being close to zero is `5e-324`. Values lower than this range are taken as zero. The `complex` data type can be used to store complex numbers. The `string` data type can be used to store character arrays in more compact form. Finally, the `boolean` data type can store a value either as `True` or `False`. For more information on Python data types and how they are handled within its interpreter, please see [2].

7.2 Analog to Digital Conversion in Embedded Systems

In order to process an analog value in an embedded system, we should convert it to digital form. Hence, the digital value will have limited range. Besides, if we intend to obtain a digital signal from its analog counterpart, then we will have limited samples as well. The ADC module within the embedded system is responsible for this operation.

The ADC module performs two operations as sampling and quantization. We will explain these in detail in this section. Then, we will focus on the STM32F4 microcontroller as the embedded system and explain all the steps within it related to the ADC operation.

7.2.1 Sampling

Sampling is the first step in analog to digital conversion. The analog value is acquired in time by sampling. In other words, the analog value is observed at a certain time instant.

The sampling operation is performed in a periodic manner most of the times. Therefore, we acquire successive samples of the analog signal in time. We will need

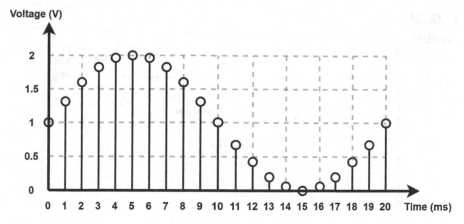

Fig. 7.3 Sampled signal

a clock signal for this operation. The most important issue here is the sampling rate in which how many samples are acquired in 1 s. This is called sampling frequency represented by Hz. The duration between two samples is called the sampling period (one over the sampling frequency).

There are theoretical limits on the required sampling frequency for a given analog signal [7]. Unfortunately, it is not possible to select a specific sampling frequency in an embedded system due to its physical hardware. We will evaluate possible values for the STM32F4 microcontroller in Sect. 7.2.3.

Let's provide a simple example on the sampling operation. Assume that the analog voltage at an electrical circuit is as in Fig. 7.1. This is in fact an analog signal with 2 V peak to peak and 1 V DC offset value. We would like to take samples of this signal at every millisecond. Hence, we will have observations of the signal by the 1 kHz sampling frequency. Then, the sampled signal becomes as in Fig. 7.3.

7.2.2   Quantization

As we sample the signal, we obtain its values at certain time instants. However, the value of each sample still has infinite range. Since we cannot store such a value in the embedded system, we should map it to finite range. Quantization performs this operation. Let's see how this can be done.

First and most important of all, there should be minimum and maximum voltage values as limits in the quantization operation. This is called the full-scale range (FSR). Second, we should allocate certain number of bits to represent the analog value. The number of allocated bits indicates the total number of levels we can use to represent the analog value. To be exact, N number of bits will result in a total of 2^N levels. Each level in quantized form corresponds to a bin having certain width in analog form. Therefore, all analog values within the bin are represented by the same

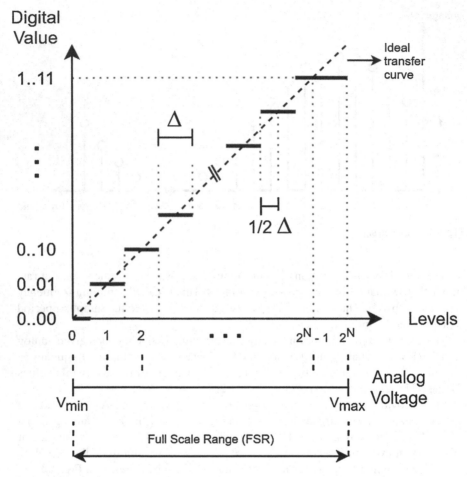

Fig. 7.4 The quantization operation

digital value. This will be the case for all bins in the quantization operation. We can represent the bin width based on FSR and number of levels as $\Delta = FSR/2^N$. The bin width is also directly related to the resolution such that a higher quantization level corresponds to better resolution with the same FSR value. We schematically show all these definitions in Fig. 7.4.

We can represent the quantization operation for the STM32F4 microcontroller as follows. The FSR for this microcontroller is set by the positive and negative reference voltage values V_{ref+} and V_{ref-}. The negative reference voltage is set to VSSA (ground voltage) for most general-purpose microcontrollers. This is also the case for the STM32F4 microcontroller. Hence, we will have $FSR = V_{ref+}$. The maximum number of bits assigned to the result of the ADC operation is 12 in the STM32F4 microcontroller. Hence, $N = 12$ here. This leads to

$$\Delta = \frac{V_{ref+}}{2^{12}} \tag{7.1}$$

The ADC operation represents the lowest analog voltage bin range, Δ, with the digital value zero. Likewise, the highest analog voltage bin range, $V_{ref+} - \Delta$, will have the corresponding digital value as 2^{N-1}. Based on these limits, we can obtain the digital (quantized) value for a given input voltage V_{in} as

$$Q = \left\lfloor \frac{V_{in}}{\Delta} \right\rfloor \tag{7.2}$$

where $\lfloor \cdot \rfloor$ indicates the floor function. We should note that when $V_{in} = V_{ref+}$ we will still have $Q = 2^{N-1}$. Although this case could have been formalized in Eq. 7.2, it would complicate the formula. Therefore, we handled it as a separate case.

Since we assign 12 bits for quantization, the quantized value can be represented in unsigned integer with 16 bits, unsigned short or uint16_t format, as mentioned in Sect. 7.1.3. Since 12 is less than 16, the most significant 4 bits will be set to zero. As a result, we have an unsigned number to represent the input analog voltage.

After obtaining the digital (quantized) value of an input voltage, we can convert it to float form by

$$Q_f = \frac{Q}{2^N} V_{ref+} \tag{7.3}$$

Please remember that even in this case, we will have quantized levels. However, each quantization level will be in float form.

Let's provide three examples on the quantization operation. In our examples, we will take $V_{ref+} = 3.3$ V and $N = 12$. First, assume that $V_{in} = \Delta/2$. The digital value corresponding to this input voltage will be 0. Hence, it can be represented in unsigned short form as 0x0000. Second, assume that $V_{in} = 3.3/2$. The digital value corresponding to this input voltage will be 2048. Hence, it can be represented in unsigned short form as 0x0800. Third, assume that $V_{in} = 3.3 - \Delta/4$. The digital value corresponding to this input voltage will be 4095. Hence, it can be represented in unsigned short form as 0x0FFF.

7.2.3   ADC Operation in the STM32F4 Microcontroller

Up to now, we explained sampling and quantization as if they are two separate operations. In practice, they are handled together by the ADC module. Besides, ADC modules in different embedded systems have their limitations. Therefore, we pick the STM32F4 microcontroller and explain working principles of its ADC module next.

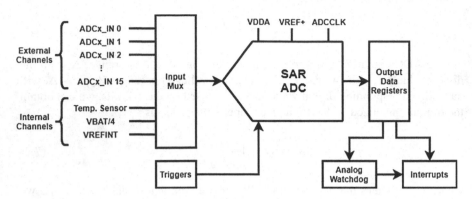

Fig. 7.5 Block diagram of the ADC module in the STM32F4 microcontroller

There are three ADC modules in the STM32F4 microcontroller called ADC1, ADC2, and ADC3. These modules can work independently. Therefore, we can configure the sampling frequency used in the ADC operation for each module separately. These modules provide 12-bit output. Hence, they are called 12-bit ADC modules. Based on the definitions in the previous section, we know that the total number of levels these ADC modules provide is 2^{12}. Finally, each ADC module works by the successive approximation register (SAR) methodology in analog to digital conversion. For further information on SAR, please see [5].

General block diagram of one ADC module in the STM32F4 microcontroller is as in Fig. 7.5. We will next explore the steps in this block diagram in detail. Therefore, we will cover input sources, triggers, modes, and sampling and clocks next. To note here, we will only cover one ADC module from this point on since the other two have exactly the same properties.

7.2.3.1 Input Sources
The ADC module in the STM32F4 microcontroller has up to 19 input channels (input sources) as being internal and external. External channels are the GPIO pins that can be used for connecting external sources. We tabulated GPIO pins of the STM32F4 microcontroller, available on the STM32F4 board, that can be used for ADC input in Sect. 2.3.2. Internal channels are connected to the internal reference voltage (Vrefint), internal temperature sensor (Vsense), and Vbat monitoring (Vbat/4) pins of the microcontroller.

The STM32F4 microcontroller ADC module also has voltage supply inputs as VDDA and VSSA. They are used to feed power to the module. The VSSA pin of the ADC module is connected to ground externally on the STM32F4 board. The VDDA pin of the ADC module is connected to VDD pin of the STM32F4 board externally with a noise filter. Hence, other digital operations do not add noise during the conversion operation. Besides, the ADC module has Vref+ and Vref− inputs. These define the FSR in the analog to digital conversion operation. The Vref− pin is connected to ground internally in the STM32F4 microcontroller. The Vref+ pin is connected to VDDA which is also connected to VDD in the STM32F4 board. Hence, the FSR for the ADC operation becomes VDD.

7.2.3.2 Triggers

The ADC operation can start in two different ways. First, the ADC module can be triggered by software via code snippet. Second, the ADC module can be triggered by hardware using a peripheral unit such as GPIO or timer. As for the software trigger, the ADC module should be set up first. Then, we can use a special register bit which starts the conversion. This bit can be accessed by the HAL library, Mbed, or MicroPython functions in the user code.

If we want to use the hardware trigger, then we should wait for an event (interrupt) to occur. The hardware trigger edge polarity can be selected as rising, falling, or both rising and falling. This means that the ADC module will be triggered when the trigger signal goes from low to high, from high to low, or in both transitions. We can use the EXTI_11 line, connected to 11th pin of the GPIO ports, for generating a hardware trigger. Also, the timer capture/compare events or timer trigger out events (interrupts) can be used for this purpose. Please consult the reference manual to see which timers and timer channels can trigger the ADC module [4].

7.2.3.3 Operating Modes

The ADC module can work in different operating modes as single-shot, continuous, or discontinuous. These modes have their configurations as single or scan (multi) channel. We provide them as a flowchart in Fig. 7.6.

The ADC module performs conversion using a single input in the single channel mode. In the scan (multi) channel mode, there can be more than one input to the ADC module. The module performs conversion using these sequentially (one after the other).

The ADC module stops after performing one conversion in the single-shot mode. If the scan channel mode is selected in this mode, then the conversion is performed for each channel. Afterward, the conversion stops. If a new conversion is required in the single-shot mode, then the module should be triggered again. In the continuous mode, the ADC module performs conversion automatically without re-triggering until the user stops it within the code. There is also discontinuous mode which can be only used if channel conversion mode is configured as scan channel. In this mode, the conversion stops after each channel while performing the conversion using multiple channels. The ADC module should be triggered to move on to the next channel.

There may be cases in which it may be required to convert another input while the ADC module is converting preselected inputs regularly. This can be accomplished by the injected mode in which the configuration of regular conversion operation is not modified. Figure 7.7 shows the injected conversion operation. Here, the regular ADC conversion is interrupted by injected ADC conversion and a new conversion from another channel (injected channel) is performed. Afterward, the ADC module keeps performing its regular conversion. To note here, if the injection operation

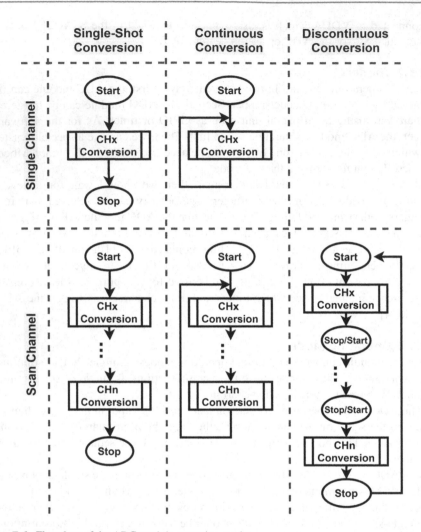

Fig. 7.6 Flowchart of the ADC module operating modes

starts in the middle of the regular ADC operation, then the present conversion value is discarded. As the module returns to the previous operation after the injected mode, a fresh conversion restarts for it.

Two ADC modules can work simultaneously or sequentially in master-slave fashion to convert analog input or inputs. This is called dual mode which can be used to increase the sampling rate of the ADC module or decrease the required time to convert multiple channels. This operation requires DMA usage.

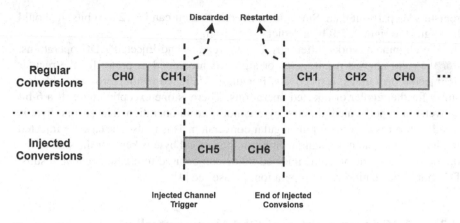

Fig. 7.7 Injected channel conversion operation

7.2.3.4 ADC Interrupts

The ADC module can generate an interrupt either at the end of conversion or by its analog watchdog property. These interrupts should be enabled separately by their respective bits. The end of conversion interrupt is generated separately for the regular or injected modes. Their corresponding callback functions are also separate.

The ADC result can be compared with two threshold values set as low and high via the analog watchdog feature. This is done automatically after setting up the ADC module. Hence, there is no need to run a separate code for the operation. If the conversion is outside the threshold values, then an interrupt can be generated. This operation is a sort of software configurable window comparator. The advantage of this operation is that it does not need an analog reference voltage. The comparison is done by the set threshold values.

7.2.3.5 Sampling and Clocks

We should use a clock signal to have periodic sampling as mentioned in Sect. 7.2.1. The ADC module has its dedicated clock, ADCCLK, for this purpose. The maximum clock speed for ADCCLK is 36 MHz.

The total conversion time for the ADC module is the sum of configured sampling period plus SAR operation duration depending on data resolution. The sampling period can be one of 3, 15, 28, 56, 84, 112, 144, or 480 ADC clock cycles. The SAR operation duration is 15 ADC clock cycles for 12-bit data resolution. This duration can be reduced to 13, 11, or 9 ADC clock cycles for 10-, 8-, and 6-bit data resolutions, respectively. As a result, the ADC module can produce 2.4 mega samples per second with 12-bit resolution. It is possible to go up to 4 mega samples per second with lower (6-bit) resolution by setting the ADCCLK to work at 36 MHz.

7.2.3.6 Obtaining the ADC Operation Result

As the ADC operation ends, the conversion result is written to the ADC_DR or ADC_JDRx register based on the regular or injected operation, respectively. These registers keep 16-bit data. Since the conversion result can be 12 to 6 bits, it should be aligned to right or left by a dedicated register.

The alignment works differently for the regular and injected ADC operations. For the right aligned result, zeros or sign bits are added as prefix for the regular or injected operations, respectively. For the left aligned result, zeros are added as suffix for the regular or injected operations. There is one exception here. If a 6-bit conversion is performed, the first byte is not used in operation. The second byte is filled with zeros as suffix for the regular conversion. This is also the case for injected mode, in which the most significant bit in the second byte is kept for the sign of the operation. As a side note, the injected mode can be used to obtain negative results. For more information on this operation, please see [4].

7.3   ADC Setup in the STM32F4 Microcontroller

We have seen that there are several options to perform the ADC operation in the STM32F4 microcontroller. We will consider them in this section. To do so, we will benefit from the available resources in C, C++, and MicroPython languages.

7.3.1   ADC Setup via C Language

We will benefit from STM32CubeMX to set up the ADC module properties in C language. At this stage, we assume that a project has already been created in STM32CubeIDE as given in Sect. 3.1.3. In order to set up ADC via the STM32F4 microcontroller, first open the "Device Configuration Tool" window by clicking the ".ioc" file in the project explorer window. Then, click on one of ADC1, ADC2, or ADC3 under the "Analog" section from the left menu of STM32CubeMX. Then, configuration can be done in this section as shown in Fig. 7.8. We will provide details of this operation next.

7.3.1.1 Setting Up Input Sources
If the internal temperature sensor, Vbat or Vrefint channel, will be selected as the ADC input source, then enable the channel under the "Analog - ADCx" section as in Fig. 7.9. Here, x can be 1, 2, or 3 to denote ADC1, ADC2, and ADC3, respectively.

If an external source will be connected to the ADC module, then click on the GPIO pin with ADC input capability on the STM32F4 microcontroller in the STM32CubeMX interface. Afterward, select the GPIO type as "ADCx_INy" in the "Device Configuration Tool" window as in Fig. 7.10. To note here, y can be any value between 0 and 15. Finally, enable the channel under the "Analog - ADCx" section.

7.3.1.2 Setting Up Operating Modes
Once the input source is selected, ADCx settings can be configured from the opened "Parameter Settings" tab in the "Configuration" menu as in Fig. 7.11. By default,

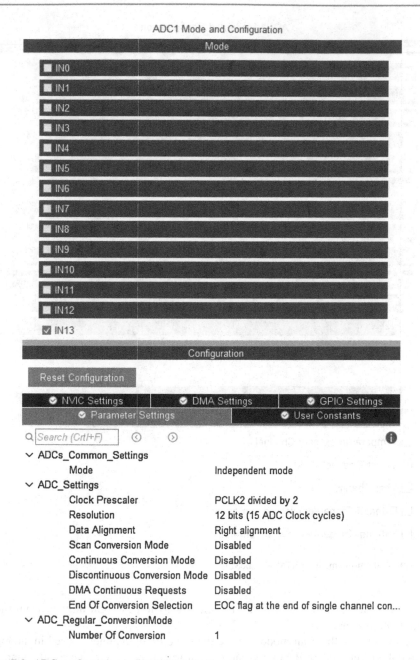

Fig. 7.8 ADC configuration in STM32CubeMX

the ADC module is configured to work in single-shot conversion mode triggered by software in 12-bit right aligned data mode. Data width can be selected as 12,

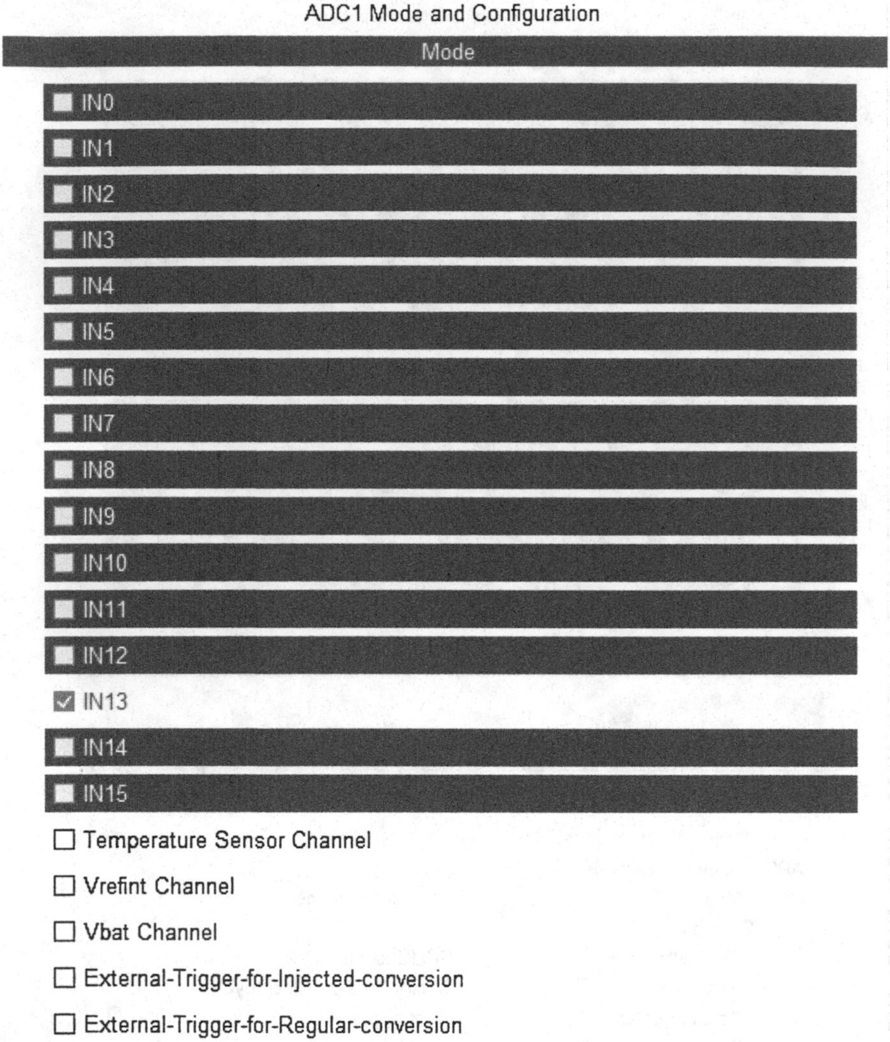

ADC1 Mode and Configuration

Mode
■ IN0
■ IN1
■ IN2
■ IN3
■ IN4
■ IN5
■ IN6
■ IN7
■ IN8
■ IN9
■ IN10
■ IN11
■ IN12
☑ IN13
■ IN14
■ IN15

☐ Temperature Sensor Channel

☐ Vrefint Channel

☐ Vbat Channel

☐ External-Trigger-for-Injected-conversion

☐ External-Trigger-for-Regular-conversion

Fig. 7.9 ADC mode menu in STM32CubeMX

10, 8, or 6 bits. Data alignment can be selected as left or right aligned from the configuration menu.

We can enable the scan mode if more than one input channel is used in analog to digital conversion. To do so, we should first select more than one input channel as ADC input. The selected and configured pins should all belong to the same ADC module such as ADC1_IN1 or ADC1_IN2. Then, the "Number of Conversion" option under the "ADC_Regular_ConversionMode" setting should be set to the number of selected input channels in the configuration menu as in Fig. 7.11.

Fig. 7.10 ADC input source
configuration in
STM32CubeMX

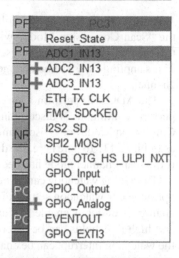

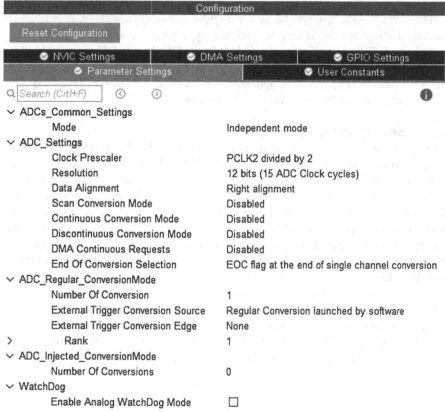

Fig. 7.11 ADC configuration menu in STM32CubeMX

Afterward, there will be multiple "Rank" options in the configuration menu and the "Scan Conversion Mode" will be enabled automatically. In each rank option, the channel number should be selected in the order that the conversion will occur. The sampling time can be changed under the "Rank" options individually for each channel.

The ADC will perform a single-shot conversion by default. If we would like to perform continuous or discontinuous conversion, we should enable the "Continuous Conversion Mode" or "Discontinuous Conversion Mode" in the configuration menu as in Fig. 7.11. To note here, the discontinuous conversion mode can only be enabled if the scan channel mode has been selected.

The analog watchdog can be enabled using the "Enable Analog Watchdog Mode" option under watchdog settings in the configuration menu as in Fig. 7.11. Once the analog watchdog is enabled, additional options will appear in the menu. Here, low and high thresholds can be entered to the appropriate places in digital form. Also, the watchdog interrupt can be enabled using the "Interrupt Mode" option in the same settings.

The default ADC conversion is for regular conversion without injection. If needed, the injected mode can be enabled by entering a nonzero value to the "Number of Conversions" section under the "ADC_Injected_ConversionMode" setting in the configuration menu as in Fig. 7.11. Once the injected mode is enabled this way, additional options will appear in the configuration menu. Afterward, the injection channel can be configured similar to a regular channel.

7.3.1.3 Setting Up ADC Interrupts

If the ADC interrupt is not required, then this step can be skipped. Otherwise, the ADC interrupt should be enabled from the NVIC tab in the configuration menu as in Fig. 7.11. The interrupt priority can be changed under the "System Core - NVIC" section. To note here, it is suggested to lower the ADC interrupt priority since the HAL library functions use the SysTick interrupt to create precise 1 ms ticks for internal operations and the HAL_Delay function. The SysTick interrupt has the highest priority among peripheral units. Hence, assigning a lower priority to the ADC interrupt will not affect it. Otherwise, precision of the HAL_Delay function may degrade.

7.3.1.4 Setting Up Triggers

The trigger source can be selected from the "External Trigger Conversion Source" option under the "ADC_Regular_ConversionMode" setting in the configuration menu as in Fig. 7.11. The default value for this setup is "Regular Conversion." If a hardware trigger will be used, then one of the EXTI line 11 or timer source with capture/compare or trigger out event options should be selected. To select the EXTI line 11, the 11th pin of any port should be configured as "ADCx_EXTI11" on the "Device Configuration Tool" window first. The trigger edge should also be selected as rising, falling, or rising and falling edge from the "External Trigger Conversion Edge" option under the configuration menu as in Fig. 7.11. If a timer will be used for the triggering operation, then the corresponding timer and its output event should be configured as explained in Sect. 6.3. Then, the trigger should be selected as the respective timer event from the "External Trigger Conversion Edge" option under the configuration menu as in Fig. 7.11.

7.3.1.5 Setting Up Clock Options
The last step is configuring the clock for the ADC module. Here, we assume that the clock configuration for the STM32F4 microcontroller has been done as explained in Chap. 6. We can set the ADC clock divider as 2, 4, 6, or 8 from the clock prescaler option in the configuration menu as in Fig. 7.11. The clock configuration of the microcontroller may not be suitable for the ADC module. For such cases, click on the "Clock Configuration" tab on the top menu of the "Device Configuration Tool" window. Then, click on the "Resolve Clock Issues" button to solve the problem. The reader can also set the ADC clock and its source clocks manually to the desired value in the same window.

7.3.2 ADC Setup via C++ Language

We can benefit from Mbed and available functions there to adjust the ADC module properties in C++ language. To do so, we should first adjust ADC channels. This can be done by the class `AnalogIn`. Usage of this class is similar to the `DigitalIn` class introduced in Sect. 4.4.2. Therefore, we should supply a channel name to the `AnalogIn` class. Optionally, a `Vref` value can be provided to the class. If this value is not provided, then the `AnalogIn` class uses the default `Vref` source value.

Mbed does not support different ADC configurations. Therefore, we cannot use the interrupt, hardware trigger, and injected mode options in it. In fact, we can only use the ADC module in its default configuration which is the single-channel, single-shot mode regular conversion with software trigger. We can indirectly realize hardware trigger using other STM32F4 microcontroller peripheral units. As an example, we can trigger the ADC module in a timer or GPIO callback function to realize the hardware trigger. However, this mode should be used with care as the ADC function used in the callback function blocks the code. Hence, it is better to set a flag in the callback function and realize the ADC operation in the main function based on the flag value.

7.3.3   ADC Setup via MicroPython

MicroPython has an ADC class to control the ADC module. The function pyb.ADC(pin=pyb.Pin('PinName')) is used to create the ADC module. Here, pin is the microcontroller pin which can be used as analog input. The PinName can be entered as Pxy with x being the port name and y being the pin number.

7.4   ADC Usage in the STM32F4 Microcontroller

As the ADC module properties are set, then we can use it for our specific purposes. We will explore how this can be done in C, C++, and MicroPython languages in this section. We will also provide examples related to ADC usage in these languages.

7.4.1   ADC Usage in C Language

The HAL library has special functions for the ADC operation after generating the project under STM32CubeIDE. We explore these functions based on their usage area next. Then, we provide usage examples for these functions on different applications.

7.4.1.1 Starting and Stopping the Conversion Operation

The regular conversion process can be started by the function HAL_ADC_Start. If interrupt generation is required after the conversion operation, then we should use the function HAL_ADC_Start_IT. We can stop and disable the ADC module with and without interrupt usage via functions HAL_ADC_Stop_IT and HAL_ADC_Stop, respectively. Once the regular ADC conversion has started, then the code can wait in a while loop (polled) using the function HAL_ADC_PollForConversion. As the conversion is done, then the function terminates. Afterward, the converted value can be read using the function HAL_ADC_GetValue. The HAL_ADC_PollForConversion function can also be terminated by its declared timeout value. In this setting, the conversion operation will not finalize. If the analog watchdog is enabled, then the function HAL_ADC_PollForEvent can be used to retrieve its status. We provide these functions in detail next.

```
HAL_StatusTypeDef HAL_ADC_Start(ADC_HandleTypeDef *hadc)
/*
hadc: pointer to the ADC_HandleTypeDef struct
*/

HAL_StatusTypeDef HAL_ADC_Stop (ADC_HandleTypeDef *hadc)

HAL_StatusTypeDef HAL_ADC_Start_IT(ADC_HandleTypeDef *hadc)

HAL_StatusTypeDef HAL_ADC_Stop_IT (ADC_HandleTypeDef *hadc)

HAL_StatusTypeDef HAL_ADC_PollForConversion(ADC_HandleTypeDef *
    hadc, uint32_t Timeout)
/*
Timeout: Timeout value in milliseconds
*/

uint32_t HAL_ADC_GetValue(ADC_HandleTypeDef *hadc)

HAL_StatusTypeDef HAL_ADC_PollForEvent(ADC_HandleTypeDef * hadc,
    uint32_t EventType, uint32_t Timeout)
/*
EventType: the ADC event type. This parameter can be one of the
    following values.
ADC_AWD_EVENT: Analog watchdog event.
ADC_OVR_EVENT: Overrun event.
*/
```

7.4.1.2 The Injected Conversion Operation

The injected conversion operation can be started by one of the functions: HAL_ADCex_InjectedStart or HAL_ADCex_InjectedStart_IT. If the injected conversion will be used without interrupt generation, then the former function should be used. If interrupt generation is required after the ADC operation, then the latter function should be used. In a similar manner, the injected conversion can be stopped and disabled by functions HAL_ADCex_InjectedStop or HAL_ADCex_InjectedStop_IT for regular- and interrupt-based operations, respectively. Once the injected ADC conversion has started, then the code can wait in a while loop (polled) using the function HAL_ADCex_InjectedPollForConversion. As the conversion is done, then the function terminates. Afterward, the converted value can be read by the function HAL_ADCex_InjectedGetValue. The function HAL_ADCex_InjectedPollForConversion can also be terminated by its declared timeout value. In this setting, the conversion operation will not finalize. We next provide detail of all these functions.

```
HAL_StatusTypeDef HAL_ADCEx_InjectedStart(ADC_HandleTypeDef *hadc
    )
/*
hadc: pointer to the ADC_HandleTypeDef struct
*/

HAL_StatusTypeDef HAL_ADCEx_InjectedStop(ADC_HandleTypeDef *hadc)

HAL_StatusTypeDef HAL_ADCEx_InjectedStart_IT(ADC_HandleTypeDef *
    hadc)
```

```
HAL_StatusTypeDef HAL_ADCEx_InjectedStop_IT(ADC_HandleTypeDef *
    hadc)

HAL_StatusTypeDef HAL_ADCEx_InjectedPollForConversion(
    ADC_HandleTypeDef *hadc, uint32_t Timeout)

uint32_t HAL_ADCEx_InjectedGetValue(ADC_HandleTypeDef *hadc,
    uint32_t InjectedRank)
/*
InjectedRank: the converted ADC injected rank. This parameter can
    be one of the following values:
ADC_INJECTED_RANK_1: Injected channel1 selected
ADC_INJECTED_RANK_2: Injected channel2 selected
ADC_INJECTED_RANK_3: Injected channel3 selected
ADC_INJECTED_RANK_4: Injected channel4 selected
*/
```

7.4.1.3 ADC Interrupt Usage

HAL library also handles ADC end of conversion interrupt generation. For this purpose, we can start the regular or injected conversion operations by the functionHAL_ADC_Start_IT. Then, we should use the callback functions HAL_ADC_ConvCpltCallback or HAL_ADCex_InjectedConvCpltCallback, respectively. If the analog watchdog and its interrupt are enabled, then the function HAL_ADC_LevelOutOfWindowCallback is called. We should define these functions within the "main.c" file in our project. To note here, these functions are generated in weak form (without any content) automatically under the "STM32F4xx_hal_adc.c" file when a project is created. This means when the reader forgets defining the actual HAL_ADC_ConvCpltCallback or HAL_ADCex_InjectedConvCpltCallback functions, the code still runs. When the user defines his or her function, then these weak functions are ignored. We next provide detail of these functions.

```
void HAL_ADC_ConvCpltCallback(ADC_HandleTypeDef *hadc)
/*
hadc: pointer to the ADC_HandleTypeDef struct
*/

void HAL_ADCEx_InjectedConvCpltCallback(ADC_HandleTypeDef *hadc)

void HAL_ADC_LevelOutOfWindowCallback(ADC_HandleTypeDef * hadc)
```

7.4.1.4 The ADC Module Usage Examples

We provide seven examples on the usage of HAL library ADC functions. The first example, with the C code given in Listing 7.1, aims to familiarize the reader with the setup and usage of the ADC module. Here, temperature of the STM32F4 microcontroller will be measured using its internal temperature sensor. The ADC1 module will be configured in single-shot mode with software trigger.

Listing 7.1 Usage of the ADC module with internal channel in single-shot mode, the C code

```
/* USER CODE BEGIN PD */
#define VREF 3.00
#define V25 0.76
#define AVG_SLP 0.00205
/* USER CODE END PD */

/* USER CODE BEGIN PV */
float temperatureC = 0;
float temperatureV = 0;
uint16_t adcValue = 0;
/* USER CODE END PV */

/* Infinite loop */
/* USER CODE BEGIN WHILE */
while (1)
{
HAL_ADC_Start(&hadc1);
HAL_ADC_PollForConversion(&hadc1, 1000);
adcValue = HAL_ADC_GetValue(&hadc1);

temperatureV = ((float)adcValue * VREF) / 4095;
temperatureC = ((temperatureV - V25) / AVG_SLP) + 25;

HAL_Delay(1000);
/* USER CODE END WHILE */

/* USER CODE BEGIN 3 */
}
/* USER CODE END 3 */
```

In order to execute the code in Listing 7.1, create an empty project and enable the internal temperature channel of the ADC1 module. As the project is compiled and run, the temperature sensor output will be first converted to digital value and then to Celsius degrees in an infinite loop. The reader can observe changing temperature values by adding a breakpoint to the code.

The second example, with the C code given in Listing 7.2, aims to familiarize the reader with the external ADC channels and hardware triggers. To do so, we should connect a 10 kΩ potentiometer between ground, VDD, and PA5 pins of the STM32F4 board. Also, we should connect an external push button in active high pull-down configuration to pin PC11. Hence, we can use it as hardware trigger.

Listing 7.2 Usage of the ADC module with external channel in single-shot mode with hardware trigger, the C code

```
/* USER CODE BEGIN PV */
uint16_t adcValue = 0;
float divRatio = 0.00;
/* USER CODE END PV */

/* Infinite loop */
/* USER CODE BEGIN WHILE */
HAL_ADC_Start(&hadc2);
while (1)
{
HAL_ADC_PollForConversion(&hadc2, HAL_MAX_DELAY);
adcValue = HAL_ADC_GetValue(&hadc2);
divRatio = (float)adcValue / 4095;
```

```
HAL_Delay(1000);
/* USER CODE END WHILE */

/* USER CODE BEGIN 3 */
}
/* USER CODE END 3 */
```

The hardware setup for the second example is given in Fig. 7.12. Here, ADC2 channel 5 will be configured as single-channel in single-shot mode and will be triggered with EXTI_11 line using the button connected to pin PC11. In order to execute the code in Listing 7.2, create an empty project. Then, configure pins PA5 and PC11 as ADC2_IN5 and ADC2_EXTI11, respectively. Afterward, configure ADC2 channel 5 in single-shot mode. As pin PC11 is configured as EXTI line, hardware trigger will be automatically selected. After the project is compiled and run, the level of the potentiometer is calculated in the infinite loop when the onboard button is pressed. The reader can observe different potentiometer levels by placing a breakpoint to the delay line in the code.

In the third example, we repeat the second example by using a timer trigger instead of push button as hardware trigger. To do so, we use the Timer 8 update event as hardware trigger for ADC. We set the timer clock to 144 MHz and Timer 8 counter as 144. Hence, the ADC module is triggered with 1 MHz timer output. We set the ADC clock to 36 MHz and sampling time to three cycles. Hence, the ADC module can convert 2.4 mega samples per second. Also, we enable the end of conversion interrupts. As all these settings are done, we add the callback function given in Listing 7.3 to the code given in Listing 7.2. Within the code, we put the CPU

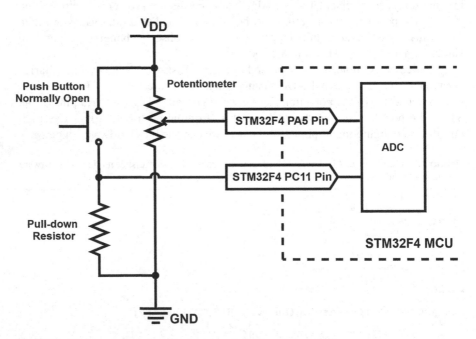

Fig. 7.12 Hardware connections for the second ADC module example

to sleep mode while the microcontroller is in idle mode or during ADC conversion. After the project is compiled and run, the ADC module will be triggered by Timer 8. The reader can observe different potentiometer levels by placing a breakpoint to the __NOP line in the callback function.

Listing 7.3 Usage of the ADC module with timer trigger and end of conversion interrupt, the C code

```c
/* USER CODE BEGIN PV */
uint16_t adcValue = 0;
float divRatio = 0.00;
/* USER CODE END PV */

/* USER CODE BEGIN 0 */
void HAL_ADC_ConvCpltCallback(ADC_HandleTypeDef *hadc)
{
adcValue = HAL_ADC_GetValue(hadc);
divRatio = (float)adcValue / 4095;
__NOP();
}
/* USER CODE END 0 */

/* USER CODE BEGIN 2 */
HAL_ADC_Start_IT(&hadc2);
HAL_TIM_Base_Start(&htim8);
HAL_PWR_EnableSleepOnExit();
HAL_PWR_EnterSLEEPMode(PWR_MAINREGULATOR_ON, PWR_SLEEPENTRY_WFI);
/* USER CODE END 2 */
```

The fourth example, with the C code given in Listing 7.4, can be used to convert multiple inputs to digital values continuously in an infinite loop. To do so, we should connect two 10 kΩ potentiometers between ground, VDD, and pins PA5 and PC3 of the STM32F4 board as shown in Fig. 7.13.

Listing 7.4 Usage of ADC module with multiple channels in continuous mode, the C code

```c
/* USER CODE BEGIN PD */
#define CONV_NUM 100
/* USER CODE END PD */

/* USER CODE BEGIN PV */
uint32_t adcValue[CONV_NUM] = {0};
float divisionLvl5[CONV_NUM / 2] = {0};
float divisionLvl13[CONV_NUM / 2] = {0};
uint32_t counter = 0, ind = 0;
/* USER CODE END PV */

/* Infinite loop */
/* USER CODE BEGIN WHILE */
HAL_ADC_Start(&hadc1);
while (1)
{
if (counter < (CONV_NUM - 1))
{
HAL_ADC_PollForConversion(&hadc1, 1000);
adcValue[counter] = HAL_ADC_GetValue(&hadc1);
counter++;

HAL_ADC_PollForConversion(&hadc1, 1000);
```

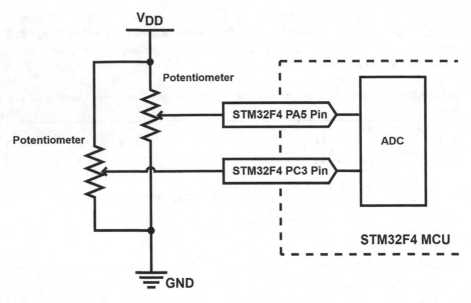

Fig. 7.13 Hardware connections for the fourth ADC module example

```
adcValue[counter] = HAL_ADC_GetValue(&hadc1);
counter++;
}
else
{
for (ind = 0; ind < CONV_NUM; ind += 2)
{
divisionLvl13[ind / 2] = (float)adcValue[ind] / 4095;
divisionLvl13[ind / 2 + 1] = (float)adcValue[ind + 1] / 4095;
}

HAL_ADC_Stop(&hadc1);
HAL_Delay(100);
}
/* USER CODE END WHILE */

/* USER CODE BEGIN 3 */
}
/* USER CODE END 3 */
```

In Listing 7.4, ADC1 external channels 5 and 13 are configured as scan channel in continuous mode and will be triggered by software. Also, the ADC clock will be configured to 36 MHz. Hence, the sampling times will be configured as 480 cycles. The converted values will be stored in the adcValue array.

In order to execute the code in Listing 7.4, create an empty project. Then, configure pins PA5 and PC3 as ADC1_IN5 and ADC1_IN13, respectively. Afterward, enable ADC1 channels 5 and 13. Then configure the ADC1 module to work in scan channel continuous mode. Afterward, adjust the ADC clock to 36 MHz and sampling time of each channel to 480 cycles. After the project is compiled and run, the level of potentiometers is calculated in infinite loop after 100 conversions are done. The reader can observe different potentiometer levels by placing a breakpoint to the delay line. The reader can also check that lowering the sampling times, such as to three cycles, will cause a problem. To be able to work at such speeds, DMA should be configured to gather data from the ADC.

In the fifth example, we repeat the fourth example with discontinuous mode. The C code for this case is given in Listing 7.5. After the project is compiled and run, the level of the potentiometers is calculated in the infinite loop after 100 conversions are done. Here, the ADC is triggered after each channel instead of gathering all conversion results with one trigger. The reader can observe different potentiometer levels by placing a breakpoint to the delay line in the code.

Listing 7.5 Usage of the ADC module with multiple channels in discontinuous mode, the C code

```
/* USER CODE BEGIN PD */
#define CONV_NUM 100
/* USER CODE END PD */

/* USER CODE BEGIN PV */
uint32_t adcValue[CONV_NUM] = {0};
float divisionLvl5[CONV_NUM / 2] = {0};
float divisionLvl13[CONV_NUM / 2] = {0};
uint32_t counter = 0, ind = 0;
/* USER CODE END PV */

/* Infinite loop */
/* USER CODE BEGIN WHILE */
while (1)
{
if (counter < (CONV_NUM - 1))
{
HAL_ADC_Start(&hadc1);
HAL_ADC_PollForConversion(&hadc1, 1000);
adcValue[counter] = HAL_ADC_GetValue(&hadc1);
counter++;

HAL_ADC_PollForConversion(&hadc1, 1000);
adcValue[counter] = HAL_ADC_GetValue(&hadc1);
counter++;

HAL_ADC_Stop(&hadc1);
}
else
{
for (ind = 0; ind < CONV_NUM; ind += 2)
{
divisionLvl13[ind / 2] = (float)adcValue[ind] / 4095;
divisionLvl13[ind / 2 + 1] = (float)adcValue[ind + 1] / 4095;
}

HAL_Delay(100);
}
/* USER CODE END WHILE */

/* USER CODE BEGIN 3 */
}
/* USER CODE END 3 */
```

The sixth example aims to familiarize the reader with the analog watchdog property. To do so, we repeat the second example in continuous mode and enable the watchdog with low and high threshold levels as 512 and 3586, respectively. We should also enable the end of conversion and watchdog interrupts and add the callback functions given in Listing 7.6 to Listing 7.2.

Listing 7.6 Usage of the ADC module with analog watchdog property, the C code

```c
/* USER CODE BEGIN PD */
#define CONV_NUM 100
/* USER CODE END PD */

/* USER CODE BEGIN PV */
uint32_t adcValue[CONV_NUM] = {0};
uint16_t counter = 0, idx = 0;
uint32_t sum = 0;
float divRatio = 0;
uint8_t adcWatchdogFlag = 0;
/* USER CODE END PV */

/* USER CODE BEGIN 0 */
void HAL_ADC_ConvCpltCallback(ADC_HandleTypeDef *hadc)
{
adcValue[counter++] = HAL_ADC_GetValue(hadc);
if (counter == CONV_NUM)
{
HAL_ADC_Stop_IT(hadc);
}
}

void HAL_ADC_LevelOutOfWindowCallback(ADC_HandleTypeDef *hadc)
{
HAL_ADC_Stop_IT(hadc);
HAL_GPIO_WritePin(GPIOG, GPIO_PIN_13, GPIO_PIN_SET);
adcWatchdogFlag = 1;
}
/* USER CODE END 0 */

/* Infinite loop */
/* USER CODE BEGIN WHILE */
HAL_ADC_Start_IT(&hadc1);
while (1)
{
if (adcWatchdogFlag)
{
adcWatchdogFlag = 0;
HAL_Delay(2000);
HAL_GPIO_WritePin(GPIOG, GPIO_PIN_13, GPIO_PIN_RESET);
counter = 0;
HAL_ADC_Start_IT(&hadc1);
HAL_Delay(100);
}
else if (counter == CONV_NUM)
{
for (idx = 0; idx < CONV_NUM; idx++)
{
sum += adcValue[idx];
}
divRatio = (float)sum / (CONV_NUM * 4095);
sum = 0;
counter = 0;
```

```
HAL_ADC_Start_IT(&hadc1);
HAL_Delay(100);
}

HAL_PWR_EnterSLEEPMode(PWR_MAINREGULATOR_ON, PWR_SLEEPENTRY_WFI);
/* USER CODE END WHILE */

/* USER CODE BEGIN 3 */
}
/* USER CODE END 3 */
```

After the project is compiled and run, ADC is triggered when the onboard button is pressed. Once 100 conversions are done, obtained values are summed and averaged in an infinite loop. If the analog value is above or below the threshold levels, then the watchdog interrupt is triggered during the conversion operation. Then, the variable adcWatchdogFlag is set and the onboard green LED connected to pin PG13 is turned on in the watchdog callback function. The adcWatchdogFlag is checked in the infinite loop in the main function. If this variable has been set in the callback function, then the ADC is stopped and onboard green LED is turned off after 2 s. The reader can observe different potentiometer levels by placing a breakpoint to the delay line in the code. The reader can also observe working principle of the analog watchdog property via this example.

The seventh example, with the C code given in Listing 7.7, adds injected conversion to the third example. Here, we enable the internal temperature sensor channel. Then, we enable the injected conversion and select the internal temperature sensor. We also enable the hardware trigger for injected channels and connect a push button in active high pull-down configuration to pin PA15 to use it as hardware trigger. The hardware setup for the seventh example is given in Fig. 7.14.

Listing 7.7 Usage of the ADC module with injected conversion property, the C code

```
/* USER CODE BEGIN PD */
#define VREF 3.00
#define V25 0.76
#define AVG_SLP 0.00205
/* USER CODE END PD */

/* USER CODE BEGIN PV */
uint16_t adcValue = 0;
float divRatio = 0.00;

float temperatureC = 0;
float temperatureV = 0;
uint16_t adcInjValue = 0;
/* USER CODE END PV */

/* USER CODE BEGIN 0 */
void HAL_ADC_ConvCpltCallback(ADC_HandleTypeDef *hadc)
{
adcValue = HAL_ADC_GetValue(hadc);
divRatio = (float)adcValue / 4095;
__NOP();
}

void HAL_ADCEx_InjectedConvCpltCallback(ADC_HandleTypeDef *hadc)
```

```
{
adcInjValue = HAL_ADCEx_InjectedGetValue(hadc,
    ADC_INJECTED_RANK_1);

temperatureV = ((float)adcInjValue * VREF) / 4095;
temperatureC = ((temperatureV - V25) / AVG_SLP) + 25;
__NOP();
}
/* USER CODE END 0 */

/* USER CODE BEGIN 2 */
HAL_ADC_Start_IT(&hadc1);
HAL_TIM_Base_Start(&htim8);
HAL_TIM_Base_Start(&htim1);
HAL_PWR_EnableSleepOnExit();

HAL_ADCEx_InjectedStart_IT(&hadc1);
HAL_PWR_EnterSLEEPMode(PWR_MAINREGULATOR_ON, PWR_SLEEPENTRY_WFI);
/* USER CODE END 2 */
```

After the project is compiled and run, the ADC module is triggered via Timer 8. When the push button is pressed, the injected channel is triggered. Hence, the ADC module starts converting the internal channel. Afterward, the module continues converting the regular channel. The reader can observe different potentiometer levels by placing a breakpoint to the __NOP line in the regular end of conversion callback function. The reader can also observe the temperature values by placing a breakpoint to the __NOP line in the injected end of conversion callback function.

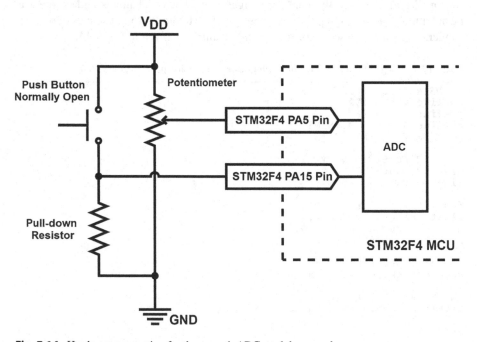

Fig. 7.14 Hardware connection for the seventh ADC module example

7.4.2 ADC Usage in C++ Language

We can benefit from the available functions in Mbed to use the ADC module via C++ language. Here, the `AnalogIn` class or one of the functions `read`, `read_u16`, or `read_voltage` will be of help. The `AnalogIn` class and `read` function provide a float value between 0 and 1.0. For example, if V_{ref} is 3 V and the ADC input voltage is 0.75 V, then the read value will be 0.25 with this function. Likewise, the function `read_voltage` returns a float value representing the input voltage, measured in volts. In order to use this function, we should set the reference voltage beforehand using the function `set_reference_voltage`. This value is the same as V_{DD} for the STM32F4 board. The `read_u16` function returns a value between 0 and `0xFFFF`.

We next provide examples on the usage of the ADC module in Mbed. Our first example, with the C++ code given in Listing 7.8, is on setting the ADC module. To execute this code in Mbed Studio, we should copy and paste it to the "main.cpp" file in our active project. As can be seen in Listing 7.8, the ADC module converts the analog input connected to pin PA5 of the STM32F4 board and prints the result to the terminal. It then sleeps for 1 s in the infinite loop. As a side note, the `platform.minimal-printf-enable-floating-point` value should be set to true within the json file of the Mbed project to print float numbers by the function `printf`.

Listing 7.8 First example on the usage of the ADC module, the C++ code

```
/*
In mbed_app.json file below settings should be set to true in
    order to print
floating numbers to console "platform.minimal-printf-enable-
    floating-point"
*/

#include "mbed.h"

#define WAIT_TIME_MS 1000
AnalogIn aIn(PA_5);

int main()
{
float aInput, volt;
uint16_t dInput;

aIn.set_reference_voltage(3);

while (true)
{
dInput = aIn.read_u16();
aInput = aIn.read();
volt = aIn.read_voltage();

printf("digital: 0x%04X \n", dInput);
printf("normalized: %.6f \n", aInput);
printf("voltage: %.6f \n\n\n", volt);

thread_sleep_for(WAIT_TIME_MS);
}
}
```

The second example, with the C++ code given in Listing 7.9, implements the first example by using a timer interrupt. Hence, the ADC module is triggered indirectly by hardware. As can be seen in Listing 7.9, the ADC converts the analog input connected to pin PA5 of the STM32F4 board at each timer interrupt and stores the result to the array divRatio. After 20 conversions, the timer interrupt is disabled by the function HwADCTrigger.detach(). Then, the result is printed to the terminal and CPU goes to sleep mode.

Listing 7.9 Second example on the usage of the ADC module, the C++ code

```
/*
In mbed_app.json file below settings should be set to true in
    order to print
floating numbers to console "platform.minimal-printf-enable-
    floating-point"
*/

#include "mbed.h"

#define CONV_NUM 20

Ticker adcTrigger;
AnalogIn aIn(PA_5);

float divRatio[CONV_NUM];
int cnt = 0, ind;
bool adcFlag = false;

void ADCTrigger()
{
adcFlag = true;
}

int main()
{
printf("ADC Conversion Started\n\n\n");

aIn.set_reference_voltage(3);
adcTrigger.attach(&ADCTrigger, 200ms);

while (true)
{

if (adcFlag && cnt < CONV_NUM)
{
adcFlag = false;
divRatio[cnt++] = aIn.read();
}

if (cnt >= CONV_NUM)
{
adcTrigger.detach();
for (ind = 0; ind < cnt; ind++)
{
printf("normalized: %.2f \n", divRatio[ind]);
}
}

sleep();
}
}
```

The third example, with the C++ code given in Listing 7.10, shows how an internal channel can be used in the ADC operation. To do so, we benefit from the internal temperature sensor of the STM32F4 microcontroller as intTemp(ADC_TEMP). As the code runs, internal temperature of the STM32 microcontroller is obtained and printed in an infinite loop every 0.1 s.

Listing 7.10 Third example on the usage of the ADC module, the C++ code

```cpp
#include "mbed.h"

#define WAIT_TIME_MS 100

AnalogIn intTemp(ADC_TEMP);

int main()
{
while (true)
{
int value = intTemp.read_u16();
printf("Analog value read %u \n", (unsigned int)value);

thread_sleep_for(WAIT_TIME_MS);
}
}
```

7.4.3   ADC Usage in MicroPython

The function ADC.read() can be used to read analog value from a pin in 12-bit resolution. The function ADC.read_timed(buf, timer) can be used to perform multiple reads with a precise sampling rate. Here, conversion results will be stored in the buf array. Size of this array determines the number of ADC conversions. Sampling rate is set by the selected timer module. However, this is a blocking function and code execution will be halted until ADC conversions are complete. Hence, ADC conversions can be realized inside a timer callback function to perform the non-blocking version of the same operation. Finally, read_core_temp(), read_core_vbat(), read_vref(), and read_core_vref() functions can be used to read internal temperature sensor, external battery, reference, and core reference voltage values, respectively. In order to use these functions, related ADC channels must be initialized beforehand by the function pyb.ADCAll(resolution, mask). Here, resolution is the ADC resolution and mask is the value used to select the desired ADC channel.

We next provide examples on the usage of the ADC module in MicroPython. In the first example, ADC module converts the analog input connected to pin PA5 of the STM32F4 board to respected voltage value. The result is printed on the terminal. The MicroPython code for this example is given in Listing 7.11

Listing 7.11 First example on the usage of the ADC module, the MicroPython code

```
import pyb

def main():
    aIn = pyb.ADC(pyb.Pin('PA5'))
    while True:
        dInput = aIn.read()
        volt = 3.0 * dInput / 4095
        print("digital:", dInput, "\n")
        print("voltage:", volt, "\n\n\n")
        pyb.delay(1000)

main()
```

In the second example, ADC conversion is triggered using timer interrupt in non-blocking mode. ADC converts the analog input connected to pin PA5 of the STM32F4 board at 200 ms sampling intervals and stores the results to the array divRatio. After 20 conversions, the timer interrupt is disabled by the function Timer1.deinit() and the results are printed to the terminal as voltage values. The MicroPython code for this example is given in Listing 7.12.

Listing 7.12 Second example on the usage of the ADC module, the MicroPython code

```
import pyb
from array import array

CONV_NUM = const(20)
timer1 = pyb.Timer(1, freq=5, mode=pyb.Timer.UP)
divRatio = array('I', (0 for i in range(100)))
adcFlag = 0
printFlag = 0
cnt = 0

def ADCTrigger(timer):
    global adcFlag
    global cnt
    global printFlag
    adcFlag = 1
    cnt = cnt + 1
    if cnt >= CONV_NUM:
        timer1.deinit()
        printFlag = 1

def main():
    global adcFlag
    global cnt
    global printFlag
    aIn = pyb.ADC(pyb.Pin('PA5'))
    timer1.callback(ADCTrigger)
    while True:
        if (adcFlag == 1) and (cnt < CONV_NUM):
            adcFlag = 0
            divRatio[cnt] = aIn.read()
        if printFlag == 1:
            for ind in range(cnt):
                print("voltage:", 3.0 * divRatio[ind] / 4095, "\n
                ")
            printFlag = 0
```

```
main()
```

In the third example, the ADC module is used to obtain the temperature value from the internal temperature sensor of the STM32F4 microcontroller. Here, the temperature is obtained in Celsius degrees and printed on the terminal in an infinite loop every 0.1 s. The MicroPython code for this example is given in Listing 7.13.

Listing 7.13 Third example on the usage of the ADC module, the MicroPython code

```
import pyb

def main():
    adcTemp = pyb.ADCAll(12, 0x70000)
    while True:
        value = adcTemp.read_core_temp()
        print(value)
        pyb.delay(100)

main()
```

7.5 Digital to Analog Conversion in Embedded Systems

Digital to analog conversion (DAC) aims to obtain the analog representation of a given digital value. We will cover two methods to perform this operation. The first method is called zero-order hold (ZOH) which can be performed if the embedded system has a DAC module to be used for this purpose. The second method is based on pulse width modulation (PWM). This method does not require any specific module except the timer which should be available in most embedded systems. Since the ZOH and PWM depend on physical hardware of the embedded system, we will consider how these operations are done in the STM32F4 microcontroller in a separate section.

7.5.1 Zero-Order Hold

The zero-order hold (ZOH) operation, performed by the DAC module, works by a simple principle such that a constant voltage level is fed to output for a given digital input value. This voltage is kept at output for a predefined time. When another digital value comes to the DAC module, the output voltage changes accordingly. There should be associated pins for the module within the embedded system to handle analog voltage values.

We can explain working principles of the DAC module with the help of the concepts introduced in the ADC module. The DAC module works in a certain voltage range determined by V_{ref+} in the STM32F4 microcontroller. The DAC module has a 12-bit digital input. Hence, the STM32F4 microcontroller has a 12-bit DAC module. When the minimum digital value (zero) is fed to it, output of the module will be 0 V. Likewise, when the maximum digital value ($2^N - 1$) is fed to

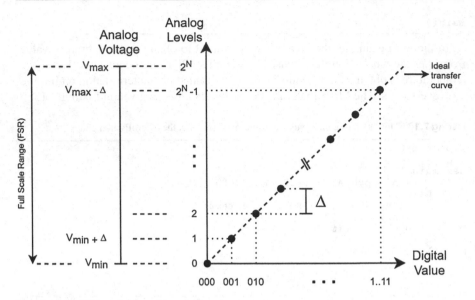

Fig. 7.15 DAC transfer function

the DAC module, its output will be $V_{ref+} - \Delta$ where Δ is given in Eq. 7.1. For other digital input values, we can calculate the analog value from the DAC module as

$$V_{out} = \Delta \times d \tag{7.4}$$

where d is the digital value fed to the DAC module. It is evident from Eq. 7.4 that we have limited analog output values since d is digital. We schematically explained these concepts in Fig. 7.15.

If we feed successive digital values to the DAC module, then we will have a staircase like analog signal at output. We provide such a signal in Fig. 7.16. This signal can be smoothed out further by a low-pass filter connected to output of the DAC module.

7.5.2 Pulse Width Modulation

The embedded system may not have a dedicated DAC module to convert a digital value to analog form. For such cases, we can benefit from pulse width modulation (PWM). PWM provides fixed amplitude pulses with varying width. These pulses are formed periodically such that there is a base periodic signal composed of pulses with constant amplitude, with value V_{DD}, and varying width. The important point here is that period of the base periodic signal is low. In other words, frequency of this signal is high. When such a high-frequency signal is fed to an analog system, it cannot detect such high-frequency changes. Instead, the analog system observes

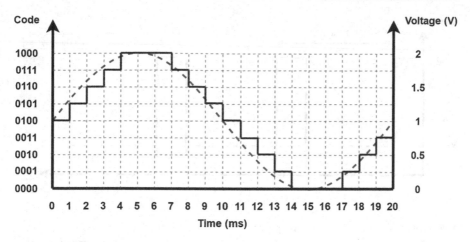

Fig. 7.16 DAC signal output example

the average voltage value within one period of the PWM signal. We can explain this process as follows. The analog system may not respond to high-frequency changes. Hence, it acts as a low-pass filter. As a result, we can change the effective voltage value observed by the analog system by changing the pulse width.

We can formalize the analog voltage value, V_{avg}, obtained by PWM as

$$V_{avg} = \frac{t_{on}}{t_{period}} \times V_{DD} = D \times V_{DD} \tag{7.5}$$

where t_{on} is the "on" duration of the pulse and t_{period} is the period of the base signal.

The ratio, t_{on}/t_{period}, is called the duty cycle, D, of the PWM signal. This value can be changed to obtain the desired analog voltage. To note here, t_{period} will be fixed for the analog system at hand such that it cannot detect high-frequency changes. The t_{on} value can be modified to change the duty cycle. We will see how to modify both t_{period} and t_{on} values in the STM32F4 microcontroller in the next section.

We provide three scenarios on obtaining different analog voltage values with different duty cycles in Fig. 7.17. In this figure, a small portion of PWM signals are provided. As can be seen in this figure, V_{avg} changes linearly with the selected duty cycle. The duty cycle can also change on the fly. Hence, we can have a varying analog voltage value at output.

We can generate three different PWM signals as left aligned, right aligned, and centered. We provide them in Fig. 7.18. As can be seen in this figure, when the signal is left aligned, it always starts with the V_{DD} value in one period. The signal then goes to 0 V after t_{on} duration. When the signal is right aligned, it always starts with 0 V within one period. Then, it goes to V_{DD} and stays there for t_{on} duration till the end of period. Finally, we can have the centered PWM signal such that it always

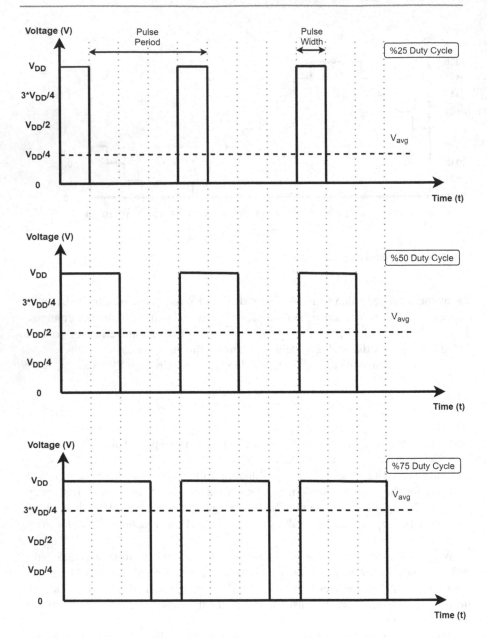

Fig. 7.17 PWM signals with different duty cycles

starts and ends with 0 V. Within the period, the signal goes to the V_{DD} value and stays there with t_{on} duration.

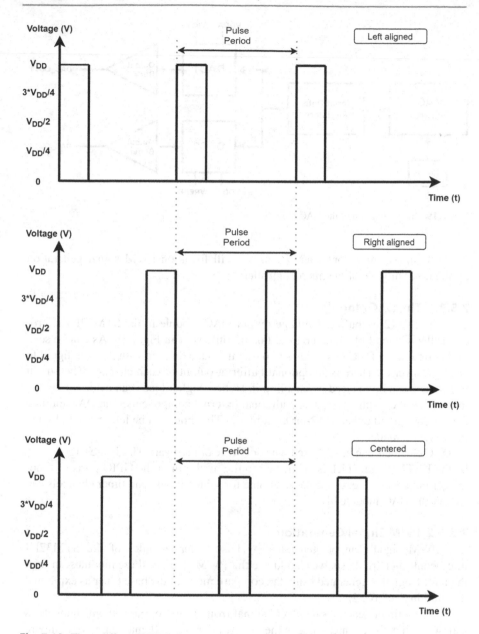

Fig. 7.18 Three different PWM alignment options

7.5.3 DAC Operation in the STM32F4 Microcontroller

We will explore how the DAC operation can be performed by the STM32F4 microcontroller in this section. To do so, we will start with the DAC module

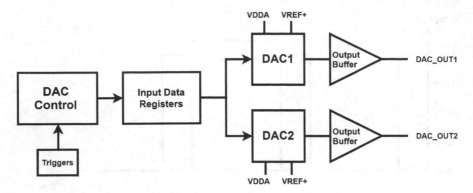

Fig. 7.19 Block diagram of the DAC module

available in the microcontroller. Then, we will focus on PWM signal generation by the timer module of the microcontroller.

7.5.3.1 The DAC Module

There is one 12-bit buffered voltage output DAC module in the STM32F4 microcontroller. General block diagram of this module is as in Fig. 7.19. As can be seen in this figure, the DAC module has two output channels each with its own digital to analog converter. There is an optional buffer at output of each channel. The output impedance can be reduced by enabling it. Hence, higher loads can be driven by the microcontroller without needing additional external components. The DAC module can be configured either to 12- or 8-bit mode. The input can be left or right aligned in the 12-bit mode.

DAC channels can be triggered with software or hardware. Hardware triggers can be the EXTI_9 line which is connected to the ninth pin of the GPIO ports or timer trigger out events. Please see the reference for which timers and timer channels can trigger the DAC module [4].

7.5.3.2 PWM Signal Generation

The PWM signal can be generated by base timer modules of the STM32F4 microcontroller. To do so, we should use the PWM setup of these modules. In fact, the PWM signal is generated using the compare mode of the base timer as explained in detail in Sect. 6.2.2.

The base timer generates the PWM signal from its appropriate output channels as follows. When the counter value in the base timer reaches 0, maximum, or compare value defined for the channel, its output can be changed. This leads to PWM signal generation based on PWM mode 1 and 2. When the base timer is counting up, output is high, while the counter is lower than the compare value in PWM mode 1. The output is low while the counter is higher than the compare value. This is just the opposite in PWM mode 2.

Table 7.1 PWM signal frequency and duty cycle formulas based on alignment

Alignment	Frequency f_{PWM}	Duty cycle D
Left	$\dfrac{f_{Timer}}{ARRx+1}$	$\dfrac{CCRx+1}{ARRx+1}$
Center	$\dfrac{f_{Timer}}{2 \times (ARRx+1)}$	$1 - \dfrac{CCRx+1}{ARRx+1}$
Right	$\dfrac{f_{Timer}}{ARRx+1}$	$1 - \dfrac{CCRx+1}{ARRx+1}$

We can explain this operation in more detail as follows. When PWM mode 1 is selected and base counter is counting up, then the PWM signal at OCxREF is at logic level 1 as long as TIMx_CNT < TIMx_CCRx. Else it becomes logic level 0. If the compare value in TIMx_CCRx is greater than the auto-reload value (in TIMx_ARR), then OCxREF is held at logic level 1. If the compare value is 0, then OCxREf is held at logic level 0. When PWM mode 1 is selected with the base counter counting down, the reference signal OCxREF is at logic level 0 as long as TIMx_CNT > TIMx_CCRx. Else, it becomes logic level 1. If the compare value in TIMx_CCRx is greater than the auto-reload value in TIMx_ARR, then OCxREF is held at logic level 1.

The PWM signal can be left or right aligned (in edge) or in centered mode. In the left edge aligned mode and the timer counting in upward direction, the rising edge occurs when the counter value resets to 0 and falling edge occurs when the counter value reaches the compare value. In the right edge aligned mode, the rising edge occurs when the counter value reaches the compare value, and falling edge occurs when the counter value reaches the maximum count. In the centered mode, timer can count only in up/down mode. Here, rising and falling edges both occur when the counter value reaches the compare value. In PWM mode 1 and centered mode, falling edge occurs while the counter is counting in the upward direction, and falling edge occurs while in downward direction. In addition to these modes, output polarity for the signal can be selected. For more information on this option, please see [4].

Based on the alignment of the PWM signal, we can obtain different frequency and duty cycle values. We tabulate these in Table 7.1. In this table, ARRx is the auto-reload and CCRx is the compare value of the base timer. For f_{Timer}, please see Sect. 6.2.1.

7.6 DAC Setup in the STM32F4 Microcontroller

As explained in previous sections, we can generate analog voltage from the STM32F4 microcontroller either by using the DAC module or PWM signal. We will consider how these can be done in C, C++, and MicroPython languages in this section. Let's start with the C language-based setup.

7.6.1 DAC Setup via C Language

DAC setup via C language is based on STM32CubeMX usage with the support of available HAL library functions. This gives us freedom in setting up the DAC module and generating the PWM signal. Next, we consider these two options in separate sections.

7.6.1.1 DAC Module Setup

We assume a project has already been formed and ready to be modified under STM32CubeMX for the DAC module usage. To configure the DAC module, click on the GPIO pin with DAC output capability on the STM32F4 microcontroller and select the GPIO type as "DAC_Outy" on the "Device Configuration Tool" window as in Fig. 7.20. To note here, y can be 1 or 2. Afterward, enable the channel using one of "OUT1 Configuration" or "OUT2 Configuration" option under the "Analog-DAC" section as in Fig. 7.21. Once the output pin and channel is configured, DAC settings can be configured from the opened "Parameter Settings" tab in the "Configuration" menu in the same section as in Fig. 7.22.

Fig. 7.20 DAC output source configuration in STM32CubeMX

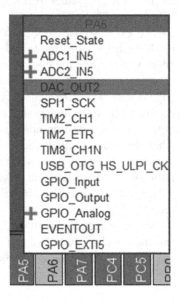

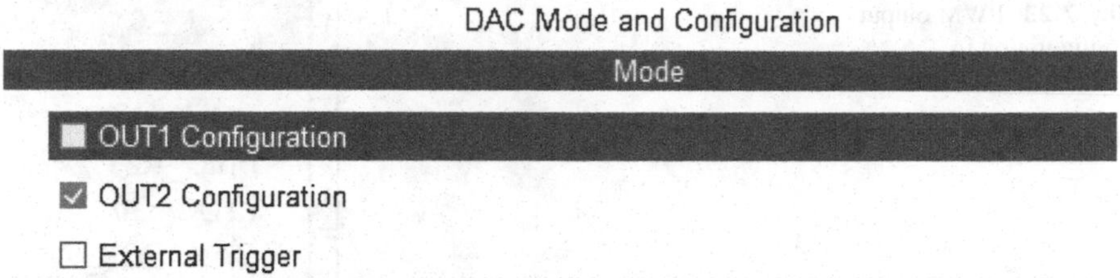

DAC Mode and Configuration

Mode
■ OUT1 Configuration
☑ OUT2 Configuration
☐ External Trigger

Fig. 7.21 DAC mode menu in STM32CubeMX

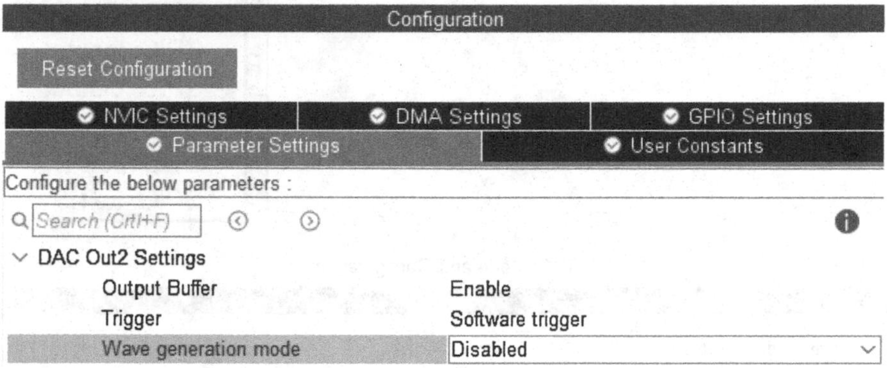

Configuration		
Reset Configuration		

● NVIC Settings	● DMA Settings	● GPIO Settings
● Parameter Settings	● User Constants	

Configure the below parameters :

Q Search (Crtl+F)   ⊘   ⊙   ⓘ

∨ DAC Out2 Settings
Output Buffer	Enable
Trigger	Software trigger
Wave generation mode	Disabled ∨

Fig. 7.22 DAC configuration menu in STM32CubeMX

The output buffer can be enabled or disabled using the "Output Buffer" option in the configuration menu. The DAC trigger should also be selected from the "Trigger" option in the configuration menu. If a software trigger will be used, then "Software Trigger" should be selected from "Trigger" option in configuration menu as in Fig. 7.22. If a hardware trigger will be used, then the EXTI line 9 or timer source with "Trigger Out" event should be selected from "Trigger" option in configuration menu as in Fig. 7.22. To select the EXTI line 9, the ninth pin of any port should be configured as "DAC_EXTI9" on the "Device Configuration Tool" window first. Then, the trigger should be selected as "External line 9" from the "Trigger" option under the configuration menu as in Fig. 7.22. If a timer will be used as hardware trigger, then the timer and its output event should be configured first as explained in Sect. 6.3. Then, the trigger should be selected as respective timer event from the "Trigger" option under the configuration menu.

7.6.1.2 PWM Signal Setup

We can use STM32CubeMX to configure the PWM signal. Here, we assume that a project has already been created beforehand as explained in Chap. 3. Then, click on the GPIO pin with timer output capability on the STM32F4 microcontroller and select the GPIO type as "TIMx_CHy" on the "Device Configuration Tool" window as in Fig. 7.23. To note here, x can be 1 to 5 or 8 to 14 and y can be any value

Fig. 7.23 PWM output
configuration in
STM32CubeMX

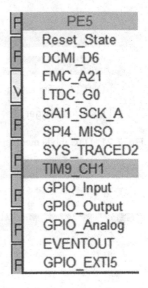

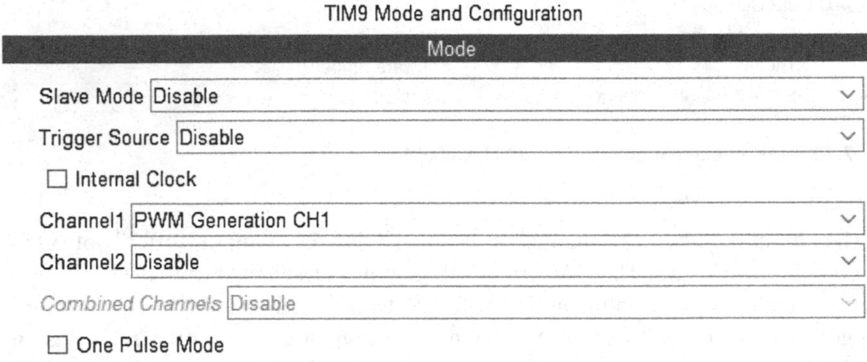

TIM9 Mode and Configuration

Mode	
Slave Mode	Disable ⌄
Trigger Source	Disable ⌄
☐ Internal Clock	
Channel1	PWM Generation CH1 ⌄
Channel2	Disable ⌄
Combined Channels	Disable ⌄
☐ One Pulse Mode	

Fig. 7.24 PWM mode selection in STM32CubeMX

between 1 and 4 depending on the timer. Afterward, enable the channel using the
"PWM Generation CHy" option under the "Timers - TIMx" section as in Fig. 7.24.

We should configure the timer clock first as explained in detail in Sect. 6.3. Once
the output pin, channel, and clock are configured, PWM settings can be configured
from the opened "Parameter Settings" tab in the "Configuration" menu in the same
section as in Fig. 7.25. Here, we can divide the timer clock using the internal
clock divider option. We should set the counter mode as up or down. Then, we
should select the period and duty cycle using the counter period and pulse in the
"Configuration" menu as in Fig. 7.25. PWM mode and output polarity can also be
selected from the same menu.

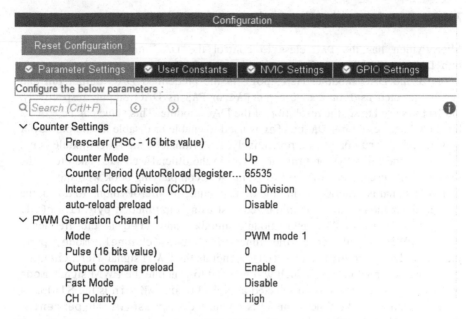

Fig. 7.25 PWM configuration settings in STM32CubeMX

7.6.2 DAC Setup via C++ Language

We can benefit from Mbed and available functions there to adjust the DAC module and PWM signal properties. In order to set up the DAC module, we should first adjust the related DAC pins. This can be done by the class `AnalogOut`. Usage of this class is similar to the class definition `AnalogIn` introduced in Sect. 7.3.2. We should supply a pin name we would like to use to this class. Afterward, the analog value can be set using the class `AnalogOut` or one of the functions `write` or `write_u16`. The `AnalogOut` class and `write` function sets a float value between 0 and 1.0. As an example, if the reference voltage value is 3 V and we want to set the output voltage to 0.75 V, then the value within these functions will be $0.75/3 = 0.25$. The function `write_u16` sets value between 0 and `0xFFFF`. Here, `0xFFFF` corresponds to V_{ref+}. Therefore, we should use `0x8000` in the function if we want to get an output voltage $V_{ref+}/2$.

In order to set up the PWM signal, we should first adjust the related pins. This can be done by the class `PwmOut`. We should supply a pin name for this class. Afterward, the period of PWM signal should be set using one of the functions `period`, `period_ms`, or `period_us`. These functions set the period in seconds, milliseconds, and microseconds, respectively. Duty cycle of the PWM signal can be set using the class `PwmOut` or one of the functions `pulsewidth`, `pulsewidth_ms`, or `pulsewidth_us`. Here, the `PwmOut` class and the function `pulsewidth` set the duty cycle in seconds. The function `pulsewidth_ms` sets the duty cycle in milliseconds. The function `pulsewidth_us` sets the duty cycle in microseconds.

7.6.3   DAC Setup via MicroPython

MicroPython has the DAC class to control the DAC module. The function `pyb.DAC(port, bits=DACbits, buffering=DACbuffer)` can be used to initialize the DAC module. Here, `port` is the predefined DAC output pin and can be selected as 1 or 2 to use pin PA4 or PA5 as DAC output, respectively. `DACbits` is used to set the resolution of the DAC module. This value can be 8 or 12 for the related resolution. `DACbuffer` is used to enable or disable the output buffer by setting it to `True` or `False`, respectively. Unfortunately, the DAC module is not enabled in the MicroPython firmware given in the official web site. Therefore, the reader should use the firmware provided in the accompanying book web site.

PWM signal is generated using the timer module in MicroPython. To do so, the desired timer module must be initialized first using the function `pyb.Timer(id, freq)`. Here, `id` is the timer module number and `freq` is the frequency of the PWM signal. Then, the function `channel(channel, mode, pin, pulse_width_percent)` can be used to generate the PWM signal. Here, `channel` is the timer channel number which will be used to generate the PWM signal. `mode` is the timer mode and must be selected as `pyb.Timer.PWM`. `pin` is the STM32F4 microcontroller pin the timer channel is connected to. `pulse_width_percent` is the duty cycle of the PWM signal.

7.7   DAC Usage in the STM32F4 Microcontroller

As the DAC module or PWM signal is set up, then we can use them for our specific purposes. Here, we provide how to do this in C, C++, and MicroPython languages. We also provide examples on sample usage scenarios.

7.7.1   DAC Usage in C Language

We handle the DAC usage in C language in two separate parts based on the DAC module and PWM signal generation. For both cases, we benefit from the available HAL library functions. We also provide usage examples on DAC usage in this section.

7.7.1.1 DAC Module Usage

HAL library has special functions to handle the DAC module. Here, we will only mention the general DAC usage in an infinite loop. The DAC output value can be set using the function `HAL_DAC_SetValue`. The regular conversion operation can be started using the function `HAL_DAC_Start`. The DAC module can be stopped and disabled using the function `HAL_DAC_Stop`. There is no conversion interrupt function for DAC. However, we can use the GPIO or timer interrupt to set the output value while using the DAC module with hardware triggers. We provide detailed information on all these functions next.

```
HAL_StatusTypeDef HAL_DAC_Start (DAC_HandleTypeDef * hdac,
    uint32_t Channel)
/*
hdac: pointer to a DAC_HandleTypeDef struct
Channel: the selected DAC channel and can be one of the following
    values
DAC_CHANNEL_1: DAC Channel1 selected
DAC_CHANNEL_2: DAC Channel2 selected
*/

HAL_StatusTypeDef HAL_DAC_Stop (DAC_HandleTypeDef * hdac,
    uint32_t Channel)

HAL_StatusTypeDef HAL_DAC_SetValue (DAC_HandleTypeDef * hdac,
    uint32_t Channel, uint32_t Alignment, uint32_t Data)
/*
Alignment: specifies the data alignment and can be one of the
    following values.
DAC_ALIGN_8B_R: 8bit right data alignment selected
DAC_ALIGN_12B_L: 12bit left data alignment selected
DAC_ALIGN_12B_R: 12bit right data alignment selected
Data: to be loaded in the selected data holding register
*/
```

7.7.1.2 PWM Signal Usage

HAL library also has special functions to handle PWM signal generation. Here, we will only mention general PWM usage in an infinite loop. The PWM output can be enabled using the function HAL_TIM_PWM_Start.The PWM output can be stopped and disabled using the function HAL_TIM_PWM_Stop. The duty cycle can be set using the CCRx value which can be accessed while the timer is running. This can be done using the macro __HAL_TIM_SET_COMPARE. We provide detailed information on all these functions next.

```
HAL_StatusTypeDef HAL_TIM_PWM_Start (TIM_HandleTypeDef * htim,
    uint32_t Channel)
/*
htim: pointer to a TIM_HandleTypeDef struct
Channel: TIM Channels to be enabled This parameter can be one of
    the following values:
TIM_CHANNEL_1: TIM Channel 1 selected
TIM_CHANNEL_2: TIM Channel 2 selected
TIM_CHANNEL_3: TIM Channel 3 selected
TIM_CHANNEL_4: TIM Channel 4 selected
*/

HAL_StatusTypeDef HAL_TIM_PWM_Stop (TIM_HandleTypeDef * htim,
    uint32_t Channel)

__HAL_TIM_SET_COMPARE(__HANDLE__, __CHANNEL__, __COMPARE__)
/*
__HANDLE__: pointer to a TIM_HandleTypeDef struct
__CHANNEL__: TIM Channels to be enabled This parameter can be one
    of the following values:
__COMPARE__: specifies the Capture Compare register (CCRx) new
    value.
*/
```

7.7.1.3 DAC Usage Examples

We provide three examples on the usage of DAC functions. The first example, with the C code given in Listing 7.14, aims to familiarize the reader with the setup and usage of the DAC module. Within the code, the DAC operation is triggered with software in an infinite loop. In order to execute the code in Listing 7.14, create an empty project and configure pin PA5 as DAC_OUT2. Afterward, enable the buffer and select the software trigger. As the project is compiled and run, brightness of the LED (connected to pin PA5) will change every second.

Listing 7.14 Usage of DAC module with software trigger, the C code

```
/* USER CODE BEGIN PV */
uint32_t i = 2048;
/* USER CODE END PV */

/* Infinite loop */
/* USER CODE BEGIN WHILE */
while (1)
{
i += 128;
if (i > 4095)
i = 2048;

HAL_DAC_SetValue(&hdac, DAC_CHANNEL_2, DAC_ALIGN_12B_R, i);
HAL_DAC_Start(&hdac, DAC_CHANNEL_2);
HAL_Delay(1000);
/* USER CODE END WHILE */

/* USER CODE BEGIN 3 */
}
/* USER CODE END 3 */
```

To execute the code in Listing 7.14, we should connect an external LED and resistor between ground and pin PA5 of the STM32F4 board. The hardware connections are as in Fig. 7.26.

In the second example, we repeat the first example. This time we use a push button as hardware trigger instead of software trigger. The C code for this example is given in Listing 7.15. In addition to the external LED connected to pin PA5, we should also connect a push button in active high pull-down configuration to pin PG9 and use it as hardware trigger.

Listing 7.15 Usage of the DAC module with GPIO hardware trigger, the C code

```
/* USER CODE BEGIN PV */
uint32_t i = 2048;
/* USER CODE END PV */

/* USER CODE BEGIN 0 */
void HAL_GPIO_EXTI_Callback(uint16_t GPIO_Pin)
{
i += 128;
if (i > 4095)
i = 2048;
HAL_DAC_SetValue(&hdac, DAC_CHANNEL_2, DAC_ALIGN_12B_R, i);
}
/* USER CODE END 0 */
```

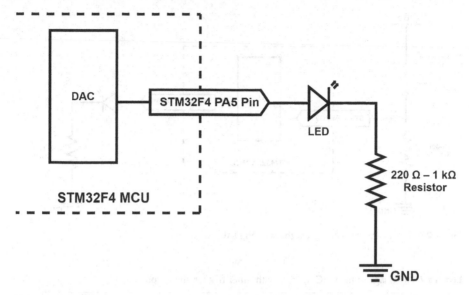

Fig. 7.26 Hardware connections for the first DAC example

```
/* USER CODE BEGIN 2 */
HAL_DAC_SetValue(&hdac, DAC_CHANNEL_2, DAC_ALIGN_12B_R, i);
HAL_DAC_Start(&hdac, DAC_CHANNEL_2);

HAL_PWR_EnableSleepOnExit();
HAL_PWR_EnterSLEEPMode(PWR_MAINREGULATOR_ON, PWR_SLEEPENTRY_WFI);
/* USER CODE END 2 */
```

The hardware setup for the second example is as in Fig. 7.27. In order to execute the code in Listing 7.15, create an empty project. Then, configure pin PA5 as DAC_OUT2 and pin PG9 as DAC_EXTI9. Afterward, enable the buffer and select the trigger as external line 9. Finally, enable the EXTI line[9:5] interrupt under NVIC. As the project is compiled and run, brightness of the LED can be changed by every button press.

In the third example, we repeat the first example by using a timer update event as hardware trigger instead of software trigger. The C code for this example is given in Listing 7.16. In order to execute the code in Listing 7.16, create an empty project. Then, configure pin PA5 as DAC_OUT2. Afterward, enable the buffer and select the trigger as "Timer 6 Trigger Out" event. Finally, configure Timer 6 to generate an update event at every second and enable TIM6 interrupt under NVIC. As the project is compiled and run, brightness of the LED can be changed every second without CPU intervention.

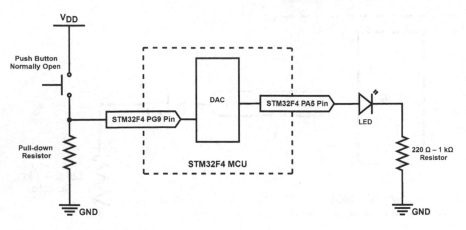

Fig. 7.27 Hardware connections for the second DAC example

Listing 7.16 Usage of the DAC module with timer trigger, the C code

```
/* USER CODE BEGIN PV */
uint32_t i = 2048;
/* USER CODE END PV */

/* USER CODE BEGIN 0 */
void HAL_TIM_PeriodElapsedCallback(TIM_HandleTypeDef *htim)
{
i += 128;
if (i > 4095)
i = 2048;
HAL_DAC_SetValue(&hdac, DAC_CHANNEL_2, DAC_ALIGN_12B_R, i);
}
/* USER CODE END 0 */

/* USER CODE BEGIN 2 */
HAL_TIM_Base_Start_IT(&htim6);
HAL_DAC_SetValue(&hdac, DAC_CHANNEL_2, DAC_ALIGN_12B_R, i);
HAL_DAC_Start(&hdac, DAC_CHANNEL_2);

HAL_PWR_EnableSleepOnExit();
HAL_PWR_EnterSLEEPMode(PWR_MAINREGULATOR_ON, PWR_SLEEPENTRY_WFI);
/* USER CODE END 2 */
```

We next provide an example on PWM signal generation. To do so, we should connect an external LED and resistor between the ground and PE5 pins of the STM32F4 board as shown in Fig. 7.28. The C code for this example is given in Listing 7.17. In order to execute the code, create an empty project. Then, configure pin PE5 as TIM8_CH3. Afterward, configure the timer clock to 48 MHz, set the prescaler 0, and counter period to 255. Hence, the generated PWM signal frequency will be 187.5 kHz. As the project is compiled and run, brightness of the LED will change every second.

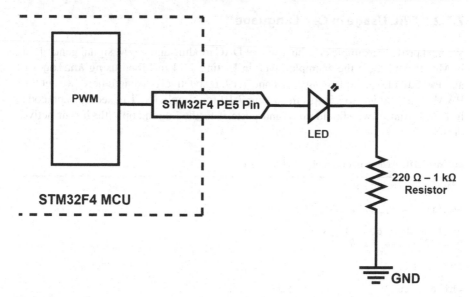

Fig. 7.28 Hardware setup for the PWM signal generation example

Listing 7.17 PWM signal generation, the C code

```
/* USER CODE BEGIN PV */
uint32_t i = 0;
/* USER CODE END PV */

/* USER CODE BEGIN 2 */
HAL_TIM_PWM_Start(&htim9, TIM_CHANNEL_1);
/* USER CODE END 2 */

/* Infinite loop */
/* USER CODE BEGIN WHILE */
while (1)
{
i += 16;
if (i > 0xFF)
i = 0;

__HAL_TIM_SET_COMPARE(&htim9, TIM_CHANNEL_1, i);

HAL_Delay(1000);
/* USER CODE END WHILE */

/* USER CODE BEGIN 3 */
}
/* USER CODE END 3 */
```

7.7.2　DAC Usage in C++ Language

We next provide examples on the usage of DAC module and PWM signal generation in Mbed. We repeat the example given in Listing 7.14 in Mbed using `AnalogOut` and `PwmOut` classes in Listings 7.18 and 7.19, respectively. To note here, we set the PWM signal period to 256 μs in Listing 7.19 for simplicity. To execute each code in Mbed Studio, we should copy and paste it to the "main.cpp" file in our active project.

Listing 7.18　Analog out example, the C++ code

```cpp
#include "mbed.h"

#define WAIT_TIME_MS 1000

AnalogOut dacOut(PA_5);
uint32_t i = 0x8000;

int main()
{
while (true)
{
i += 2048;
if (i > 0xFFFF)
i = 0x8000;
dacOut.write_u16(i);

thread_sleep_for(WAIT_TIME_MS);
}
}
```

Listing 7.19　PWM out example, the C++ code

```cpp
#include "mbed.h"

#define WAIT_TIME_MS 1000

PwmOut pwmOut(PC_8);
uint32_t i = 0;

int main()
{
pwmOut.period_us(256);

while (true)
{
i += 16;
if (i > 0XFF)
i = 0;
pwmOut.pulsewidth_us(i);

thread_sleep_for(WAIT_TIME_MS);
}
}
```

We next provide an example on the usage of the ticker module to trigger the DAC module. Here, voltage values for a sinusoidal signal with 128 elements are created.

Then, this array is fed to the DAC output with 8 Hz frequency. Hence, frequency of the sinusoidal signal will be 8/128 Hz. Again, the output can be observed by connecting a multimeter to pin PA5. We provide the corresponding C++ code in Listing 7.20

Listing 7.20 Triggering the DAC module by the ticker module, the C++ code

```cpp
#include "mbed.h"

#define PI 3.14159265358979323846

Ticker timerTicker;
AnalogOut dacOut(PA_5);

float sineVoltages[128];
int dacFlag = 0;

void setDAC()
{
dacFlag = 1;
}

int main()
{
int i;
timerTicker.attach(&setDAC, 125ms);
for (i = 0; i < 128; i++)
{
sineVoltages[i] = 1.65 + 1.65 * sin(2 * PI * i / 128);
}
i = 0;

while (true)
{
if (dacFlag == 1)
{
dacOut.write(sineVoltages[i] / 3.3);
i++;
if (i == 128)
i = 0;
dacFlag = 0;
}
}
}
```

7.7.3 DAC Usage in MicroPython

The function DAC.write(value) can be used to feed analog value from a selected pin in MicroPython. Here, value can be selected between 0 and 255 if the DAC resolution is set to 8 bits. It can be selected between 0 and 4095 if the DAC resolution is set to 12 bits. DAC output can also be generated using DMA with the function DAC.write_timed(data, freq, mode). Here, data is the array that holds DAC values. freq is the frequency of the DAC module and Timer 6 is automatically used to create this frequency. mode is the DMA mode (to be explained in Chap. 9) and can be selected as DAC.NORMAL or DAC.CIRCULAR.

We provide three examples on the usage of DAC operation and PWM signal generation in MicroPython. In the first example, the DAC module is used to generate analog voltage at pin PA5. Initially, voltage of the DAC output is set to 1.5 V. Then, it is increased by approximately 0.094 V every second. If the output voltage reaches 3 V, then it is reset to 1.5 V again. Output of the DAC module can be observed by a LED connected to pin PA5. The MicroPython code for this example is given in Listing 7.21.

Listing 7.21 Analog output example, the MicroPython code

```
import pyb

def main():
    i = 2048

    dacOut = pyb.DAC(2, bits=12, buffering=True)

    while True:
        i = i + 128
        if i > 4095:
            i = 2048
        dacOut.write(i)
        pyb.delay(1000)

main()
```

In the second example, a PWM signal is fed from pin PC8. Channel3 of Timer 3 is connected to pin PC8. Hence, it is configured to generate 1 kHz PWM signal with initial duty cycle equal to zero. Then, the duty cycle is increased by 1% every second. The output can be observed by connecting an LED to pin PC8. The MicroPython code for this example is given in Listing 7.22.

Listing 7.22 PWM output example, the MicroPython code

```
import pyb

def main():
    i = 0

    timer3 = pyb.Timer(3, freq=1000)
    timer3Ch3 = timer3.channel(
        3, pyb.Timer.PWM, pin=pyb.Pin.board.PC8,
            pulse_width_percent=0)

    while True:
        i = i + 1
        if i > 100:
            i = 0
        timer3Ch3.pulse_width_percent(i)
        pyb.delay(1000)

main()
```

In the third example, the DAC module is used with DMA to generate a sine wave at pin PA5 in circular mode. Here, the sine wave is created with 128 elements for one period. Amplitude of the sine wave changes between 0 and 3 V. The DAC module is triggered with 8 Hz via Timer 2. Hence, 128 element output (in one period) is given to output in 16 s and frequency of sine wave is 1/16 Hz. Finally, these values are converted to analog voltage and fed to output using the DMA module in circular mode. The output can be observed by connecting an LED to pin PA5. The MicroPython code for this example is given in Listing 7.23.

Listing 7.23 Triggering the DAC module by the timer module, the MicroPython code

```python
import pyb

from array import array
from math import pi, sin

def convertDACVoltage2Value(voltage):
    value = int(4095 * voltage / 3.0)
    return value

def main():
    sineVoltages = array('f', 1.5 + 1.5 * sin(2 * pi * i / 128)
        for i in range(128))
    DACValues = array('H', (0 for i in range(128)))

    for i in range(len(DACValues)):
        DACValues[i] = convertDACVoltage2Value(sineVoltages[i])
    dacOut = pyb.DAC(2, bits=12, buffering=True)
    dacOut.write_timed(DACValues, pyb.Timer(2, freq=8), mode=
        dacOut.CIRCULAR)

main()
```

7.8 Application: ADC and DAC Operations in the Robot Vacuum Cleaner

In this chapter, we perform battery level measurement and display and distance sensing and control the speed and sweep rate of our robot via ADC, DAC, and PWM modules. To do so, we use the ADC module to measure the battery level of the robot. We use the DAC module to display the measured value on an LED bar. We use PWM to control the speed of the motors connected to wheels of the robot. Likewise, we benefit from PWM to adjust the sweeping rate in different modes. Finally, we add an analog distance sensor in front of the robot to check possible obstacles. This will lead to controlling the speed of the robot. We provided the equipment list, circuit diagram, detailed explanation of design specifications, and peripheral unit settings in a separate document in the folder "Chapter7\EoC_applications" of the accompanying supplementary material. In the same folder, we also provided the complete C, C++, and MicroPython codes for the application.

## 7.9	Summary of the Chapter

Digital input and output concepts introduced in Chap. 4 are valuable in interacting of the microcontroller by the outside world. However, this interaction can be expanded further by adding analog signal processing capability to the microcontroller. We introduced how such an operation can be done in this chapter. Therefore, we started with the in-depth explanation of analog and digital values. Then, we focused on the steps applied in the analog to digital conversion operation. Hence, the embedded system can convert the input analog value to the corresponding digital value. Afterward, we introduced the ADC module in the STM32F4 microcontroller as a representative of such modules in embedded systems. We also provided the setup and usage examples of this module in C, C++, and MicroPython languages. We next covered the digital to analog conversion operation. Hence, the embedded system can convert the digital input value to the corresponding analog value. We introduced the DAC module in the STM32F4 microcontroller and PWM method to perform this operation in the STM32F4 microcontroller as a representative of such modules in embedded systems. We also provided the setup and usage examples of the DAC module and PWM in C, C++, and MicroPython languages. We expanded our previously introduced real-world example by adding ADC and DAC modules to it as the end of chapter application. As microcontrollers become more powerful, they have been used in processing analog signals. Hence, understanding how ADC and DAC modules can be used in such applications is extremely important. The concepts introduced in this chapter form a foundation for such usages.

Problems

7.1. Why cannot an analog value be processed in a digital system?

7.2. Summarize data types in C and C++ languages with their range (specific to the Arm® Cortex™-M microcontroller).

7.3. Explain the sampling theorem (or Nyquist-Shannon sampling theorem) in mathematical terms.

7.4. Assume that analog voltage values 0.128, 0.2, and 2.753 V are fed to the ADC module in the STM32F4 microcontroller with $V_{ref+} = 3.3$ V. What will be the quantized output value for each analog input voltage?

7.5. How does SAR work in the ADC module of the STM32F4 microcontroller?

7.6. Design a system that controls its own supply with the STM32F4 microcontroller. The VDD pin of the microcontroller will be regulated and kept constant by a voltage regulator with 3 V output. Input of the voltage regulator will be fed from more than one source. Switching and connection between sources will be done by controllable switching elements (such as power multiplexer or analog switch IC). Controllable switching elements can be controlled with GPIO pins. The first source will be a rechargeable battery with a nominal 3.7 V output. The output of the battery can vary between 3.3 and 4.1 V. The second source will be a solar panel with maximum 5 V output. The panel output can vary between 0 and 5 V. The setup for the system will be as follows. If the solar panel has an output voltage higher than 3.3 V, then the voltage regulator will be connected to the solar panel using the first controllable switching element. If the solar panel has an output voltage less than 3.3 V, then the voltage regulator will be connected to the battery using the second controllable switching element. If the output voltage of the solar panel is higher than the voltage of the battery, then the battery will be connected to the solar panel using a third controllable switching element. The green and red LEDs connected to pins PG13 and PG14 of the STM32F4 board will indicate the source the system is fed. When the green LED is on, this indicates that the STM32F4 microcontroller is powered by the solar panel and battery is not charged. When the red LED is on, this indicates that the STM32F4 microcontroller is powered by the battery only. When both LEDs are on, this indicates that the STM32F4 microcontroller is powered by the solar panel and battery is charged. Form the project using:

(a) C language under STM32CubeIDE.
(b) C++ language under Mbed.
(c) MicroPython.

7.7. Form a project with the following specifications. Every time the user button on the STM32F4 board is pressed, temperature value should read from the STM32F4 microcontroller. Use the appropriate ADC module in operation. Add power management option to your solution such that the STM32F4 microcontroller waits in an appropriate low power mode when not used. Form the project using:

(a) C language under STM32CubeIDE.
(b) C++ language under Mbed.
(c) MicroPython.

7.8. Modify the project in Problem 7.7 such that the temperature reading operation is done in every 10 s. Use an appropriate timer module in operation.

7.9. Form a project with the following specifications. Every time the button connected to pin PC11 is pressed, the voltage value at pin PA5 should be read from the STM32F4 microcontroller. The voltage there is formed by a voltage source and voltage divider circuit. The voltage divider is formed by a potentiometer as shown in Fig. 7.12. Hence, the pin gets voltage values between 0 and 3.3 V. Use the appropriate ADC module and ADC interrupt in operation. Add power management option to your solution such that the STM32F4 microcontroller waits in an appropriate low power mode when not used. Form the project using:

(a) C language under STM32CubeIDE.
(b) C++ language under Mbed.
(c) MicroPython.

7.10. Design a musical tone generator using the STM32F4 microcontroller with the following specifications. There should be a function which generates the note using sine waves for a given duration. Then, the main function should play the notes sequentially as C4, D4, E4, F4, G4, A4, and B4 for three seconds each. Use Timers and DAC modules for this operation. Note: The frequencies are 261.6 Hz, 293.6 Hz, 329.6 Hz, 349.2 Hz, 392.0 Hz, 440.0 Hz, and 493.8 Hz for notes C4, D4, E4, F4, G4, A4, and B4, respectively. The given frequency values can be generated up to 10% error. You can use a 3 V piezo buzzer to generate a tone signal. You can use pin PA4 pin of the STM32F4 microcontroller to connect the buzzer. Form the project using:

(a) C language under STM32CubeIDE.
(b) C++ language under Mbed.
(c) MicroPython.

7.11. Add a new module to Problem 7.9 such that the LED, connected to pin PE5 of the STM32F4 microcontroller as shown in Fig. 7.28, will change its brightness proportional or inversely proportional to the input voltage level. Use the PWM mode separately for this operation. Every time the user button on the STM32F4 board is pressed, direction will be changed. Form the project using:

(a) C language under STM32CubeIDE.
(b) C++ language under Mbed.
(c) MicroPython.

References

1. Arm: ARM Compiler: armcc User Guide, arm dui0472k edn (2014)
2. Python Community. https://docs.python.org/3/library/datatypes.html. Accessed 4 June 4 2021
3. Mbed. https://os.mbed.com/handbook/C-Data-Types. Accessed 4 June 2021
4. STMicroelectronics. STM32F405/415, STM32F407/417, STM32F427/437 and STM32F429/439 advanced Arm-based 32-bit MCUs, rm0090 rev 19 ed. (2021)
5. Ünsalan, C., Gürhan, H.D.: Programmable Microcontrollers with Applications: MSP430 LaunchPad with CCS and Grace. McGraw-Hill, New York (2013)
6. Ünsalan, C., Gürhan, H.D., Yücel, M.E.: Programmable Microcontrollers: Applications on the MSP432 LaunchPad. McGraw-Hill, New York (2018)
7. Ünsalan, C., Yücel, M.E., Gürhan, H.D.: Digital Signal Processing Using Arm Cortex-M based Microcontrollers: Theory and Practice. Arm Education Media, Cambridge (2018)

Digital Communication

8

8.1 Background on Digital Communication

There are several definitions used in digital communication. Therefore, we provide them as background in this section. We will use these definitions in the following sections to explain working principles of a selected digital communication type.

8.1.1 Data, Frame, and Field

We will start with the definitions based on the transmitted data and related concepts. We should first mention that some definitions may have different meanings for different digital communication types. Hence, there is no unique definition which can be applied to all digital communication types. In order to reach a consensus and explain digital communication concepts more coherently, we pick the most reasonable ones in this section and use them as they are from this point on.

We will call the message to be transmitted as data. This can be 1 byte or 1 kilobyte. Due to the used communication type, we may not be able to transmit data at once. Hence, it should be segmented. Each segment will be called data chunk.

All digital communication types, to be explained in this chapter, have their own standards while transmitting data. In general, we can call the structure, containing data and other supplementary bits, transmitted to the receiver as frame (also called packet). We will extensively use the frame definition in this book.

Supplementary Information The online version contains supplementary material available at (https://doi.org/10.1007/978-3-030-88439-0_8).

C. Ünsalan et al., *Embedded System Design with Arm Cortex-M Microcontrollers*, https://doi.org/10.1007/978-3-030-88439-0_8

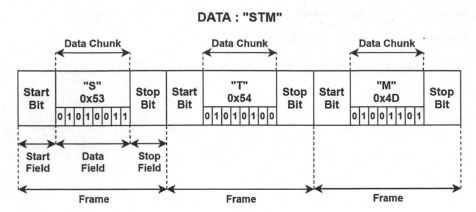

Fig. 8.1 Generic example explaining data and related concepts

Bits and bytes within one frame are grouped. We call each group as field. As an example, I^{2}C has start, address, read/write, data, and stop fields. A field can be formed by 1, 8, 32, or 1024 bits. This depends on the digital communication type. As a side note, a field formed by 1 bit is simply called bit. For example, instead of the start field composed of 1 bit, we will use the start bit naming instead.

We next provide a generic example in Fig. 8.1 to explain the concepts introduced in this section. Here, we transfer the characters "STM" represented in ASCII format by a pseudo digital communication method. As can be seen in this figure, we have data, data chunks, frames, and fields.

8.1.2 Serial and Parallel Data Transfer

There are two different data transfer types as parallel and serial. Parallel data transfer is used to achieve high transfer speeds. One such application is receiving data from a digital camera or feeding data to an LCD. We will use both operations in the following chapters. The disadvantage of parallel data transfer is that it requires several pins to operate. Serial data transfer provides an alternative such that it only needs one or two pins. This allows communicating with more than one device at the same time. The disadvantage of serial data transfer is that it is slow since data should be transmitted sequentially. We provide a simple demonstration to explain the parallel and serial data transfer in Fig. 8.2. Here, we transfer 8 bits in serial and parallel form.

8.1.3 Synchronous and Asynchronous Data Transfer

Data transfer in digital communication can be synchronous or asynchronous. In synchronous data transfer, the transmitter and receiver are synchronized by a

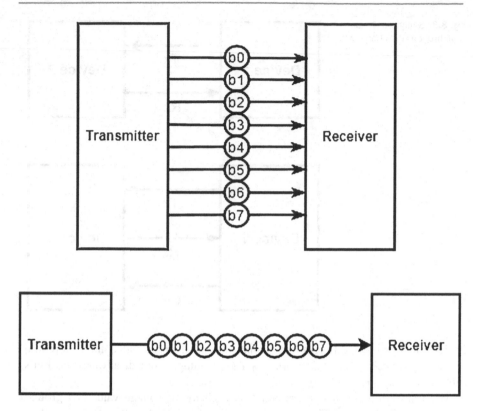

Fig. 8.2 Parallel and serial data transfer

common clock signal. This synchronization signal can be sent along with data. The asynchronous data transfer does not need such a synchronization signal. Instead, start and stop bits for each data packet are sent. However, the transmitter and receiver should know and agree upon the transfer speed beforehand.

We provide a simple demonstration to explain the synchronous and asynchronous data transfer in Fig. 8.3. Here, we transfer data between devices 1 and 2 via pins Tx, Rx, and Clk. The Clk pin carries the common clock signal between the transmitter and receiver. In the first setting, there is no Clk connection, or common clock, between both devices. Hence, asynchronous data transfer is achieved. In the second setting, there is a common clock signal carried by the Clk pin between the transmitter and receiver to form synchronous data transfer.

8.1.4 Signal Representation and Line Formations

We introduced how a bit is represented in voltage levels in Chap. 4. We also represent signals to be transmitted or received as voltage levels in digital communication.

Fig. 8.3 Synchronous and asynchronous data transfer

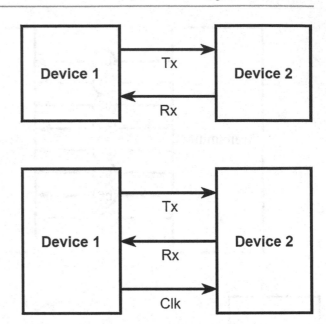

Besides, there are two different line formations called single-ended and differential to transfer signals. Each formation has its advantages and disadvantages. Let's summarize them next.

The transferred signal is represented as a changing voltage value wrt ground in the single-ended line. The advantage of this formation is that it requires less number of wires and less complex circuitry. The disadvantage of the single-ended line is that the transferred signal is affected by noise and line length in long range. Therefore, it is used in short-range data transfer most of the times. For the STM32F4 microcontroller, voltage levels used in single-ended line signal transfer are compatible with the standards introduced in Chap. 4. Hence, they can be generated by GPIO pins of the microcontroller.

Two different lines are used in the differential line formation. The transferred signal is represented by the difference of two opposite voltage values in these lines. As an example, the first and second lines get 3 V and 0 V, to represent logic level 1. The receiver takes the difference of these two signals and gets 3 V. Based on this voltage value, it decides that the signal represents logic level 1. The opposite voltage values are transmitted for logic level 0 representation. Hence, the first and second lines get 0 and 3 V, respectively. The receiver takes the difference of these two values and gets −3 V. Based on this voltage value, the receiver decides that this signal represents logic level 0. The advantage of the differential line usage is that it is not affected much by noise or range as in single-ended line. Hence, the differential line is used in long-range data transfer most of the times. The disadvantage of differential line is the need for extra line. Besides, the voltage levels used in operation exceed the ones available in the GPIO pins of the STM32F4 microcontroller. Hence, extra circuitry may be needed to use the differential line signal transfer.

There must be a common ground between the connected devices in the single-ended line. Although we did not show this common ground explicitly in the connection diagrams in this chapter, the reader should take this property into account while forming the connection. Having a common ground is not mandatory for the differential line formation. It depends on the communication type.

8.1.5   Data Encoding Types

There are different data encoding types for digital communication. These can be summarized as non return to zero, return to zero, non return to zero inverted, return to zero inverted, and Manchester coding. Each data encoding type has its usage areas. We introduce them next. As a side note, we will base our definitions on the single-ended line representation in this section.

8.1.5.1 Non Return to Zero

The first data encoding type is non return to zero (NRZ). In this method, logic level 1 to be transmitted is represented by a signal with high voltage value such as 3 V. The logic level 0 is represented by a signal with low voltage value such as 0 V. The name non return to zero is given for the following reason. Assume that the data bit to be transmitted is logic level 1. Hence, the signal goes to high voltage. This value is kept till the next data bit arrives. Hence, it does not return to the old value. If the next coming data bit is logic level 0, then the signal goes to low voltage. If the next data to be transmitted is again logic level 1, then the signal stays at high voltage. Some sources call NRZ as NRZ-L such that the suffix L indicates the level-based operation.

8.1.5.2 Return to Zero

The second data encoding type is return to zero (RZ). In this method, data representation has a changing voltage level within one clock period. As an example, if the data bit to be transmitted is logic level 1, then the voltage level is high at the first half of the clock period and low at the second half of the clock period. Therefore, the signal turns back to its initial low value after the high value. As a result, a clock-like signal is generated. If the data bit to be transmitted is logic level 0, then the voltage level stays at low value. This type of encoding can be obtained by logically ANDing the clock signal and data bits to be transmitted.

8.1.5.3 Non Return to Zero Inverted

The third data encoding type is non return to zero inverted (NRZI) which is in fact a subset of NRZ. However, NRZI does not correspond to the actual inverse operation. Instead, data is represented by an edge signal (low to high or high to low transition) instead of a voltage level in NRZI. If the data bit to be transmitted is logic level 1, then a low to high or high to low transition occurs. The transmission type is selected by the previous value of the signal. As an example, if successive data bits have value logic level 1, then a low to high and high to low transition occur repeatedly. If the data bit to be transmitted has value logic level 0, then the voltage level stays at its previous level. In other words, if successive logic level zeros come, then the signal always stays at the same level.

8.1.5.4 Return to Zero Inverted

The fourth data encoding type is return to zero inverted (RZI). This is also a subset of RZ such that it is the actual inverted operation for the RZ type. In other words, if the data bit to be transmitted is logic level 0, then the signal will have high voltage value in the first half of the clock signal. It will have the low voltage value in the second half of the clock signal. If the data bit to be transmitted is logic level 1, then the signal stays at low voltage level.

8.1.5.5 Manchester Coding

The fifth data encoding type is Manchester coding. In this method, data bits are represented by edges (low to high or high to low transition) as in NRZI. If the data bit to be transmitted is logic level 1, then a high to low voltage transition occurs in the signal. Therefore, it will be similar to the clock signal. If the data bit to be transmitted is logic level 0, then a low to high voltage transition occurs in the signal. Therefore, it will be similar to inverse of the clock signal. This type of encoding can be obtained by logically XORing data bits with the clock signal.

8.1.5.6 Demonstrating Different Data Encoding Types

We next provide an example to demonstrate all data encoding types. To do so, we picked an 8-bit data packet to be transmitted as 10100111. Based on this sequence, we provide the signals generated by the mentioned data encoding types in Fig. 8.4.

As can be seen in Fig. 8.4, when successive bits come as 101, the signal has high, low, and high voltage levels in NRZ data encoding. When the following bit values are 00, the corresponding signal stays at low voltage level. When, the following bit values are 111, the signal voltage level goes to high value and stays there. Next, we can observe the NRZI data encoding type. As mentioned before, the generated signal with this method is not the inverse of the one generated by NRZ. Instead, it has its own signal pattern for the selected 8-bit data packet. We can summarize this operation as follows. For the 101 part of data bits, the signal has a high to low transition in the first logic level 1 bit. Then, it stays at low voltage level for the logic level 0 bit. When the next logic level 1 comes, a low to high voltage transition occurs in the signal. As two successive logic level 0 bits come, the signal stays in high voltage level. When three successive logic level 1 bits come, then the signal has three successive transitions as high to low, low to high, and high to low.

We can also analyze the signals generated by RZ data encoding type in Fig. 8.4. As can be seen in this figure, when the data bit to be transmitted is logic level 1, the signal acts as the clock signal. When a logic level 0 comes, the signal stays at low voltage level. The RZI data encoding type acts just in the reverse manner such that when logic level 0 comes, the signal acts as the clock signal. When logic

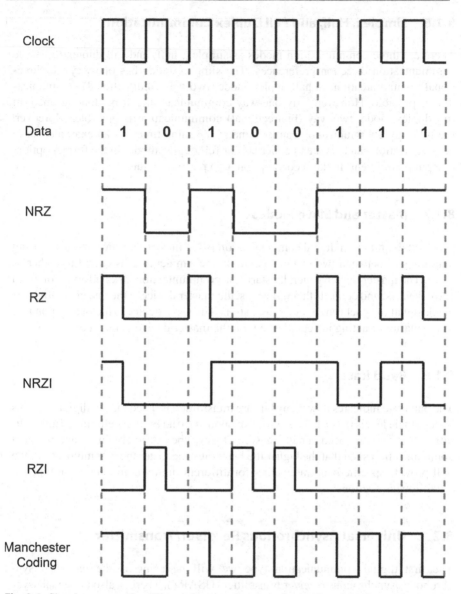

Fig. 8.4 Signals generated by different data encoding types for the given 8-bit data packet

level 1 comes, the signal stays at the low voltage level. Finally, Manchester coding generates a signal such that it always changes. To be more specific, when the first logic level 1 comes, the signal acts as the clock signal. When there is a change in successive data bits, then the signal switches from high to low or low to high value.

8.1.6   Simplex, Half, and Full Duplex Communication

There are three communication modes as simplex, half, and full duplex to handle communication direction preferences. The simplex mode has one-way (unidirectional) communication. In half duplex mode, two-way (bidirectional) communication is possible. However, only one-way communication can be done at once. In full duplex mode, two-way (bidirectional) communication is possible. Moreover, the two-way communication can be done at the same time. As an example, serial communication can be done in either half or full duplex mode. In the former option, one pin is sufficient. In the second option, two pins are required.

8.1.7   Master and Slave Modes

The next definition in digital communication is the master and slave modes. During data transfer between two or more devices, one can act as master and the other as slave. The master device generally starts the communication, asks information from slave devices, and controls them. Besides, the master device is the one generating the clock signal in synchronous communication. The slave device is the one responding to the master or acting in accordance with the master device commands.

8.1.8   Baud Rate

The baud rate indicates how many bits are transmitted in 1 second for digital systems having logic levels 0 and 1. This also corresponds to the bit rate definition. Both definitions indicate the speed of data transfer. This will be extremely important in digital communication such that the higher the baud rate, the faster the communication. We will provide specific baud rate values for different digital communication types in the following sections.

8.2   Universal Asynchronous Receiver/Transmitter

The first digital communication type we will be using is the universal synchronous/asynchronous receiver/transmitter (USART). There is also universal asynchronous receiver/transmitter (UART). We will be using the latter in asynchronous mode. Therefore, it will act as UART. Hence, we will only focus on UART from this point on. As a side note, USART works in synchronous mode with an extra clock pin. In this setting, it acts as SPI in master setting. We will not deal with this option in this book.

8.2.1 UART Working Principles

UART is generally preferred when fast data transfer speed is not needed. Hence, it is most of the times used for debugging and programming a microcontroller. Besides, it is used to communicate two microcontrollers. If two microcontrollers on different boards need to communicate with UART, a physical layer should be available such as RS232 or RS485. These will eliminate noise and power loss due to long-range signal transfer.

We can summarize UART based on the definitions in Sect. 8.1 as follows. Data transfer in UART can be done in frames. Each frame is composed of the start bit, data, parity bit, and stop bit(s). Data can be composed of 7, 8, or 9 bits. Start and stop bits are available in each frame. However, the parity bit can be used if needed. UART uses NRZ data encoding. In general, UART uses the TTL voltage levels to represent signals. UART can work in either full duplex, half duplex, or simplex modes. In the full duplex mode, there are two pins called Rx (receive) and Tx (transmit). In the half duplex mode, the Tx pin is used for both as Tx and Rx. In the simplex mode, Tx pin of one device is connected to Rx pin of the other device. The remaining pins are not connected. The master and slave definitions do not apply to UART. There is no theoretical limit for the baud rate in UART. However, a limit applies based on the used microcontroller and its associated clock.

We next provide general setup for UART communication between two devices in Fig. 8.5. As can be seen in this figure, each device has two pins as Rx (receive) and Tx (transmit). They are connected such that one Rx pin is connected to Tx and vice versa.

8.2.2 UART Modules in the STM32F4 Microcontroller

The STM32F4 microcontroller has four UART modules called UART4, UART5, UART7, and UART8. The STM32F4 microcontroller also has four USART modules called USART1, USART2, USART3, and USART6. These modules can be used in UART mode.

We provide block diagram of a generic USART/UART module for the STM32F4 microcontroller in Fig. 8.6. In this figure, the clock pin will not be used when UART is active. Here, the clock is provided by APB1 peripheral clock (PCLK1). There

Fig. 8.5 General setup for UART communication between two devices

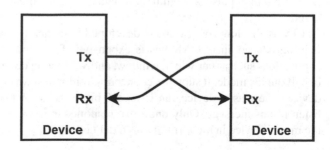

Fig. 8.6 USART/UART
block diagram in the
STM32F4 microcontroller

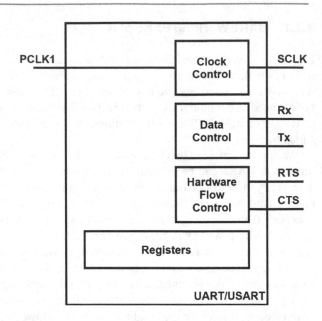

are two pins for data transfer as Tx and Rx. The module has hardware flow control
support which requires two extra pins as request to send (RTS) and clear to send
(CTS). UART does not need these pins. Finally, there is a clock output pin for
USART.

The USART1 module is connected to ST-LINK. Hence, it can communicate
with PC through virtual COM port. As a side note, this port is generated by USB
which is not an actual RS232 port. Through it, we can transfer data between the
microcontroller and PC as if we are communicating with a real communication port.

The USART1 and USART6 can reach baud rates up to 11.25 Mbits/s. The other
modules can reach a maximum baud rate of 5.625 Mbits/s. All supported baud rate
values for UART and USART modules are given in the datasheet [7]. The reader
can reach the necessary detailed information from there.

The USART and UART modules in the STM32F4 microcontroller support IrDA
and LIN protocols. IrDA is the infrared light-based communication method. The
STM32F4 microcontroller only supports the IrDA SIR ENDEC method which uses
the RZI data encoding. As a reminder, when data to be transmitted is logic level 0,
an infrared light pulse is submitted. If data to be transmitted is logic level 1, nothing
is sent.

LIN is the low-cost protocol developed specifically for automotive industry.
It is introduced since CAN bus is expensive. Therefore, it is used in noncritical
applications such as car wiper, window, and ventilating control. LIN protocol works
in half duplex mode. It supports connecting multiple slave devices to a single master
device. The master device can issue a 6 bit message. All connected slave devices
listen to this message. Only one slave responds to it and sends back a message to
the master device in the form of 2 to 6 data bytes.

All USART modules in the STM32F4 microcontroller support smartcard protocol used in communicating with chip cards. This protocol is similar to UART. The STM32F4 microcontroller supports the ISO 7816-3 standard for this purpose. According to this standard, each data packet is composed of 1 start bit, 8 data bits, 1 parity bit, and 1.5 stop bits. The 1.5 stop bits stand for sending the stop bit for 1.5 clock periods. The smartcard operation is half duplex such that only the Tx pin is used in communication.

8.2.3 UART Setup in the STM32F4 Microcontroller

We will consider UART setup in this section. Therefore, we will start with the C language based setup. This will be followed by C++ and MicroPython based setup operations.

8.2.3.1 Setup via C Language

We will benefit from STM32CubeMX to setup the UART/USART module properties in C language. At this stage, we assume that a project has already been created under STM32CubeIDE as given in Sect. 3.1.3. In order to set up UART/USART via STM32CubeMX, first open the "Device Configuration Tool window by clicking the ".ioc" file in the project explorer window. Then, configuration can be done in this section as shown in Fig. 8.7. We will provide details of this operation next.

In order to use the UART/USART module, Rx and Tx pins should be configured first. To do so, click on the GPIO pin with UART/USART capability on the STM32F4 microcontroller in the STM32CubeMX interface. Afterward, select the GPIO type as "USARTx_RX," "USARTx_TX," "UARTy_RX," or "UARTy_TX" in the "Device Configuration Tool" window as in Fig. 8.7. To note here, x can be one of 1, 2, 3, or 6 and y can be one of 4, 5, 7, or 8. Once the Rx and Tx pins are selected, click the set USART or UART module under the "Connectivity" section

Fig. 8.7 UART/USART configuration in STM32CubeMX

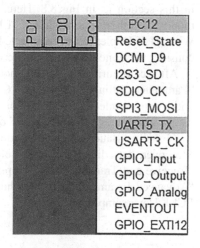

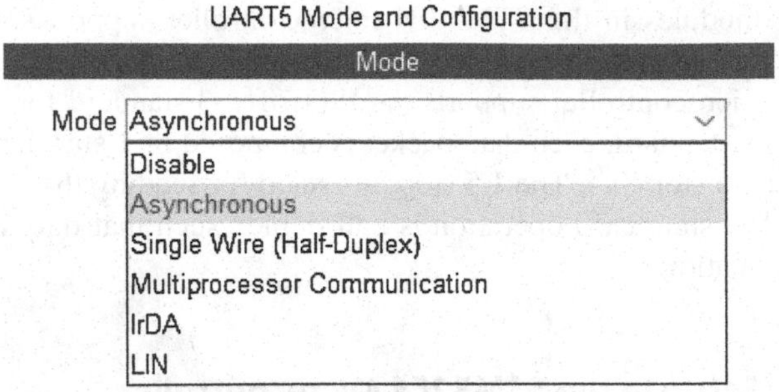

UART5 Mode and Configuration

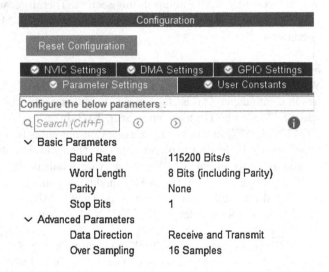

Fig. 8.8 UART/USART mode menu in STM32CubeMX

Fig. 8.9 UART/USART
configuration menu in
STM32CubeMX

from the left menu of STM32CubeMX. Then, mode configuration can be done
in this section as in Fig. 8.8. Here, select the "Asynchronous" option for UART
mode. The other options "IrDa," "LIN," "Smartcard," "Synchronous," and "Single
Wire (Half Duplex)" can be selected for other UART/USART modes. We will not
consider these modes in this chapter.

After configuring the UART mode, its parameters can be set from the opened
"Parameter Settings" tab in the "Configuration" menu as in Fig. 8.9. Here, data
width can be selected as "8-bit" or "9-bit" including the parity bit. The parity bit
can be selected as "None," "Even," or "Odd." The stop bit can be selected as "1" or
"2." Finally, "Baud Rate" can be edited in terms of bits/s under "Basic Parameters."
Moreover, data direction can be selected as "Receive Only," "Transmit Only," or
"Receive and Transmit." Oversampling ratio can be selected as "8" or "16" under
"Advanced Parameters." To note here, baud rate can take the maximum value as
PCLK1/oversampling ratio.

The UART module can operate in two modes as blocking (polling) or non-blocking (interrupt). In the blocking operation mode, the module waits (polling) until the transmit or receive operation has finished once it starts sending or receiving data in a loop. However, this mode blocks other routines while it waits. A timeout duration can be configured to terminate the operation if transmission or reception is not finished in due time. In the non-blocking operation mode, the module sends or receives data using first in, first out (FIFO) buffers. As the transmit or receive operation ends, an interrupt is generated. This operation does not block the code routine. That is why this mode is called non-blocking.

If an interrupt is not required, then this step can be skipped. Otherwise, the UART interrupt should be enabled from the NVIC tab in the configuration menu as in Fig. 8.9. The interrupt priority can be changed under the "System Core - NVIC" section. To note here, it is suggested to lower the UART interrupt priority since the HAL system uses the SysTick interrupt to create precise 1 ms ticks for internal operations and the HAL_Delay function. The SysTick interrupt has the highest priority among peripheral units. Hence, assigning a lower priority to the UART interrupt will not affect it. Otherwise, precision of the HAL_Delay function may degrade.

8.2.3.2 Setup via C++ Language

We can benefit from Mbed and available functions there to adjust the UART/USART module properties in C++ language. To do so, we should first adjust the Rx and Tx pins. This can be done by the class BufferedSerial or UnbufferedSerial. Usage of these classes is similar to the class DigitalIn introduced in Sect. 4.4.2. Therefore, we should supply Tx and Rx pin names to the class BufferedSerial or UnbufferedSerial. Optionally, we can provide the Baud Rate value to the classes. If this value is not provided, then the BufferedSerial or UnbufferedSerial class uses the default Baud Rate value which is 9600 bits/s.

The difference between BufferedSerial and UnbufferedSerial classes is as follows. The BufferedSerial class uses an intermediary buffer to send and receive data to transmit multiple bytes. The UnbufferedSerial class does not use such a buffer to transmit or receive data. Therefore, it is recommended to use the BufferedSerial class in terms of CPU load as all buffering operations are handled by HAL library functions. The BufferedSerial class uses circular buffers with different size. This size can be changed in the "mbed.json" file using the parameters drivers.uart-serial-txbuf-size and drivers.uart-serial-rxbuf-size. The default buffer size here is 256 bytes.

The BufferedSerial and UnbufferedSerial class parameters can be configured after initialization. The configurable parameters are baud rate, data length (7 or 8 bits), parity bit (odd, even, none), and stop bit (1 or 2 bits). Moreover, the operation mode can be selected as blocking or non-blocking. A callback function can be called whenever an interrupt occurs in non-blocking mode.

8.2.3.3 Setup via MicroPython

MicroPython has the UART class to perform UART operations. The UART module is created using the function pyb.UART(bus). Here, bus represents the UART module number. After the UART module is created, it can be initialized using the function init(baudrate, bits, parity, stop, timeout, flow). Here, baudrate is used to set the baud rate. The bits is used to set the number of data bits. It can be selected as 7, 8 (default), or 9. The parity is used to set the parity which can be selected as None, 0 (Even), or 1 (Odd). It is set to None by default. The stop is used to select the number of stop bits which can be selected as 1 (default) or 2. The timeout is used to define the timeout value in milliseconds. This is the waiting time for receiving or transmitting the first data. The flow is used to select the type of flow control. We will not explain flow control in this book. Hence, it can be left undefined which selects no flow control automatically. Unfortunately, only USART1 is enabled in the MicroPython firmware given in the official web site. Moreover, USART1 is used by REPL. Hence, it cannot be used to communicate with PC. The reader can use the MicroPython firmware provided in the book web site. In this firmware, UART5 is enabled with pins PD2 and PC12 set as Tx and Rx, respectively.

8.2.4 UART Usage in the STM32F4 Microcontroller

As the UART module is set up in the previous section, we will focus on its usage in this section. Therefore, we will benefit from C, C++, and MicroPython languages.

8.2.4.1 Usage via C Language

We can benefit from the HAL library functions to use the UART/USART module in C language after setting its properties as explained in Sect. 8.2.3. In this section, we only provide the functions related to UART mode. We explain these functions in detail next. Then, we provide usage examples of these functions in different applications.

There are two functions for blocking (polling) mode operation of the UART module. The function HAL_UART_Transmit can be used to send data. Likewise, the function HAL_UART_Receive can be used to receive data in blocking mode. We next provide the detail of these functions.

```
HAL_UART_Transmit(UART_HandleTypeDef *huart, uint8_t *pData,
    uint16_t Size, uint32_t Timeout)
/*
huart: pointer to the UART_HandleTypeDef struct
pData: pointer to data buffer
Size: size of data elements to be sent
Timeout: timeout duration
*/

HAL_UART_Receive(UART_HandleTypeDef *huart, uint8_t *pData,
    uint16_t Size, uint32_t Timeout)
/*
Size: size of data elements to be received
*/
```

There are two functions for non-blocking (interrupt) mode operation of UART module. The function `HAL_UART_Transmit_IT` can be used to send data in non-blocking mode. Likewise, the function `HAL_UART_Receive_IT` can be used to receive data in non-blocking mode.

If the UART interrupt has been enabled and data is sent or received, then the callback function `HAL_UART_TxCpltCallback` or `HAL_UART_RxCpltCallback` is called. We should define these functions within the "main.c" file in our project. To note here, these functions are generated in weak form (without any content) automatically under the file "STM32F4xx_hal_uart.c" when a project is created. This means when the reader forgets defining the actual `HAL_UART_TxCpltCallback` or `HAL_UART_RxCpltCallback` function, the code still runs. When the user defines his or her function, then these weak functions are ignored. We next provide detail of these functions.

```
HAL_UART_Transmit_IT(UART_HandleTypeDef *huart, uint8_t *pData,
    uint16_t Size)
/*
huart: pointer to the UART_HandleTypeDef struct
pData: pointer to data buffer
Size: size of data elements to be sent
*/

HAL_UART_Receive_IT(UART_HandleTypeDef *huart, uint8_t *pData,
    uint16_t Size)
/*
Size: size of data elements to be received
*/

void HAL_UART_TxCpltCallback (UART_HandleTypeDef *huart)

void HAL_UART_RxCpltCallback (UART_HandleTypeDef *huart)
```

Let's provide examples on the usage of UART/USART modules via C language. The first example, with the C code given in Listing 8.1, aims to familiarize the reader with the setup and usage of the UART/USART module. Here, a string and counter value is sent to PC from the STM32F4 microcontroller via ST-LINK virtual COM port every 2 s. The green LED connected to pin PG13 of the STM32F4 board is toggled once data is transmitted. In this example, the UART module is configured in blocking mode. On the PC side, we should use an application, such as Tera Term, to receive data sent from the STM32F4 microcontroller.

Listing 8.1 Usage of the UART/USART module in blocking operation mode, the C code

```
/* USER CODE BEGIN Includes */
#include "stdio.h"
/* USER CODE END Includes */

/* USER CODE BEGIN PV */
uint8_t serData[] = " STM32F4 UART PC Example\r\n";
uint8_t decNum[6];
uint16_t counter = 0;
/* USER CODE END PV */

/* Infinite loop */
```

```
/* USER CODE BEGIN WHILE */
while (1)
{
sprintf(decNum, "%d", counter++);
HAL_UART_Transmit(&huart1, decNum, (uint16_t)sizeof(decNum),
    1000);
HAL_UART_Transmit(&huart1, serData, (uint16_t)sizeof(serData),
    1000);
HAL_GPIO_TogglePin(GPIOG, GPIO_PIN_13);
HAL_Delay(2000);
/* USER CODE END WHILE */

/* USER CODE BEGIN 3 */
}
/* USER CODE END 3 */
```

In order to execute the code in Listing 8.1, create an empty project. Then, configure the pins PA9 and PA10 as USART1_TX and USART1_RX, respectively. Afterward, enable the asynchronous mode of the USART1 module; set the baud rate to 115200 bits/s, data word length to 8, parity to none, stop bits to 1, data direction to transmit only, and oversampling to 16. As the project is compiled and run, the string will be transmitted to the host PC via UART and the green LED will be toggled. Do not forget to use the terminal program to observe the sent string in the host PC.

The second example given in Listing 8.2 repeats the first example given in Listing 8.1 in non-blocking mode. This time when the character 'g' is received, the green LED connected to pin PG13 of the STM32F4 board is toggled. When the character 'r' is received, the red LED connected to pin PG14 of the STM32F4 board is toggled. For execution, replace the code given in Listing 8.1 with the one in Listing 8.2. Then, enable the USART1 interrupt. As the project is compiled and run, the code will wait to receive data from the UART module. To test the code, send the characters 'r' and 'g' from the host PC to the STM32F4 microcontroller. Hence, we can observe that the red and green LEDs toggle.

Listing 8.2 Usage of the UART/USART module in non-blocking operation mode, the C code

```
/* USER CODE BEGIN PV */
uint8_t serRxData;
/* USER CODE END PV */

/* USER CODE BEGIN 0 */
void HAL_UART_RxCpltCallback(UART_HandleTypeDef *huart)
{
if (serRxData == 'g')
HAL_GPIO_TogglePin(GPIOG, GPIO_PIN_13);
else if (serRxData == 'r')
HAL_GPIO_TogglePin(GPIOG, GPIO_PIN_14);

__NOP();
HAL_UART_Receive_IT(&huart1, &serRxData, 1);
}
/* USER CODE END 0 */

/* USER CODE BEGIN 2 */
HAL_UART_Receive_IT(&huart1, &serRxData, 1);
/* USER CODE END 2 */
```

The third example sends and receives 1 byte of data using loopback property of the UART module. We will perform this operation at every button press of the STM32F4 board. Here, the Rx pin of the UART5 module should be connected to the Tx pin of the same module using an external jumper wire. Hence, it can receive the sent data. The code blocks for sending and receiving data are executed in the same microcontroller in the loopback operation. Hence, the code can be easily debugged without using an additional board. The C code for this example is given in Listing 8.2. Here, the UART5 module will be configured in non-blocking mode. Then, the related interrupts are enabled. The green LED on the STM32F4 board will toggle once data is transmitted. The red LED on the STM32F4 board will toggle once data is received.

Listing 8.3 Usage of the UART/USART module in non-blocking loopback operation mode, the C code

```
/* USER CODE BEGIN PV */
uint8_t serTxData = 0;
uint8_t serRxData;
/* USER CODE END PV */

/* USER CODE BEGIN 0 */
void HAL_GPIO_EXTI_Callback(uint16_t GPIO_Pin)
{
if (GPIO_Pin == GPIO_PIN_0)
{
HAL_UART_Transmit_IT(&huart5, &serTxData, 1);
serTxData++;
}
}

void HAL_UART_TxCpltCallback(UART_HandleTypeDef *huart)
{
HAL_GPIO_TogglePin(GPIOG, GPIO_PIN_13);
}

void HAL_UART_RxCpltCallback(UART_HandleTypeDef *huart)
{
HAL_GPIO_TogglePin(GPIOG, GPIO_PIN_14);
__NOP();
HAL_UART_Receive_IT(&huart5, &serRxData, 1);
}
/* USER CODE END 0 */

/* USER CODE BEGIN 2 */
HAL_UART_Receive_IT(&huart5, &serRxData, 1);
/* USER CODE END 2 */
```

In order to execute the code in Listing 8.3, create an empty project; configure the pins PC12 and PD2 as UART5_TX and UART5_RX, respectively. Then, connect the pin PC12 to PD2 using a jumper cable. Afterward, enable asynchronous mode of the UART5 module; set the baud rate to 9600 bits/s, data word length to 8, parity to none, stop bits to 1, data direction to receive and transmit, and oversampling to 16. Do not forget to enable the UART5 module interrupt. As the project is compiled and run, the string will be transmitted and received once the onboard button (connected to pin PA0) is pressed. The reader can observe the received byte by placing a breakpoint to the __NOP line in the function HAL_UART_RxCpltCallback.

8.2.4.2 Usage via C++ Language

We can benefit from the available functions in Mbed to use the UART/USART module via C++ language. Here, the functions of BufferedSerial or UnbufferedSerial class will be of help. The function baud sets the baud rate of the UART/USART module. The function format sets the transmission format used by the serial port. The functions read and write can be used to receive and transmit data, respectively. Moreover, the function readable can be used to determine whether there is a character available to be read in blocking mode. The function set_blocking can be used to select the blocking or non-blocking mode. The function attach can be used to attach a callback function whenever a UART/USART receive interrupt is generated in non-blocking mode. To note here, the interrupt functions can only be attached with the class UnbufferedSerial. As the event in the non-blocking mode is enabled, BufferedSerial class handles the receive and transmit operation internally. We provide the detailed explanation of the mentioned functions next.

```
BufferedSerial (PinName tx, PinName rx, int baud=
    MBED_CONF_PLATFORM_DEFAULT_SERIAL_BAUD_RATE)
/*
tx: transmit pin
rx: receive pin
baud: baud rate
*/

UnbufferedSerial (PinName tx, PinName rx, int baud=
    MBED_CONF_PLATFORM_DEFAULT_SERIAL_BAUD_RATE)

void baud (int baudrate)

void format (int bits=8, Parity parity=SerialBase::None, int
    stop_bits=1)
/*
bits: number of bits in a word, it can be one of 5, 6, 7, or 8
parity: parity bit used, it can be one of SerialBase::None,
    SerialBase::Odd, SerialBase::Even, SerialBase::Forced1 or
    SerialBase::Forced0
stop_bits: number of stop bits, it can be one of 1 or 2
*/

ssize_t write (const void *buffer, size_t length)
/*
buffer: buffer to write from
length: number of bytes to write
*/

ssize_t read (void *buffer, size_t length)
/*
buffer: buffer to read into
length: number of bytes to read
*/

bool readable ()
```

```
bool writable ()

virtual int set_blocking (bool blocking)
/*
blocking true for blocking mode, false for nonblocking mode
*/

void attach (Callback< void()> func, IrqType type=RxIrq)
/*
func: pointer to a void function
type: which serial interrupt to attach the member function to, it
    can be one of RxIrq or TxIrq
*/
```

We next provide examples on the usage of the UART/USART module in Mbed. Our first example, with the C++ code given in Listing 8.4, sends data from the STM32F4 microcontroller to the host PC. To execute this code in Mbed Studio, we should copy and paste it to the "main.cpp" file in our active project. As can be seen in Listing 8.4, the baud rate is set to 115200 bits/s first. Then, the code sends a fixed string and counter value to the host PC and sleeps for 2 s in an infinite loop.

Listing 8.4 First example on the usage of the UART/USART module, the C++ code

```
#include "mbed.h"

#define WAIT_TIME_MS 2000

DigitalOut greenLED(LED1);
static BufferedSerial serialPort(USBTX, USBRX) ;

uint8_t serData[] = "STM32F4 UART PC Example\n\r";

int main(void)
{
serialPort.set_baud(115200);

while (1) {
greenLED = !greenLED;
serialPort.write(serData, sizeof(serData));
thread_sleep_for(WAIT_TIME_MS);
}
}
```

The second example, with the C++ code given in Listing 8.5, repeats the example given in Listing 8.1. This time when the character 'g' is received, the green LED connected to pin PG13 of the STM32F4 board is toggled. When the character 'r' is received, the red LED connected to pin PG14 of the STM32F4 board is toggled. To execute this code in Mbed Studio, we should copy and paste it to the "main.cpp" file in our active project.

Listing 8.5 Second example on the usage of the UART/USART module, the C++ code

```
#include "mbed.h"

DigitalOut greenLED(LED1);
DigitalOut redLED(LED2);
```

```
static UnbufferedSerial serialPort(USBTX, USBRX);
uint8_t serRxData;

void UARTRxISR()
{
serialPort.read(&serRxData, 1);
if (serRxData == 'g')
greenLED = !greenLED;
else if (serRxData == 'r')
redLED = !redLED;
}

int main()
{
serialPort.set_blocking(false);
serialPort.attach(&UARTRxISR);
__enable_irq();

while (true);
}
```

The third example, with the C++ code given in Listing 8.6, implements the loopback property of the UART module such that the transmitted data is received by the same microcontroller. Hence, the program flow can be debugged without additional hardware. To do so, connect the pin PC12 to PD2 of the STM32F4 board using a jumper cable. Then, copy and paste the code given in Listing 8.6 to the "main.cpp" file in the active project and run it. Here, UnbufferedSerial class is used to generate a UART receive interrupt. When the onboard button (connected to pin PA0) is pressed, a counter value is sent. Once this value is received, the callback function is executed.

Listing 8.6 Third example on the usage of the UART/USART module, the C++ code

```
#include "mbed.h"

DigitalOut greenLED(LED1);
DigitalOut redLED(LED2);
InterruptIn userButton(BUTTON1);

static UnbufferedSerial serialPort(PC_12, PD_2, 9600);

uint8_t serTxData = 0;
uint8_t serRxData;

void getButtons()
{
serialPort.write(&serTxData, 1);
serTxData++;
greenLED = !greenLED;
}

void UARTRxISR()
{
redLED = !redLED;
serialPort.read(&serRxData, 1);
}

int main()
{
```

```
serialPort.set_blocking(false);
serialPort.attach(&UARTRxISR);
userButton.fall(&getButtons);
__enable_irq();

while (true);
}
```

8.2.4.3 Usage via MicroPython

In MicroPython, there are two functions write(buf) and writechar(char) for UART write operations. The function write(buf) is used to send buf byte array. The function writechar(char) is used to send 1 byte data. For the UART read operation, there are four functions as read([nbytes]), readchar(), readinto(buf[nbytes]), and readline(). The function read([nbytes]) is used to read number of bytes specified by nbytes. If timeout occurs beforehand, the function returns the read bytes. If nbytes is not defined, it is used to read as many bytes as possible until timeout occurs. This function returns read bytes or None if there is no data after timeout. The function readchar() is used to read 1 byte data. It returns read byte or -1 on timeout. The function read([nbytes]) works the same as the function readinto(buf[nbytes]). Only the read data is stored directly to the buf byte array. This function returns number of bytes stored in buf or None if there is no data after timeout. The function readline() is used to read bytes until receiving a newline character. If timeout occurs before the newline character is read, all available data is returned. If there is no data after timeout, it returns None. Finally, the function any() is used to return number of bytes waiting. It returns 0 if there is no received data.

We next provide examples on the usage of the UART/USART module in MicroPython. In the first example, we send a fixed string to the PC every 2 s in an infinite loop. The green LED connected to pin PG13 is toggled to indicate data transfer is done. The MicroPython code for this example is given in Listing 8.7.

Listing 8.7 First example on the usage of the UART/USART module, the MicroPython code

```
import pyb
serData = "STM32F4 UART PC Example\r\n"

def main():
    greenLED = pyb.Pin('PG13', mode=pyb.Pin.OUT_PP)
    uart5 = pyb.UART(5)
    uart5.init(115200, timeout=100)
    while True:
        uart5.write(serData)
        greenLED.value(not greenLED.value())
        pyb.delay(2000)

main()
```

In the second example, the UART5 module is used in blocking mode to check the receiving character. The green LED connected to pin PG13 is toggled if the received character is 'g'. The red LED connected to pin PG14 is toggled if the received character is 'r'. The MicroPython code for this example is given in Listing 8.8.

Listing 8.8 Second example on the usage of the UART/USART module, the MicroPython code

```
import pyb

def main():
    uart5 = pyb.UART(5)
    uart5.init(115200, timeout=100)
    greenLED = pyb.Pin('PG13', mode=pyb.Pin.OUT_PP)
    redLED = pyb.Pin('PG14', mode=pyb.Pin.OUT_PP)
    while True:
        input = uart5.readchar()
        if input == 0x67:
            greenLED.value(not greenLED.value())
        elif input == 0x72:
            redLED.value(not redLED.value())

main()
```

In the third example, we use the loopback property of the UART module. Here, Tx and Rx pins of UART module are connected to each other. Hence, the transmitted data is received by the same microcontroller. Therefore, connect the pins PC12 and PD2 using a jumper cable. Each time onboard push button is pressed, data is transmitted and received via the UART5 module. Green and red LEDs toggle to track the transmit and receive operations, respectively. The MicroPython code for this example is given in Listing 8.9.

Listing 8.9 Third example on the usage of the UART/USART module, the MicroPython code

```
import pyb

serTxData = 0
uart5 = pyb.UART(5)
uart5.init(115200, timeout=100)
greenLED = pyb.Pin('PG13', mode=pyb.Pin.OUT_PP)
redLED = pyb.Pin('PG14', mode=pyb.Pin.OUT_PP)

def getButtons(id):
    global serTxData
    uart5.writechar(serTxData)
    serTxData = serTxData + 1
    greenLED.value(not greenLED.value())

def main():
    userButtonInt = pyb.ExtInt('PA0', pyb.ExtInt.IRQ_RISING, pyb.
        Pin.PULL_DOWN, getButtons)
    while True:
        if uart5.any():
            redLED.value(not redLED.value())
            serRxData = uart5.readchar()
            print(serRxData)

main()
```

8.3 Serial Peripheral Interface

The second digital communication type we will be introducing is the serial peripheral interface (SPI). We will next focus on its working principles. Then, we will consider the setup and usage of the SPI module specific to the STM32F4 microcontroller.

8.3.1 SPI Working Principles

SPI is generally used for high-speed serial data transfer. Hence, it is suitable for transferring data with large size. We can summarize SPI properties based on the definitions in Sect. 8.1 as follows. SPI has an 8 or 16 bit frame consisting only data. Hence, no start, stop, or parity bits are used in the frame. SPI uses NRZ data encoding. SPI can work in half or full duplex mode. Moreover, if we do not receive data from the slave in the full duplex mode (we do not use one of the MISO or MOSI pins, to be explained next), actually it works in simplex mode. SPI can work in either master or slave mode based on the application. The master device generates the clock signal. More than one slave device can be connected to the communication line. In this setting, the SPI master chip select (CS) pins should be used to select which slave to be communicated to. In this setup, one dedicated CS pin should be used for each slave device. If one slave device is used in data transfer, its CS pin can be set to ground or V_{DD} based on its being active low or high, respectively. There is no limit for data transmission speed in SPI. As in UART, baud rate directly depends on the microcontroller and its clock used in operation.

In order to explain working principles of SPI, we should take a look at the connection diagram between the transmitter and receiver given in Fig. 8.10. As can be seen in this figure, there is one master and one slave device in operation. There are four pins in operation as master in/slave out (MISO), master out/slave in (MOSI), serial clock (SCLK), and not slave select (NSS). The same named pins are connected. However, they work in reverse order as follows. When the device is in master mode, MISO works for data input, MOSI for data output, SCLK as clock output, and NSS as slave select output (in active low mode). When the device is in slave mode, MISO works for data output, MOSI for data input, SCLK as clock input, and NSS as slave select input (in active low mode).

Data transfer between the master and slave devices occurs in serial mode between MOSI and MISO pins in Fig. 8.10. The most significant bit is sent first in data transfer. Moreover, communication is initiated by the master device. When the master device transmits data to a slave device via the MOSI pin, the slave device responds via the MISO pin. This implies full duplex communication with both data out and data in, synchronized with the same clock signal (which is provided by the master device via the SCK pin).

Fig. 8.10 SPI connection
diagram between the
transmitter and receiver

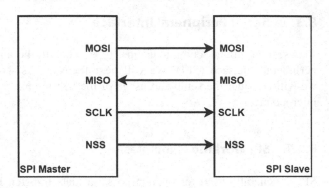

8.3.2 SPI Modules in the STM32F4 Microcontroller

The STM32F4 microcontroller has six SPI modules as SPI1, SPI2, SPI3, SPI4, SPI5, and SPI6. Each module can work either in master or slave mode. They can also work in half or full duplex modes.

SPI modules use the peripheral clock as the clock signal. The master device has its own clock prescaler so that the clock signal can be divided by 2, 4, 8, 16, 32, 64, 128, or 256. The resulting clock signal will decide on the baud rate. To be more specific, SPI1, SPI4, SPI5, and SPI6 modules use the APB2 peripheral clock (PCLK2). Hence, the maximum baud rate can be 45 Mbits/s. The SPI2 and SPI3 modules use the APB1 peripheral clock (PCLK1). This may result in the maximum baud rate to 22.5 Mbits/s.

We provide block diagram of a generic SPI module available in the STM32F4 microcontroller in Fig. 8.11. As can be seen in this figure, there is a shift register in the module. It has an essential role in data transfer. The communication and CRC controller blocks are responsible for reading and writing data coming from software to FIFO buffer modules. Moreover, they are responsible for interrupt generation. If the SPI module is selected as master, the NSS logic module becomes responsible for setting the output and selecting the slave device. If the SPI module is selected as slave, then the NSS logic module activates the read/write operation. The NSS logic module also disables the shift register if it is not selected. This is done to eliminate data confusion if more than one slave device is present in operation. The baud rate generator module is responsible for generating the SPI clock signal when the module works as master. The register block keeps the modification bits responsible for setting up all properties of the SPI module.

We can explain working principles of the SPI module as follows. Shift registers in the master and slave devices feed 1 bit to their output at each clock pulse generated by the master device. Since the master output pin is connected to the slave input pin, the bit fed from the master device shift register is received by the shift register in the slave device. Likewise, the master input is connected to the slave output. Hence, the bit fed from the shift register of the slave device is fed to the shift register in the master device. As the master device generates eight clock pulses, 1 data byte is

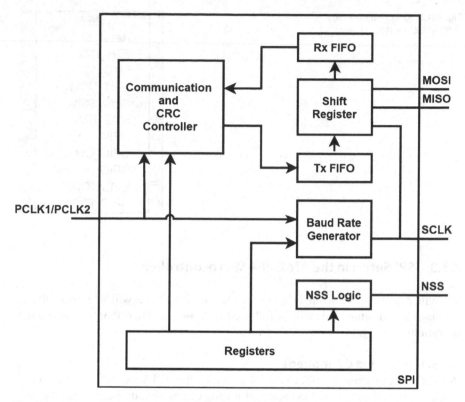

Fig. 8.11 Block diagram of the SPI module in the STM32F4 microcontroller

transferred between the master and slave devices. As the transfer operation ends, the master or slave device moves data in the shift register to the Rx FIFO buffer. Hence, data read cycle is complete. If the master or slave device will transfer new data, it is first written to the Tx FIFO buffer. As the data in the shift register is moved to the Rx FIFO buffer, this data is stored in the shift register.

SPI modules in the STM32F4 microcontroller support hardware cyclic redundancy check (CRC). This allows using an SD card via SPI. As a side note, the SD card sends 2 bytes of CRC information in its data packets. This should be decoded first. Since the SPI module in the STM32F4 microcontroller handles this operation, the SD card can be used by it.

The SPI module in the STM32F4 microcontroller also supports I^2S audio interface. To be more specific, there are I^2S modules available in the microcontroller multiplexed by SPI2 and SPI3 modules. SPI is the default mode in these modules. I^2S can be selected by software. We will introduce I^2S in Sect. 8.7.

Fig. 8.12 SPI configuration
in STM32CubeIDE

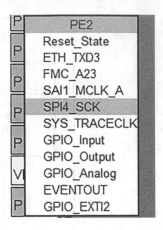

8.3.3 SPI Setup in the STM32F4 Microcontroller

We will consider SPI setup in this section. Therefore, we will start with the C
language based setup. This will be followed by C++ and MicroPython based setup
operations.

8.3.3.1 Setup via C Language

We will benefit from STM32CubeMX to setup the SPI module properties in C
language. At this stage, we assume that a project has already been created under
STM32CubeIDE as given in Sect. 3.1.3. In order to setup SPI via STM32CubeIDE,
first open the "Device Configuration Tool" window by clicking the ".ioc" file in the
project explorer window. Then, configuration can be done as shown in Fig. 8.12. We
will provide details of this operation next.

In order to use the SPI module, MISO, MOSI, and SCK (and NSS if it will
be used) pins should be configured first. To do so, click on the GPIO pin with
SPI capability in the STM32F4 microcontroller in the STM32CubeMX interface.
Afterward, select the GPIO type as "SPIx_MISO," "SPIx_MOSI," "SPIx_SCK," or
"SPIx_NSS" in the "Device Configuration Tool" window as in Fig. 8.12. To note
here, x can be 1, 2, 3, 4, 5, or 6 to denote SPI1, SPI2, SPI3, SPI4, SPI5, or SPI6,
respectively. Once the SPI pins are selected and configured, click the set SPI module
under the "Connectivity" section from the left menu of STM32CubeMX. Then,
mode configuration can be done as shown in Fig. 8.13. Here, the mode can be one of
"Full-Duplex Master," "Full-Duplex Slave," "Half-Duplex Master," "Half-Duplex
Slave," "Receive-Only Master," "Receive-Only Slave," "Transmit-Only Master," or
"Transmit-Only Slave." Also, the hardware NSS signal can be enabled by clicking
its checkbox.

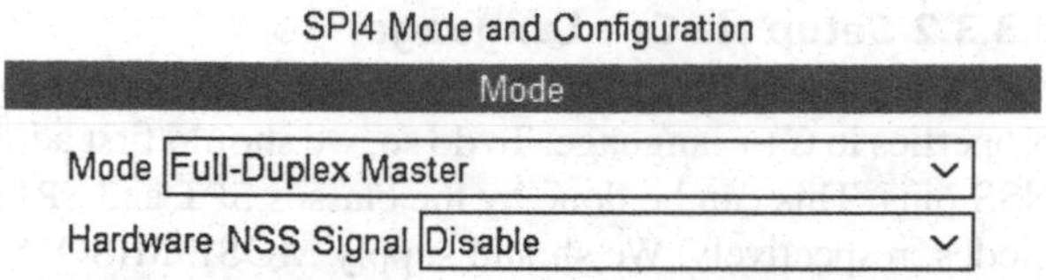

Fig. 8.13 SPI mode menu in
STM32CubeIDE

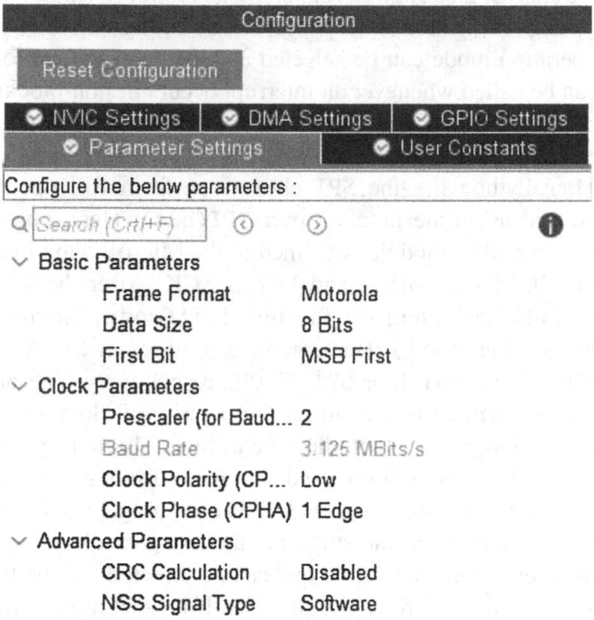

SPI4 Mode and Configuration

Mode

Mode | Full-Duplex Master   ⌄

Hardware NSS Signal | Disable   ⌄

Fig. 8.14 SPI configuration
menu in STM32CubeIDE

Configuration

Reset Configuration

⊘ NVIC Settings | ⊘ DMA Settings | ⊘ GPIO Settings
⊘ Parameter Settings | ⊘ User Constants

Configure the below parameters :

Q | Search (Crt+F)   ⊘   ⊘   ⓘ

∨ Basic Parameters
        Frame Format   Motorola
        Data Size   8 Bits
        First Bit   MSB First
∨ Clock Parameters
        Prescaler (for Baud... 2
        Baud Rate   3.125 MBits/s
        Clock Polarity (CP... Low
        Clock Phase (CPHA)   1 Edge
∨ Advanced Parameters
        CRC Calculation   Disabled
        NSS Signal Type   Software

After configuring the SPI mode, its parameters can be set from the opened
"Parameter Settings" tab in the "Configuration" menu as in Fig. 8.14. Here, frame
format can be selected as one of "Motorola" or "TI," data size can be selected as one
of "8-bit" or "16-bit," and the first bit can be selected as "MSB First" or "LSB First"
under "Basic Parameters." The "TI" format can be only selected if hardware NSS
signal is enabled. The clock prescaler for baud rate, clock polarity, and phase can be
configured under "Clock Parameters." The clock prescaler is available only for the
master device. CRC calculation can be enabled under "Advanced Parameters."

Similar to UART, SPI can operate in two operation modes as blocking (polling)
or non-blocking (interrupt). If interrupt generation is not required, then this step
can be skipped. Otherwise, the SPI interrupt should be enabled from the NVIC tab
in the configuration menu as in Fig. 8.14. The interrupt priority can be changed
under the "System Core - NVIC" section. To note here, it is suggested to lower
the SPI interrupt priority since HAL functions use the SysTick interrupt to create
precise 1 ms ticks for internal operations and the HAL_Delay function. The SysTick
interrupt has the highest priority among peripheral units. Hence, assigning a lower
priority to the SPI interrupt will not affect it. Otherwise, precision of the HAL_Delay
function may degrade.

8.3.3.2 Setup via C++ Language

We can benefit from Mbed and available functions there to adjust the SPI module properties in C++ language. To do so, we should first adjust MOSI, MISO, SCK, and NSS pins. This can be done by the classes SPI and SPISlave for master and slave modes, respectively. We should supply MOSI, MISO, SCK, and NSS pin names to the class SPI or SPISlave. Parameters for both classes, such as frequency, data size, clock polarity, and clock phase, can be configured after initialization. Moreover, the operation mode can be selected as blocking or non-blocking. A callback function can be called whenever an interrupt occurs in non-blocking mode.

8.3.3.3 Setup via MicroPython

MicroPython has the SPI class to perform SPI operations. The SPI module is created using the function pyb.SPI(bus). Here, bus represents the SPI module number. SPI5 module is defined in the MicroPython firmware with the pins PF9 as MOSI, PF8 as MISO, and PF7 as SCK. After the SPI module is created, it can be initialized using the function init(mode, baudrate, polarity, phase, bits, firstbit). Here, mode is used to set the SPI module as master or slave using SPI.MASTER or SPI.SLAVE, respectively. The baudrate is the SCK clock rate. polarity is used to set the polarity of clock line during idle state. It can be selected as 0 or 1 to set idle state as low or high, respectively. The phase is used to set clock edges to sample and shift data. If it is set to 0, data is sampled at the rising clock edge and shifted at the falling clock edge. If it is set to 1, data is sampled at the falling clock edge and shifted at the rising clock edge. The bits is used to set the number of bits transferred and can be selected as 8 or 16. The firstbit is used to set the transfer direction of the bits. If it is set as SPI.MSB, data bits are transferred starting from the most to least significant bit. If it is set as SPI.LSB, data bits are transferred starting from the least to most significant bit.

8.3.4 SPI Usage in the STM32F4 Microcontroller

As the SPI module is setup in the previous section, we will focus on its usage here. Therefore, we will benefit from C, C++, and MicroPython languages.

8.3.4.1 Usage via C Language

We can benefit from the HAL library functions to use the SPI module in C language after setting its properties in STM32CubeMX as explained in Sect. 8.3.3. There are three functions for blocking (polling) mode operation of SPI module. The function HAL_SPI_Transmit can be used to send data. The function HAL_SPI_Receive can be used to receive data in blocking mode. The function HAL_SPI_TransmitReceive combines the former two functions and transmits and receives data at the same time in blocking mode. We next provide detail of these functions.

```
HAL_SPI_Transmit(SPI_HandleTypeDef *hspi, uint8_t *pData,
    uint16_t Size, uint32_t Timeout)
/*
hspi: pointer to the  SPI_HandleTypeDef struct
pData: pointer to data buffer
Size: size of data to be sent
Timeout: timeout duration
*/

HAL_SPI_Receive(SPI_HandleTypeDef * hspi, uint8_t *pData,
    uint16_t Size, uint32_t Timeout)
/*
Size: size of data to be received
*/

HAL_SPI_TransmitReceive(SPI_HandleTypeDef *hspi, uint8_t *pTxData
    , uint8_t *pRxData, uint16_t Size, uint32_t Timeout)
/*
pTxData: pointer to the transmission data buffer
pRxData: pointer to the reception data buffer
Size: size of data elements to be sent and received
*/
```

There are three functions for non-blocking (interrupt) mode operation of SPI module. The function HAL_SPI_Transmit_IT can be used to send data. The function HAL_SPI_Receive_IT can be used to receive data in non-blocking mode. The function HAL_SPI_TransmitReceive_IT combines the former two functions and transmit and receive data at the same time in non-blocking mode.

If the SPI interrupt is enabled, and data is sent or received, then the callback function HAL_SPI_TxCpltCallback or HAL_SPI_RxCpltCallback is called. Also, the callback function HAL_SPI_TransmitReceive_IT is called when data is sent and received at the same time. We should define these functions within the "main.c" file in our project. To note here, these functions are generated in weak form (without any content) automatically under the file "STM32F4xx_hal_spi.c" when a project is created. This means when the reader forgets defining the actual function HAL_SPI_TxCpltCallback, HAL_SPI_RxCpltCallback, or HAL_SPI_TransmitReceive_IT, the code still runs. When the user defines his or her function, then these weak functions are ignored. We next provide detail of these functions.

```
HAL_SPI_Transmit_IT(SPI_HandleTypeDef * hspi, uint8_t * pData,
    uint16_t Size)
/*
hspi: pointer to the SPI_HandleTypeDef struct
pData: pointer to data buffer
Size: size of data to be sent
*/

HAL_SPI_Receive_IT(SPI_HandleTypeDef * hspi, uint8_t * pData,
    uint16_t Size)
/*
Size: size of data to be received
*/
```

```
HAL_SPI_TransmitReceive_IT(SPI_HandleTypeDef *hspi, uint8_t *
    pTxData, uint8_t *pRxData, uint16_t Size)
/*
pTxData: pointer to the transmission data buffer
pRxData: pointer to the reception data buffer
Size: size of data to be sent and received
*/

HAL_SPI_TxCpltCallback(SPI_HandleTypeDef *hspi)

HAL_SPI_RxCpltCallback(SPI_HandleTypeDef *hspi)

HAL_SPI_TxRxCpltCallback(SPI_HandleTypeDef *hspi)
```

Let's provide examples on the usage of SPI module via C language. The first example, with the C code given in Listing 8.10, aims to familiarize the reader with the setup and usage of the SPI module. Here, SPI4 and SPI5 modules will be configured as master and slave, respectively. There are two counter values in the code. The first one will be sent to the slave from master. Upon receiving this data by the slave, the green LED connected to pin PG13 of the STM32F4 board will toggle if mod 3 of the received counter value is equal to zero. Afterward, the second counter will be sent to master from slave. Upon receiving this data by the master, the red LED connected to pin PG14 of the STM32F4 board will toggle if mod 5 of the received counter value is equal to zero. This operation will repeat every 2 s.

Listing 8.10 Usage of the SPI module in blocking operation mode, the C code

```
/* USER CODE BEGIN PV */
uint8_t masterTxData = 0;
uint8_t slaveTxData = 0;
uint8_t masterRxData;
uint8_t slaveRxData;
/* USER CODE END PV */

/* Infinite loop */
/* USER CODE BEGIN WHILE */
while (1)
{
masterTxData++;
slaveTxData++;

HAL_SPI_TransmitReceive(&hspi4, &masterTxData, &masterRxData, 1,
    1000);
HAL_SPI_TransmitReceive(&hspi5, &slaveTxData, &slaveRxData, 1,
    1000);

if ((slaveRxData % 3) == 0)
HAL_GPIO_TogglePin(GPIOG, GPIO_PIN_13);

if ((masterRxData % 5) == 0)
HAL_GPIO_TogglePin(GPIOG, GPIO_PIN_14);

HAL_Delay(2000);
/* USER CODE END WHILE */

/* USER CODE BEGIN 3 */
}
```

```
/* USER CODE END 3 */
```

In order to execute the code in Listing 8.10, create an empty project; configure the pins PE2, PE4, PE5, and PE6 as SPI4_SCK, SPI4_NSS, SPI4_MISO, and SPI4_MOSI, respectively. Likewise, the pins PF6, PF7, PF8, and PF9 should be set as SPI5_NSS, SPI5_SCK, SPI5_MISO, and SPI5_MOSI, respectively. Then, connect the pin PE2 to PF7, PE4 to PF6, PE5 to PF8, and PE6 to PF9 with jumper cables. Afterward, enable the full duplex master mode of SPI4 and the full duplex slave mode of SPI5, and enable the hardware NSS pins. Set the PCLK2 clock to 1.6 MHz as shown in Chap. 6. Then, set the prescaler values of SPI4 to 16 to get 100 kHz baud rate. Finally, set frame format to Motorola, data size to 8 bits, first bit to MSB first, clock polarity to low and clock phase to 1 edge, and disable the CRC calculation. As the project is compiled and run, the counter values will be transmitted from master to slave and slave to master, and LEDs will toggle accordingly. The second example, with the C code given in Listing 8.10, repeats the first example given in Listing 8.10 in non-blocking mode. When the transmission and reception is complete, the LEDs are toggled in callback functions. The pin PA0, in which the push button on the STM32F4 board is connected to, is configured in interrupt mode. It triggers the SPI transmission when pressed. In order to execute the code in Listing 8.11, first enable the SPI interrupts in the first example. Then, replace the "main.c" file with code given in Listing 8.11. As the project is compiled and run, the counter values will be transmitted from master to slave and slave to master, and LEDs will toggle when the onboard push button is pressed.

Listing 8.11 Usage of the SPI module in non-blocking operation mode, the C code

```
/* USER CODE BEGIN PV */
uint8_t masterTxData = 0;
uint8_t slaveTxData = 0;
uint8_t masterRxData;
uint8_t slaveRxData;
/* USER CODE END PV */

/* USER CODE BEGIN 0 */
void HAL_GPIO_EXTI_Callback(uint16_t GPIO_Pin)
{
if (GPIO_Pin == GPIO_PIN_0)
{
masterTxData++;
slaveTxData++;

HAL_SPI_TransmitReceive_IT(&hspi4, &masterTxData, &masterRxData,
    1);
HAL_SPI_TransmitReceive_IT(&hspi5, &slaveTxData, &slaveRxData, 1)
    ;
__NOP();
}
}

void HAL_SPI_TxRxCpltCallback(SPI_HandleTypeDef *hspi)
{
if (hspi->Instance == SPI4)
{
if ((masterRxData % 5) == 0)
```

```
HAL_GPIO_TogglePin(GPIOG, GPIO_PIN_14);
__NOP();
}
else if (hspi->Instance == SPI5)
{
if ((slaveRxData % 3) == 0)
HAL_GPIO_TogglePin(GPIOG, GPIO_PIN_13);
__NOP();
}
}
/* USER CODE END 0 */
```

8.3.4.2 Usage via C++ Language

We can benefit from the available functions in Mbed to use the SPI module via C++ language. Here, the classes SPI and SPISlave will be of help. The function frequency sets the baud rate of the SPI module. The function format sets the data width, clock polarity, and clock phase of SPI. We next provide the detail of these functions.

```
SPI(PinName mosi, PinName miso, PinName sclk, PinName ssel=NC)
/*
mosi: SPI master out, slave in pin.
miso: SPI master in, slave out pin.
sclk: SPI clock pin.
ssel: SPI chip select pin.
*/

SPISlave(PinName mosi, PinName miso, PinName sclk, PinName ssel)

frequency(int hz=1000000)
/*
hz: clock frequency in Hz.
*/

format(int bits, int mode = 0)
/*
bits: number of bits per SPI frame, it can be any value between
    4-16
mode: clock polarity and phase mode, it can be any value between
    0-3
*/
```

The function write of the SPI class can be used to transmit and receive data in blocking mode. Moreover, the functions select and deselect can be used to select and deselect the slave device via NSS pin, respectively. We next provide detail of these functions.

```
write(int value)
/*
value: data to be sent to the SPI slave device.
*/

write(const char *tx_buffer, int tx_length, char *rx_buffer, int
    rx_length)
/*
tx_buffer: pointer to the byte-array of data to write to the
    device.
tx_length: number of bytes to write, may be zero.
```

```
rx_buffer: pointer to the byte-array of data to read from the
    device.
rx_length: number of bytes to read, may be zero.
*/
```

```
select(void)
```

```
deselect(void)
```

The functions `receive`, `read`, and `reply` of the SPISlave class can be used to receive and then transmit data in blocking mode. The SPISlave class does not support non-blocking mode in the current Mbed version. The `receive` function checks SPI if data has been received. The function `read` gets data from the receive buffer. The function `reply` fills the transmission buffer with the value to be written out as slave on the next received message from the master. We next provide detail of these functions.

```
reply(int value)
/*
value: data to be transmitted.
*/
```

```
read()
```

```
receive()
```

We next repeat the first example given in the previous section now in C++ language. We provide the corresponding code in Listing 8.12. To execute the code, first make necessary hardware connections. Then, copy and paste the code to the "main.cpp" file in the active project and run it. This code performs the same operations as its C version.

Listing 8.12 Usage of the SPI module in blocking operation mode, the C++ code

```cpp
#include "mbed.h"

#define WAIT_TIME_MS 2000

DigitalOut greenLED(LED1);
DigitalOut redLED(LED2);
SPI spiMaster(PE_6, PE_5, PE_2, PE_4);      // SPI4
SPISlave spiSlave(PF_9, PF_8, PF_7, PF_6); // SPI5

uint8_t masterTxData = 0;
uint8_t slaveTxData = 0;
uint8_t masterRxData;
uint8_t slaveRxData;

int main()
{
spiMaster.frequency(100000);
spiMaster.format(8, 0);
spiSlave.frequency(100000);
spiSlave.format(8, 0);

while (true)
```

```
{
masterTxData++;
slaveTxData++;

spiMaster.select();
spiMaster.write(masterTxData);
masterRxData = spiMaster.write(0x00);
if (spiSlave.receive())
{
slaveRxData = spiSlave.read();
spiSlave.reply(slaveTxData);
}
spiMaster.deselect();

if ((slaveRxData % 3) == 0)
greenLED = !greenLED;

if ((masterRxData % 5) == 0)
redLED = !redLED;

thread_sleep_for(WAIT_TIME_MS);
}
}
```

8.3.4.3 Usage via MicroPython

In MicroPython, the function send(send, timeout) is used to send data via SPI. Here, send is the data array which will be transferred. It can be an integer or buffer (byte) object. timeout is the timeout value in milliseconds and is set to 5000 by default. The function recv(recv, timeout) is used to receive data via SPI. Here, recv can be integer which represents the number of bytes to receive or buffer (byte) object in which the received bytes will be saved. timeout is the timeout value in milliseconds and is set to 5000 by default. Finally, the function send_recv(send, recv, timeout) is used to send and receive data via SPI at the same time. Here, send is the data array which will be transferred. It can be integer or buffer (byte) object. recv is the data array in which the received bytes will be saved. It can only be buffer (byte) object. timeout is the timeout value in milliseconds and is set to 5000 by default.

Let's provide an example on the usage of SPI module via MicroPython. In this example, we will use the loopback property of the SPI module. Here, MOSI and MISO pins of SPI module are connected to each other. Hence, the transmitted data is received by the same microcontroller. To do so, connect the pins PF8 and PF9 using a jumper cable. The MicroPython code for this example is given in Listing 8.13. As the code is executed, data is increased by one every 2 s. It is then transmitted and received by the SPI5 module. If received data is a multiple of 3, then the green LED toggles. If it is a multiple of 5, then the red LED toggles.

Listing 8.13 Example on the usage of the SPI module, the MicroPython code

```
import pyb
from struct import unpack

def main():
```

```
masterTxData = 0
masterRxData = bytearray(1)
greenLED = pyb.Pin('PG13', mode=pyb.Pin.OUT_PP)
redLED = pyb.Pin('PG14', mode=pyb.Pin.OUT_PP)
spiMaster = pyb.SPI(5)
spiMaster.init(pyb.SPI.MASTER, baudrate=2000000, polarity=0,
    phase=1)
while True:
    masterTxData += 1
    spiMaster.send_recv(masterTxData, masterRxData)
    data = unpack('<B', masterRxData)[0]
    print(data)
    if data % 3 == 0:
        greenLED.value(not greenLED.value())
    if data % 5 == 0:
        redLED.value(not redLED.value())
    pyb.delay(2000)

main()
```

8.4   Inter-integrated Circuit

The third digital communication type we will be using is the inter-integrated circuit (I^2C). We will next focus on its working principles. Then, we will consider the setup and usage of the I^2C module specific to the STM32F4 microcontroller.

8.4.1 I^2C Working Principles

I^2C is generally used for communication between a microcontroller and one or more external devices. More specifically, I^2C is suitable for setting an external device through its registers and reading data from it. It allows reaching a maximum of 128 such devices (with 7-bit address) in standard mode and 1024 devices (with 10 bit address) in extended mode. In these modes, I^2C simplifies circuit connection and address management while reaching external devices.

We can explain I^2C properties based on the definitions in Sect. 8.1 as follows. I^2C has a frame structure with start, address byte, ACK, data, and stop fields. We provide a generic frame in Fig. 8.15 to explain these concepts better. I^2C uses NRZ for data encoding. I^2C works in half duplex mode. I^2C has both the master and slave modes. The clock in synchronous communication is generated by the master device. Moreover, the communication is always started by the master device. I^2C has standard and fast data transfer modes. In the former mode, the maximum baud rate that can be reached is 100 kbits/s. In the latter mode, there are several sub options as fast-mode, fast-mode plus, and high-speed (HS) mode. Maximum baud rates that can be reached by these are 400 kbits/s, 1000 kbits/s, and 3.4 Mbits/s, respectively.

In order to explain working principles of I^2C, we should take a look at the connection diagram between the master and slave devices in Fig. 8.16. As can be seen in this figure, I^2C has two pins in operation as clock (SCL) and data (SDA). Initially, the data line is in receive (input) mode for the slave device and transmit

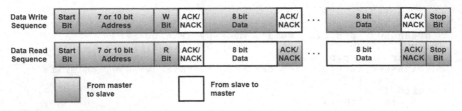

Fig. 8.15 A generic I^{2}C frame

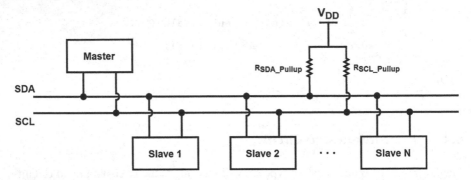

Fig. 8.16 I^{2}C connection diagram between master and slave devices

(output) mode for the master device. Then, the master device sends address of the slave device it needs to communicate with. Afterward, the master device can send or receive data to the slave. This is declared in the read/write bit sent along the address. The slave device knows what to do by checking this bit. The slave device sets its data pin accordingly, sends the acknowledge (ACK) signal, and sends data to the master if necessary.

8.4.2 I^{2}C Modules in the STM32F4 Microcontroller

The STM32F4 microcontroller has three I^{2}C modules as I2C1, I2C2, and I2C3. The first two modules are connected to the LCD and external RAM on the STM32F4 board. Therefore, they should be used with caution.

Each I^{2}C module in the STM32F4 microcontroller can work in either master or slave mode. They can work as master transmitter, master receiver, slave transmitter, and slave receiver. Each module waits in slave mode by default. Whenever a module needs to transmit data, it generates the clock signal and starts working as master. As data transmission ends, the module turns back to slave mode again. This setup allows multi-master operation such that each connected module in the line can work as both master and slave device in different times. Each slave device has a programmable address composed of 7 or 10 bits. Moreover, there may be two addresses for each device. In this setting, the module can work as if it is two different devices.

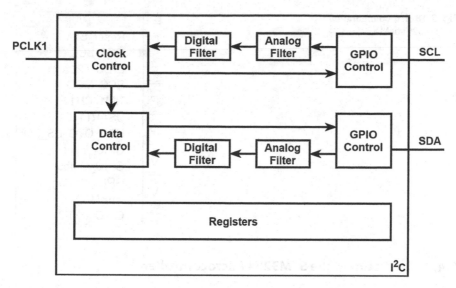

Fig. 8.17 Block diagram of an I^2C module in the STM32F4 microcontroller

I^2C modules use the APB1 peripheral clock (PCLK1). There are two speed modes as standard and fast with 100 kHz and 400 kHz clock frequency, respectively. In order to have the fast mode, the APB1 clock should be set to a value higher than 4 MHz.

We provide block diagram of a generic I^2C module available in the STM32F4 microcontroller in Fig. 8.17. As can be seen in this figure, there are two pins in the module as SCL and SDA as explained before. These pins are set in open-drain mode. Besides, they are bidirectional. Hence, each pin can work either as output or input. To perform these operations, there is a specific GPIO control block. The clock control module is responsible for clock signal generation if the I^2C module works as the master device. If the I^2C module works as a slave device, then the clock control module works to sample the incoming (data) signal. The data control block is responsible for reading and writing data. The register block keeps all the information necessary to set the I^2C module properties.

There is a maximum time limit to suppress spikes in data and clock signals I^2C standard. These time limits are 50 ηs for standard and fast plus modes. It becomes 10 ηs for the high-speed (HS) standard. There are two filter blocks to suppress these spikes in the designated time in the STM32F4 microcontroller. These are labeled as analog and digital filters in Fig. 8.17. The analog filter has fixed parameters. Hence, it can only be enabled or disabled. On the other hand, the coefficients in the digital filter can be adjusted.

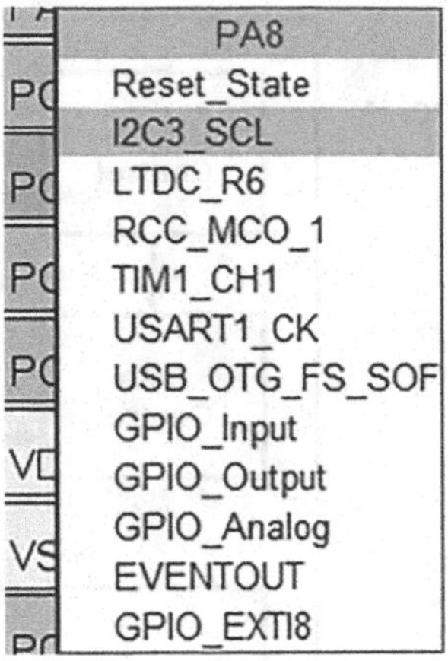

Fig. 8.18 I^{2}C configuration
in STM32CubeMX

8.4.3 I^{2}C Setup in the STM32F4 Microcontroller

We will consider I^{2}C setup in this section. Therefore, we will start with the C language based setup. This will be followed by C++ and MicroPython based setup operations.

8.4.3.1 Setup via C Language

We will benefit from STM32CubeMX to set up the I^{2}C module properties in C language. At this stage, we assume that a project has already been created under STM32CubeIDE as given in Sect. 3.1.3. In order to set up I^{2}C via STM32CubeMX, first open the "Device Configuration Tool" window by clicking the ".ioc" file in the project explorer window. Then, configuration can be done as shown in Fig. 8.18. We will provide details of this operation next.

In order to use the I^{2}C module, SCL and SDA pins should be configured first. To do so, click on the GPIO pin with I^{2}C capability in the STM32F4 microcontroller in the STM32CubeMX interface. Afterward, select the GPIO type as "I2Cx_SDA" or "I2Cx_SCLA" in the "Device Configuration Tool" window as in Fig. 8.18. To note here, x can be 1, 2, or 3 to denote I2C1, I2C2, or I2C3, respectively.

Once the I^{2}C pins are selected and configured, click the set I^{2}C module in the "Connectivity" section from the left menu of STM32CubeMX. Then, mode configuration can be done as in Fig. 8.19. Here, select the "I2C" for the I^{2}C mode. The other options "SMBus-Alert-Mode" and "SMBus-two-wire-interface" can be selected for other modes of the I^{2}C module. However, we will not deal with these modes in this chapter.

Fig. 8.19 I^{2}C mode menu in STM32CubeIDE

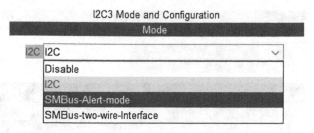

After configuring the I^{2}C mode, its settings can be configured from the opened "Parameter Settings" tab in the "Configuration" menu as in Fig. 8.20. Here, I^{2}C speed mode can be selected as one of "Standard Mode" or "Fast Mode." I^{2}C clock speed can be set in Hz under "Master Features" section. To note here, "Fast Mode" is only available if the PCLK1 is set to work above 4 MHz. The primary address length can be selected as 7 or 10 bits in the "Slave Features" section. Associated to this value, primary slave address can be set to a value between 0 and 127 or 0 and 1023, respectively. To note here, the slave address will be shifted to left by 1 bit. For example, if you set the slave address as 80 (0x50 in hexadecimal form), then the actual address will be 160 (0xA0) for the read access and 161 (0xA1) for the write access. The analog and digital filters for SDA and SC lines are configured in the "Timing Configuration" section. Here, the analog filter can be enabled or disabled. Coefficient of the digital filter can be set a value between 0 and 15, where 0 disables the filter. To note here, the analog filter is enabled and digital filter is disabled by default.

Similar to UART and SPI, I^{2}C can operate in two operation modes as blocking (polling) or non-blocking (interrupt). If interrupt generation is not required, then this step can be skipped. Otherwise, the I^{2}C interrupt should be enabled from the NVIC tab in the configuration menu as in Fig. 8.20. The interrupt priority can be changed under the "System Core - NVIC" section. To note here, it is suggested to lower the I^{2}C interrupt priority since the HAL functions use the SysTick interrupt to create precise 1 ms ticks for internal operations and the HAL_Delay function. The SysTick interrupt has the highest priority among peripheral units. Hence, assigning a lower priority to the I^{2}C interrupt will not affect it. Otherwise, precision of the HAL_Delay function may degrade.

8.4.3.2 Setup via C++ Language

We can benefit from Mbed and available functions there to adjust the I^{2}C module properties. To do so, we should first adjust SDA and SCK pins. This can be done by the classes I2C and I2CSlave for the master and slave modes, respectively. We should supply SDA and SCL pin names to these classes. Parameters of both class functions, such as frequency and slave address, can be configured after initialization. Moreover, the operation mode can be modified as blocking or non-blocking, and the callback function can be called whenever an interrupt occurs in non-blocking mode.

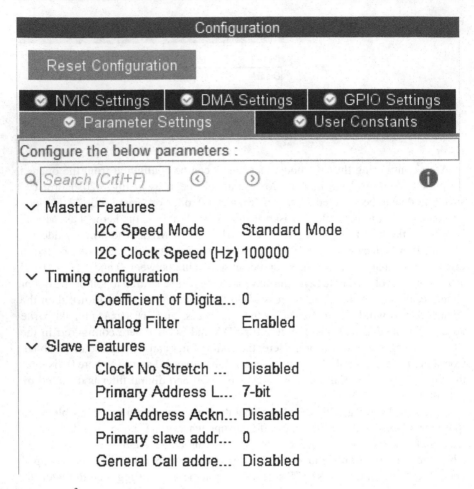

Fig. 8.20 I^2C configuration menu in STM32CubeMX

8.4.3.3 Setup via MicroPython

MicroPython has the I^2C class to perform I^2C operations. The I^2C module is created using the function pyb.I2C(bus). Here, bus represents the I^2C module number. I^2C3 module is defined in the MicroPython firmware with the pins PA8 as SCL and PC9 as SDA. After the I^2C module is created, it can be initialized using the function init(mode, addr, baudrate, gencall). Here, mode is used to set the I^2C module as master or slave using I2C.MASTER or I2C.SLAVE, respectively. The addr is the 7-bit slave address. The baudrate is the SCL clock rate. The gencall is used to enable or disable general call property. It can be enabled or disabled by setting it as True or False, respectively.

8.4.4 I²C Usage in the STM32F4 Microcontroller

As the I²C module is setup in the previous section, we will focus on its usage in this section. Therefore, we will benefit from C, C++, and MicroPython languages.

8.4.4.1 Usage via C Language

We can benefit from the HAL library functions to use the I²C module in C language after setting its properties in STM32CubeMX as explained in Sect. 8.4.3. There are seven functions for the blocking (polling) mode operation of the I²C module. The functions `HAL_I2C_Master_Transmit`, `HAL_I2C_Slave_Transmit`, and `HAL_I2C_Mem_Write` can be used to send data. Likewise, the functions `HAL_I2C_Master_Receive`, `HAL_I2C_Slave_Receive`, and `HAL_I2C_Mem_Read` can be used to receive data in blocking mode. The function `HAL_I2C_IsDeviceReady` checks whether the target device is ready for communication. We next provide detail of all these functions.

```
HAL_I2C_Master_Transmit(I2C_HandleTypeDef *hi2c, uint16_t
    DevAddress, uint8_t *pData, uint16_t Size, uint32_t Timeout)
/*
hi2c: pointer to the I2C_HandleTypeDef struct
DevAddress: target device address
pData: pointer to the data buffer
Size: size of data to be sent
Timeout: timeout duration
*/

HAL_I2C_Master_Receive(I2C_HandleTypeDef *hi2c, uint16_t
    DevAddress, uint8_t *pData, uint16_t Size, uint32_t Timeout)
/*
Size: size of data to be received
*/

HAL_I2C_Slave_Transmit(I2C_HandleTypeDef *hi2c, uint8_t *pData,
    uint16_t Size, uint32_t Timeout)
/*
Size: size of data to be sent
*/

HAL_I2C_Slave_Receive(I2C_HandleTypeDef *hi2c, uint8_t *pData,
    uint16_t Size, uint32_t Timeout)
/*
Size: size of data to be received
*/

HAL_I2C_Mem_Write(I2C_HandleTypeDef *hi2c, uint16_t DevAddress,
    uint16_t MemAddress, uint16_t MemAddSize, uint8_t *pData,
    uint16_t Size, uint32_t Timeout)
/*
MemAddress: internal memory address
MemAddSize: size of internal memory address
Size: size of data to be sent
*/

HAL_I2C_Mem_Read(I2C_HandleTypeDef * hi2c, uint16_t DevAddress,
    uint16_t MemAddress, uint16_t MemAddSize, uint8_t *pData,
    uint16_t Size, uint32_t Timeout)
/*
```

```
Size: size of data to be received
*/

HAL_I2C_IsDeviceReady(I2C_HandleTypeDef *hi2c, uint16_t
    DevAddress, uint32_t Trials, uint32_t Timeout)
/*
Trials: number of trials
*/
```

There are six functions for the non-blocking (interrupt) mode operation of the I^{2}C module. The functions HAL_I2C_Master_Transmit_IT, HAL_I2C_Slave_Transmit_IT, and HAL_I2C_Mem_Write_IT can be used to send data. Likewise, the functions HAL_I2C_Master_Receive_IT, HAL_I2C _Slave_Receive_IT, and HAL_I2C_Mem_Read_IT can be used to receive data in blocking mode.

If the I^{2}C interrupt is enabled and data is sent or received in master mode, then the callback function HAL_I2C_MasterTxCpltCallback or HAL_I2C_MasterRxCpltCallback is called. Similarly, if the I^{2}C interrupt is enabled and data is sent or received in slave mode, then the callback function HAL_I2C_SlaveTxCpltCallback or HAL_I2C_SlaveRxCpltCallback is called. Moreover, if data is sent or received in master mode using the function HAL_I2C_Mem_Write_IT or HAL_I2C_Mem_Read_IT, then the function HAL_I2C_MemTxCpltCallback or HAL_I2C_MemRxCpltCallback is called, respectively. To note here, these functions are generated in weak form (without any content) automatically in the "STM32F4xx_hal_i2c.c" file when a project is created. This means when the reader forgets defining the actual functions, the code still runs. When the user defines his or her function, then these weak functions are ignored. We next provide detail of these functions.

```
HAL_I2C_Master_Transmit_IT(I2C_HandleTypeDef *hi2c, uint16_t
    DevAddress, uint8_t *pData, uint16_t Size)
/*
hi2c: Pointer to the I2C_HandleTypeDef struct
DevAddress: target device address
pData: pointer to the data buffer
Size: size of data to be sent
*/

HAL_I2C_Master_Receive_IT(I2C_HandleTypeDef *hi2c, uint16_t
    DevAddress, uint8_t *pData, uint16_t Size)
/*
Size: size of data to be received
*/

HAL_I2C_Slave_Transmit_IT(I2C_HandleTypeDef *hi2c, uint8_t *pData
    , uint16_t Size)
/*
Size: size of data to be sent
*/

HAL_I2C_Slave_Receive_IT(I2C_HandleTypeDef *hi2c, uint8_t *pData,
    uint16_t Size)
/*
Size: size of data to be received
*/
```

```
HAL_I2C_Mem_Write_IT(I2C_HandleTypeDef *hi2c, uint16_t DevAddress
    , uint16_t MemAddress, uint16_t MemAddSize, uint8_t *pData,
    uint16_t Size)
/*
MemAddress: internal memory address
MemAddSize: size of internal memory address
Size: size of data to be sent
*/

HAL_I2C_Mem_Read_IT(I2C_HandleTypeDef * hi2c, uint16_t DevAddress
    , uint16_t MemAddress, uint16_t MemAddSize, uint8_t *pData,
    uint16_t Size)
/*
Size: size of data to be received
*/

HAL_I2C_MasterTxCpltCallback(I2C_HandleTypeDef * hi2c)

HAL_I2C_MasterRxCpltCallback(I2C_HandleTypeDef * hi2c)

HAL_I2C_SlaveTxCpltCallback(I2C_HandleTypeDef * hi2c)

HAL_I2C_SlaveRxCpltCallback(I2C_HandleTypeDef * hi2c)

HAL_I2C_MemTxCpltCallback(I2C_HandleTypeDef * hi2c)

HAL_I2C_MemRxCpltCallback(I2C_HandleTypeDef * hi2c)
```

Let's provide two examples on the usage of I^2C module via C language. Based on the property of the I^2C communication, we need two STM32F4 boards (one working as master, the other as slave) for operation. Therefore, the following examples are formed for this setup. The reader can use the same code for both master and slave devices. However, the code line #define MASTER should be uncommented for the master device. Likewise, the code line #define SLAVE should be uncommented for the slave device. Finally, we should connect the pins PB11 and PB10 of the STM32F4 board (working as master) to the pins PC9 and PA8 of the STM32F4 board (working as slave), respectively. Ground pin of both boards should be connected together.

The first example, with the C code given in Listing 8.14, aims to familiarize the reader with the setup and usage of the I^2C module. Here, I2C2 and I2C3 modules are configured as I^2C master and slave on the two boards, respectively. There are two counter values in the code. The first one will be sent to the slave from the master. Upon receiving this value by the slave, the green LED connected to pin PG13 of the STM32F4 board (working as slave) is toggled if mod 3 of the received counter value is zero. Afterward, the second counter will be sent to the master from slave. Upon receiving this value by master, the red LED connected to pin PG14 of the STM32F4 board (working as master) is toggled if mod 5 of the received counter value is zero. This operation will repeat every 2 s.

Listing 8.14 Usage of the I^2C module in blocking transmit operation mode, the C code

```
/* USER CODE BEGIN PD */
#define MASTER
```

```
//#define SLAVE
/* USER CODE END PD */

/* USER CODE BEGIN PV */
#ifdef MASTER
uint8_t masterTxData = 0;
uint8_t masterRxData;
#endif

#ifdef SLAVE
uint8_t slaveTxData = 0;
uint8_t slaveRxData;
#endif
/* USER CODE END PV */

/* Infinite loop */
/* USER CODE BEGIN WHILE */
while (1)
{
#ifdef MASTER
masterTxData++;
HAL_I2C_Master_Transmit(&hi2c2, (uint16_t)160, &masterTxData, 1,
    10000);

HAL_I2C_Master_Receive(&hi2c2, (uint16_t)160, &masterRxData, 1,
    10000);
if ((masterRxData % 5) == 0)
HAL_GPIO_TogglePin(GPIOG, GPIO_PIN_14);

HAL_Delay(2000);
#endif

#ifdef SLAVE
slaveTxData++;
HAL_I2C_Slave_Receive(&hi2c3, &slaveRxData, 1, 10000);
if ((slaveRxData % 3) == 0)
HAL_GPIO_TogglePin(GPIOG, GPIO_PIN_13);
HAL_I2C_Slave_Transmit(&hi2c3, &slaveTxData, 1, 10000);
#endif
}
/* USER CODE END WHILE */

/* USER CODE END 3 */
```

In order to execute the code in Listing 8.14, create an empty project (for each board separately); configure the PB11 and PB10 pins as I2C2_SDA and I2C2_SCL, respectively. Also, set pins PC9 and PA8 as I2C3_SDA and I2C3_SCL, respectively. Afterward, enable the I2C mode of I2C2 and I2C3 modules. Set the I^2C speed mode to standard, frequency to 100 kHz, and slave address to 80 (0x50). As the project is compiled and run, counter values will be transmitted from the master to slave and slave to master, and LEDs will toggle accordingly.

The second example, with the C code given in Listing 8.15, repeats the first example given in Listing 8.14 in non-blocking mode. When the transmission and reception is complete, the LEDs are toggled in callback functions. The pin PA0, the onboard push button on the STM32F4 board (working as master) is connected to, is configured in interrupt mode. It triggers the transmission when pressed. In order to execute the code, replace the "main.c" file with code given in Listing 8.15 for both projects (for the master and slave boards). Do not forget to enable I^2C interrupts. As the projects are compiled and run, the counter values will be transmitted from the master to slave and slave to master, and LEDs will toggle when the onboard push button on the STM32F4 board (working as master) is pressed.

Listing 8.15 Usage of the I^2C module in non-blocking operation mode, the C code

```
/* USER CODE BEGIN PD */
#define MASTER
//#define SLAVE
/* USER CODE END PD */

/* USER CODE BEGIN PV */
#ifdef MASTER
uint8_t masterTxData = 0;
uint8_t masterRxData;
#endif

#ifdef SLAVE
uint8_t slaveTxData = 0;
uint8_t slaveRxData;
#endif
/* USER CODE END PV */

/* USER CODE BEGIN 0 */
void HAL_GPIO_EXTI_Callback(uint16_t GPIO_Pin)
{
if (GPIO_Pin == GPIO_PIN_0)
{
#ifdef MASTER
masterTxData++;
HAL_I2C_Master_Transmit(&hi2c2, (uint16_t)160, &masterTxData, 1,
    1000);
HAL_I2C_Master_Receive_IT(&hi2c2, (uint16_t)160, &masterRxData,
    1);
#endif
}
}

#ifdef MASTER
void HAL_I2C_MasterRxCpltCallback(I2C_HandleTypeDef *hi2c)
{
if ((masterRxData % 5) == 0)
HAL_GPIO_TogglePin(GPIOG, GPIO_PIN_14);
}
#endif

#ifdef SLAVE
void HAL_I2C_SlaveRxCpltCallback(I2C_HandleTypeDef *hi2c)
{
if ((slaveRxData % 3) == 0)
HAL_GPIO_TogglePin(GPIOG, GPIO_PIN_13);

slaveTxData++;
HAL_I2C_Slave_Transmit(&hi2c3, &slaveTxData, 1, 1000);
HAL_Delay(100);
HAL_I2C_Slave_Receive_IT(&hi2c3, &slaveRxData, 1);
}
#endif
/* USER CODE END 0 */
```

```
/* USER CODE BEGIN 2 */
#ifdef SLAVE
HAL_I2C_Slave_Receive_IT(&hi2c3, &slaveRxData, 1);
HAL_I2C_Slave_Transmit_IT(&hi2c3, &slaveTxData, 1);
#endif
/* USER CODE END 2 */
```

8.4.4.2 Usage via C++ Language

We can benefit from the available functions under Mbed to use the I^2C module via C++ language. Here, the functions of I2C and I2CSlave classes will be of help. The function frequency of both classes can be used to configure the speed of I^2C module. We next provide detail of these functions.

```
I2C(PinName sda, PinName scl)
/*
sda: I2C data line pin
scl: I2C clock line pin
*/

I2CSlave(PinName sda, PinName scl)

frequency(int hz)
/*
hz: the bus frequency in Hertz
*/
```

The functions write and read of the I2C class can be used to transmit and receive data in blocking mode. Moreover, the functions start and stop of the I2C class can be used to send start and stop conditions to the I^2C line. We next provide detail of these functions.

```
int write(int data)
/*
data: data to write to the I2C line
*/

int write(int address, const char *data, int length, bool
    repeated = false)
/*
address: 8-bit I2C slave address [ addr | 0 ]
data: pointer to the byte-array data to send
length: number of bytes to send
repeated: repeated start if true.
*/

int read(int address, char *data, int length, bool repeated =
    false)
/*
address: 8-bit I2C slave address [ addr | 1 ]
data: pointer to the byte-array to read data into
length: number of bytes to read
*/

int read(int ack)
/*
ack: indicates if the byte is to be acknowledged (1 = acknowledge
    )
*/

start(void)
```

```
stop(void)
```

The functions `write` and `read` of the `I2CSlave` class can be used to transmit and receive data in blocking mode. The `address` function sets the slave address. We next provide detail of these functions.

```
int write(int data)
/*
data: value to write.
*/
```

```
int write(const char *data, int length)
/*
data: pointer to the buffer containing the data to be sent.
length: number of bytes to send.
*/
```

```
int read(void)
```

```
int read(char *data, int length)
/*
data: pointer to the buffer to read data into.
length: number of bytes to read.
*/
```

```
address(int address)
/*
address: the address to set for the slave (least significant bit
    is ignored).
*/
```

```
enum RxStatus{
        NoData          = 0,
        ReadAddressed  = 1,
        WriteGeneral    = 2,
        WriteAddressed = 3
    };
```

```
NoData: The slave has not been addressed.
ReadAddressed: The master has requested a read from this slave.
WriteAddressed: The master is writing to this slave.
WriteGeneral: The master is writing to all slaves.
```

We next modify the example in Listing 8.14 to work under Mbed in C++ language. The reader should follow the steps given there to form the hardware setup. Then, the C++ code given in Listing 8.16 should be used for the master and slave devices (STM32F4 boards) to execute the example.

Listing 8.16 Usage of the I²C module in blocking operation mode, the C++ code

```
#include "mbed.h"

#define MASTER
// #define SLAVE

#define WAIT_TIME_MS 2000

#ifdef MASTER
DigitalOut redLED(LED2);
I2C i2c(PB_11, PB_10); // i2c2
```

```
uint8_t masterTxData = 0;
uint8_t masterRxData = 0;
#endif

#ifdef SLAVE
DigitalOut greenLED(LED1);
I2CSlave i2c(PC_9, PA_8); // i2c3

uint8_t slaveTxData = 0;
uint8_t slaveRxData = 0;
#endif

#define SLAVEADDRESS 80

int main()
{

i2c.frequency(100000);

#ifdef SLAVE
i2c.address(SLAVEADDRESS);
#endif

while (true)
{

#ifdef MASTER
masterTxData++;
i2c.write(SLAVEADDRESS, (const char *)(&masterTxData), 1, false);
i2c.read(SLAVEADDRESS, (char *)(&masterRxData), 1);

if ((masterRxData % 5) == 0)
redLED = !redLED;

thread_sleep_for(WAIT_TIME_MS);
#endif

#ifdef SLAVE
if (i2c.receive() == i2c.NoData)
{
slaveRxData = i2c.read();
if ((slaveRxData % 3) == 0)
greenLED = !greenLED;

slaveTxData++;
i2c.write(slaveTxData);
}
#endif
}
}
```

8.4.4.3 Usage via MicroPython

In MicroPython, the function scan() is used to scan all 7-bit slave addresses from 0x01 to 0x7F. The function returns the list of addresses which responded to it. This function can only be used in master mode. The function is_ready(addr) is used to check only the slave device with address value addr instead of scanning all address values. This function can only be used in master mode. The function send(send, addr, timeout) is used to send data via I^2C. Here, send is the data array which will be transmitted. It can be an integer or buffer (byte) object. The addr is the address of the slave device and can only be used in master mode. timeout is the timeout value in milliseconds and is set to 5000 by default. The function recv(recv, addr, timeout) is used to receive data via I²C. Here, recv can be integer which represents the number of bytes to receive or buffer (byte) object in which received bytes will be saved. The addr is the address of the slave device and can only be used in master mode. timeout is the timeout value in milliseconds and is set to 5000 by default. The function mem_write(data, addr, memaddr, timeout, addr_size) is used to send data to a specific memory address of a slave device via I²C. Here, data is the data array which will be transferred. It can be an integer or buffer (byte) object. The addr is the address of the slave device. The memaddr is the memory location of the slave device. The timeout is the timeout value in milliseconds and is set to 5000 by default. The addr_size is used to select the bit size of the memaddr. It can be selected as 8 or 16. This function can only be used in master mode. The function mem_read(data, addr, memaddr, timeout, addr_size) is used to receive data from a specific memory address of a slave device via I²C. Here, data can be integer which represents the number of bytes to receive or buffer (byte) object in which the received bytes will be stored. The addr is the address of the slave device. The memaddr is the memory location of the slave device. The timeout is the timeout value in milliseconds and is set to 5000 by default. The addr_size is used to select the bit size of the memaddr. It can be selected as 8 or 16. This function can only be used in master mode.

We next provide an example on forming an I²C communication between the two STM32F4 boards, one working as master and the other as slave. We provide the codes to be used in the master and slave boards in Listings 8.17 and 8.18, respectively. The address for the slave device is set as 0xB0 Listing 8.18. The slave device reads 1 byte data from the master in 0.5 s intervals in an infinite loop. If the received data is 0x67 (the character 'g' in ASCII format), then the green LED is toggled. If the received data is 0x72 (the character 'r' in ASCII format), then the red LED is toggled. The master device sends 1 byte in an infinite loop. When the slave device receives it, the master sends the next data. Hence, the character sent from the master always switches between 'g' and 'r'.

Listing 8.17 MicroPython code for the master device in I²C communication

```
import pyb

def main():
    i2cMaster = pyb.I2C(3)
    i2cMaster.init(pyb.I2C.MASTER)
    data = b'g'
    while True:
        i2cMaster.send(data, 0xB0)
        if data == b'g':
            data = b'r'
        elif data == b'r':
            data = b'g'
        pyb.delay(500)
main()
```

Listing 8.18   MicroPython code for the slave device in I^{2}C communication

```
import pyb
from struct import unpack

def main():
    i2cSlave = pyb.I2C(3)
    i2cSlave.init(pyb.I2C.SLAVE, addr=0xB0)
    greenLED = pyb.Pin('PG13', mode=pyb.Pin.OUT_PP)
    redLED = pyb.Pin('PG14', mode=pyb.Pin.OUT_PP)
    while True:
        try:
                RxData = i2cSlave.recv(1)
        except OSError:
                pass
        data = unpack('<B', RxData)[0]
        if data[0] == 0x67:
            greenLED.value(not(greenLED.value()))
        elif data[0] == 0x72:
            redLED.value(not(redLED.value()))

main()
```

8.5   Controller Area Network

The fourth digital communication type we will be considering is the controller area network (CAN). We will next focus on its working principles. Then, we will consider the setup and usage of the CAN module specific to the STM32F4 microcontroller.

8.5.1   CAN Working Principles

CAN is the serial communication type extensively used in automotive industry. It has low error rate and does not need a main controller for operation. In CAN communication, there can be several devices connected via two cables (each having 120 Ω terminal resistance). This line is called CAN bus with properties set by the ISO 11898 standard. CAN communication can best be explained by open systems interconnection (OSI) model. OSI is formed by ISO for multilayer communication. For more information on the OSI model, please see [9]. CAN bus forms the lowest two layers of the OSI model as data link and physical layer. The data link layer is available in the microcontroller as the CAN module. The physical layer should be formed by an external transceiver which converts data to electrical CAN signal. Higher layers of OSI should be formed by the user in code.

We can explain working principles of CAN (and CAN bus) based on the definitions in Sect. 8.1. CAN has five frames called data, extended data, remote, error, and overload. Among these, data frame is the most important one. We will focus on it next. For more information on other frames, please see [5]. Data frame is formed by start of frame (SOF) bit, arbitration field, control field, data field,

CRC field, ACK, and EOF. SOF is always dominant (logic level 0). Arbitration field is composed of ID and remote transmission request (RTR) bit. ID is the 11 bit identification number of the submitted message. RTR indicates that the frame does not contain data and it will be sent as remote frame. Control field is composed of IDE, r0, and DLC. Here, IDE is the identifier extension bit. If this bit has value logic level 0, 11 bit ID is used. Otherwise, 29 bit ID is used. This indicates the extended data frame. r0 is the reserved bit set to logic level 0. DLC stands for the data length code. It is 4 bits and keeps size of the submitted data in bytes. Data field keeps the data to be submitted. It has length $N \times 8$ bits. N should be between 0 and 8. Hence, data length can be between 0 and 64 bits. If data length is zero, this indicates we have a remote frame. CRC stands for the cyclic redundancy check. The CRC field is composed of 16 bits of which 15 are for CRC and a logic level 1 bit as suffix. ACK is the acknowledge value composed of two bits. Each node sends back the ACK value to confirm data has been received. This is the case even if the node will not use data in the received frame. The ACK value will be 01 if there is no error in data reception. The ACK value will be 11 if there is error in data reception. Finally, EOF stands for the end of frame value which is composed of seven bits with logic level 1 value.

Data encoding used in CAN is NRZ with one difference. Since 0 V corresponds to logic level 1 and a voltage value higher than 1.5 V represents logic level 0 in CAN, logic levels are inverted. However, the reader should not confuse this with NRZi data encoding. Instead, we will still call data encoding in CAN as NRZ with the mentioned logic level representations. There are two separate usages of half and full duplex modes in CAN. Data transfer between the microcontroller and transceiver is in full duplex mode. This is because there is two-sided data transfer in the CAN Rx and Tx pins. On the other hand, the CAN bus itself uses half duplex mode in communication since there is one-sided data transfer in the line. The master slave option is not applicable to CAN bus. Instead, there are specific protocols called CSMA/CD for this purpose. CSMA stands for carrier-sense multiple access. CD stands for collision detection. For more information on these methods, please see [5].

The CAN bus does not carry a common clock signal. Hence, it is asynchronous. The bus speed is set to a constant value during design phase. The automotive environment has heavy noise. Moreover, data line lengths between nodes can reach 100 meters or more. Therefore, it is important for the nodes to synchronize the baud rate in communication. The first synchronization is done by the start bit at the beginning of data. This is called hard synchronization. If a synchronization shift is detected in the following operations, then synchronization is done again. For more information on this topic, please see [3].

Output voltage from the STM32F4 microcontroller is not compatible with the standard CAN bus voltage levels. Hence, there should be a transceiver connected to the microcontroller for CAN bus. In other words, the transceiver converts the signals at the CAN Tx and Rx pins of the microcontroller to CAN bus voltage levels. The CAN bus has a differential line. Let's call the lines in this setup as CAN high and CAN low. The difference of the two signals (CAN high - CAN low) will give us the actual CAN signal. Let's call this difference as CANDiff. In representing logic level 1, CAN high and low become 2.5 V. Hence, CANDiff becomes 0 V, whereas, in representing logic level 0, it is sufficient to have CANDiff to have a voltage value higher than 1.5 V. In the CAN bus standard, logic levels 0 and 1 are called dominant and recessive, respectively. For more information on them, please see [2]. Furthermore, there can be more than one device/node connected to the line. To note here, devices in the CAN bus are specifically called as nodes since they are not separate devices but parts of one system (such as an automotive).

The ISO 11898 standard specifies that the CAN bus transceiver should be able to drive a 40 m bus at 1 Mbits/s. However, this may not be satisfied by the length of used cable. Therefore, the user should take this into account while designing the CAN communication system.

We provide block diagram of a generic CAN module and CAN bus in Fig. 8.21. As can be seen in this figure, the CAN bus is composed of two wires as CAN high and low. These wires are connected by terminal resistances with 120 Ω value. There may be more than one node connected to the bus. Each node is composed of one CAN controller, CAN transceiver, and software on the microcontroller for CAN communication. CAN transceiver is connected to the CAN bus by CAN high and low pins. As a reminder, these pins form a differential pair. With this setting, the CAN transceiver converts data coming from CAN controller to CAN differential voltage. CAN controller and CAN transceiver are connected by CAN Rx and Tx pins. The CAN Rx pin transmits data coming from transceiver to data controller. The CAN Tx pin transmits data from data controller to transceiver.

The CAN module works as follows. If a node does not transmit data, it stays in recessive level. If the CAN bus is empty and no other node is transmitting data, we can send data to another node. However, the transmitter should check the bus for a possible voltage change in the line after sending each bit. There may be more than one transmission attempt at the same time. There is a specific structure for this operation. For more information on this topic, please see [5]. Finally, there is bit stuffing operation during data transfer in CAN. In this case, if five successive bits come during transmission, the following sixth bit is inverted and it is not considered in operation. The aim in this operation is to ensure that nodes can make synchronization.

Let's focus on frame handling in CAN communication. There is an inter-frame space before and after sending each frame. This ensures a time tolerance for the transmitter and receiver. The inter-frame space is formed by 11 bits composed of logic level 1 values. If a node sends a frame from the CAN bus, all nodes connected to the bus receive this frame. As a reminder, each frame has an ID section. Hence, each node checks this ID and decides on whether the frame is sent to them or not. This is done by masking and filtering operations. We can explain this operation by a pseudocode as if((ID & Mask) == (Filter & Mask)), then accept the frame; else discard the frame.

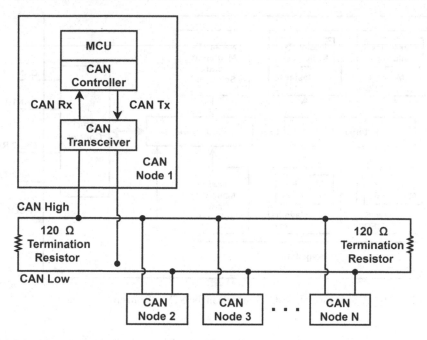

Fig. 8.21 Generic CAN modules and CAN bus connection diagram

8.5.2 CAN Modules in the STM32F4 Microcontroller

The exact name for the CAN module in the STM32F4 microcontroller is basic extended CAN, bxCAN. This module does not have PHY in itself. It only has a core section modifying data packets. The bxCAN can support only CAN protocols version 2.0 A and B. The STM32F4 microcontroller has two bxCAN modules. These are CAN1 (master bxCAN) and CAN2 (slave bxCAN). The master bxCAN can be used for managing the communication between a slave bxCAN and 512 byte SRAM. The slave bxCAN does not have direct access to SRAM. The two bxCANs share the 512 byte SRAM. To note here, the master and slave naming originates from STM. Hence, they do not correspond to the standard master and slave notations introduced in Sect. 8.1.

We provide block diagram of one bxCAN module in the STM32F4 microcontroller in Fig. 8.22. We will explain the CAN topics based on this block diagram specific to the STM32F4 microcontroller. Other microcontrollers may have different settings in their CAN modules.

In Fig. 8.22, the GPIO control block configures related GPIO pins so that the CAN module and transceiver can communicate. CAN1 and CAN2 modules have different pins used for this purpose. CAN1 and CAN2 modules share the GPIO control, CAN core, and Filters blocks. There are dedicated Rx FIFO blocks and Tx mailbox blocks for CAN1 and CAN2 modules. CAN core block is responsible

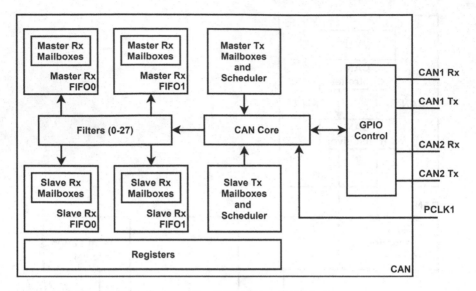

Fig. 8.22 Block diagram of the bxCAN module in the STM32F4 microcontroller

for receiving and transmitting frame, ACK and CRC control, clock generation, and synchronization. While receiving a frame from the CAN bus, the received frame is checked by the CAN core block in terms of ACK and CRC. Afterward, only the Arbitration, Control, and Data fields in the frame are sent to the Filters block for filtering and masking operation. The Filters block consists of 28 programmable filters and masks shared by CAN1 and CAN2. This block is responsible for accepting or discarding the incoming frame by checking its ID field. If the frame is accepted, then it is sent to the Rx FIFO block. There are four FIFO buffers for Rx in CAN1 and CAN2, dedicated two for each. We can direct different frames to different FIFO buffers. Each FIFO buffer consists of three Rx mailboxes used to receive incoming frame. Each mailbox is a buffer and used to handle a new incoming frame when the present one is processed. When the software reads data inside the mailbox, it is emptied automatically. If all mailboxes are full, then further incoming frames will be discarded.

While transmitting a frame to the CAN bus, the Arbitration, Control, and Data fields in the frame are obtained from Tx mailbox blocks by the CAN core block. There are six Tx mailboxes, three dedicated for CAN1 and three for CAN2 separately. There is a priority mechanism here. Hence, there is no FIFO operation. This means that the prioritized frame enters the mailbox. The ACK and CRC control bits are generated according to these fields, appended to the frame in CAN core block, and the frame is sent to the CAN bus. Afterward, the used mailbox is emptied automatically. If an error occurs during transfer, there are options such as automatic resend and discard.

The bxCAN module in the STM32F4 microcontroller has a loopback mode. This mode connects the Tx and Rx pins within the module without any need of external connection. This way, the CAN communication tests can be performed. We will use this option in the following section.

8.5.3 CAN Setup in the STM32F4 Microcontroller

We will consider CAN setup in this section. Therefore, we will start with the C language based setup. This will be followed by C++ and MicroPython language based setup operations.

8.5.3.1 Setup via C Language
We will benefit from STM32CubeMX to setup CAN module properties in C language. At this stage, we assume that a project has already been created under STM32CubeIDE as explained in Sect. 3.1.3. In order to set up CAN via STM32CubeMX, first open the "Device Configuration Tool" window by clicking the ".ioc" file in the project explorer window. Then, configuration can be done as shown in Fig. 8.23. We will provide details of this operation next.

In order to use the CAN module, Tx and Rx pins should be configured first. To do so, click on the GPIO pin with CAN capability on the STM32F4 microcontroller in STM32CubeMX interface. Afterward, select the GPIO type as "CANx_TX" or "CANx_RX" in the "Device Configuration Tool" window as in Fig. 8.23. To note here, x can be 1 or 2 to denote CAN1 or CAN2, respectively.

Once CAN pins are selected and configured, click the set CAN module under the "Connectivity" section from the left STM32CubeMX menu. Then, mode configuration can be done in this section as in Fig. 8.24. Here, the CAN module can be activated by clicking the "Activated" checkbox.

Fig. 8.23 CAN module configuration in STM32CubeMX

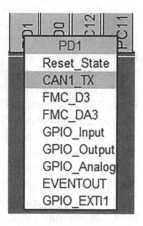

Fig. 8.24 CAN mode menu
in STM32CubeMX

Fig. 8.25 CAN
configuration menu in
STM32CubeIDE

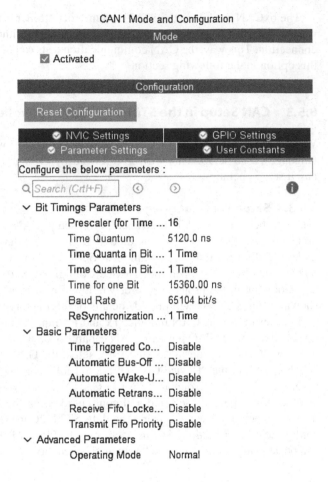

CAN1 Mode and Configuration

Mode

☑ Activated

Configuration

Reset Configuration

● NVIC Settings   ● GPIO Settings
● Parameter Settings   ● User Constants

Configure the below parameters :

🔍 Search (Crtl+F)   ⊙   ⊙   ⓘ

∨ Bit Timings Parameters
        Prescaler (for Time ... 16
        Time Quantum   5120.0 ns
        Time Quanta in Bit ... 1 Time
        Time Quanta in Bit ... 1 Time
        Time for one Bit   15360.00 ns
        Baud Rate   65104 bit/s
        ReSynchronization ...   1 Time
∨ Basic Parameters
        Time Triggered Co...   Disable
        Automatic Bus-Off ...   Disable
        Automatic Wake-U...   Disable
        Automatic Retrans...   Disable
        Receive Fifo Locke...   Disable
        Transmit Fifo Priority Disable
∨ Advanced Parameters
        Operating Mode   Normal

After activating the CAN module, its parameters can be set from the opened "Parameter Settings" tab in the "Configuration" menu as in Fig. 8.25. Here, baud rate of the CAN module can be configured using "Prescaler (for Time Quantum)," "Time Quanta in Bit Segment 1," and "Time Quanta in Bit Segment 2" options under "Bit Timing Parameters." We will use the "Automatic Retransmission" option under "Basic Parameters" and "Operating Mode" to select the CAN mode under "Advanced Parameters." The rest of the options can be left in their default settings. The "Automatic Retransmission" can be selected as "Enabled" or "Disabled." This is to set or reset resending the data packet if it fails in the first attempt. The "Operating Mode" can be selected as "Normal," "Loopback," "Silent," or "Loopback combined with Silent."

If interrupt generation is not required, then this step can be skipped. Otherwise, the CAN interrupt should be enabled from the NVIC tab in the configuration menu as in Fig. 8.25. The interrupt priority can be changed under the "System Core -

NVIC" section. To note here, it is suggested to lower the CAN interrupt priority since HAL functions use the SysTick interrupt to create precise 1 ms ticks for internal operations and the HAL_Delay function. The SysTick interrupt has the highest priority among peripheral units. Hence, assigning a lower priority to the CAN interrupt will not affect it. Otherwise, precision of the HAL_Delay function may degrade.

8.5.3.2 Setup via C++ Language
We can benefit from Mbed and available functions there to adjust the CAN module properties. To do so, we should first adjust Tx and Rx pins. This can be done by the CAN class. Hence, we should supply Tx and Rx pin names to the class. The frequency parameter can be configured after initialization. Moreover, a callback function can be called whenever an interrupt occurs.

8.5.3.3 Setup via MicroPython
MicroPython has the CAN class to handle CAN operations. The CAN module is created using the function pyb.CAN(bus). Here, bus represents the CAN module number. CAN1 and CAN2 modules are defined in the MicroPython firmware. Pins PB8 and PB9 are set as Rx and Tx for CAN1. Pins PB12 and PB13 are set as Rx and Tx for CAN2. As the CAN module is created, it can be initialized using the function init(mode, prescaler, sjw, bs1, bs2, auto_restart, baudrate). Here, the mode is used to set the CAN mode using NORMAL, LOOPBACK, SILENT, or SILENT_LOOPBACK. The prescaler, sjw, bs1, and bs2 are used together to set the baud rate. The time quanta (tq) is the basic unit of the baud rate calculation for the CAN module. It is calculated by dividing prescaler value by PCLK1 frequency. After tq is set, bittime can be calculated using the equation bittime = (sjw + bs1 + bs2)*tq. Here, sjw is the resynchronization jump width in units of the time quanta and can be selected as 1, 2, 3, or 4. The bs1 is the location of the sample point in units of the time quanta and can be selected between 1 and 1024. The bs2 is the location of the transmit point in units of the time quanta and can be selected between 1 and 16. Then, the baudrate can be calculated by 1/bittime. If baudrate is directly set using baudrate input, it overrides the prescaler, sjw, bs1, and bs2 selections. Finally, the auto_restart is used to restart the communication after bus-off state by setting it to True. If this property is not used, then it is set as False.

8.5.4 CAN Usage in the STM32F4 Microcontroller

As the CAN module is set up in the previous section, we will focus on its usage in this section. Therefore, we will benefit from C, C++, and MicroPython languages.

8.5.4.1 Usage via C Language
We can benefit from HAL library functions to use the CAN module in C language after setting its properties as explained in Sect. 8.5.3. Configuration of the CAN filter

parameter can be made by the function `HAL_CAN_ConfigFilter`. This function needs a pointer to the struct `CAN_HandleTypeDef` that contains the configuration information for the specified CAN module. The reader should define and configure this structure within code. We next provide detailed information on the configuration function and filter structure.

```
struct CAN_FilterTypeDef{
uint32_t FilterIdHigh; // specifies the filter identification
    number. MSBs for a 32-bit configuration, first one for a 16-
    bit configuration.
uint32_t FilterIdLow; // specifies the filter identification
    number. LSBs for a 32-bit configuration, second one for a 16-
    bit configuration
uint32_t FilterMaskIdHigh; // specifies the filter mask number or
    identification number, according to the mode. MSBs for a 32-
    bit configuration, first one for a 16-bit configuration.
uint32_t FilterMaskIdLow; // specifies the filter mask number or
    identification number, according to the mode. LSBs for a 32-
    bit configuration, second one for a 16-bit configuration.
uint32_t FilterFIFOAssignment; // specifies the FIFO which will
    be assigned to the filter and can be one of CAN_filter_FIFO0
    or CAN_filter_FIFO1
uint32_t FilterBank; // specifies the filter bank which will be
    initialized. For single CAN instance it can be 0-14, For dual
    CAN instances it can be 0-27
uint32_t FilterMode; // specifies the filter mode to be
    initialized and can be one of CAN_FILTERMODE_IDMASK or
    CAN_FILTERMODE_IDLIST
uint32_t FilterScale; // specifies the filter scale and can be
    one of CAN_FILTERSCALE_16BIT or CAN_FILTERSCALE_32BIT
uint32_t FilterActivation; // enable or disable the filter and
    can be one of CAN_FILTER_DISABLE or CAN_FILTER_ENABLE
uint32_t SlaveStartFilterBank; // start index of filter bank for
    the slave CAN instance for dual CAN instances
}

HAL_CAN_ConfigFilter (CAN_HandleTypeDef *hcan, CAN_FilterTypeDef
    *sFilterConfig)
/*
hcan: pointer to the CAN_HandleTypeDef struct
sFilterConfig: pointer to the CAN_FilterTypeDef struct
*/
```

Before sending or receiving data, message headers to be sent or received should be configured. The user should define and configure these headers using the structures `CAN_TxHeaderTypeDef` and `CAN_RxHeaderTypeDef` within code. Afterward, the CAN module can be started using the function `HAL_CAN_Start`. Then, a message can be added to the first free Tx mailbox. Also, the corresponding transmission request can be activated using the function `HAL_CAN_AddTxMessage`. Similarly, a CAN message can be taken from the Rx FIFO zone into the message RAM using the function `HAL_CAN_GetRxMessage`. If a transmission request is pending at the selected Tx, mailboxes can be checked using the function `HAL_CAN_IsTxMessagePending`. More information on these functions and structures are provided next.

```
struct CAN_TxHeaderTypeDef{
uint32_t StdId; // specifies the standard identifier between 0
    and 0x7FF
uint32_t ExtId; // specifies the extended identifier between 0
    and 0x1FFFFFFF
uint32_t IDE; // specifies the type of identifier for the message
    that will be transmitted and can be one of CAN_ID_STD or
    CAN_ID_EXT
uint32_t RTR; // specifies the type of frame for the message that
    will be transmitted and can be one of CAN_RTR_DATA or
    CAN_RTR_REMOTE
uint32_t DLC; // specifies the length of the frame that will be
    transmitted between 0-8
FunctionalState TransmitGlobalTime; // specifies whether the
    timestamp counter value captured on start of frame
    transmission and can be either ENABLE or DISABLE
}

struct CAN_RxHeaderTypeDef{
uint32_t StdId; // specifies the standard identifier between 0
    and 0x7FF
uint32_t ExtId; // specifies the extended identifier between 0
    and 0x1FFFFFFF
uint32_t IDE; // specifies the type of identifier for the message
    that will be transmitted and can be one of CAN_ID_STD or
    CAN_ID_EXT
uint32_t RTR; // specifies the type of frame for the message that
    will be transmitted and can be one of CAN_RTR_DATA or
    CAN_RTR_REMOTE
uint32_t DLC; // specifies the length of the frame that will be
    transmitted between 0-8
uint32_t Timestamp; // specifies the timestamp counter value
    captured on start of frame reception between 0 and 0xFFFF
uint32_t FilterMatchIndex; // specifies the index of matching
    acceptance filter element between 0 - 0xFF
}

HAL_CAN_Start(CAN_HandleTypeDef *hcan)
/*
hcan: pointer to the CAN_HandleTypeDef struct
*/

HAL_CAN_AddTxMessage (CAN_HandleTypeDef *hcan,
    CAN_TxHeaderTypeDef *pHeader, uint8_t aData, uint32_t *
    pTxMailbox)
/*
pHeader: pointer to the CAN_TxHeaderTypeDef struct.
aData: array containing the payload of the Tx frame.
pTxMailbox: pointer to a variable where the function will return
    the TxMailbox used to store the Tx message
*/

HAL_CAN_GetRxMessage(CAN_HandleTypeDef *hcan, uint32_t RxFifo,
    CAN_RxHeaderTypeDef *pHeader, uint8_t aData)
/*
RxFifo: Fifo number of the received message to be read and can be
    one of CAN_filter_FIFO0 or CAN_filter_FIFO1
pHeader: pointer to the CAN_RxHeaderTypeDef struct
aData: array where the payload of the Rx frame will be stored.
*/
```

```
HAL_CAN_IsTxMessagePending (CAN_HandleTypeDef *hcan, uint32_t
    TxMailboxes)
/*
TxMailboxes: list of Tx mailboxes to check, it can be any
    combination of CAN_TX_MAILBOX0, CAN_TX_MAILBOX1 and
    CAN_TX_MAILBOX2.
*/
```

HAL library also covers CAN end of conversion interrupt generation. Therefore, we should activate the interrupts using the function HAL_CAN_Activate Notification. Then, we should use the callback function HAL_CAN_TxMailbox0 CompleteCallback, HAL_CAN_TxMailbox1CompleteCallback, or HAL_CAN_Tx Mailbox2CompleteCallback. We can also use the functions HAL_CAN_RxFifo 0MsgPendingCallback and HAL_CAN_RxFifo1MsgPendingCallback for pending message on receive FIFO interrupts. To note here, these functions are generated in weak form (without any content) automatically in the "STM32F4xx_hal_can.c" file when a project is created. This means when the reader forgets defining the actual functions, the code still runs. When the user defines his or her function, then these weak functions are ignored. We next provide detail of these functions.

```
HAL_CAN_ActivateNotification (CAN_HandleTypeDef *hcan, uint32_t
    ActiveITs)
/*
hcan: pointer to the CAN_HandleTypeDef struct
ActiveITs: indicates which interrupts will be enabled, it can be
    any combination of CAN_Interrupts.
*/

HAL_CAN_TxMailbox0CompleteCallback (CAN_HandleTypeDef *hcan)
HAL_CAN_TxMailbox1CompleteCallback (CAN_HandleTypeDef *hcan)
HAL_CAN_TxMailbox2CompleteCallback (CAN_HandleTypeDef *hcan)
HAL_CAN_RxFifo0MsgPendingCallback (CAN_HandleTypeDef *hcan)
HAL_CAN_RxFifo1MsgPendingCallback (CAN_HandleTypeDef *hcan)
```

We next provide an example on the usage of the CAN module. The C code for this example is given in Listing 8.19. It aims to familiarize the reader with the setup and usage of the CAN module. Here, CAN1 module is configured in loopback mode. Hence, the code block for message sending and receiving parts can be executed on the same microcontroller. As a result, the code can be easily debugged. There is a counter value in the code that will be transmitted and received every 2 s. If mod 3 of the received counter value is zero, then the green LED connected to pin PG13 of the STM32F4 board is toggled. If mod 5 of the received counter value is zero, then the red LED connected to pin PG14 of the STM32F4 board is toggled.

Listing 8.19 Usage of the CAN module, the C code

```
/* USER CODE BEGIN PV */
CAN_FilterTypeDef sFilterConfig;
CAN_TxHeaderTypeDef sTxHeader;
CAN_RxHeaderTypeDef sRxHeader;
uint32_t TxMailbox;

uint8_t canTxData[8] = {0};
```

```
uint8_t canRxData[8];
/* USER CODE END PV */

/* USER CODE BEGIN 2 */
sFilterConfig.FilterBank = 0;
sFilterConfig.FilterMode = CAN_FILTERMODE_IDMASK;
sFilterConfig.FilterScale = CAN_FILTERSCALE_32BIT;
sFilterConfig.FilterIdHigh = 0x12 << 5;
sFilterConfig.FilterIdLow = 0x0000;
sFilterConfig.FilterMaskIdHigh = 0x0000;
sFilterConfig.FilterMaskIdLow = 0x0000;
sFilterConfig.FilterFIFOAssignment = CAN_RX_FIFO0;
sFilterConfig.FilterActivation = ENABLE;
sFilterConfig.SlaveStartFilterBank = 14;

HAL_CAN_ConfigFilter(&hcan1, &sFilterConfig);
HAL_CAN_Start(&hcan1);

sTxHeader.StdId = 0x12;
sTxHeader.RTR = CAN_RTR_DATA;
sTxHeader.IDE = CAN_ID_STD;
sTxHeader.DLC = 2;
sTxHeader.TransmitGlobalTime = DISABLE;
/* USER CODE END 2 */

/* Infinite loop */
/* USER CODE BEGIN WHILE */
while (1)
{
canTxData[0]++;

HAL_CAN_AddTxMessage(&hcan1, &sTxHeader, canTxData, &TxMailbox);
while (HAL_CAN_IsTxMessagePending(&hcan1, TxMailbox));
HAL_CAN_GetRxMessage(&hcan1, CAN_RX_FIFO0, &sRxHeader, canRxData)
    ;

if ((canRxData[0] % 3) == 0)
HAL_GPIO_TogglePin(GPIOG, GPIO_PIN_13);

if ((canRxData[0] % 5) == 0)
HAL_GPIO_TogglePin(GPIOG, GPIO_PIN_14);

HAL_Delay(2000);
/* USER CODE END WHILE */

/* USER CODE BEGIN 3 */
}
/* USER CODE END 3 */
```

In order to execute the code in Listing 8.19, create an empty project; configure the pins PA12 and PA11 as CAN1_TX and CAN1_RX, respectively. Then, enable the CAN1 module and set the loopback mode. Set the "Prescaler (for Time Quantum)" to 16, "Time Quanta in Bit Segment 1" to 4, and "Time Quanta in Bit Segment 2" to 2, and enable "Automatic Retransmission." As the project is compiled and run, the counter value will be transmitted and received and LEDs will toggle accordingly.

8.5.4.2 Usage via C++ Language
We can benefit from the available functions in Mbed to use the CAN module via C++ language. Here, the CAN and CANMessage classes will be of help. Under

these, the function `frequency` can be used to configure the speed of operation. The function `mode` sets mode of the CAN module. The function `filter` can be used to configure filter parameters of the CAN module. `CANMessage` creates a CAN message. The functions `write` and `read` can be used to transmit and receive CAN messages, respectively. The function `attach` can be used to attach a callback function whenever a CAN transmit or receive interrupt is generated. We next provide detail of these functions.

```
frequency(int hz)
/*
hz: the bus frequency in Hz
*/

mode (Mode mode)
/*
mode: the operation mode, it can be one of CAN::Normal, CAN::
    Silent, CAN::LocalTest, CAN::GlobalTest or CAN::SilentTest
*/

filter (unsigned int id, unsigned int mask, CANFormat format=
    CANAny, int handle=0)
/*
id: the id to filter on
mask: the mask applied to the id
format: format to filter on
handle: message filter handle
*/

CANMessage

write (CANMessage msg)
/*
msg: the CANMessage to write.
*/

read (CANMessage &msg, int handle=0)
/*
msg: the CANMessage to read to.
handle: message filter handle (0 for any message)
*/

attach (Callback< void()> func, IrqType type=RxIrq)
/*
func: pointer to a void function, or 0 to set as none
type: CAN interrupt to attach the member function
*/

  CANMessage(unsigned int _id, const unsigned char *_data,
      unsigned char _len=8, CANType _type=CANData, CANFormat
      _format=CANStandard)
/*
 _id: Message ID
 _data: Message Data
 _len: Message Data length
 _type: Type of Data, can be one of CANData or CANRemote
 _format: Data Format, can be one of CANStandard, CANExtended or
      CANAny
*/
```

We next repeat the example given in the previous section now in C++ language. We provide the corresponding code in Listing 8.20. Copy and paste the code to the "main.cpp" file in the active project under Mbed Studio and run it. This code performs the same operations as its C version.

Listing 8.20 Usage of the CAN module, the C++ code

```cpp
#include "mbed.h"

#define WAIT_TIME_MS 2000
#define filterHandle 0

DigitalOut greenLED(LED1);
DigitalOut redLED(LED2);

CAN can(PA_11, PA_12);

uint8_t canTxData = 0;

int main()
{
CANMessage canRxData;
CANMessage TxMsg(0x12, &canTxData, 1, CANData, CANStandard);

can.mode(CAN::LocalTest);
can.filter(0x12, 0xFF, CANStandard, filterHandle);

while (true)
{
if (can.write(TxMsg))
{
canTxData++;
}

if (can.read(canRxData, filterHandle))
{
if ((canRxData.data[0] % 3) == 0)
greenLED = !greenLED;
if ((canRxData.data[0] % 5) == 0)
redLED = !redLED;
}
thread_sleep_for(WAIT_TIME_MS);
}
}
```

8.5.4.3 Usage via MicroPython

In order to use the CAN module in MicroPython, CAN filters must be configured first using the function setfilter(bank, mode, fifo, params). Here, bank is the number of the filter coefficients which will be used. The mode is used to select the filter mode. It can be set as CAN.LIST16 for selecting four different 16 bit IDs, CAN.LIST32 for selecting two different 32 bit IDs, CAN.MASK16 for selecting two sets of 16 bit ID/mask pairs, or CAN.MASK32 for selecting one set of 32 bit ID/mask pair. The fifo is the number of the FIFO buffers used by the filter and can be selected as 0 or 1. Finally, the params is the array which holds the values selected according to the filter mode.

As the filter is configured, the function send(data, id, timeout) can be used to send a message to the CAN bus. Here, data is the data which will be sent. It can be an integer or buffer (byte) object. id is the id of the data which will be sent. timeout is the timeout value in milliseconds. The function recv(fifo, timeout) can be used to receive a message from the CAN bus. Here, fifo is the FIFO number which will be used. timeout is the timeout value in milliseconds and is set to 5000 by default. This function returns a four-element tuple as follows. The ID of the message is the first element. The RTR flag is the second element. The filter match index value is the third element. Finally, the array containing the received message is the fourth element.

Let's provide an example on the usage of CAN module via MicroPython. In this example, we will use the loopback property of the CAN module. Hence, we can observe the CAN module without any external hardware modification. There is a counter value in the code that will be transmitted and received every 2 s. If mod 3 of the received counter value is zero, then the green LED connected to pin PG13 of the STM32F4 board is toggled. If mod 5 of the received counter value is zero, then the red LED connected to pin PG14 of the STM32F4 board is toggled. The MicroPython code for this example is given in Listing 8.21.

Listing 8.21 Usage of the CAN module, the MicroPython code

```
import pyb
from struct import unpack

def main():
    TxMsg = 0
    greenLED = pyb.Pin('PG13', mode=pyb.Pin.OUT_PP)
    redLED = pyb.Pin('PG14', mode=pyb.Pin.OUT_PP)
    can = pyb.CAN(1)
    can.init(pyb.CAN.LOOPBACK, baudrate=100000)
    can.setfilter(0, pyb.CAN.LIST16, 0, (10, 11, 12, 13))
    while True:
        can.send(TxMsg, 12)
        TxMsg += 1
        RxMsg = can.recv(0)
        print(RxMsg[3])
        data = unpack('<B', RxMsg[3])[0]
        if data % 3 == 0:
            greenLED.value(not greenLED.value())
        if data % 5 == 0:
            redLED.value(not redLED.value())
        pyb.delay(2000)

main()
```

8.6 Universal Serial Bus

The fifth digital communication type we will be using is the universal serial bus (USB). We will next focus on its working principles. Then, we will consider the setup and usage of the USB module specific to the STM32F4 microcontroller.

8.6.1 USB Working Principles

USB is a standard aiming to connect the PC, cell phones, tablets, and peripheral devices such as mouse, keyboard, and webcam. Besides, USB also sets the standards for the cable, connector, voltage level, signal type, and protocols used in operation. USB has three major versions as USB1.x, USB2.x, and USB3.x. Each major version also has its subversions. For more information on them, please see [11].

We can explain working principles of the USB (and its line structure) based on the definitions in Sect. 8.1. Data in USB is represented in layers. The highest layer is the frame. There are transactions within a frame. These consist of packets. Each packet is composed of fields. We will explain working principles of the USB based on this layered structure in the following paragraphs.

In USB communication, one side should be master and the other slave. However, the naming for these in USB terminology is host and peripheral (device), respectively. Moreover, there is a device called hub in USB. This is used to increase the number of connected peripheral devices. Via hub, a total of 127 peripheral devices can be connected to a USB line. However, there can only be one host in operation.

A device need not to be a peripheral all the time in USB. There is a specific property developed for it called USB on the go (OTG). A good example for this operation is the cell phone. If the cell phone is connected to a PC, the PC acts as host and cell phone acts as peripheral. If a USB memory is connected to the cell phone, then the cell phone changes its mode and acts as host, and USB memory acts as peripheral. This is achieved since the cell phone supports the USB OTG standard. As a note, not all USB devices support the USB OTG property.

Data transfer speeds in USB are set by standards and these cannot be changed. Besides, there are modes in USB versions (specific to USB 2.x and USB 1.x) as low speed (LS), full speed (FS), and high speed (HS). Based on these, data transfer speeds in terms of bit rates are tabulated in Table 8.1. Since the STM32F4 microcontroller supports the USB2.0 version, we will just focus on it.

Output voltage from the microcontroller (such as STM32F4) may not be compatible with the standard USB line voltage levels. Hence, there should be a transceiver (specifically called PHY) connected to the microcontroller. In other words, the transceiver converts signals at the microcontroller output to actual USB line voltage levels. As a side note, the STM32F4 microcontroller has an embedded PHY within itself.

USB uses a differential line with line names D+ and D−. The STM32F4 microcontroller reference manuals use the DP and DM naming for these, respectively. Data encoding used in operation is NRZI. Besides encoding, D+ and D− line

Table 8.1 USB2.x and USB1.x bit rates

Mode	Acronym	Bit rate (Mbits/s)	USB version
Low speed	LS	1.5	USB 1.0
Full speed	FS	12	USB 1.1
High speed	HS	480	USB 2.0

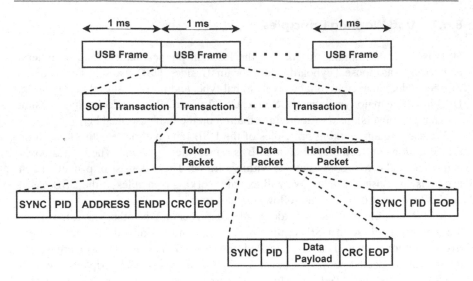

Fig. 8.26 USB frame

voltage values are used to define states. For more information on these, please see [4, 11]. The USB line works in half duplex mode.

We can explain working principles of USB based on its layered structure as in Fig. 8.26. USB frame is started at every millisecond in USB LS and FS. In USB HS, this timing becomes 125 μs. A peripheral device (such as mouse) cannot send data by itself. Instead, it can send data as a response to a frame started by the host.

As can be seen in Fig. 8.26, each frame consists of one start of frame (SOF) token and more than one transaction. Each transaction is composed of three packets as token, data, and handshake. The data packet is optional. Each packet is composed of different fields as SYNC, PID, ADDR, ENDP, PAYLOAD, CRC, and EOP. The SYNC field is 8 and 32 bits for USB LS and USB HS, respectively. This field is responsible for clock synchronization between the host and peripheral devices. The PID stands for packet ID. This field is 8 bits and defines type of the packet. ADDR is the 7-bit address field. Through it, the receiver is set if the hub option is used in communication. ENDP is the 4-bit endpoint address. There are virtual endpoints in each peripheral device defining the function of the USB device. PAYLOAD is the transmitted data which can be between 0 and 1023 bytes. Token packets have 5 bit CRC. Data packets have 16 bit CRC. EOP represents the end of packet.

Now, let's focus on packets. Token packets are available in every transaction. They are always sent by the host. This is the first packet transmitted in transaction. The aim of this packet is to start the transaction, address the device and endpoint, and define what will be sent or received in the transaction. There are four tokens as IN, OUT, SETUP, and SOF. Data packets keep the transferred data. Handshake must be the last packet in transaction. Please see [4] for more information on how a data or handshake packet is formed.

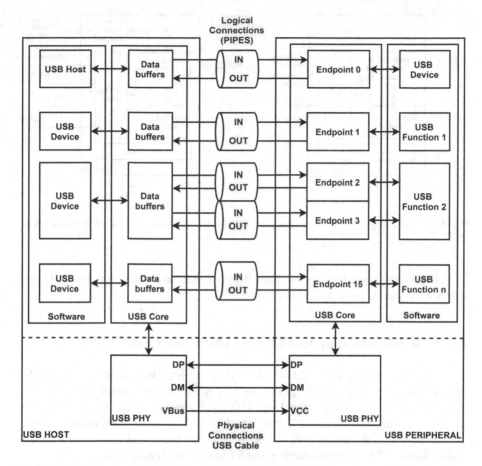

Fig. 8.27 USB connection diagram between the host and peripheral

We next provide the general USB connection diagram between a host and peripheral device in Fig. 8.27. As can be seen in this figure, the USB line is composed of two data lines and one VBUS line. Data lines can be used in both differential and single-ended form.

The host and peripheral blocks are composed of one USB core (controller) and USB PHY. The latter is used to adjust voltage level between the USB core and USB line. Each peripheral has at most 16 endpoints (16 in, 16 out). Endpoints can be grouped within themselves. These groups are called transfers. There can be control, interrupt, isochronous, and bulk transfer. The control transfer is used to understand the peripheral device type and configure it. This is done at the beginning of operation. The interrupt transfer is used to send or receive data for peripheral devices such as mouse and keyboard. In these, data has high priority but low size. As a side note, this is not an actual interrupt operation. The isochronous transfer is used for audio and video data. Bulk transfer is used in devices such as USB memory

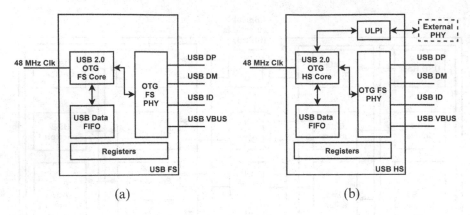

Fig. 8.28 Block diagram of USB modules in the STM32F4 microcontroller. (**a**) USB FS. (**b**) USB HS

and printers. In these, data size is large. Besides, there are descriptors in USB. We will not consider them in this book.

8.6.2   USB Modules in the STM32F4 Microcontroller

The STM32F4 microcontroller has two USB modules. One of them is the USB FS and the other is USB HS. Both modules are compatible with the USB 2.0 standard. They also support the USB OTG property. As a reminder, USB FS supports transmission speed of 12 Mbits/s. USB HS supports transmission speed of 480 Mbits/s. Moreover, both modules support low-speed (LS) 1.5 Mbits/s transmission. PHY is also available in the STM32F4 microcontroller.

The USB FS and USB HS modules have their own FIFO RAMs. These are used to receive and transmit data. The USB FS module FIFO RAM has size of 1.25 kB. The USB HS module FIFO RAM has size of 4 kB. Each FIFO RAM can hold multiple packets.

We provide block diagram of the USB FS and USB HS modules of the STM32F4 microcontroller in Fig. 8.28. In this figure, DP and DM pins are for data transfer. VBUS is for 5 V power supply. ID pin is used to check whether the device is host or peripheral in the OTG operation. PHY converts the USB line electrical signal to standard logic levels and transfers them to the USB core module. As all frame operations are done in the USB core module, data is transferred to the related FIFO RAM. USB FS and USB HS modules work at 48 MHz clock speed. The USB HS module also supports external PHY support besides the internal PHY.

The USB FS and USB HS modules in the STM32F4 microcontroller have two modes as host and peripheral. In the host mode, 5 V voltage should be powered to the connected peripheral device. This is called VBUS. The host is responsible for this operation done by the external charge pump circuit. In the host mode, USB FS

Fig. 8.29 USB configuration
in STM32CubeMX

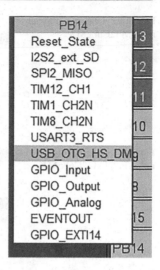

and HS support 8 and 12 connection channels. This means eight different functions can be supported or data can be received or sent in USB FS module. The number of functions becomes 12 in the USB HS module. In the peripheral mode, USB FS module supports one bidirectional (three in, three out) endpoint. This means the module can support one microphone, one headphone, and their control keys at the same time. USB HS module supports one bidirectional (five in, five out) endpoint.

8.6.3   USB Setup in the STM32F4 Microcontroller

We will consider USB setup in this section. Therefore, we will start with the C language based setup. This will be followed by MicroPython based setup operations.

8.6.3.1 Setup via C Language

We will benefit from STM32CubeMX to setup the USB HS and USB FS module properties in C language. At this stage, we assume that a project has already been created under STM32CubeIDE as given in Sect. 3.1.3. In order to setup USB via STM32CubeMX, first open the "Device Configuration Tool" window by clicking the ".ioc" file in the project explorer window. Then, configuration can be done in this section as shown in Fig. 8.29. We will provide details of this operation next.

In order to use the USB HS or USB FS module, DM and DP pins should be configured first. To do so, click on the GPIO pin with USB capability of the STM32F4 microcontroller in the STM32CubeMX interface. Afterward, select the GPIO type as "USB_OTG_FS_DM," "USB_OTG_FS_DP," "USB_OTG_HS_DM," or "USB_OTG_HS_DP," in the "Device Configuration Tool" window as in Fig. 8.29.

Once the DP and DM pins are selected, click the appropriate option ("USB_OTG_FS" or "USB_OTG_HS") under the "Connectivity" section from the left STM32CubeMX menu. Then, mode configuration can be done in this

Fig. 8.30 USB mode menu
in STM32CubeMX

Fig. 8.31 USB configuration
menu in STM32CubeMX

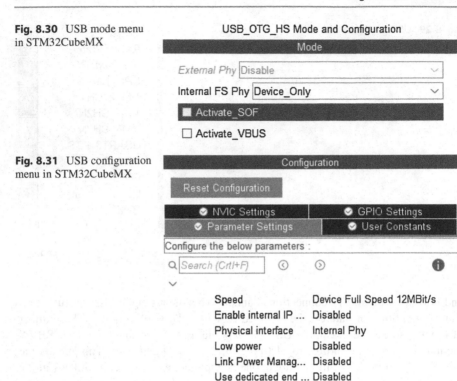

section as shown in Fig. 8.30. Here, select the mode as "Device only." The other modes can be "Host only" or "OTG." We will not consider these modes in this chapter. To note here, USB_OTG_FS uses internal PHY and USB_OTG_HS can use internal or external PHY. Hence, we will only use the internal PHY in this chapter.

After configuring the USB module in device mode, its parameters can be set from the opened "Parameter Settings" tab in the "Configuration" menu as in Fig. 8.31. At this stage, we will use the default settings which will be sufficient for us.

We should also set the clock properties for the USB to work. To do so, the "48MHz clocks" clock output should be active. This clock is fed by the main PLL. Hence, the HSE oscillator should also be activated. To do so, select the "Crystal/Ceramic Resonator" as high-speed oscillator (HSE) under the STM32CubeMX RCC window. Afterward, the main PLL input should be selected as HSE in the clock configuration window. Related to all these settings, we should also set the PCLK to 168 MHz.

To use the USB module, we can benefit from ST USB middleware [8]. Therefore, click on to "USB_DEVICE" under the "Middleware" section from the left STM32CubeMX menu. Then, mode configuration of USB middleware can be done

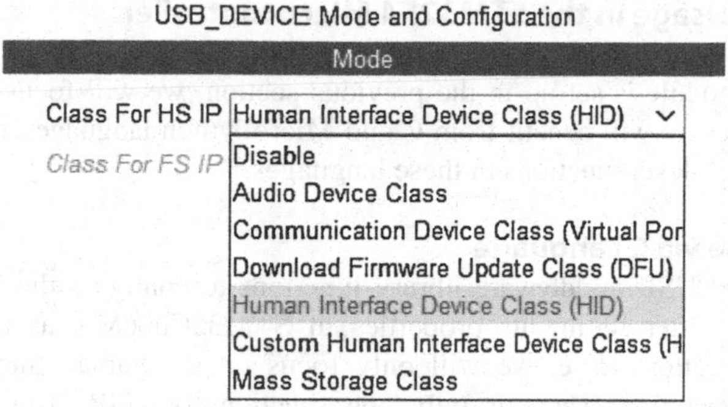

USB_DEVICE Mode and Configuration

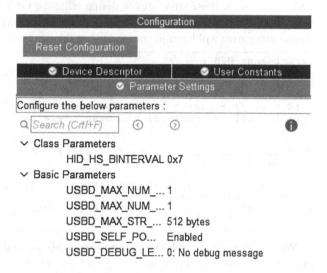

Fig. 8.32 USB middleware mode menu in STM32CubeMX

Fig. 8.33 USB middleware
configuration menu in
STM32CubeMX

in this section as shown in Fig. 8.32. Here, select "Human Interface Device Class (HID)." The other modes can be "Audio Device Class," "Communication Device Class," "Download Firmware Device Class," "Custom Human Interface Device Class," or "Mass Storage Class." We will not use these modes in this chapter.

After configuring the USB middleware in "Human Interface Device Class (HID)," its settings can be configured from the opened "Parameter Settings" tab in the "Configuration" menu as in Fig. 8.33. At this stage, we will use the default settings which will be sufficient for us.

8.6.3.2 Setup via MicroPython
MicroPython has the USB_HID class to use the USB human interface device property. Before using the USB_HID module in the code, go to the "boot.py" file and uncomment the line pyb.usb_mode('VCP+HID'). Then, the USB_HID module can be created and initialized using the function pyb.USB_HID().

8.6.4   USB Usage in the STM32F4 Microcontroller

As the USB module is set up in the previous section, we will focus on its usage next. Therefore, we will benefit from C and MicroPython languages. To note here, we will use high-level functions in these languages.

8.6.4.1 Usage via C Language

We can use the USB middleware library functions to configure the USB module in C language after setting its properties in STM32CubeMX as introduced in the previous section. Here, we will only focus on the human interface device (HID) class operation. There is only one function for USB HID operation as USBD_HID_SendReport. This function can be used to send HID data to host using USB. Also, we will use structures to define what the USB HID device is. Here, the struct mouseHID_t is used to define the USB mouse features. The button press or mouse movement will be implemented using this structure.

```
struct mouseHID_t {
uint8_t buttons; //  bit2: middle button, bit1: right button,
    bit0: left button
int8_t x; // cursor movement in x axis, relative to last position
int8_t y; // cursor movement in y axis, relative to last position
int8_t wheel; // vertical movement, relative to last position
};

USBD_HID_SendReport(USBD_HandleTypeDef *pdev, uint8_t *report,
    uint16_t len)
/*
pdev: pointer to the USBD_HandleTypeDef struct
report: pointer to report
len: length of report elements to be sent
*/
```

We next provide examples on the usage of USB module via C language. The first example, with the code given in Listing 8.22, aims to familiarize the reader with the setup and usage of the USB module. Here, the USB HS module will be configured as USB device. To do so, the USB_Device middleware will be used to implement a USB mouse. The onboard button connected to pin PA0 will be configured in interrupt mode. It will trigger the transmission when pressed. Also, digital value of this button will be sent to host PC as left mouse button click. Hence, the user can use the onboard button as the left mouse button.

Listing 8.22 Usage of the USB module, first example, the C code

```
/* USER CODE BEGIN Includes */
#include "usbd_hid.h"
/* USER CODE END Includes */

/* USER CODE BEGIN PV */
// HID Mouse
struct mouseHID_t
{
uint8_t buttons;
int8_t x;
```

```
int8_t y;
int8_t wheel;
};
struct mouseHID_t mouseHID = {0, 0, 0, 0};

extern USBD_HandleTypeDef hUsbDeviceHS;
/* USER CODE END PV */

/* USER CODE BEGIN 0 */
void HAL_GPIO_EXTI_Callback(uint16_t GPIO_Pin)
{
if (GPIO_Pin == GPIO_PIN_0)
{
if (HAL_GPIO_ReadPin(GPIOA, GPIO_PIN_0) == GPIO_PIN_SET)
mouseHID.buttons = 1;
else
mouseHID.buttons = 0;
USBD_HID_SendReport(&hUsbDeviceHS, (uint8_t *)&mouseHID, (
    uint16_t)sizeof(struct mouseHID_t));
__NOP();
}
}
/* USER CODE END 0 */
```

In order to run the code in Listing 8.22, create an empty project; configure the pins PB14 and PB15 as USB_OTG_HS_DM and USB_OTG_HS_DP, respectively. Set the internal USB FS PHY of USB_OTG_HS as "Device Only." Then, set the "Human Interface Device Class (HID)" of class for USB HS of USB_DEVICE middleware. Enable the crystal/ceramic resonator for HSE and set 48 MHz clock output. Configure the PA0 pin in external interrupt mode with falling/rising trigger detection. As the project is compiled and run, the onboard button can be used as left mouse button. The user can test the code by moving the mouse over a file on the PC desktop and then pressing the onboard button.

The second example, with the C code given in Listing 8.23, repeats the first example given in Listing 8.22. Here, mouse movements will be implemented in addition to the mouse button press action. Replace the code given in Listing 8.22 with the one in Listing 8.23. As the project is compiled and run, the onboard button can be used as left mouse button. The user will notice that the mouse will move 20 pixels in horizontal and vertical directions every second.

Listing 8.23 Usage of the USB module, second example, the C code

```
/* USER CODE BEGIN Includes */
#include "usbd_hid.h"
/* USER CODE END Includes */

/* USER CODE BEGIN PV */
// HID Mouse
struct mouseHID_t
{
uint8_t buttons;
int8_t x;
int8_t y;
int8_t wheel;
};
struct mouseHID_t mouseHID = {0, 0, 0, 0};
```

```
extern USBD_HandleTypeDef hUsbDeviceHS;
/* USER CODE END PV */

/* USER CODE BEGIN 0 */
void HAL_GPIO_EXTI_Callback(uint16_t GPIO_Pin)
{
if (GPIO_Pin == GPIO_PIN_0)
{
if (HAL_GPIO_ReadPin(GPIOA, GPIO_PIN_0) == GPIO_PIN_SET)
mouseHID.buttons = 1;
else
mouseHID.buttons = 0;
USBD_HID_SendReport(&hUsbDeviceHS, (uint8_t *)&mouseHID, (
    uint16_t)sizeof(struct mouseHID_t));
__NOP();
}
}
/* USER CODE END 0 */

/* Infinite loop */
/* USER CODE BEGIN WHILE */
while (1)
{
mouseHID.x = 20;
mouseHID.y = 20;
USBD_HID_SendReport(&hUsbDeviceHS, (uint8_t *)&mouseHID, (
    uint16_t)sizeof(struct mouseHID_t));
HAL_Delay(1000);
/* USER CODE END WHILE */

/* USER CODE BEGIN 3 */
}
/* USER CODE END 3 */
```

8.6.4.2 Usage via MicroPython

In MicroPython, the function send(data) can be used to send data over USB
HID interface. Here, data represents the data which will be sent over USB HID
interface. It can be a tuple/list of integers or a byte array. The function recv(data,
timeout) can be used to receive data over the USB HID interface. Here, data can
be set as number of data bytes which will be received, or a byte array in which
received bytes will be saved. The timeout is the timeout value in milliseconds and
is set to 5000 by default.

Let's provide an example on the usage of USB HID module via MicroPython. In
this example, the onboard button is used as a mouse left click. Pressing and releasing
the mouse left button is controlled with a flag inside the associated callback function.
The MicroPython code for this example is given in Listing 8.24.

Listing 8.24 Usage of the USB module, the MicroPython code

```
import pyb
flag = 0
userButton = pyb.Pin('PA0', mode=pyb.Pin.IN, pull=pyb.Pin.
    PULL_DOWN)

def get_button(id):
    global flag
    if userButton.value() == 1:
        flag = 1
    else:
        flag = 2

def main():
    hid = pyb.USB_HID()
    userButtonInt = pyb.ExtInt('PA0', pyb.ExtInt.
        IRQ_RISING_FALLING, pyb.Pin.PULL_DOWN, get_button)
    while True:
        if flag == 1:
            # (button status, x-direction, y-direction, scroll)
            hid.send((1, 0, 0, 0))
        elif flag == 2:
            # (button status, x-direction, y-direction, scroll)
            hid.send((0, 0, 0, 0))

main()
```

8.7 Other Digital Communication Types

There are other important but less frequently used digital communication types available on the STM32F4 microcontroller. We will provide brief information on them in this section.

8.7.1 SD Bus Interface

SD bus is used to communicate with the SD card or other devices attached to the SD card slot. SD bus interface is the communication type used for this purpose. Secure digital input and output (SDIO) is the peripheral unit in the STM32F4 microcontroller performing SD bus interface operations.

Recently SD cards have been used in embedded systems. All SD card systems support both SPI and SD bus interface. The connected device decides on which one to use. We explored SPI communication in Sect. 8.3. Here, we will only focus on SD bus interface. SD bus interface allows communicating with SD card families MMC, RS-MMC, MMCplus, MMCmobile, SecureMMC, SD, miniSD, and microSD. It also allows communicating with a camera or microphone having an SD card socket with extended specifications [1].

We can explain working principles of the SD bus interface based on the definitions in Sect. 8.1 as follows. Data types and encoding options are similar to SPI. Hence, NRZ is used in operation. Voltage level used in operation is 3.3 V. Some

Table 8.2 SD bus interface communication speeds

Card type	Max bit Clock (MHz)	Max transfer rate (Mbits/s) 1-bit wide	4-bit wide	8-bit wide
MMC	20	20	–	–
RS-MMC	20	20	–	–
MMCplus	52	52	208	832
MMCmobile	52	52	208	832
SecureMMC	20	20	–	–
SD	208	208	832	–
MiniSD	208	208	832	–
MicroSD	208	208	832	–
SDIO	50	50	200	–

cards can switch to 1.8 V after start-up. This is done to save power. SD bus interface can use one cable for serial communication. It can also communicate in parallel form with four or eight cables. Based on these, the communication is called 1-bit, 4-bit, or 8-bit wide. Normally, all SD cards support 1-bit-wide communication. If there is extra support for 4- or 8-bit-wide communication, then the SD card can switch to these modes.

SD bus interface has two different pin groups as command and data. The master device sends commands, such as read/write, via one (command) pin to the slave device. There are one, four, and eight pins for 1-, 4-, and 8-bit-wide data communication, respectively. The master device only sends data to be read or written to. Therefore, command and data pins work in half duplex mode. SD bus interface supports master and slave communication. However, they are called host and card, respectively. Communication speed in SD bus interface depends on the used SD card and number of data cables. We tabulated them in Table 8.2. As can be seen in this table, the maximum transfer rate can be 832 Mbits/s.

Frames in SD bus interface are constructed in two different forms. The first one handles the commands sent through its pin. The second one is the data sent through data pin or pins. The host device sends the command frame with 48 bits. Fields forming this frame are start bit, transmission bit, command index, argument, CRC, and end bit. The start bit is always set to logic level 0. Likewise, the transmission bit is always set to logic level 1. The command index is composed of 6 bits keeping commands. Argument is composed of 32 bits keeping the argument. CRC is composed of 7 bits. The end bit is always set to logic level 1. The SD card responds to a command with frames having different structures. The simplest one is the same as host frame. For more information on this topic, please see [1].

The host device sends the data frame in a structure as follows. Fields in the frame are start bit, data, CRC, and end. The start bit always has logic level 0 value. Data size is fixed to 512 bytes for some commands. For others, its size can change. However, the maximum size cannot exceed 512 bytes. CRC is composed of 16 bits. The end bit has always logic level 1 value.

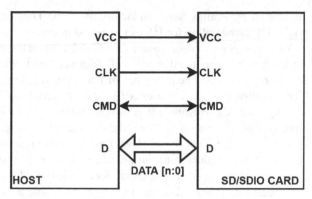

Fig. 8.34 SD bus connection diagram

The connection diagram for the SD bus is as in Fig. 8.34. As can be seen in this figure, there is one clock line called CLK. The clock is provided by the host device. There are also one command (CMD) pin and n+1 data pins. Here, n can be 7, 3, or 0 based on the communication type. The host also supports power to the card.

Data transfer in the SD bus interface can be summarized as follows. The host first sends the command to the SD card such as read, write, or erase. The SD card responds to this. When the 1-bit-wide data line is used, data is transferred synchronously with the clock as in SPI communication. When 4-bit-wide data line is used, there are two methods. In the first method, most significant 4 bits of 1 byte data are sent in the first clock pulse; the remaining least significant 4 bits of data are sent in the second clock pulse. Hence, 1 byte of data is transferred after two clock pulses. In the second method, each cable works independently. Hence, 4 bytes of data is taken. Most significant (seventh) bits of all 4 bytes are sent from four cables separately at the first clock pulse. The next sixth bits of all 4 bytes are sent from four cables separately at the second clock pulse. This operation repeats till the least significant (zeroth) bits of the 4 bytes are sent. As a summary, 4 bytes of data is transmitted after eight clock pulses. The same method applies to 8-bit-wide data line. The only difference will be the number of used parallel lines, which is eight in this case. Hence, 8 bytes of data is transmitted after eight clock pulses.

The STM32F4 microcontroller has one SDIO module. Through it, we can use SD cards supporting protocols MultiMediaCard system specification version 4.2, SD memory card specifications version 2.0, SD I/O card specification version 2.0, and CE-ATA digital protocol Rev1.1. The STM32F4 microcontroller supports 48 MHz clock speed for these operations. The module also supports 1-, 4-, and 8-bit-wide data transfer.

8.7.2   Inter-IC Sound

Inter-IC sound (I^2S) is extensively used for audio data transfer between ADC/DAC, codec integrated circuits (IC), and the microcontroller or DSP chips. It has been introduced by Philips Semiconductors in 1986. Then, it is revised in 1996. The original documentation for I^2S can be found in [6].

We can explain working principles of I^2S based on the definitions in Sect. 8.1. We should first mention that this protocol is derived from SPI. Hence, they have several common properties. The difference of I^2S from SPI is that, in the former, the clock signal is always active. I^2S uses NRZ for data encoding. I^2S can work in full duplex or simplex mode. However, it is used in simplex mode most of the times. In I^2S, there is master slave setup. Here, the module generating the clock signal is called master. The other IC becomes slave. Besides, there is also transmitter, receiver, and controller. The module transmitting data is called transmitter. The module receiving data is called the receiver. The controller decides on who will transmit data and who will receive it. This module need not to be available all the times. Based on these, there may be different combinations such as master transmitter, master receiver, or master controller.

There is no maximum bit rate value for operation. However, the bit rate is most of the times related to the sampling frequency of audio signals which is 44.1, 48, 96, or 192 kHz. The bit length for these operations can be 16, 24, or 32 bits for mono audio signals. For stereo audio signals, these values are doubled. Based on these, the stereo data will have clock speed of $2 \times 48000 \times 64 = 6.144$ MHz if the used sampling frequency is 48 kHz and the bit length is 64 bits.

Data in I^2S is sampled at rising edge of the clock. There is also an accompanying word select (WS) signal which selects the stereo channel to be transmitted. According to the initial standards, the left channel (channel 1) of the stereo signal is transmitted when WS has logic level 0 value. The right channel (channel 2) of the stereo signal is transmitted when WS has logic level 1 value. The WS signal can have period of 32 or 64 clock pulses. Besides, there is no start, stop, ACK, or other values in the frame. Only data is transmitted.

We provide general connection diagram for three different I^2S scenarios in Fig. 8.35. As can be seen in this figure, there are four pins in each module as MCK, CLK, WS, and SD. These stand for master clock, clock, word select, and serial data respectively. The master device generates the MCK for the slave to work and clock signal for data to be transmitted. All slave devices may not need a MCK signal. Besides, the transmitter module generates the SD and WS signals and sends them to the receiver.

The STM32F4 microcontroller has two I^2S modules as I2S2 and I2S3. These are multiplexed by SPI modules. We refer the reader to our book on digital signal processing on the usage of these modules in a similar microcontroller [10].

8.8   Application: Digital Communication for the Robot Vacuum Cleaner

We add remote control ability to our robot vacuum cleaner in this chapter. Therefore, we first use a stand-alone board as remote controller with Bluetooth connection. As the second option, we develop a remote controller application to run on

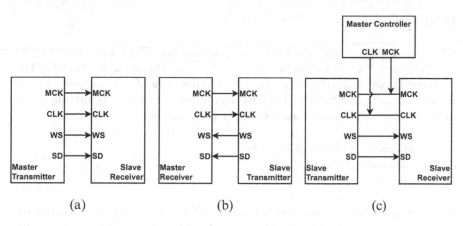

Fig. 8.35 I^{2}S connection diagram for three different settings. (**a**) Master transmitter. (**b**) Master receiver. (**c**) Master controller

Android mobile phones. We provided the equipment list, circuit diagram, detailed explanation of design specifications, and peripheral unit settings in a separate document in the folder "Chapter8\EoC_applications" of the accompanying supplementary material. In the same folder, we also provided the complete C, C++, and MicroPython codes for the application.

8.9 Summary of the Chapter

Embedded systems have evolved such that Internet of things (IoT) has become one of the hot topics in today's world. Digital communication is the backbone of IoT. Hence, we considered several modules to perform digital communication in embedded systems. To be more specific, we introduced UART, SPI, I^{2}C, CAN, USB, SDIO, and I^{2}S modules available in the STM32F4 microcontroller. Although we introduced them for the STM32F4 microcontroller, we first explained the general definitions in digital communication. Then, we introduced each communication type starting from a general perspective. Then, we provided its setup and usage for the STM32F4 microcontroller. While doing so, we benefit from the C, C++, and MicroPython languages. Hence, the reader can observe how a digital communication module works in real life. Related to this, we expanded our previously introduced real-world example by adding digital communication property to it as the end of chapter application. To sum up, one of the building blocks of IoT is digital communication. The concepts introduced in this chapter offer valuable information to understand it. Therefore, the reader should master them to get familiar with IoT.

Problems

8.1. Form a project with the following specifications. There are two STM32F4 boards at hand. We will use the button on the first board to toggle the green LED on the second board. Use, UART, SPI, and I²C modules to transfer the toggle command from one board to other. The STM32F4 microcontroller on both boards should be in low power mode when not used. Form the project using:

(a) C language under STM32CubeIDE.
(b) C++ language under Mbed.
(c) MicroPython.

8.2. Expand the setup in Problem 8.1 as follows. As the second board receives the toggle command, it turns back an acknowledge bit to the first board. Hence, the green LED on it turns on.

8.3. Form a project with the following specifications. There are two STM32F4 boards at hand. We will use the ADC module on the first board to measure temperature on the STM32F4 microcontroller. This measured value will be transferred to the second board. If the temperature value is greater than a predefined threshold, the red LED on the second board will turn on. Otherwise, the green LED on the second board will turn on. Use UART, SPI, I²C, and CAN modules to transfer the ADC value from one board to other. The STM32F4 microcontroller on both boards should be in low power mode when not used. Form the project using:

(a) C language under STM32CubeIDE.
(b) C++ language under Mbed.
(c) MicroPython.

8.4. Expand the setup in Problem 8.3 as follows. As the second board receives the ADC value, it turns back an acknowledge bit to the first board. Hence, the green LED on it turns on.

References

1. SD Association: https://www.sdcard.org/. Accessed: June 4, 2021
2. Griffith, J.: https://e2e.ti.com/blogs_/b/industrial_strength/posts/the-inner-workings-of-a-can-bus-driver. Accessed: June 4, 2021
3. ISO: Road vehicles – Controller area network (CAN) – Part 1: Data link layer and physical signalling, ISO 11898-1:2015 edn. (2015)
4. Murphy, R.: USB 101: An Introduction to Universal Serial Bus 2.0. Cypress, an57294 edn.
5. Pazul, K.: Controller Area Network (CAN) Basics. Microchip, an713 edn. (1999)
6. Philips Semiconductors: I2S bus specification (1996)
7. STMicroelectronics: STM32F427xx STM32F429xx, docid024030 rev 10 edn. (2018)
8. STMicroelectronics: STM32Cube USB Device Library, um1734 edn. (2019)

9. Ünsalan, C., Gürhan, H.D., Yücel, M.E.: Programmable Microcontrollers: Applications on the MSP432 LaunchPad. McGraw-Hill (2018)
10. Ünsalan, C., Yücel, M.E., Gürhan, H.D.: Digital Signal Processing using Arm Cortex-M based Microcontrollers: Theory and Practice. Arm Education Media (2018)
11. USB: https://www.usb.org/. Accessed: June 4, 2021

Memory Operations

9

9.1 Memory Working Principles

This section introduces how memory works in Arm® microcontrollers, including STM32F4. Therefore, we will first explain the bus architecture. Then, we will focus on RAM working principles. Finally, we will explore the flash memory and its structure.

9.1.1 Bus Architecture

As briefly introduced in Chap. 2, bus is composed of parallel wires in which data and instructions are transferred between microcontroller units. Arm® microcontrollers use advanced microcontroller bus architecture (AMBA) for this purpose. AMBA controls all data transfer between microcontroller units. Here, the unit which initiates data transfer is called master. The unit which responds to master is called slave.

There are two different bus structures in AMBA. These are called advanced high-performance bus (AHB) and advanced peripheral bus (APB). AHB supports high-speed and high bandwidth operations. Therefore, it is generally used by CPU, DMA modules, and external and internal memory units. AHB supports pipelined operations and burst transfer. Burst transfer is used to realize multiple data transactions at once. This is especially useful for communicating with slave units with high initial access latency. APB supports low bandwidth operations. This

---

Supplementary Information The online version contains supplementary material available at (https://doi.org/10.1007/978-3-030-88439-0_9).

C. Ünsalan et al., *Embedded System Design with Arm Cortex-M Microcontrollers*, https://doi.org/10.1007/978-3-030-88439-0_9

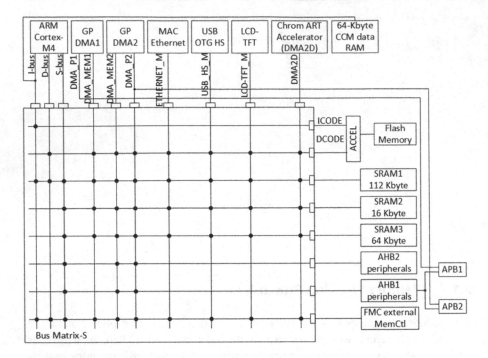

Fig. 9.1 The STM32F4 bus matrix connections [3]

leads to low power consumption. Communication between most peripheral units is performed via this interface.

The STM32F4 microcontroller has a multilayer bus matrix as in Fig. 9.1. This bus matrix allows connecting several master and slave units. To be more specific, the STM32F4 microcontroller has 10 master and 8 slave units.

As can be seen in Fig. 9.1, there are three bus types connecting to the bus matrix as instruction bus (I-bus), data bus (D-bus), and system bus (S-bus). The instruction bus is used by the CPU to fetch instructions from flash memory, SRAM, or external memory (wherever the code is running from). The prefetch operation in the adaptive real-time accelerator, to be introduced in Sect. 9.1.4, is also performed via the instruction bus. The CPU controls code or data in memory locations by the data bus. The system bus allows the CPU to reach peripheral units or SRAM. Besides, DMA buses connect the related DMA master devices to desired peripheral unit and memory blocks. Hence, DMA can be used for providing high-speed data transfer between peripheral units to memory, memory to peripheral units, and memory to memory without the CPU intervention. We will evaluate these operations in detail in Sect. 9.4.

9.1.2 Memory in General

The memory in the STM32F4 microcontroller is based on Harvard architecture [3]. Hence, data and instructions are transferred through their own dedicated data and instruction bus. The number of possible memory locations is determined by the address bus size. There are 2^k addressable memory locations for k address bits. AHB and APB of the STM32F4 microcontroller have 32 bit address bus. Hence, there are a total of 4 GByte possible usable address locations. The partitioning of these address locations is given in Sect. 2.1.2. Besides, each address location can keep 1 byte data as will be explained next. Hence, the maximum data size that can be stored in memory is 4 GByte.

Memory is composed of data storage blocks. Each block has its own address value and content. As the name implies, address of a memory block indicates its unique location in memory. The content is the value stored in that address. The minimum memory block size for the STM32F4 microcontroller is 1 byte. In other words, the minimum addressable memory block can have size of 1 byte. Hence, these memory blocks are called byte-addressable. For RAM, the addressable memory size can be 1 byte, 2 bytes (half-word), and 4 bytes (word). For the flash memory, the addressable memory size can be 1 byte, 2 bytes (half-word), and 4 bytes (word) for the writing operation. We can also have 8 byte (double word) addressable memory size in flash memory for reading operation. Since the minimum addressable memory location is 1 byte, half-word and word data are stored in the memory in little endian format [3].

9.1.3 RAM

SRAM of the STM32F4 microcontroller is divided into different blocks as tabulated in Table 9.1. In this table, CCM stands for core coupled memory. Backup SRAM can be used as internal EEPROM in standby mode. Read and write operations to SRAM are performed at any clock speed without delay.

As can be seen in Table 9.1, SRAM1, SRAM2, and SRAM3 form a total of 192 kB SRAM used by global and static variables, heap, and stack. We will talk about global and static variables in Sect. 9.2.4 in detail.

Heap is a specific RAM region in which dynamic memory allocation can be performed. We will explain how to use heap in C and C++ languages in detail in

Table 9.1 SRAM blocks and their properties

Name	Size (kB)	Start address	Accessed by
SRAM1	112	0x20000000	All AHB master units
SRAM2	16	0x2001C000	All AHB master units except D-bus and I-bus
SRAM3	64	0x20020000	All AHB master units except D-bus and I-bus
CCM data RAM	64	0x10000000	D-bus
Backup SRAM	4	0x40024000	All AHB master units

Sect. 9.2. Heap allocation and usage is automatically done in MicroPython. The user can check heap usage by the command `mem_info` there.

Stack is the specific memory region initially located at the end of usable SRAM. Here local variables and temporary data are stored during code execution. We will talk about local variables in Sect. 9.2.4 in detail. Data is moved to and removed from stack in last-in first-out (LIFO) order. This means last data moved to stack is removed first. Stack memory allocation and deallocation is done automatically in C and C++ languages. However, stack should be used carefully to prevent stack overflow. This overflow occurs when variables added to the stack exceed its free area during code execution. For such cases, either existing variables or critical code segments in nearby RAM regions are overwritten. This may cause serious problems. As in heap, stack allocation and usage is automatically done in MicroPython. The user can check stack usage by the command `mem_info` there.

9.1.4 Flash Memory

The STM32F4 microcontroller has 2 MB flash memory divided into two banks, each having 1 MB space. Each bank is further divided into 12 sectors. Hence, sectors from 0 to 11 are in bank 1 and sectors from 12 to 23 are in bank 2. For each bank, first four sectors are 16 kB, fifth sector is 64 kB, and remaining sectors are 128 kB. Starting address of flash memory (beginning of sector 0) is `0x08000000` and end address (end of sector 23) is `0x081FFFFF`. Flash memory erasing operation can be done at sector level, bank level, or whole (mass flash erase).

The CPU has higher clock speed compared to flash memory. Therefore, when CPU performs an operation connected to flash memory, it should wait for the end of operation. This is called wait state. To be more specific, when reading data or code from flash memory, there is a relation between HCLK frequency, supply voltage level, and number of flash wait states. If the supply voltage is kept constant, then the wait state increases with respect to the HCLK frequency. As an example, when the HCLK frequency is in the range 0–30 MHz, the wait state will be zero clock cycle. When the HCLK frequency is in the range 150–180 MHz, the wait state becomes five clock cycles. Hence, increasing the clock speed does not directly increase performance. To overcome this problem, the STM32F4 microcontroller has the adaptive real-time (ART) accelerator unit. Normally, 128-bit read operations can be performed sequentially and flash is stalled during number of wait states between each read. The ART accelerator uses an instruction prefetch queue and branch cache. Hence, latency between successive reads may be prevented. As a result, 0 wait state code execution from flash memory may be achieved at CPU clock frequencies up to 180 MHz.

The STM32F4 microcontroller also has other memory regions besides its 2 MB flash memory. These are tabulated in Table 9.2. In this table, OTP stands for one-time programmable. System memory region is used to boot the device in system memory boot mode. As the name implies, OTP bytes can be programmed only once and cannot be rewritten again. Option bytes region is used to configure read and

Table 9.2 Other flash
memory regions

Name	Size (bytes)	Start address
System memory	32,768	0x1FFF0000
OTP-bytes	528	0x1FFF7800
Option bytes, bank 1	16	0x1FFFC000
Option bytes, bank 2	16	0x1FFEC000

write protection, brown-out reset (BOR) level, watchdog module, and dual bank boot mode. When the device is in standby or stop mode, it can also be used to adjust the reset operation.

Some microcontrollers have memory protection unit (MPU) which adds secure access to SRAM and flash memory blocks. Unfortunately, the STM32F4 microcontroller does not have such a unit. Still, memory read/write protection can be performed by option bytes. There are three protection levels as read protection (RDP), write protection, and proprietary code readout protection (PCROP).

RDP protects the selected flash memory block against copying and checking via debugging tools. There are three levels (as 0, 1, and 2) for this purpose. The default mode is level 0. There is no protection here. When level 1 is activated, the user cannot perform read, write, or erase operations on flash memory or backup SRAM via any debug interface. If level 1 is disabled and level 0 is activated afterward, mass erase operation can be done on flash memory. Hence, it can be reprogrammed. Level 2 acts like level 1. However, all option bytes are frozen and cannot be changed when level 2 is activated. Hence, this is a nonreversible operation. The reader should use this option with caution. The reader should apply power-on reset while the debugger is active to activate a level change. RDP can be set by bits 15 to 8 of 2 bytes starting from the memory address 0x1FFFC000. Writing 0xAA to this region activates level 0. Writing 0xCC to this region activates level 2. Writing any other value to this region activates level 1.

Write protection adds protection to write or erase operations to each flash sector separately. PCROP works in a similar manner with additional protection for the end user code and debugger-based read operation. The main difference of PCROP is that it does not allow a different protection method for the overall flash. In other words, if PCROP is selected for a flash sector, other sectors can only be protected by it if needed. Write protection or PCROP can be selected by the 15th bit of the 2 bytes starting from the memory address 0x1FFFC008. By default, this bit is reset. Hence, write protection has been selected. If this bit is set, PCROP is selected. Bits 0–11 of the 2 bytes can be used to enable or disable the selected protection (write protection or PCROP) for flash bank 1 sectors 0–11. If the 15th bit is reset (write protection is selected), resetting a bit among bits 0–11 activates write protection for the corresponding sector. Let's assume that we reset bit 15 and bit 6 and set the remaining bits. This way, we apply write protection to sector 6 of flash bank 1. No protection is applied to other sectors. If the 15th bit is set, we should also set the bit among bits 0–11 to activate PCROP for the corresponding sector. Let's assume that we set bit 15 and bit 6 and reset the remaining bits. This way, we apply PCROP to sector 6 of flash bank 1. No protection is applied to other sectors. The same operations can be done to flash bank 2 sectors 12–23 by modifying the bits 11–0 of the 2 bytes starting from the memory address `0x1FFEC008`.

9.2 Memory Management in C and C++ Languages

We will consider memory management in C and C++ languages in this section. Since both languages share common properties, we handle them together. We will start memory management with RAM and flash memory modification. Then, we will focus on pointers and their usage for memory management. Afterward, we will handle local and global variables in C and C++ languages.

9.2.1 RAM Partitioning

We can modify RAM usage, more specifically heap and stack memory, after opening a new project under STM32CubeIDE. To do so, we should open the linker script file "STM32F429ZITX_FLASH.ld" located at the top level of the project folder. In this file, the highest address of the user mode stack (end of SRAM) is defined by the code line `_estack = ORIGIN(RAM) + LENGTH(RAM)`. Stack is used in decreasing order starting from this address. Code lines `_Min_Heap_Size` and `_Min_Stack_Size` are used to define the minimum heap and stack size, respectively.

The user can allocate memory from heap using the function `malloc()` or `calloc()` within RAM. Allocated memory can be freed using the function `free()` during code execution. Memory management of the heap must be controlled by the user. If unused allocated memory is not freed, heap will eventually fill and memory leakage will occur. Another problem encountered when using heap is fragmentation. Memory blocks in heap are allocated and freed not in the same order. As a result, the required memory space in heap can be available in fragmented form. Therefore, it may not be used.

Global and static variables are placed in SRAM starting from the memory address `0x20000000`. Heap is placed after this placement. As a reminder, stack is placed at the end of SRAM region. The user should make sure that the memory space used in all these regions should not exceed the total SRAM size in the microcontroller.

9.2.2 Memory Modification

A code block can be placed to a desired memory region using the attribute property of STM32CubeIDE. To do so, we should open the linker script file "STM32F429ZITX_FLASH.ld" located at the top level of the project folder. There, we can modify the existing memory declaration

```
MEMORY
{
  CCMRAM (xrw) : ORIGIN = 0x10000000, LENGTH = 64K
  RAM (xrw) : ORIGIN = 0x20000000, LENGTH = 192K
  FLASH (rx) : ORIGIN = 0x8000000, LENGTH = 2048K
}
```

as follows to specify a new memory region.

```
MEMORY
{
  CCMRAM (xrw) : ORIGIN = 0x10000000, LENGTH = 64K
  RAM (xrw) : ORIGIN = 0x20000000, LENGTH = 192K
  FLASH (rx) : ORIGIN = 0x8000000, LENGTH = 1024K
  FLASH_CODE (rx) : ORIGIN = 0x8100000, LENGTH = 1024K
}
```

Then, we should place the code given below between ".data" and ".bss" sections. Hence, all sections named as ".flash_code*" in the code will be placed in the constructed FLASH_CODE memory region.

```
.flash_code :
{
*(.flash_code*);
} > FLASH_CODE
```

There are two main reasons to place a code to a specific memory location. The first one is related to MPU of a microcontroller. As a reminder, this unit ensures that the code written to a specific memory region cannot be reached by another program. Unfortunately, the STM32F4 microcontroller does not have MPU. The second reason for placing a code to a specific memory location is speeding up its operation. This is done by placing the code to a designated place in RAM. As a side note, this code block will be retained from flash memory after every reset operation.

We next provide a usage example for placing a code segment to specific memory location. To do so, we form the function called sum in our C code. Then, we place it to the memory location defined as FLASH_CODE. We use the sum function to add 2–3 as follows.

```
/* USER CODE BEGIN 0 */
__attribute__((section(".flash_code"))) int sum(int a, int b)
{
 return a+b;
}

/* USER CODE END 0 */

/* USER CODE BEGIN 2 */
int c;
c = sum(2,3);
/* USER CODE END 2 */
```

As the code is executed, we can see that the variable c will have value 5 as expected. Moreover, if we visit the address 0x8100000 in the "Memory" window, we can see that our sum function has been placed in the FLASH_CODE memory region.

9.2.3   Pointer-Based Operations

Pointers and pointer arithmetic allow reaching and modifying a specific memory address in C and C++ languages. This is one of the most powerful and possibly confusing properties of C language. Since we are using pointers in the STM32F4 microcontroller, we can grasp the basic idea behind them more easily as we can observe the memory map of the microcontroller via STM32CubeIDE. This will be of great help in understanding the usage of pointers and pointer arithmetic. Next, we provide several usage areas of pointers in C language. We can revisit pointer-based operations in C++ language in Mbed Studio as well. Since C and C++ languages handle the pointer in a similar manner, we provide the modified codes for C++ language as supplementary material in the accompanying book web site.

9.2.3.1 Pointer to a Variable

We provide a pointer definition and usage in Listing 9.1. In this code, we define an integer variable a and assign value 3 to it. In fact, when we make such a definition and assignment, the variable is kept in a specific memory address. In order to reach this memory address, we define a pointer aPointer with integer type. This pointer can keep memory address of the variable a by the code line aPointer=&a. For our case, this memory address is 0x20000000 as can be seen in the "Expressions" window in Fig. 9.2 after the code is executed in STM32CubeIDE. We can indirectly reach and modify this memory address content by the code line *aPointer=5. Hence, the a value is modified as can be seen in Fig. 9.2.

Listing 9.1   Pointer to a variable example, the C code

```
int a = 3;
int *aPointer;

int main(void)
{
aPointer = &a;
*aPointer = 5;

while (1);
}
```

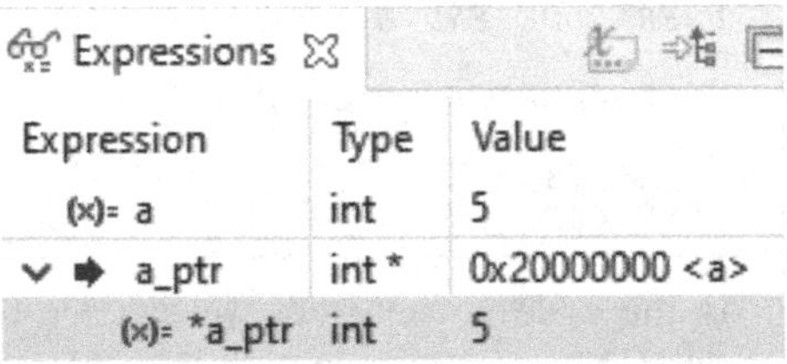

Fig. 9.2 "Expressions" window for the pointer usage example

Expression	Type	Value
(x)= a	int	5
⌄ ➡ a_ptr	int *	0x20000000 \<a\>
(x)= *a_ptr	int	5

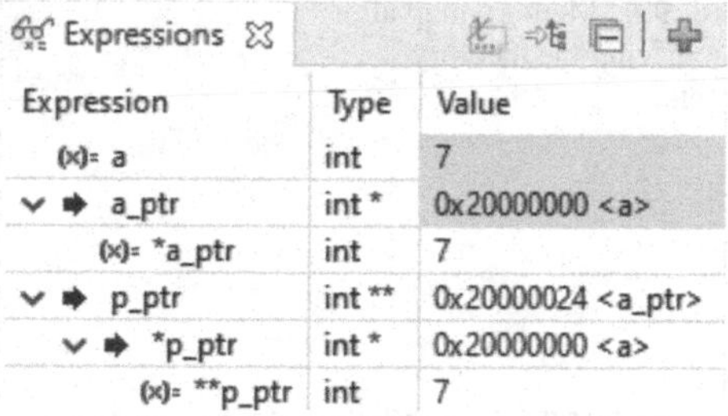

Fig. 9.3 "Expressions" window for the pointer to pointer usage example

Expression	Type	Value
(x)= a	int	7
⌄ ➡ a_ptr	int *	0x20000000 <a>
(x)= *a_ptr	int	7
⌄ ➡ p_ptr	int **	0x20000024 <a_ptr>
⌄ ➡ *p_ptr	int *	0x20000000 <a>
(x)= **p_ptr	int	7

9.2.3.2 Pointer to a Pointer

The pointer is itself another variable kept in memory. We can reach this memory address by using pointer to a pointer. This method is also called double pointer in literature. To note here, this is an advanced and powerful operation. We only use it for reaching the pointer address in this book.

We provide the pointer to a pointer (double pointer) usage in Listing 9.2. In this code, we extend our previous example in Listing 9.1 such that we define the pointer to a pointer by the code line int **pPointer and equate it to the address of aPointer by the code line pPointer = &aPointer. As we run the code in STM32CubeIDE, we can observe that the aPointer is kept in the memory address 0x20000024 as can be seen in the "Expressions" window in Fig. 9.3. In Listing 9.2, we modify the value of our variable a by the pointer to pointer usage as **pPointer=7.

Listing 9.2 Pointer to pointer usage example, the C code

```
int a = 3;
int *aPointer;
int **pPointer;

int main(void)
{
aPointer = &a;

pPointer = &aPointer;

*aPointer = 5;

**pPointer = 7;

while (1);
}
```

9.2.3.3 Reaching a Specific Memory Address by Pointers

Pointers can be used to reach a specific memory address in the STM32F4 microcontroller. We provide two methods for this purpose in this section. The first method is based on pointer arithmetic and casting operation. We provide such an example in Listing 9.3. In this code, we modify the memory addresses 0x20000020 and

Fig. 9.4 Memory map after
modifying the memory
content

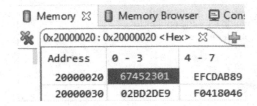

0x20000024 via pointer arithmetic and casting operation. We provide the memory
map after this operation in Fig. 9.4. As can be seen in this figure, the memory content
has been modified as expected.

Listing 9.3 Reaching a specific memory address using pointer arithmetic and casting operation,
the C code

```c
unsigned int *memLoc;

int main()
{
memLoc = (unsigned int *)0x20000020;

*(memLoc) = 0x01234567;
*(memLoc + 1) = 0x89ABCDEF;

while (1);
}
```

The second method in reaching a memory address is using the define statement.
We provide such an example in Listing 9.4. Here, we modify the memory addresses
0x20000020 and 0x20000024 via define statement. The memory map after this
operation will again be as in Fig. 9.4.

Listing 9.4 Reaching a specific memory address using the define statement, the C code

```c
#define memLoc1 (*(unsigned long *)0x20000020)
#define memLoc2 (*(unsigned long *)0x20000024)

int main(void)
{

memLoc1=0x1234567;
memLoc2=0x89ABCDEF;

while (1);
}
```

The second method allows obtaining the unique identification number (ID) of the
STM324 microcontroller. As a side note, each STM microcontroller has its unique
ID composed of 96 bits assigned to it. We provide the sample code for this operation
in Listing 9.5. The obtained unique ID for our microcontroller can be observed both
in the "Expressions" window and "Memory map" as in Fig. 9.5.

Fig. 9.5 Obtaining unique ID of the STM32F4 microcontroller. (**a**) Expressions window. (**b**) Memory map

Listing 9.5 Obtaining unique ID of the STM32F4 microcontroller, the C code

```c
#define ID1 (*(unsigned long *)0x1FFF7A10)
#define ID2 (*(unsigned long *)0x1FFF7A14)
#define ID3 (*(unsigned long *)0x1FFF7A18)

int main(void)
{

while (1);
}
```

9.2.3.4 Pointers and Arrays

When an array is defined in C language, it is in fact treated as pointer. We provide one such example in Listing 9.6. In this code, we first define an integer array, a, with entries {1, 2, 3, 4, 5}. We then assign the array to the pointer with the code line aPointer=a. We next reach the second element of the array and modify it by the code line *aPointer = 0. We can observe the effect of this operation in the "Expressions" window in Fig. 9.6. This window also provides another insight on the array usage. As can be seen there, the array a is kept by its starting address only. Hence, we can use it as a pointer.

Listing 9.6 Array usage as pointer example, the C code

```c
int a[5] = {1, 2, 3, 4, 5};
int *aPointer;

int main(void)
{

aPointer = a;
aPointer += 1;
*aPointer = 0;

while (1);
}
```

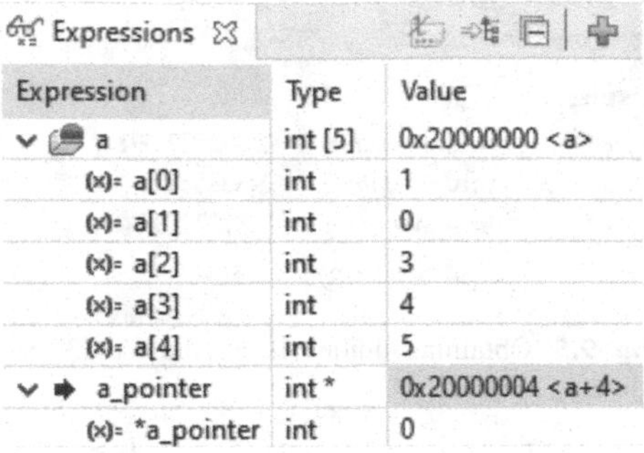

Fig. 9.6 "Expressions" window for the array as pointer example

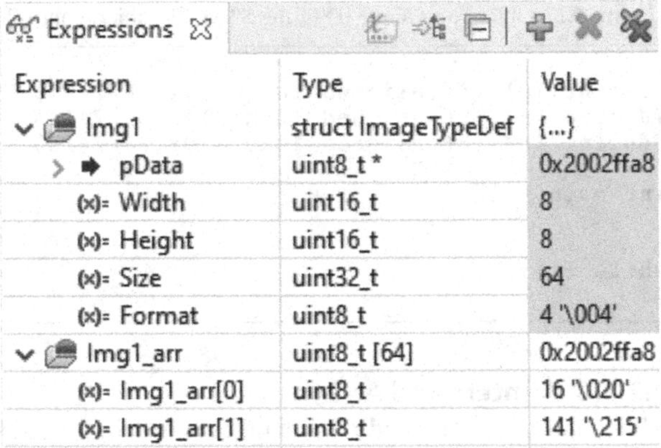

Expression	Type	Value
∨ 📁 a	int [5]	0x20000000 <a>
(x)= a[0]	int	1
(x)= a[1]	int	0
(x)= a[2]	int	3
(x)= a[3]	int	4
(x)= a[4]	int	5
∨ ➡ a_pointer	int *	0x20000004 <a+4>
(x)= *a_pointer	int	0

Fig. 9.7 "Expressions" window for the structure example

Expression	Type	Value
∨ 📁 Img1	struct ImageTypeDef	{...}
> ➡ pData	uint8_t *	0x2002ffa8
(x)= Width	uint16_t	8
(x)= Height	uint16_t	8
(x)= Size	uint32_t	64
(x)= Format	uint8_t	4 '\004'
∨ 📁 Img1_arr	uint8_t [64]	0x2002ffa8
(x)= Img1_arr[0]	uint8_t	16 '\020'
(x)= Img1_arr[1]	uint8_t	141 '\215'

9.2.3.5 Pointer to a Structure

Structure (struct) is a powerful tool in representing data in grouped form. We will be using structures in digital image processing operations. We can use pointers in reaching a specific field in a structure.

We provide an example on structure usage in Listing 9.7. Here, we define an image struct, `ImageTypeDef`, with five fields as `*pdata`, `width`, `height`, `size`, and `format`. Then, we create an image struct as `img1` and assign values to its elements. We also provide how a pointer is used to reach specific field elements in the structure in the same example. We provide the "Expressions" window in Fig. 9.7 as we run the code in STM32CubeIDE.

Listing 9.7 Structures and pointer usage on them, the C code

```
#include <stdint.h>

struct ImageTypeDef
{
uint8_t *pData;
uint16_t width;
uint16_t height;
uint32_t size;
uint8_t format;
};
```

```
int main(void)
{

struct ImageTypeDef img1;
uint8_t img1Arr[64];

img1.pData = img1Arr;
img1.width = 8;
img1.height = 8;
img1.size = 64;
img1.format = 4;
img1.pData[5] = 8;

while (1);
}
```

9.2.3.6 Function Call by Reference

Pointers allow passing array addresses directly to a function. This is function call by reference which allows modifying a specific array element in the function. Here, we do not transfer all array elements to the function and receive them back afterward. This becomes extremely important while processing long arrays such as images or audio signals.

We provide two examples on the usage of function call by reference next. In the first example, given in Listing 9.8, the function `replace` is used to replace the second element of any array with value 0. Here, only starting address of the array is passed to the function instead of passing all its elements.

Listing 9.8 The first function call by reference example, the C code

```
#include <stdint.h>
char a[5] = {1, 2, 3, 4, 5};
char *aPointer;

void replace(char *pointer);

int main()
{
aPointer = a;
replace(aPointer);

while (1);
}

void replace(char *pointer)
{
pointer += 1;
*pointer = 0;
}
```

In the second example, given in Listing 9.9, the function `squareFunc` is used to take square of the array elements. Here, only the starting address and size of the array are fed to the function instead of passing all array elements. Size of the array is fed to the function since it must know the number of array elements to operate on.

Listing 9.9 The second function call by reference example, the C code

```
char a[5] = {1, 2, 3, 4, 5};
char *aPointer;

void squareFunc(char *pointer, int size);

int main()
{
aPointer = a;
squareFunc(aPointer, sizeof(a));

while (1);
}

void squareFunc(char *pointer, int size)
{
int i;

for (i = 0; i < size; i++)
{
*(pointer + i) *= *(pointer + i);
}
}
```

9.2.3.7 Function Pointers

Pointers can also be used to call functions. This is different from previous applications such that we call a code script to be executed by an assigned pointer to it. There are two main reasons for the usage of such a method. The first usage area is when a callback function is associated with its interrupt source in Mbed. A simple example for this usage is the code line button.rise(&toggle()). Here, toggle()) is the callback function associated with the button rise interrupt. We have seen such examples starting from Chap. 5. The second usage area is forming a function which can perform the same operation on different variable types.

We provide a function pointer usage example in Listing 9.10. This code consists of a function to modify array elements as desired. We feed start address of the array, size of the array, size of each array element, and the function (dummy) to perform the desired change as a pointer. The (dummy) function is called and operated on the given data. However, we do not know what this (dummy) function is until it is called. In our example, we set this (dummy) function as the one taking square of the array elements.

Listing 9.10 Function pointer usage example, the C code

```
/* USER CODE BEGIN PV */
float arr[5] = {1.7, 2.8, 3.6, 4.5, 5.2};
/* USER CODE END PV */

/* USER CODE BEGIN 0 */
void change(void *arr, size_t arrSize, size_t elemSize, void (*
    dummy)(const void *))
{
char *ptr = (char *)arr;
```

```
int i;
for (i = 0; i < arrSize; i++)
{
dummy(ptr + i * elemSize);
}
}

void squareFunc(const void *x)
{
*(float *)x = (*(float *)x) * (*(float *)x);
}
/* USER CODE END 0 */

/* USER CODE BEGIN 2 */
change(arr, 5, sizeof(float), squareFunc);
/* USER CODE END 2 */
```

9.2.4 Local, Global, and Static Variables

We can define a variable either as local or global in C and C++ languages. As the name implies, a global variable is available to all code blocks. However, a local variable is only available to the function it is defined in. These two variables are kept in different memory locations in RAM. Global variables are kept starting from the lowest possible memory address (0x20000000) in SRAM. As a new global variable is added, the memory address is incremented and the new variable is saved there. On the other hand, local variables are kept in the highest memory address in the stack. For STM32F4 microcontroller, this address is initially 0x2002FFFF which is the end of SRAM. Based on the definition of stack, when a new local variable comes, it is saved to the recent memory address. The address value is decremented afterward. Hence, local variables are saved from top to bottom in SRAM in stack. It is important to note that the C and C++ languages take main() as a function. Hence, a variable defined within the main function is also treated as local. Therefore, a global variable should be defined before the main() declaration.

We should also explain how a static variable works in C and C++ languages. Such a variable is created with the static keyword and kept in RAM during code execution. Even if a static variable is declared as local, it is not kept in stack. Instead, the static variable is placed at the beginning of RAM as with global variables. The importance of the static variable is as follows. It can be defined in a function. Hence, it will be local to that function. However, whenever the function is called more than once, the value of the static variable is not lost unlike other local variables.

We can observe how global and local variables are stored in memory by the help of pointer usage. We provide one simple example in C language in Listing 9.11. Here, the variable a is defined as global and the variable b is defined as local. We provide the "Expressions" window after executing the code in Fig. 9.8. As can be seen in this window, the two variables are kept in different memory locations. We can observe this effect by pointer usage.

Fig. 9.8 "Expressions" window for the local and global variables example

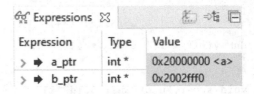

Listing 9.11 Local and global variables, the C code

```c
//Global variable
int a = 1;
int *aPointer;

int main(void)
{
//Local variable
int b = 2;
int *bPointer;

aPointer = &a;
bPointer = &b;

while (1);
}
```

We can observe the local and global variable handling in C++ language under Mbed Studio. Since C and C++ languages handle local and global variables in a similar manner, we provide the modified codes for C++ language as supplementary material in the accompanying book web site.

9.3   Memory Management in MicroPython

As explained before, MicroPython is an efficient implementation of Python language which can be used in microcontrollers with limited memory space. Therefore, memory management is one of the biggest problems encountered while using MicroPython. We provide how memory management can be done in MicroPython in this section. Besides, we introduce local and global variables in MicroPython.

9.3.1   RAM Management During Compilation Stage

MicroPython firmware is placed in flash memory of the microcontroller. However, importing additional Python scripts uses RAM during compilation and execution steps. During compilation, a Python script is converted to bytecode and stored in RAM until compilation is done. This means importing multiple modules may cause memory fault even before the execution process. There are two methods to prevent this.

9.3.1.1 Frozen Bytecode Usage

The first method in avoiding memory fault is using frozen bytecode. In this method, modules to be used in the code are compiled with MicroPython. Hence, they become part of the MicroPython firmware to be embedded in the microcontroller flash memory. We will explain how to use the frozen bytecode method in Chap. 15.

9.3.1.2 Precompiling Scripts

The second method in avoiding memory fault is precompiling Python scripts to bytecodes separately and creating individual ".mpy" files. Then, these files can be embedded in the microcontroller flash memory. We will explain how to use the precompiling scripts method in Chap. 15.

9.3.2 Effective RAM Usage During Code Execution

After compiling the MicroPython code and execution of the bytecode starts, RAM must be used. Since RAM space is limited, we must focus on effective (or minimal) RAM usage at this step. MicroPython offers several methods to perform this operation. We provide some of these methods next.

9.3.2.1 Constant Usage

The first method in decreasing RAM usage can be done by the const() function. This function works in a similar manner as with the define statement in C language. Before explaining how the const() function works, let's focus on how a variable is handled in MicroPython. When a variable is defined in MicroPython, both the variable and the necessary bytecode for it are stored in RAM. When the const() function is used, only value of the variable is stored in RAM. It is directly used during code execution. Hence, we should keep fixed variables in the code by the const() function. This way, the compiler uses value of the constant directly and saves bytecode which means less RAM space. We provide an example for the const() function in Listing 9.12.

Listing 9.12 The const() function usage in MicroPython

```
from micropython import const
SIZE = const(10)
_SIZE = const(10)

x = SIZE
y = _SIZE
```

In Listing 9.12, we declare the SIZE as constant in two different ways. First, the SIZE constant does not use space in bytecode. However, since it is defined as global, it still uses RAM space. This way, SIZE can be used in other modules as well. When a constant is used with underscore before it, it means this constant is only available for the current module. This is the second usage as _SIZE in our MicroPython script.

Here, this constant does not use RAM space. However, it is limited to the module it is defined in.

9.3.2.2 Constant Data Structures

The second method in decreasing RAM usage can be done by constant data structures. If there is constant data which does not change during code execution, it can be located in a Python module as `bytes` object and frozen as bytecode. Hence, the compiler does not copy these objects to RAM since they are constant (immutable) variables. This process can be applied in other immutable object types such as `string`, `float`, or `integer`. Only `tuple` object type cannot be used in this way, although it is an immutable object. For more information on mutable and immutable objects, please see [1].

9.3.2.3 Using Pre-allocated Buffers

The third method in decreasing RAM usage is by pre-allocated buffers. When obtaining data from a microcontroller peripheral unit (such as UART, I^2C, SPI, or USB), a new buffer is created every time data is obtained. However, this may cause a fragmentation problem in RAM. Instead, a pre-allocated buffer can be used such that the same buffer is used every time the peripheral unit is used. We have provided a usage example for this setup in Listing 8.13.

9.3.2.4 Garbage Collector

The fourth method in decreasing RAM usage is associated with the garbage collector. As explained in Sect. 9.1.3, heap is used to allocate dynamic memory. Therefore, when an object is created in MicroPython, necessary RAM space is allocated for it from heap. When the object is no longer needed, it is called garbage. The garbage collector recognizes this object and puts the allocated memory space back to the heap.

Garbage collector runs automatically under two conditions. The first one is the failed allocation attempt. When an allocation attempt is failed, garbage collector runs and allocation process is retried. The second condition is by a RAM threshold. If free heap memory is less than a predefined threshold, garbage collector runs automatically.

Garbage collector can also be triggered manually using the function `gc.collect()`. Heap fragmentation can be reduced by using this function at periodic time intervals instead of automatic garbage collection. This manual garbage collection process is important for real-time applications for the following reason. Automatic garbage collection may take several milliseconds. On the other hand, this duration can be divided into smaller time durations using a periodic manual garbage collector. Besides, if we wait for the garbage collector to work automatically, the heap memory may be fragmented more during this time. This negatively affects the heap usage.

Garbage collector can be enabled or disabled using the functions `gc.enable()` and `gc.disable()`, respectively. Related to this, the function `gc.threshold` (`[threshold_value]`) can be used to define the heap allocation threshold described before. There are also two functions to monitor free and allocated heap space as gc.mem_free() and gc.mem_alloc(), respectively. Finally, the function mem_info() can be used to monitor the overall heap condition.

9.3.3 Local and Global Variables

MicroPython also has local and global variables. By default, any variable declaration will lead to a local variable. The reader can use the keyword global to declare a variable to be global. We provide one such usage example for these cases in Listing 9.13. In this code, the variable x is declared as global within two functions and used as such. The variable y is declared as local. Hence, when asked to print its value outside its scope, it will generate an error.

Listing 9.13 Local and global variable usage in MicroPython

```
x = 3

def multiply():
    global x
    x = x * 2
    y = 1
    print("y is equal to ", y)
def add():
    global x
    x = x + 2

print("x is equal to ", x)
multiply()
print("x is equal to ", x)
add()
print("x is equal to ", x)
print("y is equal to ", y)
```

9.4 Direct Memory Access

For some applications, large amount of data may be received from a peripheral unit, sent to a peripheral unit or copied from one memory location to another. If the CPU is dedicated for these operations, it cannot perform other tasks at the same time. Hence, code execution efficiency reduces. Therefore, most microcontrollers have direct memory access (DMA) controllers for such memory operations to keep CPU free for other tasks. We will explore the DMA controller available in the STM32F4 microcontroller in this section.

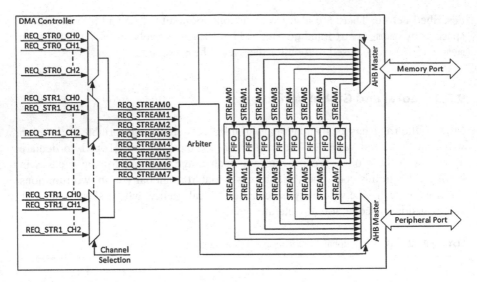

Fig. 9.9 Block diagram of the DMA controller

9.4.1   The DMA Controller in the STM32F4 Microcontroller

The STM32F4 microcontroller has two independent DMA controllers (DMA1 and DMA2) to transfer data efficiently from a peripheral unit to memory, memory to a peripheral unit, or memory to memory. These DMA controllers handle data transfer independent of the CPU. Hence, the CPU can execute other tasks or be kept in low power mode.

Block diagram of the DMA controller in the STM32F4 microcontroller is as in Fig. 9.9. As can be seen in this figure, the DMA controller has eight different streams (stream 0 to stream 7). Stream is a DMA block which can be modified independently. Hence, a different transfer type, buffer type, or other DMA properties can be set for each stream separately.

All of the eight streams in the DMA controller can be used at once. Selection between streams is handled by the arbiter according to stream priorities. There are four different software stream priorities as very high, high, medium, and low. If two streams have the same software priority, then hardware priorities are used. Here, stream 0 has the highest hardware priority. Stream 7 has the lowest hardware priority. Each stream has eight different channels (channel 0 to channel 7). A channel decides on which DMA source will be connected to the stream. The channel to be used is decided beforehand and more than one channel cannot be used for a stream. We provide the complete list for each channel in Tables 9.3 and 9.4 for DMA1 and DMA2, respectively.

Table 9.3 DMA1 sources for all channels

Stream/ Channel	Stream 0	Stream 1	Stream 2	Stream 3	Stream 4	Stream 5	Stream 6	Stream 7
Channel 0	SPI3_RX	-	SPI3_RX	SPI2_RX	SPI2_TX	SPI3_TX	-	SPI3_TX
Channel 1	I2C1_RX	-	TIM7_UP	-	TIM7_UP	I2C1_RX	I2C1_TX	I2C1_TX
Channel 2	TIM4_CH1	-	I2S3_EXT_RX	TIM4_CH2	I2S2_EXT_TX	I2S3_EXT_TX	TIM4_UP	TIM4_CH3
Channel 3	I2S3_EXT_RX	TIM2_UP TIM2_CH3	I2C3_RX	I2S2_EXT_RX	I2C3_TX	TIM2_CH1	TIM2_CH2 TIM2_CH4	TIM2_UP TIM2_CH4
Channel 4	UART5_RX	USART3_RX	UART4_RX	USART3_TX	UART4_TX	USART2_RX	USART2_TX	UART5_TX
Channel 5	UART8_TX	UART7_TX	TIM3_CH4 TIM3_UP	UART7_RX	TIM3_CH1 TIM3_TRIG	TIM3_CH2	UART8_RX	TIM3_CH3
Channel 6	TIM5_CH3 TIM5_UP	TIM5_CH4 TIM5_TRIG	TIM5_CH1	TIM5_CH4 TIM5_TRIG	TIM5_CH2	-	TIM5_UP	-
Channel 7	-	TIM6_UP	I2C2_RX	I2C2_RX	USART3_TX	DAC1	DAC2	I2C2_TX

Table 9.4 DMA2 sources for all channels

Stream/Channel	Stream 0	Stream 1	Stream 2	Stream 3	Stream 4	Stream 5	Stream 6	Stream 7
Channel 0	SPI3_RX	–	SPI3_RX	SPI2_RX	SPI2_TX	SPI3_TX	–	SPI3_TX
Channel 1	I2C1_RX	–	TIM7_UP	–	TIM7_UP	I2C1_RX	I2C1_TX	I2C1_TX
Channel 2	TIM4_CH1	–	I2S3_EXT_RX	TIM4_CH2	I2S2_EXT_TX	I2S3_EXT_TX	TIM4_UP	TIM4_CH3
Channel 3	I2S3_EXT_RX I2S3_EXT_RX	TIM2_UP TIM2_CH3	I2C3_RX I2C3_RX	I2S2_EXT_RX I2S2_EXT_RX	I2C3_TX I2C3_TX	TIM2_CH1 TIM2_CH1	TIM2_CH2 TIM2_CH4	TIM2_UP TIM2_CH4
Channel 4	UART5_RX	USART3_RX	UART4_RX	USART3_TX	UART4_TX	USART2_RX	USART2_TX	UART5_TX
Channel 5	UART8_TX UART8_TX	UART7_TX UART7_TX	TIM3_CH4 TIM3_UP	UART7_RX UART7_RX	TIM3_CH1 TIM3_TRIG	TIM3_CH2 TIM3_CH2	UART8_RX UART8_RX	TIM3_CH3 TIM3_CH3
Channel 6	TIM5_CH3 TIM5_UP	TIM5_CH4 TIM5_TRIG	TIM5_CH1 TIM5_CH1	TIM5_CH4 TIM5_TRIG	TIM5_CH2 TIM5_CH2	– –	TIM5_UP TIM5_UP	– –
Channel 7	–	TIM6_UP	I2C2_RX	I2C2_RX	USART3_TX	DAC1	DAC2	I2C2_TX

9.4.2   DMA Features

DMA can be used in direct or FIFO mode. Data transfer is initiated by each DMA request in direct mode. Transfer width can be selected as byte, half-word, or word and must be the same for source and destination. Data is first saved in a 16 byte FIFO buffer in FIFO mode. Data is transferred after it reaches the selected threshold level which can be as 1/4, 1/2, 3/4, or full. Let's assume that we set the threshold level as 1/4. This means that data transfer occurs when data size reaches 16/4 bytes in the FIFO buffer. Here, 4 bytes are transferred at once. In FIFO mode, data transfer width for source and destination can be selected differently. This becomes advantageous for applications such as DCMI module usage as explored in Chap. 14. Here, the source (digital camera) sends data in byte format. The destination (DCMI module) requires the data to be in half-word format. The FIFO mode handles this conversion automatically.

DMA can also be used in normal or circular mode. The operation ends as the required number of DMA transfers are done in the normal mode. If needed, it can be restarted again. In the circular mode, the DMA operation restarts automatically from the first address as the required number of DMA transfers is done. This becomes very advantageous in applications performing the same operation repetitively such as digital signal processing. There is also a double buffer mode available in circular mode. In this mode, DMA controller swaps between two buffers automatically after each data transfer. There is also an address increment option which can be used to automatically increase address of source or destination separately.

Size of data to be transferred can be set by the DMA controller or peripheral unit. If the DMA controller is used, then size of data that will be transferred can be selected between 1 and 65,535 (as the total number of data transfer). As an example, if data width is selected as byte, then maximum size of data becomes 65,535 bytes. If size of data that will be transferred is unknown, the peripheral unit is responsible for finishing the transfer. There is also a burst option used to send group of data at once. Burst transfer size can be selected as 4, 8, or 16. To note here, the burst transfer size is not the size of data that will be transferred. It indicates the data size to be transferred at once. As an example, if data width is selected as byte and burst size is selected as 4, then 4 byte data is transferred at once.

9.4.3   DMA Interrupts

The DMA controller can generate up to five different interrupt requests for each stream. These are half-transfer reached, transfer complete, transfer error, FIFO error, and direct mode error. Half-transfer reached and transfer complete interrupts are used for tracking the DMA transfer. The transfer error, FIFO error, and direct mode error interrupts are used for error management. Transfer error interrupt is generated if a bus error occurs during DMA read/write access. Transfer error interrupt is generated if the FIFO buffer overrun/underrun condition is detected or burst size

Fig. 9.10 DMA mode and
configuration window

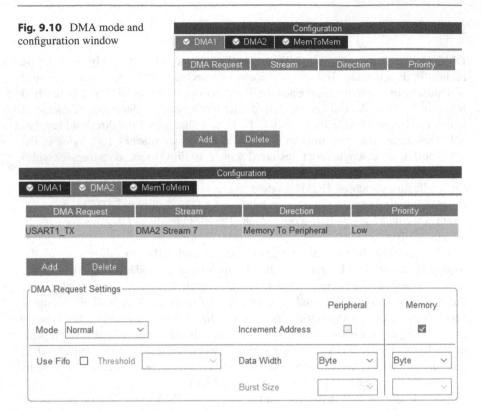

Fig. 9.11 DMA mode and configuration window after source selection

is larger than the FIFO buffer level. Direct mode error interrupt is generated if a
new DMA request occurs before existing data is transferred.

9.4.4 DMA Setup in the STM32F4 Microcontroller

We will setup the DMA only in C language in this section. In order to configure the
DMA under STM32CubeMX, go to the "System Core" dropdown menu in "Pinout
& Configuration" window, and select the desired DMA. Opening "DMA Mode and
Configuration" window will be as in Fig. 9.10. In order to add a DMA source there,
select DMA1 or DMA2, click "Add" button, and select the desired source from the
opened "Select" dropdown list. Here, Tables 9.3 and 9.4 will be of help. In order
to add the DMA source, it must be enabled beforehand as explained in previous
chapters.

After the DMA source is selected, the "DMA Mode and Configuration" window
will be changed as in Fig. 9.11. If there is more than one DMA channel for the
selected source, it can be selected from the "Stream" dropdown list. Transfer

direction can be selected from the "Direction" dropdown list. Priority for the DMA channel can be selected from the "Priority" dropdown list. DMA mode can be selected as "Normal" or "Circular" from the "Mode" dropdown list. In order to use FIFO, "Use FIFO checkbox" must be checked. Then transfer threshold size can be from the "Threshold" dropdown list. If FIFO mode is selected, burst transfer size can also be selected from the "Burst Size" dropdown lists. Auto address increment property for peripheral or memory can be enabled or disabled by selecting or deselecting related checkboxes. Data width can be selected as byte, half-word, or word for peripheral and memory from the related "Data Width" dropdown lists. As the DMA is activated for the desired source, related DMA interrupt is enabled automatically.

DMA can also be configured from the "Configuration" section of the peripheral unit. There is a "DMA Settings" window for this purpose which is identical to the "DMA Mode and Configuration" window. Same configuration steps can be applied in this window as well.

"MemToMem" selection from the "DMA Mode and Configuration" window is used for DMA transfer from memory to memory. It can be seen that there is only one DMA source as MEMTOMEM after the "Add" button is clicked. However, "DMA Stream" can be selected from "DMA2 Stream0" to "DMA2 Stream6" from the "Stream" dropdown list. Memory to memory transfer direction, normal mode, and FIFO usage are preset. However, all other configurations can be selected as desired. The DMA interrupt is not enabled automatically for this transfer type. Hence, the user should go to the "System Core" dropdown menu and enable DMA interrupt from the NVIC window.

9.4.5 DMA Usage in the STM32F4 Microcontroller

As DMA properties are set, the next step is its usage. We provide usage examples in C language for this purpose in this section. To do so, we assume that a project has already been created in STM32CubeIDE. After setting the desired DMA properties in STM32CubeMX, related DMA initialization code is created under the function MX_DMA_Init. This function is called automatically in the "main.c" file. Then, the desired peripheral unit with DMA can be used with functions special to that unit. We provide three such examples for ADC and UART modules as follows.

```
HAL_ADC_Start_DMA(ADC_HandleTypeDef* hadc, uint32_t* pData,
    uint32_t Length)
/*
hadc: pointer to the ADC_HandleTypeDef struct.
pData: pointer to array which ADC conversion results will be
    saved by DMA controller.
Length: size of the array.
*/
HAL_UART_Transmit_DMA(UART_HandleTypeDef *huart, uint8_t *pData,
    uint16_t Size)
/*
huart: pointer to the UART_HandleTypeDef struct.
```

```
pData: pointer to array which holds the data transmitted by DMA
    controller.
Size: size of the array.
*/

HAL_UART_Receive_DMA(UART_HandleTypeDef *huart, uint8_t *pData,
    uint16_t Size)
/*
huart: pointer to the UART_HandleTypeDef struct.
pData: pointer to array which received data will be saved by DMA
    controller.
Size: size of the array.
*/
```

There are also DMA callback functions specific to peripheral units. Some of these functions are as follows. The function `ADC_DMAError` is used to handle the ADC DMA error. The function `UART_DMATxHalfCplt` is used for tracking half of UART DMA transmission. The function `UART_DMAReceiveCplt` is used for tracking end of UART DMA receive operation. More information on these functions are given next.

```
ADC_DMAError(DMA_HandleTypeDef *hdma)
/*
hdma: pointer to the DMA_HandleTypeDef struct.
*/
UART_DMATxHalfCplt(DMA_HandleTypeDef *hdma)

UART_DMAReceiveCplt(DMA_HandleTypeDef *hdma)
```

There are also additional functions for the memory to memory mode as follows. The functions `HAL_DMA_Start` and `HAL_DMA_Start_IT` are used to start DMA without and with interrupt, respectively. The functions `HAL_DMA_Abort` and `HAL_DMA_Abort_IT` are used to abort them, respectively. Finally, the function `HAL_DMA_RegisterCallback` is used for assigning user functions as DMA interrupt callback functions. More information on these functions are given next.

```
HAL_DMA_Start(DMA_HandleTypeDef *hdma, uint32_t SrcAddress,
    uint32_t DstAddress, uint32_t DataLength)
/*
hdma: pointer to the DMA_HandleTypeDef struct.
SrcAddress: source memory buffer address
DstAddress: destination memory buffer address
DataLength: DMA transfer size
*/

HAL_DMA_Start_IT(DMA_HandleTypeDef *hdma, uint32_t SrcAddress,
    uint32_t DstAddress, uint32_t DataLength)

HAL_DMA_Abort(DMA_HandleTypeDef *hdma)

HAL_DMA_Abort_IT(DMA_HandleTypeDef *hdma)

HAL_DMA_RegisterCallback(DMA_HandleTypeDef *hdma,
    HAL_DMA_CallbackIDTypeDef CallbackID, void (* pCallback)(
    DMA_HandleTypeDef *_hdma))
/*
CallbackID: Identifier to select callback type. It can be
    HAL_DMA_XFER_CPLT_CB_ID for transfer complete,
```

```
HAL_DMA_XFER_HALFCPLT_CB_ID for half transfer complete,
HAL_DMA_XFER_ERROR_CB_ID for error and
HAL_DMA_XFER_ABORT_CB_ID for abort process.
pCallback: pointer to user callback function with pointer to a
DMA_HandleTypeDef struct as input.
*/
```

Let's provide examples on the usage of DMA via C language. Our first example is on memory to peripheral unit transfer using UART module. Every second, "Hello World" string will be transmitted to PC via UART using DMA. The green LED, connected to pin PG13 of the STM32F4 board, will be toggled every time to track data transfer. We will use interrupt of the Timer 1 module to obtain 1 s intervals. In order to configure this module, please see the examples given in Chap. 6. We should also configure the USART1 module in "Asynchronous Mode" with baud rate 115,200 bits/s, data length as 8 bits, 1 stop bit, and no parity. The DMA is configured from the "DMA2" window as "Normal" mode with no FIFO, byte data width for both peripheral and memory, and auto increment for memory. The stream should be selected as "DMA2 Stream 7 automatically." Priority of the DMA is left as "Low" since there is no additional DMA stream used in the code. DMA 2 module stream 0 global interrupt is enabled automatically. However, the USART1 interrupt must be enabled from the "NVIC Settings" window. We provide the C code for our first example in Listing 9.14. Please place the given code parts to appropriate places in the "main.c" file.

Listing 9.14 UART data transfer using the DMA module

```
/* USER CODE BEGIN PV */
uint8_t arr[13] = "Hello World\n\r";
/* USER CODE END PV */

/* USER CODE BEGIN 0 */
void HAL_TIM_PeriodElapsedCallback(TIM_HandleTypeDef *htim)
{
HAL_GPIO_TogglePin(GPIOG, GPIO_PIN_13);
HAL_UART_Transmit_DMA(&huart1, arr, 13);
}
/* USER CODE END 0 */

/* USER CODE BEGIN 2 */
if (__HAL_TIM_GET_FLAG(&htim1, TIM_FLAG_UPDATE) != RESET)
__HAL_TIM_CLEAR_FLAG(&htim1, TIM_FLAG_UPDATE);
HAL_TIM_Base_Start_IT(&htim1);
/* USER CODE END 2 */
```

The second example is on peripheral unit to memory transfer using the ADC module. Here, consecutive 16 ADC conversion operations are done by the help of DMA. Then, average of the obtained samples is calculated as voltage. If the average voltage level is higher than 2.5 V, the green LED connected to pin PG13 of the STM32F4 board is turned on. It will be turned off otherwise. We will trigger the ADC module with the Timer 2 module. Here, the sampling frequency will be set as 1 kHz. In order to configure the ADC module, please see the examples given in Chap. 7. The DMA is configured from the "DMA2" window as "Circular" mode with no FIFO, half-word data width for both peripheral and memory, and auto increment for memory. The stream is selected as DMA 2 Stream 0. Priority of the DMA is left as "Low" since there is no additional DMA stream used in the code. DMA 2 module stream 0 global interrupt is enabled automatically. However, the ADC interrupt must be enabled manually from the "NVIC Settings" window. Finally, DMA "Continuous Request" must be enabled from the "ADC 1 Parameter Settings" window. We provide the C code for our second example in Listing 9.15. Please place the given code parts to appropriate places in the "main.c" file.

Listing 9.15 Obtaining ADC data using the DMA module

```
/* USER CODE BEGIN PV */
uint16_t data[16];
/* USER CODE END PV */

/* USER CODE BEGIN 0 */
float ConvertToVoltage(uint16_t in)
{
return (float)in * 3.3 / 4095;
}

void HAL_ADC_ConvCpltCallback(ADC_HandleTypeDef *hadc)
{
int i;
float average = 0;
for (i = 0; i < 16; i++)
{
average += ConvertToVoltage(data[i]);
}
average /= 16;
if (average > 2.5)
HAL_GPIO_WritePin(GPIOG, GPIO_PIN_13, GPIO_PIN_SET);
else
HAL_GPIO_WritePin(GPIOG, GPIO_PIN_13, GPIO_PIN_RESET);
}
/* USER CODE END 0 */

/* USER CODE BEGIN 2 */
HAL_ADC_Start_DMA(&hadc1, (uint32_t *)data, 16);
HAL_TIM_Base_Start(&htim2);
/* USER CODE END 2 */
```

We perform memory to memory data transfer in our third example. Here, the string "stm32f4discovery" will be transferred from one memory location to another using DMA. The green LED, connected to pin PG13 of the STM32F4 board, will be turned on to indicate end of data transfer. The DMA is configured from the "MemToMem" window as "Circular" mode half FIFO threshold, byte data width for both source and destination memory and single burst size. Auto increment for both source and destination memory are also enabled. Finally, DMA 2 stream 0 is selected. DMA 2 module stream 0 global interrupt is enabled from the "NVIC Settings" window. Normal mode with FIFO is selected automatically for memory to memory transfer. We provide the C code for our third example in Listing 9.16. Please place the given code parts to appropriate places in the "main.c" file.

Listing 9.16 Transferring data from memory to memory using DMA module

```
/* USER CODE BEGIN PV */
uint8_t srcArray[16] = "stm32f4discovery";
uint8_t dstArray[16];
/* USER CODE END PV */

/* USER CODE BEGIN 0 */
static void DMATransferComplete(DMA_HandleTypeDef *DmaHandle)
{
HAL_GPIO_WritePin(GPIOG, GPIO_PIN_13, GPIO_PIN_SET);
}
/* USER CODE END 0 */

/* USER CODE BEGIN 2 */
HAL_DMA_RegisterCallback(&hdma_memtomem_dma2_stream0,
    HAL_DMA_XFER_CPLT_CB_ID, DMATransferComplete);

HAL_DMA_Start_IT(&hdma_memtomem_dma2_stream0, (uint32_t)srcArray,
    (uint32_t)dstArray, 16);
/* USER CODE END 2 */
```

9.5 Flexible Memory Controller

Available memory on the STM32F4 microcontroller may not be sufficient for some applications. External memory chips (as RAM or flash memory) may be needed for such cases. The flexible memory controller (FMC) module on the STM32F4 microcontroller is the dedicated unit used for this purpose. Through it, we can easily communicate with an external memory chip. As a side note, the FMC module can also be used to communicate with different LCD interfaces. However, we will only consider FMC module usage for reaching and using the SDRAM module available on the STM32F4 board.

9.5.1 FMC Working Principles

The address space allocated for FMC usage in the STM32F4 microcontroller is divided into banks. Each bank can be used to communicate with a different memory type. As can be seen in Fig. 9.12, two different banks are allocated for SDRAM communication. This allows using two different RAM chips at the same time.

Each bank in FMC has size of 256 MB. Data transfer size within these banks can be 8, 16, or 32 bits. The maximum clock speed supported by FMC is half of the HCLK. Hence, it is half of the maximum clock speed in the microcontroller. This indicates that reaching an external RAM location will be slower compared to reaching RAM on the STM32F4 microcontroller.

In order to use the SDRAM on the STM32F4 board by the FMC module, we should adjust several timing values. We summarize them as follows. Load mode register to active delay (tMRD) is the delay between load mode register command

and activate or refresh command. Exit self-refresh delay (tXSR) is the delay between self-refresh command and activate command. Self-refresh time (tRAS) is used to set minimum self-refresh period. SDRAM common row cycle delay (tRC) is the delay between refresh command and activate command or between two consecutive refresh commands. Write recovery time (tWR) is the delay between write command and precharge command. SDRAM common row precharge delay (tRP) is the delay between precharge command and another command. Row to column delay (tRCD) is the delay between activate command and read/write command.

9.5.2   FMC Setup in the STM32F4 Microcontroller

We can use the FMC module in C, C++, and MicroPython languages. Therefore, we provide the setup procedure for the module in this section.

9.5.2.1 Setup via C Language

We have two options to set up the FMC module in C language. The first option is based on HAL library functions. The second option is based the board support package (BSP) usage. We will next consider them separately.

Setup via HAL Library Functions

We can obtain the timing values for our SDRAM on the STM32F4 board from its datasheet [2]. Then, we can make necessary calculations to obtain the integer values to be entered to the related STM32CubeMX interface as follows. In our SDRAM,

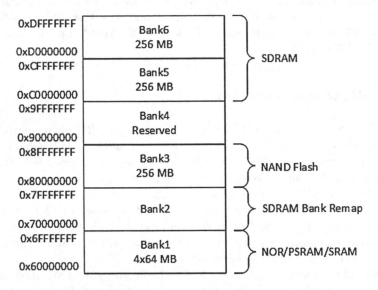

Fig. 9.12 FMC address banks for the SDRAM

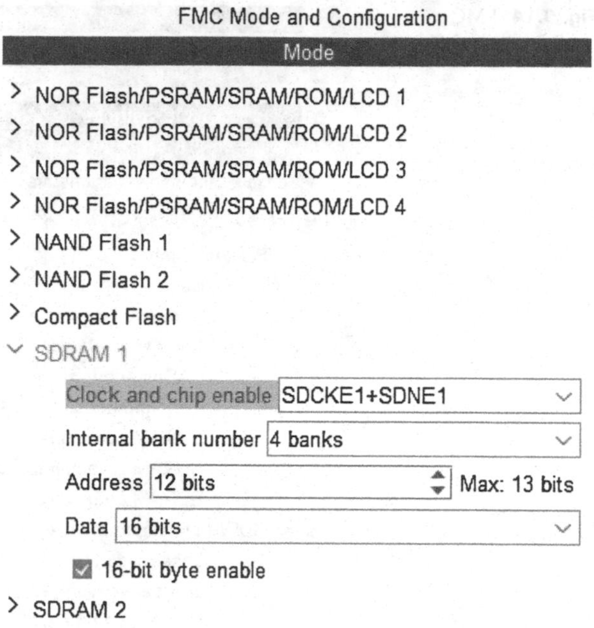

Fig. 9.13 FMC mode and configuration window

FMC Mode and Configuration

Mode

> NOR Flash/PSRAM/SRAM/ROM/LCD 1

> NOR Flash/PSRAM/SRAM/ROM/LCD 2

> NOR Flash/PSRAM/SRAM/ROM/LCD 3

> NOR Flash/PSRAM/SRAM/ROM/LCD 4

> NAND Flash 1

> NAND Flash 2

> Compact Flash

∨ SDRAM 1

Clock and chip enable	SDCKE1+SDNE1 ∨
Internal bank number	4 banks ∨
Address	12 bits ⬍ Max: 13 bits
Data	16 bits ∨

☑ 16-bit byte enable

> SDRAM 2

one memory clock cycle can be calculated as 11.11 ηc $(1/(90 * 10^{-6})$ s). Hence, tMRD should be set to two memory clock cycles. tXSR should be set to seven memory clock cycles. tRAS should be set to four memory clock cycles. tRC should be set to six memory clock cycles. tWR should be set to two memory clock cycles. tRP should be set to two memory clock cycles. Finally, tRCD should be set to two memory clock cycles.

Next, we assume that a project has already been created under STM32CubeIDE. In order to configure the FMC module for SDRAM under STM32CubeMX, go to the "Connectivity" dropdown menu in "Pinout & Configuration" window and select "FMC." Opening "FMC Mode and Configuration" window will be as in Fig. 9.13. In order to enable the FMC for SDRAM, go to SDRAM 1 or SDRAM 2, and select "SDCKE0+SDNE0" or "SDCKE1+SDNE1" from the "Clock and chip enable" dropdown list to select one of the clock and chip enable pin pairs. The number of internal SDRAM banks can be selected as 2 or 4 from the "Internal bank number" dropdown list. Number of address bits should be entered as 11, 12, or 13 to the "Address box." Number of data bits should be selected as 8 or 16 from the "Data" dropdown list. Finally, 16 bit byte enable checkbox is used to enable "FMC_NBL0" and "FMC_NBL1" pins. These are used to control data mask pins of the 16 bit SDRAM. Low byte and/or high byte of the data can be enabled or disabled using "FMC_NBL0" and "FMC_NBL1" pins, respectively.

Remaining SDRAM configurations are realized in the "Configuration" window as in Fig. 9.14. Here, SDRAM bank is selected automatically according to the selected clock and chip enable bits. Bank 1 is selected when SDCKE0+SDNE0 is set. Bank 2 is selected when SDCKE1+SDNE1 is set. Number of column and

Fig. 9.14 FMC
configuration window

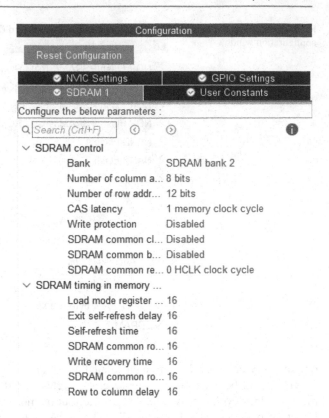

row address bits are selected from related dropdown lists. Column access strobe (CAS) latency can be selected as one, two, or three memory clock cycles from the related dropdown list. CAS latency describes the time duration between the moment an instruction is given for a column and the moment data is available. Write protection of SDRAM can be enabled/disabled from the related dropdown list. SDRAM common clock dropdown list is used to set SDRAM clock frequency. It can be selected as "2 HCLK" or "3 HCLK" clock cycles to set frequency as HCLK/2 or HCLK/3, respectively. Burst read mode can be enabled/disabled from the "SDRAM common burst read" dropdown list. SDRAM common read pipe delay dropdown list is used to specify additional data read delay after CAS latency. It can be selected as zero, one, or two HCLK clock cycles. There are also some time-related configurations under "SDRAM timing in memory clock cycles section" as described in Sect. 9.5.1.

Setup via BSP Functions

We can also benefit from the board support package (BSP) drivers provided by STMicroelectronics for configuring the FMC. BSP drivers are introduced by STMicroelectronics to simplify the usage of external modules via predefined functions. These use HAL library functions in the background. Unfortunately, BSP

Fig. 9.15 Configuring STM32CubeMX for the FMC module usage

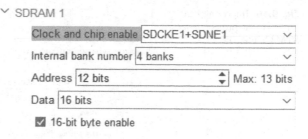

drivers require all onboard module connections to be activated in STM32CubeIDE. As an example, if we want to use a module using FMC on the board, we should also activate I^2C, SPI, LTDC, and SDRAM in STM32CubeIDE to use BSP drivers.

In order to use BSP drivers for the FMC module, we should first open a new project; press the ".ioc" file to open the "Device Configuration Tool" in STM32CubeIDE. Since all configurations are done by BSP drivers, we just have to enable the used peripheral devices in STM32CubeMX. To be more specific, we should select mode of the peripheral in its "mode and configuration window" as follows. We should set "I2C: I2C" under the I2C3 module; "Mode: Full-Duplex Master" under the SPI5 module; "Activated: checkbox" checkbox under the DMA2D module; and "Display Type: RGB565 (16bits)" under the LTDC module. SDRAM1 configuration under FMC should be as in Fig. 9.15. Besides, we should set the system clock to 180 MHz under the clock configuration tab by entering HCLK (MHz) 180. We should also set the LCD-TFT clock to 6 MHz by entering 6 (MHz).

As we save the ".ioc" file, the "main.c" file will be generated automatically. We should add BSP drivers to the project. To do so, right-click on the project under "Project Explorer." Click "Import," select "File System" under "General," and click "Next." We should add the complete address of the most recent STM32F429 folder as `Users User_Name STM32Cube Repository STM32Cube_FW_F4_V1.26.1`. Then, we should select the folders "Drivers/BSP/Components," "Drivers/BSP/STM32F429I-Discovery," and "Utilities/Fonts" as in Fig. 9.16.

Next, we should right-click on the project in "Project Explorer" to open its properties. Afterward, we should click on "Paths and Symbols" under "C/C++ General" in the opening window. We should press the "Add..." button under "Includes" section on the right. Then, we should select the added folder "Drivers/BSP" from the Workspace->Project_Name files as in Fig. 9.17. Afterward, we should select "Add to all configurations" and "Add to all languages" and press "OK." Finally, we should add the below C code script to the appropriate place in the "main.c" file.

```
/* USER CODE BEGIN Includes */
#include "STM32F429I-Discovery/stm32f429i_discovery.h"
#include "STM32F429I-Discovery/stm32f429i_discovery_sdram.h"
/* USER CODE END Includes */
```

Fig. 9.16 Importing BSP drivers to the project under STM32CubeIDE

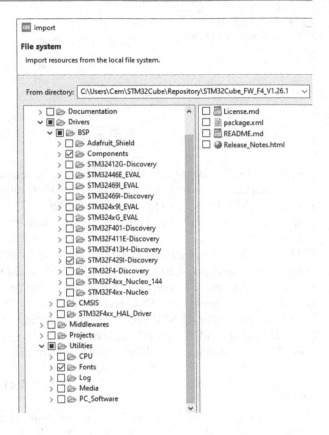

9.5.2.2 Setup via C++ Language

We can also benefit from the BSP drivers to use the FMC module in C++ language. We have two options at this stage. We will explain each option next. First, we should open a new project under Mbed Studio.

To apply the first option, we should right-click on the project and click on the "add library" button. Afterward, we should include the BSP library via the link http://os.mbed.com/teams/Embedded-System-Design-with-ARM-Cortex-M/code/BSP_DISCO_F429ZI/. As the "Next" button is pressed, the interface asks for "Branch" or "Tag." We can keep the default value. STMicroelectronics officially formed the BSP FMC wrapper library for Mbed. In this library, they formed the functions as class members. We forked this library to use with the latest Mbed version (Mbed OS6). To use this library, we should include the BSP library via the link http://os.mbed.com/teams/Embedded-System-Design-with-ARM-Cortex-M/code/SDRAM_DISCO_F429ZI/. As the "Next" button is pressed, the interface asks for "Branch" or "Tag." We can keep the default value. Then, we can use the BSP drivers by adding the code line #include "SDRAM_DISCO_F429ZI.h" to our main file.

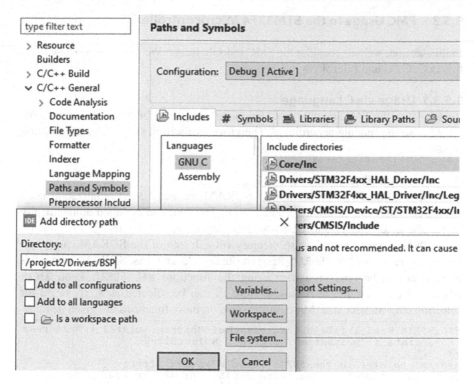

Fig. 9.17 Including BSP drivers to the project under STM32CubeIDE

To apply the second option, we will not make any specific modifications in project properties. We should only add the folders "Libraries\Cpp\BSP_DISCO_F429ZI" and "Libraries\Cpp\SDRAM_DISCO_F429ZI" directly to the project folder. To note here, these codes are available in the accompanying book web site. Then, we can use the BSP drivers by adding the code line #include "SDRAM_DISCO_F429ZI.h" to our main file.

9.5.2.3 Setup via MicroPython

MicroPython has no class to perform SDRAM operations. Therefore, we created a sdram class in our custom MicroPython firmware. The SDRAM module can be created and initialized using the function init(). All GPIO, FMC, and SDRAM module configurations are done automatically within this function. Timing configurations and initialization sequence for SDRAM module are also realized in this function. After calling this function, the SDRAM module will be ready to be used.

9.5.3   FMC Usage in the STM32F4 Microcontroller

In this section, we will consider the FMC module usage. We next provide how to do this in C, C++, and MicroPython languages.

9.5.3.1 Usage via C Language

As we setup the FMC module in C language via HAL library or BSP functions, we can use the module accordingly. Therefore, we will next consider each option separately.

Usage via HAL Library Functions

After setting the FMC properties for SDRAM, related FMC initialization code is created under the function `MX_FMC_Init`. The function is called automatically in the "main.c" file. Then, SDRAM can be configured by using the function `HAL_SDRAM_SendCommand`. Also memory refresh rate of the SDRAM can be set using the function `HAL_SDRAM_ProgramRefreshRate`. As SDRAM is configured, data can be read or written using the functions `HAL_SDRAM_Read_Xb` or `HAL_SDRAM_Write_Xb`, respectively. Here, X can be selected as 8, 16, or 32 for selecting data transfer size. More information on these functions are given next.

```
HAL_SDRAM_Read_Xb(SDRAM_HandleTypeDef *hsdram, uint32_t *pAddress
    , uintX_t *pDstBuffer, uint32_t BufferSize)
/*
hsdram: pointer to the SDRAM\_HandleTypeDef struct.
pAddress: pointer to beginning SDRAM address for read process
DstAddress: pointer to destination buffer
BufferSize: size of the destination buffer
*/

HAL_SDRAM_Write_Xb(SDRAM_HandleTypeDef *hsdram, uint32_t *
    pAddress, uintX_t *pDstBuffer, uint32_t BufferSize)
/*
hsdram: pointer to the SDRAM_HandleTypeDef struct.
pAddress: pointer to beginning SDRAM address for write process
DstAddress: pointer to source buffer
BufferSize: size of the source buffer
*/

HAL_SDRAM_ProgramRefreshRate(SDRAM_HandleTypeDef *hsdram,
    uint32_t RefreshRate)
/*
hsdram: pointer to the SDRAM_HandleTypeDef struct.
RefreshRate: Refresh rate of SDRAM
*/

HAL_SDRAM_SendCommand(SDRAM_HandleTypeDef *hsdram,
    FMC_SDRAM_CommandTypeDef *Command, uint32_t Timeout)
/*
hsdram: pointer to the SDRAM_HandleTypeDef struct.
Timeout: timeout duration
Command: SDRAM command struct given below.
typedef struct
{
  uint32_t CommandMode;                      // SDRAM Command. It can
        be FMC_SDRAM_CMD_NORMAL_MODE, FMC_SDRAM_CMD_CLK_ENABLE,
        FMC_SDRAM_CMD_PALL_MODE, FMC_SDRAM_CMD_AUTOREFRESH_MODE,
```

```
    FMC_SDRAM_CMD_LOAD_MODE, FMC_SDRAM_CMD_SELFREFRESH_MODE, or
    FMC_SDRAM_CMD_POWERDOWN_MODE.
  uint32_t CommandTarget;                // SDRAM Bank number. It
    can FMC_SDRAM_CMD_TARGET_BANK1 or
    FMC_SDRAM_CMD_TARGET_BANK2.
  uint32_t AutoRefreshNumber;            // Number of consecutive
    auto refresh command issued in auto refresh mode.
  uint32_t ModeRegisterDefinition;       // Value of the Mode
    Register of SDRAM
}FMC_SDRAM_CommandTypeDef;
*/
```

Let's provide an example on the usage of the FMC module for SDRAM. In this example, 16×32 bit data is written to the onboard SDRAM. Then, the data is read back. If the written and read data match, the green LED connected to pin PG13 of the STM32F4 board is turned on. Otherwise, the red LED connected to pin PG14 is turned on. In this example, HCLK is set to 180 MHz using the external HSE crystal.

The SDRAM on the STM32F4 board has SDCKE1 and SDNE1 pins connected to clock end chip enable pins. Hence, they must be selected from STM32CubeMX. This SDRAM has 4 banks, 12 address pins, and 16 data pins. Related parts are configured accordingly. FMC_NBL0 and FMC_NBL1 pins are connected physically to the SDRAM. Hence, 16-bit byte enable checkbox is selected even this property is not used in code. SDRAM has 12 row bits and 8 column bits. CAS latency can be programmed as 2 or 3. We will set it to 2. Burst or write protection will not be used. SDRAM clock is set to 90 MHz by setting SDRAM common clock to 2 HCLK. SDRAM uses a pipelined architecture. Hence, there is no delay between read commands such that SDRAM common read pipe delay is set to 0 HCLK. Afterward, SDRAM timing configurations are realized as explained in Sect. 9.5.2.1. We formed an initialization function for this purpose.

The SDRAM bank 2 is addressed between 0xD0000000 and 0xDFFFFFFF. We select 0xD0000E00 as the starting address of SDRAM. Before using the SDRAM, we should initialize it. Initialization steps for the SDRAM are as follows. After power-up, the memory clock is stabilized with DQM and CKE high for 100 µs. This step can be realized by setting Command-Mode to FMC_SDRAM_CMD_CLK_ENABLE. CommandTarget is set to FMC_SDRAM_CMD_TARGET_BANK2 and it will not change through the initialization process. Auto refresh and mode registers are not used at this step. Hence, AutoRefreshNumber and ModeRegisterDefinition are set to 0. Then, command is sent to SDRAM and system is delayed for 1 ms. All banks are precharged to leave them in idle state. This step can be realized by setting CommandMode to FMC_SDRAM_CMD_PALL. Then, the command is sent to SDRAM. At least two AUTO REFRESH cycles are performed. This step can be realized by setting CommandMode to FMC_SDRAM_CMD_AUTOREFRESH_MODE and AutoRefreshNumber is set to 3. Then, this command is sent to SDRAM. Finally, the mode register is programmed. Mode register has 12 bits. Bits 11 and 10 are reserved and must be set to logic level 0. Bit 9 is used to specify the write burst mode. Bits 8 and 7 are used to specify the operating mode. Standard operation is selected by resetting these bits and only this mode can be used.

Bits from 6 to 4 are used to set CAS latency. They must be set to 0b011 for three and 0b010 for two memory clock cycles. Bit 3 is used to specify the burst type. Bits from 2 to 0 are used to specify the burst length. This step can be realized by setting CommandMode to FMC_SDRAM_CMD_LOAD_MODE and ModeRegisterDefinition to 0b000000100000 since burst is not enabled and CAS is set to 2. Then, this command is sent to SDRAM.

After initialization, SDRAM refresh rate is set. According to the SDRAM datasheet, 4096 refresh cycles are realized every 64 ms. Hence, input of the function HAL_SDRAM_ProgramRefreshRate is calculated as (90000000×0.064/4096). Then, SDRAM is ready for write/read operations. We provide the C code for our example in Listing 9.17. Please place the given code parts to appropriate places in the "main.c" file.

Listing 9.17 Using SDRAM with FMC module

```
/* USER CODE BEGIN PV */
FMC_SDRAM_CommandTypeDef myCommand;

uint32_t sourceArr[16] = {
0x01020304, 0x05060708, 0x090A0B0C, 0x0D0E0F10,
0x11121314, 0x15161718, 0x191A1B1C, 0x1D1E1F20,
0x21222324, 0x25262728, 0x292A2B2C, 0x2D2E2F30,
0x31323334, 0x35363738, 0x393A3B3C, 0x3D3E3F40};
uint32_t receiveArr[16];
uint32_t refreshRate;
uint8_t errorFlag = 0;
uint32_t i;
/* USER CODE END PV */

/* USER CODE BEGIN 0 */
void SDRAM_Init(void)
{
myCommand.CommandMode = FMC_SDRAM_CMD_CLK_ENABLE;
myCommand.CommandTarget = FMC_SDRAM_CMD_TARGET_BANK2;
myCommand.AutoRefreshNumber = 0;
myCommand.ModeRegisterDefinition = 0;
HAL_SDRAM_SendCommand(&hsdram1, &myCommand, 100);
HAL_Delay(1);
myCommand.CommandMode = FMC_SDRAM_CMD_PALL;
HAL_SDRAM_SendCommand(&hsdram1, &myCommand, 100);
myCommand.CommandMode = FMC_SDRAM_CMD_AUTOREFRESH_MODE;
myCommand.AutoRefreshNumber = 3;
HAL_SDRAM_SendCommand(&hsdram1, &myCommand, 100);
myCommand.CommandMode = FMC_SDRAM_CMD_LOAD_MODE;
myCommand.ModeRegisterDefinition = 0b000000100000;
HAL_SDRAM_SendCommand(&hsdram1, &myCommand, 100);
}
/* USER CODE END 0 */

/* USER CODE BEGIN 2 */
SDRAM_Init();
refreshRate = (uint32_t)(90000000 * 0.064 / 4096);
HAL_SDRAM_ProgramRefreshRate(&hsdram1, refreshRate);
HAL_SDRAM_Write_32b(&hsdram1, (uint32_t *)(0xD0000E00), sourceArr
    , 16);
HAL_SDRAM_Read_32b(&hsdram1, (uint32_t *)(0xD0000E00), receiveArr
    , 16);
```

```
for (i = 0; i < 16; i++)
{
if (sourceArr[i] != receiveArr[i])
{
errorFlag = 1;
break;
}
}
if (errorFlag == 1)
HAL_GPIO_WritePin(GPIOG, GPIO_PIN_14, GPIO_PIN_SET);
else
HAL_GPIO_WritePin(GPIOG, GPIO_PIN_13, GPIO_PIN_SET);
/* USER CODE END 2 */
```

We next repeat our first FMC example via pointer usage. To do so, we should set up a new project with the same settings as in the first one. We will only modify the C code in Listing 9.17 such that instead of using the functions HAL_SDRAM_Write_32b and HAL_SDRAM_Read_32b, we will benefit from pointers. Hence, we will read and write data to memory as if we are reaching an array. We provide the modified code in Listing 9.18.

Listing 9.18 Using SDRAM with FMC module with pointer usage

```
/* USER CODE BEGIN PV */
FMC_SDRAM_CommandTypeDef myCommand;

uint32_t sourceArr[16] = {
0x01020304, 0x05060708, 0x090A0B0C, 0x0D0E0F10,
0x11121314, 0x15161718, 0x191A1B1C, 0x1D1E1F20,
0x21222324, 0x25262728, 0x292A2B2C, 0x2D2E2F30,
0x31323334, 0x35363738, 0x393A3B3C, 0x3D3E3F40};
uint32_t receiveArr[16];
uint32_t refreshRate;
uint8_t errorFlag = 0;
uint32_t i;
uint32_t *memDataAddress;
/* USER CODE END PV */

/* USER CODE BEGIN 0 */
void SDRAM_Init(void)
{
myCommand.CommandMode = FMC_SDRAM_CMD_CLK_ENABLE;
myCommand.CommandTarget = FMC_SDRAM_CMD_TARGET_BANK2;
myCommand.AutoRefreshNumber = 0;
myCommand.ModeRegisterDefinition = 0;
HAL_SDRAM_SendCommand(&hsdram1, &myCommand, 100);
HAL_Delay(1);
myCommand.CommandMode = FMC_SDRAM_CMD_PALL;
HAL_SDRAM_SendCommand(&hsdram1, &myCommand, 100);
myCommand.CommandMode = FMC_SDRAM_CMD_AUTOREFRESH_MODE;
myCommand.AutoRefreshNumber = 3;
HAL_SDRAM_SendCommand(&hsdram1, &myCommand, 100);
myCommand.CommandMode = FMC_SDRAM_CMD_LOAD_MODE;
myCommand.ModeRegisterDefinition = 0b000000100000;
HAL_SDRAM_SendCommand(&hsdram1, &myCommand, 100);
}
/* USER CODE END 0 */

/* USER CODE BEGIN 2 */
SDRAM_Init();
```

```
refreshRate = (uint32_t)(90000000*0.064/4096) ;
HAL_SDRAM_ProgramRefreshRate (&hsdram1 , refreshRate);

memDataAddress = (uint32_t*)0xD0000E00;

for(i = 0; i < 16; i++){
*(memDataAddress + i)= sourceArr[i];
}

for(i = 0; i < 16; i++){
receiveArr[i] = *(memDataAddress + i);
}

for(i = 0; i < 16; i++){
if(sourceArr[i] != receiveArr[i]){
errorFlag = 1;
break;
}
}
if(errorFlag == 1) HAL_GPIO_WritePin(GPIOG, GPIO_PIN_14,
    GPIO_PIN_SET);
else HAL_GPIO_WritePin(GPIOG, GPIO_PIN_13 , GPIO_PIN_SET);
/* USER CODE END 2 */
```

Usage via HAL Library Functions

We will next benefit from the available BSP driver functions to use the SDRAM via FMC module in C language. There are three functions for the blocking (polling) mode operation of the SDRAM module. The function BSP_SDRAM_Init is used to initialize the onboard SDRAM. The functions BSP_SDRAM_WriteData and BSP_SDRAM_ReadData are used to write data to and read data from the SDRAM, respectively. We next provide these functions in detail.

```
uint8_t BSP_SDRAM_Init(void)
uint8_t BSP_SDRAM_WriteData(uint32_t uwStartAddress, uint32_t *
    pData, uint32_t uwDataSize)
/*
uwStartAddress: write start address of SDRAM.
pData: pointer of the data to be written
uwDataSize: size of the data to be written
*/

uint8_t BSP_SDRAM_ReadData(uint32_t uwStartAddress, uint32_t *
    pData, uint32_t uwDataSize)
/*
uwStartAddress: read start address of SDRAM.
pData: pointer of the data to be read
uwDataSize: size of the data to be read
*/
```

We next repeat the example Listing 9.17 now by using BSP functions. We provide the modified C code for this purpose in Listing 9.19.

Listing 9.19 Using SDRAM with FMC module via BSP functions

```
/* USER CODE BEGIN PD */
#include "STM32F429I-Discovery/stm32f429i_discovery.h"
#include "STM32F429I-Discovery/stm32f429i_discovery_sdram.h"
/* USER CODE END PD */

/* USER CODE BEGIN PV */
uint32_t sourceArr[16] = {
0x01020304, 0x05060708, 0x090A0B0C, 0x0D0E0F10,
0x11121314, 0x15161718, 0x191A1B1C, 0x1D1E1F20,
0x21222324, 0x25262728, 0x292A2B2C, 0x2D2E2F30,
0x31323334, 0x35363738, 0x393A3B3C, 0x3D3E3F40};
uint32_t receiveArr[16];
uint8_t errorFlag = 0;
uint32_t i;
/* USER CODE END PV */

/* USER CODE BEGIN 2 */
BSP_SDRAM_Init();
BSP_SDRAM_WriteData((uint32_t)(0xD0000E00), sourceArr, 16);
BSP_SDRAM_ReadData((uint32_t)(0xD0000E00), receiveArr, 16);
for (i = 0; i < 16; i++)
{
if (sourceArr[i] != receiveArr[i])
{
errorFlag = 1;
break;
}
}
if (errorFlag == 1)
HAL_GPIO_WritePin(GPIOG, GPIO_PIN_14, GPIO_PIN_SET);
else
HAL_GPIO_WritePin(GPIOG, GPIO_PIN_13, GPIO_PIN_SET);
/* USER CODE END 2 */
```

To execute the C code in Listing 9.19, we should open a new project under STM32CubeIDE. Then, we should apply the settings in Sect. 9.5.2.1. We should also add the BSP driver files. As we debug and run the code, 16×32 bit data will be written to the onboard SDRAM. Then, it is read back. If the written and read data match, the green LED connected to pin PG13 of the STM32F4 board will turn on. Otherwise, the red LED connected to pin PG14 will turn on.

We can also repeat the example in Listing 9.18 using BSP functions. To do so, we formed the C code in Listing 9.20. In this code, first the FMC is initialized. Then, we use pointer operations to write data to the onboard SDRAM. Then, it is read back. We can debug and run the code as in the previous example.

Listing 9.20 Using SDRAM with FMC module via BSP functions with pointer usage

```
/* USER CODE BEGIN Includes */
#include "STM32F429I-Discovery/stm32f429i_discovery.h"
#include "STM32F429I-Discovery/stm32f429i_discovery_sdram.h"
/* USER CODE END Includes */

/* USER CODE BEGIN PV */
uint32_t sourceArr[16] = {
0x01020304, 0x05060708, 0x090A0B0C, 0x0D0E0F10,
0x11121314, 0x15161718, 0x191A1B1C, 0x1D1E1F20,
```

```
0x21222324, 0x25262728, 0x292A2B2C, 0x2D2E2F30,
0x31323334, 0x35363738, 0x393A3B3C, 0x3D3E3F40};
uint32_t receiveArr[16];
uint8_t errorFlag = 0;
uint32_t i;
uint32_t *memDataAddress;
/* USER CODE END PV */

/* USER CODE BEGIN 2 */
BSP_SDRAM_Init();

//  BSP_SDRAM_WriteData((uint32_t)(0xD0000E00), sourceArr, 16);
//  BSP_SDRAM_ReadData((uint32_t)(0xD0000E00), receiveArr, 16);
memDataAddress = (uint32_t*)0xD0000E00;

for(i = 0; i < 16; i++){
*(memDataAddress + i)= sourceArr[i];
}

for(i = 0; i < 16; i++){
receiveArr[i] = *(memDataAddress + i);
}

for (i = 0; i < 16; i++)
{
if (sourceArr[i] != receiveArr[i])
{
errorFlag = 1;
break;
}
}
if (errorFlag == 1)
HAL_GPIO_WritePin(GPIOG, GPIO_PIN_14, GPIO_PIN_SET);
else
HAL_GPIO_WritePin(GPIOG, GPIO_PIN_13, GPIO_PIN_SET);
/* USER CODE END 2 */
```

9.5.3.2 Usage via C++ Language

The official STMicroelectronics BSP SDRAM library for Mbed can be reached from
https://os.mbed.com/teams/ST/code/SDRAM_DISCO_F429ZI/. As mentioned in
Sect. 9.5.2.2, we forked this library to use with the most recent Mbed OS version.
The BSP driver functions in this library are formed as class members. We provide
details of the class members next.

```
SDRAM_DISCO_F429ZI()
/*
Class constractor. Initializes the SDRAM device.
*/

void SDRAM_DISCO_F429ZI::Init(void)
/*
Initializes the SDRAM device.
*/

void SDRAM_DISCO_F429ZI::ReadData(uint32_t uwStartAddress,
    uint32_t *pData, uint32_t uwDataSize)
/*
Reads an mount of data from the SDRAM memory in polling mode.
uwStartAddress : Read start address
```

```
pData : Pointer to data to be read
uwDataSize: Size of read data from the memory
*/

void SDRAM_DISCO_F429ZI::WriteData(uint32_t uwStartAddress,
    uint32_t *pData, uint32_t uwDataSize)
/*
Writes an mount of data to the SDRAM memory in polling mode.
uwStartAddress : Write start address
pData : Pointer to data to be written
uwDataSize: Size of written data from the memory
*/
```

The library contains the Init function for initialization. The constructor of the class SDRAM_DISCO_F429ZI also performs the initialization operation. The class function ReadData reads an amount of data from the SDRAM memory. The class function WriteData writes an amount of data to the SDRAM memory.

We next repeat the two examples in the previous section in C++ language. To do so, we formed the C++ codes in Listings 9.21 and 9.22. These codes perform the same operations as with the ones in Listings 9.17 and 9.18, respectively.

Listing 9.21 Using SDRAM with FMC module

```
#include "SDRAM_DISCO_F429ZI.h"
#include "mbed.h"
#include <cstdio>

DigitalOut greenLED(LED1);
DigitalOut redLED(LED2);

SDRAM_DISCO_F429ZI sdram;

uint32_t sourceArr[16] = {
0x01020304, 0x05060708, 0x090A0B0C, 0x0D0E0F10,
0x11121314, 0x15161718, 0x191A1B1C, 0x1D1E1F20,
0x21222324, 0x25262728, 0x292A2B2C, 0x2D2E2F30,
0x31323334, 0x35363738, 0x393A3B3C, 0x3D3E3F40};

uint32_t receiveArr[16];

bool errorFlag = false;

int main()
{
redLED = 0;
greenLED = 0;

sdram.WriteData(0xD0000E00, sourceArr, 16);
sdram.ReadData(0xD0000E00, receiveArr, 16);

for (int i = 0; i < 16; i++)
{
if (sourceArr[i] != receiveArr[i])
{
errorFlag = true;
break;
}
}
```

```
if (errorFlag)
redLED = 1;
else
greenLED = 1;

while (true)
{
thread_sleep_for(1000);
}
}
```

Listing 9.22 Using SDRAM with FMC module with pointer usage

```
#include "SDRAM_DISCO_F429ZI.h"
#include "mbed.h"
#include <cstdio>

DigitalOut greenLED(LED1);
DigitalOut redLED(LED2);

SDRAM_DISCO_F429ZI sdram;

uint32_t sourceArr[16] = {
0x01020304, 0x05060708, 0x090A0B0C, 0x0D0E0F10,
0x11121314, 0x15161718, 0x191A1B1C, 0x1D1E1F20,
0x21222324, 0x25262728, 0x292A2B2C, 0x2D2E2F30,
0x31323334, 0x35363738, 0x393A3B3C, 0x3D3E3F40};

uint32_t receiveArr[16];

bool errorFlag = false;
uint32_t i;
uint32_t *memDataAddress;

int main()
{
redLED = 0;
greenLED = 0;

memDataAddress = (uint32_t*)0xD0000E00;

for(i = 0; i < 16; i++){
*(memDataAddress + i)= sourceArr[i];
}

for(i = 0; i < 16; i++){
receiveArr[i] = *(memDataAddress + i);
}

for (int i = 0; i < 16; i++)
{
if (sourceArr[i] != receiveArr[i])
{
errorFlag = true;
break;
}
}

if (errorFlag)
redLED = 1;
else
```

```
greenLED = 1;

while (true)
{
thread_sleep_for(1000);
}
}
```

9.5.3.3 Usage via MicroPython

There are three functions as `write8b(pAddress, pSrcBuffer)`, `write16b` `(pAddress, pSrcBuffer)`, and `write32b(pAddress, pSrcBuffer)` to write data to SDRAM in our custom MicroPython firmware. These functions are used to write 8, 16, and 32 bit data packages to the SDRAM, respectively. In all these functions, `pAddress` is a list with two elements representing the beginning of the 32 bit SDRAM address for write process. The first element of the `pAddress` is the high 16 bit of 32 bit SDRAM address. The second element of the `pAddress` is the low 16 bit of 32 bit SDRAM address. The `pSrcBuffer` is the list which holds data to be written as 8, 16, or 32 bit elements according to the function. The function `read8b(pAddress, BufferSize)` is used to read 8-bit data from the SDRAM. The `BufferSize` is used to determine the number of bytes which will be read from the SDRAM.

We next provide an example on the usage of MicroPython functions to read and write data to the SDRAM on the STM32F4 board. Here, three 32-bit data are written to the SDRAM starting from a specific address. Then, the same data is read from the SDRAM as 12 bytes and printed in a loop. We provide the MicroPython code for our example in Listing 9.23.

Listing 9.23 Using SDRAM with FMC module, the MicroPython code

```
import sdram

sdram.init()
sdram.write32b([0xD000, 0x0E00], [0x01020304, 0x05060708, 0
    x090A0B0C], 3)
data = sdram.read8b([0xD000, 0x0E00], 12)

for i in range(12):
    print(data[i])
```

9.6 Application: Memory-Based Operations in the Robot Vacuum Cleaner

We handle the map generation and sweeping algorithms to be used by our robot vacuum cleaner in this chapter. Moreover, we store the location, sweeping mode, and other important information to SDRAM as the robot is turned off. We provided the equipment list, circuit diagram, detailed explanation of design specifications,

and peripheral unit settings in a separate document in the folder "Chapter9\
EoC_applications" of the accompanying supplementary material. In the same
folder, we also provided the complete C, C++, and MicroPython codes for the
application.

9.7   Summary of the Chapter

Memory is scarce in embedded systems. Therefore, memory management requires
special consideration in them. Due to the importance of the topic, we focused
on memory and memory management in this chapter. To do so, we started with
general definitions for RAM and flash memory. Then, we explored how memory
management can be done in C, C++, and MicroPython languages. We also evaluated
DMA and FMC as dedicated hardware modules for RAM usage. We also expanded
our previously introduced real-world example by adding memory management
property to it as the end of chapter application. Being scarce, memory space (both
as RAM and flash memory) should be used with caution. This chapter provides
methods to perform this. Hence, the reader can develop better embedded code by
mastering them.

Problems

9.1. What do AMBA, AHB, and APB stand for? Where are they used?

9.2. Explain memory data storage blocks for RAM and flash memory.

9.3. How can code and data be protected in a microcontroller? How is this done in
the STM32F4 microcontroller?

9.4. Obtain the ID of your STM32F4 microcontroller by forming a project using:

(a) C language under STM32CubeIDE.
(b) C++ language under Mbed.

9.5. Form a project with the following specifications. There is an array of unsigned
characters with size 160×120. Fill this array with random (unsigned character)
numbers ranging from 0 to 255. We would like to calculate the occurrence of
each unsigned character value in the array. Hence, form an unsigned integer array
with 256 elements. We can call this array as histogram. Use the function call by
reference method to calculate the occurrence of each unsigned character value in
the array. Form the project using:

(a) C language under STM32CubeIDE.
(b) C++ language under Mbed.

9.6. Form a project to implement a state machine using function pointers. There are five states in the machine as A, B, C, D, and E. B is the initial state and state machine runs as B -> C -> E -> A -> D -> B. State transitions occur by the user button press on the STM32F4 board. Form the project using:

(a) C language under STM32CubeIDE.
(b) C++ language under Mbed.

9.7. Form a project with the following specifications. There should be a function in the C code. We would like to count how many times this function has been called during code execution. How can this be done by a static variable? Form the project using:

(a) C language under STM32CubeIDE.
(b) C++ language under Mbed.

9.8. Form a project in MicroPython with the following specifications. In the code, there will be a function which adds, subtracts, or multiplies two numbers with the select input. ADD, SUBTRACT, and MULTIPLY are set as constants and they will be used in the function to select the operation.

9.9. Form a project in MicroPython with the following specifications. In the code, a pre-allocated buffer is created with 12 elements. This buffer is used to obtain the string "Hello World!" from PC via UART communication.

9.10. The function `gc.mem_free()` returns the free heap size in MicroPython. Form MicroPython project such that the function `gc.mem_free()` is called in a for loop 20 times. What happens to free unallocated memory every time this function is called? Now call the function `gc.collect()` after each time the function `gc.mem_free()` is called. What has changed?

9.11. Form a project in STM32CubeIDE with the following specifications. In the project, the ADC module is used with DMA property to obtain data from the internal temperature sensor. The ADC module is triggered with Timer 3 module such that the sampling frequency is set to 100 Hz. After acquiring 100 samples, their average is sent to PC via UART module.

9.12. Expand the project in Problem 9.5 such that the array has size 640 × 480. Here, we should store this array in the external RAM on the STM32F4 board. Therefore, we should activate and use the FMC module.

References

1. Python Community: https://docs.python.org/3/library/stdtypes.html. Accessed 4 June 2021
2. ISSI: 1 Meg Bits x 16 Bits x 4 Banks (64-MBIT) Synchronous Dynamic RAM, rev. g edn. (2014)
3. STMicroelectronics: STM32F405/415,   STM32F407/417,   STM32F427/437   and STM32F429/439 advanced Arm-based 32-bit MCUs, rm0090 rev 19 edn. (2021)

Real-Time Operating Systems

10

10.1 Fundamentals of RTOS

Codes we have formed up to now fall into the bare-metal category to indicate that they are directly executed on the microcontroller. This option becomes unfeasible when different parts (or functions) of the code try to reach a resource (such as shared data, code, or peripheral unit) at the same time. For such cases, there should be an organizer controlling flow of the code to ensure efficient usage of microcontroller resources. RTOS is introduced for this purpose. This is done by adding an extra code layer to handle all resource management and organization operations. Hence, extra code and RAM usage requirements emerge. However, these overheads become acceptable compared to the advantages offered by RTOS. To note here, codes written by RTOS follow a different structure compared to bare-metal programming.

10.1.1 RTOS Components

Forming code with RTOS is not as straightforward as in bare-metal programming. The main reason is that RTOS depends on scheduling operations for efficient usage of microcontroller resources. This is done by RTOS components specialized for different operations. The most important component for us is the kernel which is in fact a scheduler adjusting code flow. Besides, the kernel controls resource sharing between different parts of the code and allows communication between them. To

Supplementary Information The online version contains supplementary material available at (https://doi.org/10.1007/978-3-030-88439-0_10).

C. Ünsalan et al., *Embedded System Design with Arm Cortex-M Microcontrollers*, https://doi.org/10.1007/978-3-030-88439-0_10

perform all these operations, kernel depends on dividing the code into blocks called tasks.

A task is similar to a function with one difference. The task has its own stack (RAM space). The kernel is responsible for scheduling and executing tasks. This is done by assigning priority levels to them. Besides, other kernel components such as event, mutex, and semaphore can be used in operation. Kernel also depends on queue and mail for the communication between tasks. We will explain all these components in detail in the following sections.

10.1.2 RTOS Working Principles

We will provide an overview of RTOS working principles in this section. To do so, let's start with general layout of a generic RTOS code in Fig. 10.1. As can be seen in this figure, the code is divided into blocks called tasks. The kernel controls and organizes task execution based on its priority and state. There is also an idle task inherent in the generated code. RTOS has this component such that when no other task is being executed at a specific time instant, the CPU executes the idle task. Besides, RTOS provides communication option between tasks. If task operations are modified through this communication, then the event, mutex, or semaphore should be used. If only data transfer is needed between tasks, then queue or mail should be used by the kernel.

Now, let's focus on how RTOS works. As we compile and start executing the code, the kernel starts working by checking the state and priority level of each task. If a task is ready to be executed and has the highest priority, then it is executed.

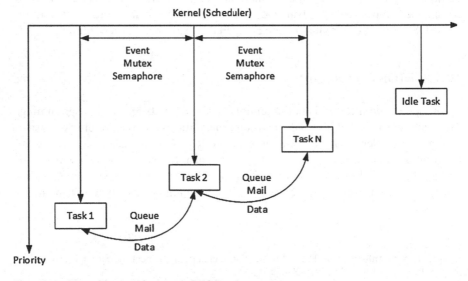

Fig. 10.1 General layout of a generic RTOS code

The code to be executed within the task should be wrapped by an infinite loop. The reason for is that, each task will be visited more than once in operation. Within the infinite loop, there should be a function which yields execution of the current task and sets the task in wait state. Hence, the kernel calls the next task in ready state to be executed. If no other task is available to be executed at a given time instant, then the kernel calls the idle task. This operation goes indefinitely in the same order. As can be seen here, each task is called based on its priority and state in RTOS. The aim here is the effective usage of microcontroller resources between different tasks. As a reminder, resource means data, code, or peripheral unit. We will explain the task and its properties in detail in Sect. 10.3.

The kernel can start or stop executing a specific task by the help of event, mutex, or semaphore. Hence, task execution order can be controlled and altered besides its priority level. We will explain the event usage in detail in Sect. 10.4. Likewise, we will focus on the mutex and semaphore in Sect. 10.5.

Data transfer may be needed between tasks in operation. We can use queue or mail for this purpose. We will explain these operations in detail in Sect. 10.6.

All kernel operations in RTOS are done at specific time instants called kernel ticks. Kernel tick is generated by SysTick (system timer) introduced in Sect. 6.2.4. The kernel checks status of the currently executed and remaining tasks at each kernel tick. Based on this, a new task may be started or the existing one may continue operating. This is done automatically within the kernel.

FreeRTOS has specific functions to perform time-based operations in connection with kernel. We will handle these topics in Sect. 10.7. Finally, RTOS needs memory management in operation. We will focus on this topic in Sect. 10.8.

The user can follow RTOS operations while executing a code by advanced debug features or available software such as Percepio Tracealyzer or Segger SystemView. Unfortunately, these programs are not free. Hence, we did not cover them in this book. We suggest the reader to consult one such program when detailed RTOS code analysis is needed.

10.2   FreeRTOS and Mbed OS

We will use FreeRTOS and Mbed OS throughout the chapter. Hence, we will briefly introduce them in this section. Besides, we will provide how to set up a project with FreeRTOS and Mbed OS.

10.2.1 FreeRTOS

FreeRTOS is a popular real-time operating system for microcontrollers. It is distributed freely under the MIT open-source license. Detailed information on FreeRTOS can be obtained from its web site https://www.freertos.org. The reader can also consult resources for RTOS components and their usage from the mentioned web site.

We will use FreeRTOS under STM32CubeIDE. Arm® formed an extra layer on top of FreeRTOS to be compatible with its CMSIS-based operations. As a result, RTOS and CMSIS operations do not interfere. As of writing this book, the formed extra layer was having the version "CMSIS_V2."

10.2.2 FreeRTOS Project Setup in STM32CubeIDE

We can use FreeRTOS in an STM32CubeIDE project by the help of STM32CubeMX. To do so, we should open a new project under STM32CubeIDE as explained in Sect. 3.1.5. Then, we should enable FreeRTOS under the "Middleware" in the "Category" section of the "Pinout and Configuration" tab. Next, we should select the "Mode" as "CMSIS_V2" from the "FreeRTOS Mode and Configuration" section on the right as in Fig. 10.2. We will extensively use this window while setting other FreeRTOS properties in the following sections.

As explained in Sect. 10.1, RTOS operations are synchronized based on kernel ticks generated by SysTick. Hence, STM32CubeMX warns the user to pick a HAL time base source other than SysTick. We can change this source from the "System Source" section of the "Pinout and Configuration" tab. Here, we should modify the "Timebase Source" under "Sys Mode and Configuration" under the "Sys" option. It is generally suggested to use Timer 6 (TIM6) for this operation. We also applied this setup in our projects in this chapter.

Fig. 10.2 FreeRTOS mode and configuration selection

Fig. 10.3 FreeRTOS file
layout from the project
explorer window

As we generate the code from the RTOS setup, all related files will be added
to our project. We can check their layout from the project explorer window as in
Fig. 10.3. As can be seen in this window, FreeRTOS sources are placed under the
folder "Third_Party->FreeRTOS." FreeRTOS functions for RTOS components can
be found in related C files. For example, all functions related to tasks can be found in
the file "tasks.c." All CMSIS-RTOS functions can be found in the file "cmsis_os2.c"
which is placed under the folder "CMSIS_RTOS_V2."

We can observe the "main.c" file contents added specific for FreeRTOS usage
in Listing 10.1. Let's briefly analyze this file. First of all, handle and attributes

structures for RTOS components are placed in section Private variables.
Default task handle and its attributes structure including its name, priority, and stack
size have already been placed in this section. After all peripherals are initialized,
RTOS scheduler is initialized using the function osKernelInitialize. Then,
RTOS components are created in associated sections. For example, functions used
to create mutexes are placed between lines USER CODE BEGIN RTOS_MUTEX and
USER CODE END RTOS_MUTEX. After all RTOS components are created, the kernel
is started using the function osKernelStart. Finally, functions which will be
executed for each task are created. For example, the function StartDefaultTask
is created for the default task.

Listing 10.1 Contents of the "main.c" file in a FreeRTOS-based project

```
/* USER CODE END Header */
/* Includes
   --------------------------------------------------------------------
   */
#include "cmsis_os.h"

/* Private variables
   ------------------------------------------------------------*/
/* Definitions for defaultTask */
osThreadId_t defaultTaskHandle;
const osThreadAttr_t defaultTask_attributes = {
.name = "defaultTask",
.stack_size = 128 * 4,
.priority = (osPriority_t)osPriorityNormal,
};

/* Private function prototypes
   ------------------------------------------------*/
void StartDefaultTask(void *argument);

/* Init scheduler */
osKernelInitialize();

/* USER CODE BEGIN RTOS_MUTEX */
/* add mutexes, ... */
/* USER CODE END RTOS_MUTEX */

/* USER CODE BEGIN RTOS_SEMAPHORES */
/* add semaphores, ... */
/* USER CODE END RTOS_SEMAPHORES */

/* USER CODE BEGIN RTOS_TIMERS */
/* start timers, add new ones, ... */
/* USER CODE END RTOS_TIMERS */

/* USER CODE BEGIN RTOS_QUEUES */
/* add queues, ... */
/* USER CODE END RTOS_QUEUES */

/* Create the thread(s) */
/* creation of defaultTask */
defaultTaskHandle = osThreadNew(StartDefaultTask, NULL, &
    defaultTask_attributes);

/* USER CODE BEGIN RTOS_THREADS */
/* add threads, ... */
```

```
/* USER CODE END RTOS_THREADS */

/* USER CODE BEGIN RTOS_EVENTS */
/* add events, ... */
/* USER CODE END RTOS_EVENTS */

/* Start scheduler */
osKernelStart();

/* We should never get here as control is now taken by the
     scheduler */
/* Infinite loop */

/* USER CODE BEGIN Header_StartDefaultTask */
/**
* @brief  Function implementing the defaultTask thread.
* @param  argument: Not used
* @retval None
*/
/* USER CODE END Header_StartDefaultTask */
void StartDefaultTask(void *argument)
{
/* USER CODE BEGIN 5 */
/* Infinite loop */
for(;;)
{
osDelay(1);
}
/* USER CODE END 5 */
}
```

10.2.3 Mbed OS

We will use Mbed OS under Mbed Studio. This operating system is based on
CMSIS-RTOS RTX. Here, all kernel components are defined as objects and they
are treated as such during code execution. The reader can get more information on
Mbed OS from its web site https://os.mbed.com/docs/mbed-os/.

10.2.4 First Mbed OS Project in Mbed Studio

We can form our first Mbed OS project by an available template as in Sect. 3.2.4. To
do so, we should pick the "mbed-os-example-blinky" project under Mbed Studio.
For completeness, we provide the "main.cpp" file of this project in Listing 10.2.
This will be our base for the remaining Mbed OS projects in this book.

Listing 10.2 The "main.cpp" file of the "mbed-os-example-blinky" project in Mbed Studio

```
#include "mbed.h"

#define BLINKING_RATE 500ms

int main()
{
```

```
DigitalOut greenLED(LED1);

while (true)
{
greenLED = !greenLED;
ThisThread::sleep_for(BLINKING_RATE);
}
}
```

While compiling the code in Listing 10.2, Mbed Studio includes all related OS components to the project. Hence, as we modify the code in future applications, we can benefit from them. As we run the code, the green LED on the STM32F4 board will start blinking. Different from the bare metal example in Sect. 3.2.4, this operation is done by RTOS components.

10.3   Task and Thread

Task is the main building block in RTOS. Before going further, we should mention that we have tasks in FreeRTOS and threads in Mbed OS. The main difference between the task and thread is their usage of memory space in operation. Since we are dealing with embedded systems in this book, this difference is not visible. Besides, we could not find a solid reference clearly indicating the difference between the task and thread in embedded systems. As a side note, the thread and task are clearly separated in TI RTOS [1] since Texas Instrument assigned specific meaning to each. In this book, we will always use the task term in FreeRTOS projects to be compatible with literature. Likewise, we will use the thread term in Mbed OS projects. While explaining general concepts, we will always use the task term.

10.3.1 Task Working Principles

To fully grasp RTOS operations, we should focus on task working principles. To do so, we should understand task states, task priorities, and idle task. Before going further, we should mention that each task has a code block wrapped by infinite loop. However, this does not mean that when the CPU starts executing the task, it will stay there forever. On the contrary, within the infinite loop, there should be a function which yields privilege of the current task and let it wait. Hence, the kernel calls the next ready task to be executed. To do so, the kernel should check two key values. The first one is the task state. The second one is the task priority. Based on these, either the next waiting (or blocked) task is called. Or the existing task is executed again if it has higher priority. We will consider the task state and priority next.

10.3.1.1 Task States
While the kernel is active, a task can be in four different states. We summarize them according to the FreeRTOS and Mbed OS naming as follows. When the task is being executed, it is in the "Running" state. There can be only one task in this state at any

time. The task to be executed next is in the "Ready" state. This can happen either by the current tasks yielding its privilege to other tasks or letting the running task to the waiting state. If a task waits for necessary resources to become available, we call it to be in the "Waiting" state. If a task has not been created yet, it will be in the "Inactive" state. Based on the defined states, the kernel synchronizes RTOS operations and changes the state of each task dynamically.

10.3.1.2 Task Priorities

Task priority defines which task should be given precedence in execution. There are 48 priority levels (excluding the priority level of the idle task) in FreeRTOS and Mbed OS. These levels are grouped as osPriorityLow, osPriorityBelowNormal, osPriorityNormal, osPriorityAboveNormal, osPriorityHigh, and osPriorityRealtime. The priority in these groups go from lowest to highest as listed. Each group has eight levels within itself. The priority in each group also goes from lowest to highest from the first to the last level.

The task with higher priority will always preempt the one with lower priority. Hence, it will take the precedence in execution. If the running and ready tasks have the same priority, then the kernel will wait for the running task to end its operation. As the running task yields, the ready task is executed.

10.3.1.3 Idle Task

Based on the structure of RTOS operations, the kernel calls a task based on its state and priority. The question arises. What happens when no task is to be called at certain time instant? For such cases, RTOS consists of an idle task having lowest priority (set by the system). Hence, when no other task is being executed, the idle task is called. Any task called afterward preempts the idle task during its execution.

FreeRTOS allows assigning operations to the idle task. Hence, the user can modify the idle task content. As an example, we can let the CPU to low power mode in the idle task under FreeRTOS. We will show how this can be done in Sect. 10.3.2.

Mbed OS has a fixed idle task (called idle loop) such that the system goes to low power mode when it is executed. Hence, we can benefit from both power saving option and let the CPU to wait in the idle task. We can also assign an operation to the idle task under Mbed OS. We will show how this can be done in Sect. 10.3.3.

10.3.2 Task in FreeRTOS

We will summarize how tasks are handled in FreeRTOS in this section. We will provide functions related to task modification and execution as well. To show how these concepts can be realized, we will also provide task usage examples.

10.3.2.1 Task Setup and Task Functions

As we form a project following the steps in Sect. 10.2.2, the default task is generated. We can observe its properties under the "Configuration" part of the "FreeRTOS

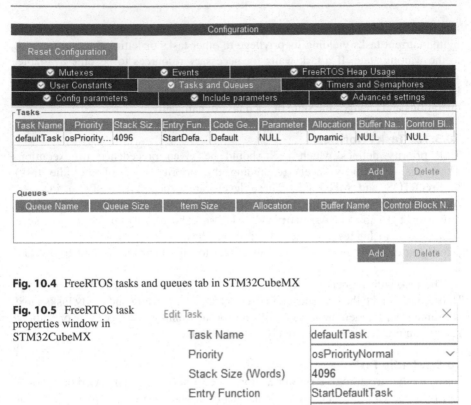

Fig. 10.4 FreeRTOS tasks and queues tab in STM32CubeMX

Fig. 10.5 FreeRTOS task
properties window in
STM32CubeMX

Mode and Configuration" section in STM32CubeMX interface. There, we should visit the "Tasks and Queues" tab as in Fig. 10.4. As can be seen in this figure, STM32CubeMX does not allow deleting the default task. We can define other tasks from the same window. We will do so in the following sections.

We can observe and modify properties of a task in the "Tasks and Queues" tab by double-clicking on it. A new window appears as in Fig. 10.5. Here, the reader can change the name, priority, stack size (to be explained in detail in Sect. 10.8), entry function (for the task), and other properties of the task.

As we make modifications in STM32CubeMX and save project files, the following code segments will be added to three sections of the "main.c" file of the project.

```
/* Definitions for defaultTask */
osThreadId_t defaultTaskHandle;
const osThreadAttr_t defaultTask_attributes = {
  .name = "defaultTask",
  .stack_size = 128 * 4,
  .priority = (osPriority_t) osPriorityNormal,
};

  /* Create the thread(s) */
  /* creation of defaultTask */
  defaultTaskHandle = osThreadNew(StartDefaultTask, NULL, &
      defaultTask_attributes);
/* USER CODE BEGIN Header_StartDefaultTask */
/**
  * @brief  Function implementing the defaultTask thread.
  * @param  argument: Not used
  * @retval None
  */
/* USER CODE END Header_StartDefaultTask */
void StartDefaultTask(void *argument)
{
  /* USER CODE BEGIN 5 */
  /* Infinite loop */
  for(;;)
  {
    osDelay(1);
  }
  /* USER CODE END 5 */
}
```

In the first section, the task handle osThreadId_t is created with the name
defaultTaskHandle. Also, task attributes structure including its name, stack size,
and priority is created. In the second section, the task is created with its handle and
attributes structure using the function osThreadNew. defaultTaskHandle will be
used to call the created task afterward. In the third section, the StartDefaultTask
function can be used to define what to do under the default task (or other tasks to be
added in the following sections). Here, the reader can observe that the task has an
infinite loop. We explained the rationale for this setup in Sect. 10.3.1.

We can summarize the functions related to task management in FreeRTOS as
follows. The function osThreadNew is used to create a new task. The task must
be created after the kernel is initialized with the function osKernelInitialize.
The function osThreadGetName is used to obtain the name of the task. The
function osThreadGetId is used to obtain the id of the active task. The function
osThreadGetState is used to obtain the current state of the task. The function
osThreadSetPriority is used to set a new priority level to the task. The
function osThreadGetPriority is used to obtain the priority level of the task.
The function osThreadYield is used to pass the control to the next task in ready
state. The function osThreadSuspend is used to suspend execution of the task. The
function osThreadResume is used to resume execution of the task. The functions
osThreadDetach and osThreadJoin are used to return all resources used by
the task to the system. Tasks can either be detached or joinable. The function
osThreadDetach is used to change the attribute of the task to detached. Then,

all resources used by this task are returned to the system when it is terminated. On the other hand, joinable tasks are kept alive until one explicitly calls the function osThreadJoin. The function osThreadExit is used to terminate the calling task. The function osThreadTerminate is used to remove the task from the list of active tasks. If the current state of the task is running, it is terminated and the execution continues with the next task in ready state. The function osThreadGetStackSize is used to obtain the stack size of the task. The function osThreadGetStackSpace is used to obtain the unused stack size of the task. The function osThreadGetCount is used to obtain the number of active tasks. The function osThreadEnumerate is used to obtain the number of enumerated tasks. Except the function osThreadGetId, other functions cannot be used inside an interrupt callback function. We next provide detailed information on the mentioned task functions.

```
osThreadId_t osThreadNew(osThreadFunc_t func, void *argument,
    const osThreadAttr_t *attr)
/*
func: task function.
argument: pointer to the destination ImageTypeDef struct.
attr: pointer to the struct containing task attributes. Set it to
    NULL in order to use default values.
*/

const char* osThreadGetName(osThreadId_t thread_id)
/*
thread_id: id which identifies the thread.
*/

osThreadId_t osThreadGetId(void)

osThreadState_t osThreadGetState(osThreadId_t thread_id)

osStatus_t osThreadSetPriority(osThreadId_t thread_id,
    osPriority_t priority)
/*
priority: new priority level for task
*/

osPriority_t osThreadGetPriority(osThreadId_t thread_id)

osStatus_t osThreadYield(void)

osStatus_t osThreadSuspend(osThreadId_t thread_id)

osStatus_t osThreadResume(osThreadId_t thread_id)

osStatus_t osThreadDetach(osThreadId_t thread_id)

osStatus_t osThreadJoin(osThreadId_t thread_id)

__NO_RETURN void osThreadExit(void)

osStatus_t osThreadTerminate(osThreadId_t thread_id)

uint32_t osThreadGetStackSpace(osThreadId_t thread_id)

uint32_t osThreadGetCount(void)
```

```
uint32_t osThreadEnumerate(osThreadId_t *thread_array, uint32_t
    array_items)
/*
thread_array: pointer to array for retrieving thread IDs.
array_items: maximum number of items in array for retrieving
    thread IDs.
*/
```

There are two delay functions in FreeRTOS as `osDelay` and `osDelayUntil`. The function `osDelay` waits for a time period specified in kernel ticks. For value one, the system waits until the next kernel tick occurs. The difference of `osDelay` from the previously used `HAL_Delay` is as follows. When `osDelay` is called, state of the running task is set to waiting and it yields to the next ready task. Hence, the kernel can call it. State of the task which executed the `osDelay` function will be set to ready automatically after given amount of kernel ticks has elapsed. The `osDelayUntil` function waits until the given absolute time value in kernel ticks. If the given time value is smaller than the current kernel tick value, the task returns to the ready state immediately. We next provide detailed information on the mentioned delay functions.

```
osStatus_t osDelay(uint32_t ticks)
/*
ticks: Number of kernel ticks value.
*/

osStatus_t osDelayUntil(uint32_t ticks)
/*
ticks: absolute time value in kernel ticks.
*/
```

We should also provide common macros used in FreeRTOS functions. These have been defined in the "cmsis_os2.h" file under the FreeRTOS folder of the project. Here, `osWaitForever` is used to define the timeout value for waiting forever. `osFlagsWaitAny` is used to wait for any flag and `osFlagsWaitAll` is used to wait for all flags. `osFlagsNoClear` is used not to clear flags which have been specified to wait for. `osWaitForever` is generally used with the wait function `osDelay`. `osFlagsWaitAny`, `osFlagsWaitAll`, and `osFlagsNoClear` are used with the specific task wait function `osThreadFlagsWait`. We provide detailed information on the mentioned macros next.

```
#define osWaitForever 0xFFFFFFFFU

#define osFlagsWaitAny 0x00000000U

#define osFlagsWaitAll 0x00000001U

#define osFlagsNoClear 0x00000002U
```

10.3.2.2 Task Usage Examples

We next provide three examples on the usage of tasks in FreeRTOS. In the first example, given in Listing 10.3, we create a new task named `myTask02` apart from the default task. Priority of this task is set to `osPriorityAboveNormal`, which is higher than the priority of the default task. When the code is executed, myTask02 is called every 0.5 s. However, myTask02 and default tasks become ready every second. Since myTask02 has higher priority, it is executed first. Then, the defaultTask is executed as soon as myTask02 yields. In order to see the effect of the priority better, the reader can change osDelay(500) to HAL_Delay(500) inside the function StartTask02. Besides, the function osThreadYield() should also be added after the HAL_Delay function. As we run the modified code, we will observe that defaultTask cannot be executed since myTask02 has higher priority and takes control immediately after it yields.

Listing 10.3 The first task example in FreeRTOS

```
/* USER CODE BEGIN PV */
uint32_t cnt = 0;
/* USER CODE END PV */

void StartDefaultTask(void *argument)
{
/* USER CODE BEGIN 5 */
/* Infinite loop */
for (;;)
{
HAL_GPIO_TogglePin(GPIOG, GPIO_PIN_14);
cnt = 1;
printf("defaultTask \n\r");
osDelay(1000);
}
/* USER CODE END 5 */
}

void StartTask02(void *argument)
{
/* USER CODE BEGIN StartTask02 */

printf("Thread id:  0x%x\n", osThreadGetId());
printf("Thread name:  %s\n", osThreadGetName(myTask02Handle));
printf("Thread state:  %i\n", osThreadGetState(myTask02Handle));
printf("Thread priority: %i\n", osThreadGetPriority(
    myTask02Handle));
printf("Thread used stack size:  %lu\n", osThreadGetStackSpace(
    myTask02Handle));
/* Infinite loop */
for (;;)
{
HAL_GPIO_TogglePin(GPIOG, GPIO_PIN_13);
cnt = 2;
printf("myTask02 \n\r");
osDelay(500);
}
/* USER CODE END StartTask02 */
}
```

The code in Listing 10.3 is formed such that the green LED on the STM32F4 board is toggled, cnt variable is set to 2, and the string "myTask02" is printed by the task myTask02 every 0.5 s. The red LED on the STM32F4 is toggled, cnt variable is set to 1, and the string "defaultTask" is printed by the task defaultTask every second. By using the cnt variable and printf function, the reader can analyze execution of these tasks through ITM properties as explained in Sect. 3.4.1.1. The id, name, state, priority, and used stack space of myTask02 are also obtained and printed at the beginning of its execution.

To note here, when we form a FreeRTOS project with default settings, the function USB init is called at the beginning of the StartDefaultTask. This delays execution of this task 0.5 s in the first run. Therefore, the reader should disable the line calling the function USB init when the defaultTask is called.

In the second example, given in Listing 10.4, we modify the idle task of the project instead of adding a new task. In order to modify and use the idle task in FreeRTOS, go to "Configuration" part of the "FreeRTOS Mode and Configuration" section in CubeMX interface. There, we should visit the "Config Parameters" tab and enable the "USE_IDLE_HOOK" option under the "Hook function related definitions." We should modify the function void vApplicationIdleHook(void) automatically defined under the "freertos.c" file. This is the function called by idle task. Hence, we should add the code to be executed by the idle task to this function.

Listing 10.4 The second task example in FreeRTOS

```
void vApplicationIdleHook(void)
{
HAL_GPIO_WritePin(GPIOG, GPIO_PIN_14, GPIO_PIN_RESET);
}

void StartDefaultTask(void *argument)
{
/* USER CODE BEGIN 5 */
/* Infinite loop */
for (;;)
{
HAL_GPIO_WritePin(GPIOG, GPIO_PIN_14, GPIO_PIN_SET);
HAL_Delay(500);
osDelay(500);
}
/* USER CODE END 5 */
}
```

In Listing 10.4, the defaultTask turns on the red LED on the STM32F4 board, halts the process for 0.5 s with the function HAL_Delay, and then yields for 0.5 s with the function osDelay. The red LED on the STM32F4 board is turned off in the idle loop. As a result, the red LED toggles every 0.5 s.

In the third example, given in Listing 10.5, we add Timer 2 to the code given in Listing 10.3. We set period of the Timer 2 to 1 s. In the Timer 2 callback function, the cnt variable is set to 3 and the string "timer2" is printed. When the code is executed, myTask02 is called every 0.5 s. However, myTask02 and defaultTask become ready every second. At the same time, Timer 2 interrupt is also triggered. Here, Timer 2 interrupt has the highest priority and is executed first. Then, myTask02 is called. Finally, the defaultTask is called as soon as myTask02 yields. We should modify ITM properties as explained in Sect. 3.4.1.1 to observe printing results in this example.

Listing 10.5   The third task example in FreeRTOS

```
/* USER CODE BEGIN 2 */
HAL_TIM_Base_Start_IT(&htim2);
/* USER CODE END 2 */

void HAL_TIM_PeriodElapsedCallback(TIM_HandleTypeDef *htim)
{
/* USER CODE BEGIN Callback 0 */

/* USER CODE END Callback 0 */
if (htim->Instance == TIM6)
{
HAL_IncTick();
}
/* USER CODE BEGIN Callback 1 */
else if (htim->Instance == TIM2)
{
cnt = 3;
printf("timer2 \n\r");
}
/* USER CODE END Callback 1 */
}
```

If a task function which is not interrupt-safe is used inside the Timer 2 callback function, then the system crashes as soon as this function is executed. The reader can try this scenario by printing name of the task myTask02 inside the Timer 2 callback function. The only interrupt-safe task function is osThreadGetId. The reader can try this by printing id of the task myTask02 inside the Timer 2 callback function. He or she can see that the system runs without problem.

10.3.3 Thread in Mbed OS

As Mbed OS starts running, four threads become active. These are the ISR/scheduler, idle, timer, and main application threads. Each thread is responsible for its own operations. We should emphasize an important property here. As can be seen, the main function is taken as thread under Mbed OS. Hence, we can use it as the default thread in RTOS operations.

10.3.3.1 Thread Functions

We next summarize thread-related functions under Mbed OS as follows. The function Thread is used to create the thread with given parameters. The function start is used to start thread execution. The function join is used to wait for the thread to terminate. The function terminate is used to terminate execution of the thread and remove it from the list of active threads. The function set_priority is used to set the priority of the thread. The function get_priority is used to obtain the priority of the thread. The function flags_set is used to set the specified flags for the thread. The function get_state is used to obtain the current state of the thread. The function stack_size is used to obtain the total stack memory size of the thread. The function free_stack is used to obtain the unused stack memory size of the thread. The function used_stack is used to obtain the used stack memory size of the thread. The function max_stack is used to obtain the maximum stack memory usage of the thread to date. The function get_name is used to obtain the name of the active thread. The function get_id is used to obtain the id of the active thread. Finally, the function ~Thread is used to destruct the thread. Except the functions flags_set, get_name, and get_id, the rest cannot be used inside an interrupt callback function. We provide detailed information on the mentioned thread functions next.

```
Thread(osPriority priority=osPriorityNormal, uint32_t stack_size=
    OS_STACK_SIZE, unsigned char *stack_mem=nullptr, const char *
    name=nullptr)
/*
priority: initial priority level of the thread.
stack_size: stack size of the thread.
stack_mem: pointer for the stack area of the thread.
name: name of the thread.
*/

osStatus start(mbed::Callback< void()> task)
/*
task: function to be executed.
*/

osStatus join()

osStatus terminate()

osStatus set_priority(osPriority priority)
/*
priority:  new priority level for thread.
*/

osPriority get_priority()

uint32_t flags_set(uint32_t flags)
/*
flags: specifies the flags of the thread that should be set.
*/

State get_state()

uint32_t stack_size()

uint32_t free_stack()

uint32_t used_stack()

uint32_t max_stack()

const char* get_name()

osThreadId_t get_id()

virtual ~Thread()
```

There is also the ThisThread namespace under Mbed OS. It can be used to control the running thread. We will use it occasionally in the following sections. There may be limited functions associated with ThisThread. Hence, we strongly suggest the reader to check the Mbed OS web site for this purpose.

There is the `sleep_for` function associated with `ThisThread`. It is used for sleeping the CPU for a specified time period given by parameter `rel_time`. This function cannot be used inside an interrupt callback function. Finally, we should mention that most of the macros defined for FreeRTOS can also be used in Mbed OS.

10.3.3.2 Thread Usage Examples

In this section, we provide three examples on the usage of threads in Mbed OS. In the first example, given in Listing 10.6, we create an additional thread named `thread1` apart from the default main thread. Priority of `thread1` is set to `osPriorityAboveNormal`, which is higher than the default main thread. As the code is executed, the green LED on the STM32F4 board is toggled, and the string "thread1" is printed by the function `thread1Fn` every 0.5 s. The red LED on the STM32F4 board is toggled and the string "Main thread" is printed inside the default main thread every second.

Listing 10.6 The first thread example in Mbed OS

```
#include "mbed.h"

DigitalOut greenLED(LED1);
DigitalOut redLED(LED2);

Thread thread1(osPriorityAboveNormal);

void thread1Fn()
{
printf("Thread id:   0x%x\n", thread1.get_id());
printf("Thread name:  %s\n", thread1.get_name());
printf("Thread state:  %i\n", thread1.get_state());
printf("Thread priority: %i\n", thread1.get_priority());
printf("Thread total stack size: %u\n", thread1.stack_size());
printf("Thread used stack size:  %u\n", thread1.used_stack());
while (true)
{
greenLED = !greenLED;
printf("thread1 \n");
ThisThread::sleep_for(500ms);
}
}

int main()
{
thread1.start(thread1Fn);
while (true)
{
redLED = !redLED;
printf("Main thread \n");
ThisThread::sleep_for(1000ms);
}
}
```

When the code in Listing 10.6 is executed, `thread1` is started by the function `thread1Fn` at the beginning of the default main thread. The function `thread1Fn` is executed every 0.5 s. However, `thread1` and default main thread should be called every second. Since thread1 has higher priority, it is executed first. The default main thread is executed as soon as thread1 yields. In order to see the effect of priority better, the reader should change ThisThread::sleep_for(500ms) to wait_us(500000) inside the function thread1Fn. It can be seen that default main thread cannot be executed since thread1 has higher priority and takes the control immediately after it yields.

In the second example, given in Listing 10.7, we modify the operations in the idle loop (thread) of the project. Here, the default main thread turns on the red LED on the STM32F4 board, halts the process for 0.5 s with the function wait_us, then yields for 0.5 s with the function ThisThread::sleep_for. The red LED is turned off in the idle loop. As a result, the red LED toggles every 0.5 s.

Listing 10.7   The second thread example in Mbed OS

```
#include "mbed.h"

DigitalOut redLED(LED1);

void newIdleLoop()
{
redLED = 0;
}

int main()
{

Kernel::attach_idle_hook(newIdleLoop);

while (true)
{
redLED = 1;
wait_us(500000);
ThisThread::sleep_for(500ms);
}
}
```

In the third example, given in Listing 10.8, we add a ticker (named Ticker1) to the code given in Listing 10.6. We set period of Ticker1 to 1 s and set its callback function as ticker1Cb. In this function, both red and green LEDs on the STM32F4 board are toggled. When the code is executed, thread1 is called every 0.5 s. However, thread1, default main thread, and Ticker1 interrupt coincide at every second. Since the Ticker1 interrupt has the highest priority, its callback function ticker1Cb is called first. Then, thread1 is called. Finally, the default main thread is called as soon as thread1 yields.

Listing 10.8   The third thread example in Mbed OS

```
#include "mbed.h"

DigitalOut greenLED(LED1);
DigitalOut redLED(LED2);

Thread thread1(osPriorityAboveNormal);
Ticker ticker1;
```

```
void thread1Fn()
{
while (true)
{
greenLED = !greenLED;
printf("thread1 \n");
ThisThread::sleep_for(500ms);
}
}

void ticker1Cb()
{
redLED = !redLED;
greenLED = !greenLED;
}

int main()
{
thread1.start(thread1Fn);
ticker1.attach(&ticker1Cb, 1s);
while (true)
{
redLED = !redLED;
printf("thread2 \n");
ThisThread::sleep_for(1000ms);
}
}
```

If any thread function which is not interrupt-safe is used inside the callback function ticker1Cb, the system crashes as soon as this function is executed. The reader can try this scenario by calling the function get_priority inside the callback function ticker1Cb. On the other hand, interrupt-safe thread functions can be used inside the callback function ticker1Cb without any problem. The reader can try this scenario by calling the function get_id inside the function ticker1Cb.

10.4 Event

During RTOS operations, we may want to change state of a task by another task. This can be done by using an event. Let's start with explaining working principles of the event.

10.4.1 Event Working Principles

The event can be used in such a way that it leads one task to waiting or ready state by another task. This is done by setting and checking predefined trigger values (or flags). Hence, one task sets a specific flag to be checked by other tasks. When the task waiting for that specific flag receives it, it acts accordingly either going to the waiting or ready state.

In order to show working principles of the event, we provide its general usage layout in Fig. 10.6. As can be seen in this figure, the first task generates an event flag

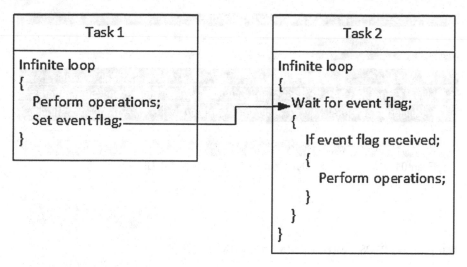

Fig. 10.6 General layout of event usage

and sends it to other tasks. Another task should be waiting for this flag in waiting state. Hence, when the flag is received, that task goes to the ready state. As a result, one task triggers another by passing the predefined event flag.

10.4.2 Event in FreeRTOS

FreeRTOS has predefined functions and setup options under STM32CubeIDE to benefit from the event. We provide these options next.

10.4.2.1 Event Setup and Event Functions

We can add an event to our FreeRTOS project in STM32CubeMX by the "Configuration" part of the "FreeRTOS Mode and Configuration" section. There, we should visit the "Events" tab as in Fig. 10.7. We can observe properties of the event by double-clicking on it.

As we make modifications in STM32CubeMX and save project files, the following code segments will be added to the two sections of the "main.c" file of the project.

```
/* Definitions for evt */
osEventFlagsId_t evtHandle;
const osEventFlagsAttr_t evt_attributes = {
  .name = "evt"
};

  /* Create the event(s) */
  /* creation of evt */
  evtHandle = osEventFlagsNew(&evt_attributes);
```

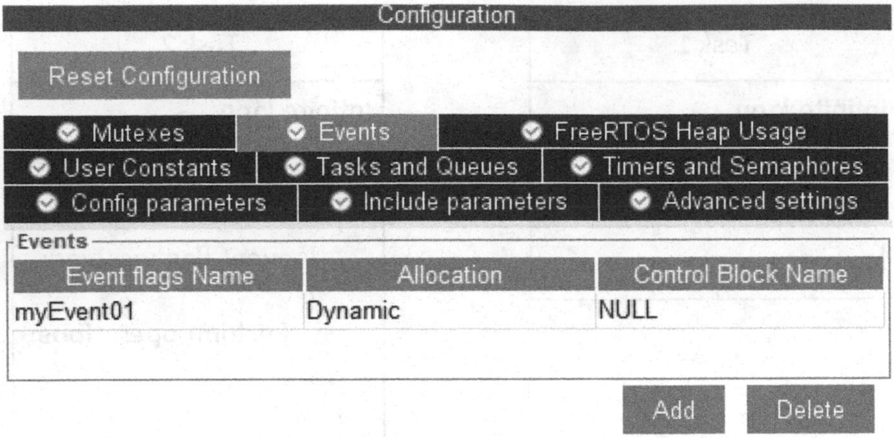

Fig. 10.7 FreeRTOS events tab in STM32CubeMX

In the first section, the event handle `osEventFlagsId_t` is created with the name `evtHandle`. Also, the event attributes structure which includes its name is created. In the second section, the event is created with its handle and attributes structure using the function `osEventFlagsNew`. `evtHandle` will be used to call the created event afterward.

The functions related to events in FreeRTOS can be summarized as follows. The function `osEventFlagsNew` is used to create a new event flags structure. The event must be created after the kernel is initialized. The function `osEventFlagsSet` is used to set the event flags. The function `osEventFlagsClear` is used to clear the event flags. The function `osEventFlagsGet` is used to obtain the current event flags. The function `osEventFlagsWait` is used to suspend execution of the running task until any or all event flags are set. The function `osEventFlagsDelete` is used to delete the event flags object. The function `osEventFlagsGetName` is used to obtain the name of the event flags object. Except `osEventFlagsNew`, `osEventFlagsDelete`, and `osEventFlagsGetName`, the remaining functions can be used inside an interrupt callback function. We next provide detailed explanation of the event functions in FreeRTOS.

```
osEventFlagsId_t osEventFlagsNew(const osEventFlagsAttr_t *attr)
/*
attr: pointer for event attributes struct
*/

uint32_t osEventFlagsSet(osEventFlagsId_t ef_id, uint32_t flags)
/*
ef_id: event flags id.
flags: event flags.
*/

uint32_t osEventFlagsClear(osEventFlagsId_t ef_id, uint32_t flags
   )
```

```
uint32_t osEventFlagsGet(osEventFlagsId_t ef_id)

uint32_t osEventFlagsWait(osEventFlagsId_t ef_id, uint32_t flags,
    uint32_t options, uint32_t timeout)
/*
options: event flag options.
timeout: event timeout value.
*/

osStatus_t osEventFlagsDelete(osEventFlagsId_t ef_id)

const char* osEventFlagsGetName(osEventFlagsId_t ef_id)
```

10.4.2.2 Event Usage Examples

In this section, we provide two examples on the usage of events in FreeR-TOS. In the first example, given in Listing 10.9, we create an additional task named myTask02 apart from the default task. The priority of this task is set to osPriorityAboveNormal, which is higher than the priority of the default task. As the code is executed, the green LED on the STM32F4 board is toggled, cnt variable is set to 1, event flag is cleared and then set to FLAG2, and the string "defaultTask" is printed by the default task every 0.5 s if the event flag is FLAG1. The red LED on the STM32F4 board is toggled, cnt variable is set to 2, event flag is cleared and then set to FLAG1, and the string "myTask02" is printed by the task myTask02 every 0.5 s if the event flag is FLAG2.

Listing 10.9 The first event example in FreeRTOS

```
/* USER CODE BEGIN PD */
#define FLAG1 0x01
#define FLAG2 0x10
/* USER CODE END PD */

/* USER CODE BEGIN PV */
uint32_t cnt = 0;
/* USER CODE END PV */

void StartDefaultTask(void *argument)
{
/* USER CODE BEGIN 5 */
/* Infinite loop */
for (;;)
{
osEventFlagsWait(myEvent01Handle, FLAG1, osFlagsWaitAny,
    osWaitForever);
HAL_GPIO_TogglePin(GPIOG, GPIO_PIN_13);
cnt = 1;
osEventFlagsClear(myEvent01Handle, FLAG1);
osEventFlagsSet(myEvent01Handle, FLAG2);
printf("defaultTask \n\r");
osDelay(500);
}
/* USER CODE END 5 */
}

void StartTask02(void *argument)
{
```

```
/* USER CODE BEGIN StartTask02 */
osEventFlagsSet(myEvent01Handle, FLAG1);
/* Infinite loop */
for (;;)
{
osEventFlagsWait(myEvent01Handle, FLAG2, osFlagsWaitAny,
    osWaitForever);
HAL_GPIO_TogglePin(GPIOG, GPIO_PIN_14);
cnt = 2;
osEventFlagsClear(myEvent01Handle, FLAG2);
osEventFlagsSet(myEvent01Handle, FLAG1);
printf("myTask02 \n\r");
osDelay(500);
}
/* USER CODE END StartTask02 */
}
```

When the code in Listing 10.9 is executed, the event flag is set to FLAG1 at the beginning of the function StartTask02. The task myTask02 has higher priority than default task. Therefore, it must be called first. However, both tasks wait for the event flag which is initially set to FLAG1. Hence, the default task is called first. It toggles the green LED on the STM32F4 board and sets the event flag to FLAG2. As soon as the event flag is set to FLAG2, myTask02 gets the precedence. Therefore, it is called before the default task yields since it has higher priority. As a result, the red LED on the STM32F4 board is toggled; the event flag is set to FLAG1; the string "myTask02" is printed. Then, myTask02 yields. Finally, the string "defaultTask" is printed, the default task yields, and both tasks wait for 0.5 s. We should modify ITM properties as explained in Sect. 3.4.1.1 to observe printing results in this example.

In the second example, given in Listing 10.10, we modify the code in Listing 10.5 by an additional task and event. The name of the new task is myTask03. Its priority is set to osPriorityHigh. We set the period of Timer 2 to 1 s. The event flag is set to FLAG1 inside the Timer 2 callback function. When the flag is set as such, the task myTask03 is called, cnt variable is set to 3, the string "timer2" is printed, and the event flag is cleared. Then, myTask02 and default tasks are called according to their priority level. We should modify ITM properties as explained in Sect. 3.4.1.1 to observe printing results in this example.

Listing 10.10 The second event example in FreeRTOS

```
/* USER CODE BEGIN PV */
uint32_t cnt = 0;
/* USER CODE END PV */

/* USER CODE BEGIN 2 */
HAL_TIM_Base_Start_IT(&htim2);
/* USER CODE END 2 */

void StartDefaultTask(void *argument)
{
/* USER CODE BEGIN 5 */
/* Infinite loop */
for (;;)
{
HAL_GPIO_TogglePin(GPIOG, GPIO_PIN_14);
cnt = 1;
```

```
printf("defaultTask \n\r");
osDelay(1000);
}
/* USER CODE END 5 */
}

void StartTask02(void *argument)
{
/* USER CODE BEGIN StartTask02 */
osEventFlagsSet(myEvent01Handle, FLAG1);
/* Infinite loop */
for (;;)
{
HAL_GPIO_TogglePin(GPIOG, GPIO_PIN_13);
cnt = 2;
printf("myTask02 \n\r");
osDelay(500);
}
/* USER CODE END StartTask02 */
}

void StartTask03(void *argument)
{
/* USER CODE BEGIN StartTask03 */
/* Infinite loop */
for (;;)
{
osEventFlagsWait(myEvent01Handle, FLAG1, osFlagsWaitAny,
    osWaitForever);
cnt = 3;
printf("timer2 \n\r");
osEventFlagsClear(myEvent01Handle, FLAG1);
osThreadYield();
}
/* USER CODE END StartTask03 */
}

void HAL_TIM_PeriodElapsedCallback(TIM_HandleTypeDef *htim)
{
/* USER CODE BEGIN Callback 0 */
/* USER CODE END Callback 0 */
if (htim->Instance == TIM6) {
HAL_IncTick();
}
/* USER CODE BEGIN Callback 1 */
else if (htim->Instance == TIM2)
{
osEventFlagsSet(myEvent01Handle, FLAG1);
}
/* USER CODE END Callback 1 */
}
```

We already know that an interrupt callback function must be kept as short as possible. Therefore, we only set the event flag inside the callback function and trigger another task to perform interrupt-related operations inside the task in Listing 10.10. This setup can be used for all callback functions. As a side note, the function osEventFlagsSet is interrupt-safe and can be used inside the Timer 2 callback function.

10.4.3 Event in Mbed OS

We will handle event-related functions in Mbed OS and their usage in active projects in this section. Let's start with event-related functions.

10.4.3.1 Event Functions

Mbed OS has several functions related to events. The function EventFlags is used to create and initialize the EventFlags object. The function set is used to set the event flags. The function clear is used to clear the event flags. The function get is used to obtain the current event flags. The functions wait_all, wait_all_for, and wait_all_until are used to wait until all of the specified event flags are set. The functions wait_any, wait_any_for, and wait_any_until are used to wait until any of the specified event flags are set. Finally, the function ~EventFlags is used to destruct the EventFlags object. The functions EventFlags, ~EventFlags, wait_all_until, and wait_any_until cannot be used inside an interrupt call-back function. The functions clear, get, and set can be used inside an interrupt callback function. The remaining functions can be used inside a callback function if their timeout value is set to zero. We provide detailed explanation of the event functions under Mbed OS next.

```
EventFlags()

EventFlags(const char *name)
/*
name: name of the EventFlags.
*/

uint32_t set(uint32_t flags)
/*
flags: the flags which will be set.
*/

uint32_t clear(uint32_t flags=0x7fffffff)
/*
flags: the flags which will be cleared.
*/

uint32_t get()

uint32_t wait_all(uint32_t flags=0, uint32_t millisec=
    osWaitForever, bool clear=true)
/*
flags: the flags to wait for.
millisec: timeout value.
clear: variable to clear specified event flags after waiting for
    them. Event flags are cleared if it is set to true, and they
    are not cleared if it is set to false.
*/

uint32_t wait_all_for(uint32_t flags, Kernel::Clock::duration_u32
    rel_time, bool clear=true)
/*
rel_time: timeout value.
*/
```

```
uint32_t wait_all_until(uint32_t flags, Kernel::Clock::time_point
    abs_time, bool clear=true)
/*
abs_time: timeout value.
*/

uint32_t wait_any(uint32_t flags=0, uint32_t millisec=
    osWaitForever, bool clear=true)

uint32_t wait_any_for(uint32_t flags, Kernel::Clock::duration_u32
    rel_time, bool clear=true)

uint32_t wait_any_until(uint32_t flags, Kernel::Clock::time_point
    abs_time, bool clear=true)

~EventFlags()
```

10.4.3.2 Event Usage Examples

In this section, we provide two examples on the usage of events in Mbed OS. In the first example, given in Listing 10.11, we create an additional thread named thread1 apart from default main thread. The priority of this thread is set to osPriorityAboveNormal, which is higher than the priority of the default main thread. As the code is executed, the green LED on the STM32F4 board is toggled, the event flag is cleared and then set to FLAG2, and the string "Main Thread" is printed by the default task every 0.5 s if event flag is FLAG1. The red LED on the STM32F4 board is toggled, the event flag is cleared and then set to FLAG1, and the string "thread1" is printed by thread1 every 0.5 s if the event flag is FLAG2.

Listing 10.11 The first event example in Mbed OS

```
#include "mbed.h"

#define FLAG1 0x01
#define FLAG2 0x10

EventFlags evt;

DigitalOut greenLED(LED1);
DigitalOut redLED(LED2);

Thread thread1;

void thread1Fn(){
evt.set(FLAG2);
while (true) {
evt.wait_any(FLAG2,osWaitForever);
greenLED = !greenLED;
evt.clear(FLAG2);
evt.set(FLAG1);
printf("thread1\r\n");
ThisThread::sleep_for(500ms);
}
}

int main()
{
thread1.start(thread1Fn);
```

```
while (true) {
evt.wait_any(FLAG1,osWaitForever);
redLED = !redLED;
evt.clear(FLAG1);
evt.set(FLAG2);
printf("Main Thread\r\n");
ThisThread::sleep_for(500ms);
}
}
```

In Listing 10.11, the event flag is set to FLAG1 at the beginning of the function thread1Fn. thread1 has higher priority than the default main thread. Hence, normally thread1 must be executed first. However, both threads wait for the event flag which is initially set to FLAG1. Hence, the default main thread is executed first. It toggles the green LED on the STM32F4 board and sets the event flag to FLAG2. As soon as the event flag is set to FLAG2, thread1 takes the precedence. Hence, it is executed before the default main thread yields since it has higher priority. As a result, the red LED on the STM32F4 board is toggled, the event flag is set to FLAG1, the string "thread1" is printed, and then thread1 yields. Finally, the string "Main thread" is printed, the default main thread yields, and both threads wait for 0.5 s.

In the second example, given in Listing 10.12, we reconsider the code in Listing 10.10 now in C++ language. Normally, the function printf cannot be used inside a callback function in Mbed OS. We can overcome this problem by using an event and thread. The function printf is used inside the thread. This thread is triggered by the event flag which is set by the callback function.

Listing 10.12 The second event example in Mbed OS

```
#include "mbed.h"

#define FLAG1 0x01

DigitalOut greenLED(LED1);
DigitalOut redLED(LED2);

Thread thread1(osPriorityAboveNormal);
Thread thread2(osPriorityHigh);
Ticker ticker1;
EventFlags evt;

void thread1Fn()
{
while (true)
{
greenLED = !greenLED;
printf("thread1 \n");
ThisThread::sleep_for(500ms);
}
}

void thread2Fn()
{
while (true)
{
evt.wait_any(FLAG1, osWaitForever);
printf("ticker1\r\n");
```

```
evt.clear(FLAG1);
}
}

void ticker1Cb()
{
evt.set(FLAG1);
}

int main()
{
thread1.start(thread1Fn);
thread2.start(thread2Fn);
ticker1.attach(&ticker1Cb, 1s);
while (true)
{
redLED = !redLED;
printf("thread2 \n");
ThisThread::sleep_for(1000ms);
}
}
```

10.5 Mutex and Semaphore

There should be an order among different tasks while reaching microcontroller resources for some applications. We can use mutex (mutual exclusion) and semaphore for this purpose. To be more specific, the mutex is a lock mechanism used between two tasks. The semaphore can be thought as the combination of an event and wait mechanism. It can either be used between two tasks (binary semaphore) or more than two tasks (counting semaphore). We will consider the mutex and semaphore in detail next.

10.5.1 Mutex Working Principles

As the name implies, mutex (mutual exclusion) makes sure that the microcontroller resources can be reached by only one task at a time. We provide the general layout of mutex usage in Fig. 10.8. As can be seen in this figure, the mutex is created and one task acquires it. In other words, the task locks the mutex. Hence, only this task can reach and control all microcontroller resources. When the task (holding the mutex) finishes its operations, it releases (unlocks) the mutex. Hence, the next task can acquire it. Then, it will have the sole ownership of microcontroller resources.

10.5.2 Semaphore Working Principles

The semaphore can be thought as the combination of an event and wait mechanism. There are binary and counting semaphores. The binary semaphore uses a binary token count (increased or decreased by different tasks) to control microcontroller

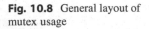

Fig. 10.8 General layout of mutex usage

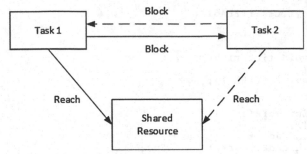

resources. When the count value is one, a task can acquire the token (decrease the count value to zero) meanwhile performs its operations. When the count value is zero, no other task can perform an operation. In other words, there is no available token. When the count value is incremented by one again, then the next task can acquire the token and perform its operations. Hence, a binary control mechanism can be formed in the same way as with mutex.

In the counting semaphore, the operation becomes more general such that more than two tasks can be coordinated. We can explain the counting semaphore by an example. Assume that we acquire 1000 samples from the microphone via ADC and DMA modules. We generate an interrupt as we perform ADC conversion. We convert each value to float form within a task (with highest priority). We apply filtering to float values in another task (with medium priority). We form a third task (with lowest priority) to convert the filtering result to integer form and feed it to the headphone via DAC module. As can be seen in this setup, the three tasks should be called in the given order for every 1000 samples acquisition. To coordinate this operation, we can form a counting semaphore with three tokens. We initially set the token count value to zero. Since there is no available token, all three tasks will be in the waiting state. Whenever the ADC interrupt is generated, three tokens are released. In other words, the token count value becomes three. Hence, each task acquires a token based on its priority level and performs the operation assigned to it. However, the tasks do not release the tokens. Hence, when all three operations are finalized, the count value will be zero. As a result, the three tasks will wait for the next ADC interrupt for available tokens. This operation goes indefinitely. From this example, we can observe that a wisely used counting semaphore can organize sequential task execution.

We provide general layout of counting semaphore usage in Fig. 10.9. Here, we pick our previous example to formalize this layout. As can be seen in this figure, at any time one task can reach microcontroller resources. The remaining tasks are blocked since there is no available token (or the count value is zero).

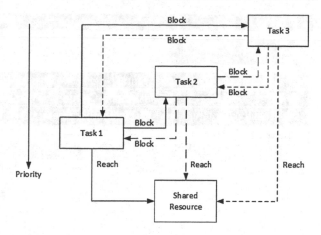

Fig. 10.9 General layout of semaphore usage

10.5.3 Mutex and Semaphore in FreeRTOS

Since mutex and semaphore share similar properties, we handle them together under FreeRTOS in this section. We will first consider mutex in terms of its setup, related functions, and usage. Then, we will consider the semaphore in the same order.

10.5.3.1 Mutex Setup and Mutex Functions

We can add a mutex to our FreeRTOS project in STM32CubeMX by the "Configuration" part of the "FreeRTOS Mode and Configuration" section. There, we should visit the "Mutexes" tab as in Fig. 10.10. We can observe properties of the mutex by double-clicking on it.

As we make modifications in STM32CubeMX and save project files, the following code segments will be added to the two sections of the "main.c" file of the project.

```
/* Definitions for Mutex1 */
osMutexId_t Mutex1Handle;
const osMutexAttr_t Mutex1_attributes = {
  .name = "Mutex1"
};

  /* Create the mutex(es) */
  /* creation of Mutex1 */
  Mutex1Handle = osMutexNew(&Mutex1_attributes);
```

In the first section, the mutex handle osMutexId_t is created with the name Mutex1Handle. Also, the mutex attributes structure which includes its name is created. In the second section, the mutex is created with its handle and attributes structure using the function osMutexNew. Mutex1Handle will be used to call the created mutex afterward.

The functions related to mutex management in FreeRTOS can be summarized as follows. The function osMutexNew is used to create a new mutex. The mutex must be created after the kernel is initialized. The function osMutexGetName is used to

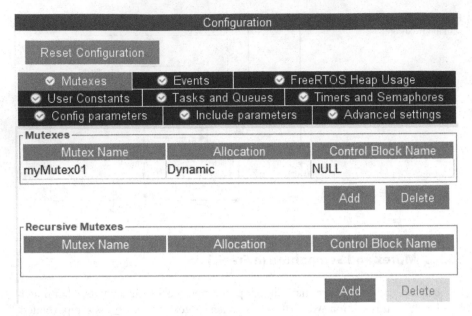

Fig. 10.10 FreeRTOS mutexes tab in STM32CubeMX

obtain the name of the mutex. The function osMutexAcquire waits for the mutex
for the given timeout value. If the mutex becomes available during this time period,
then it is taken (locked). The function osMutexRelease is used to release (unlock)
the mutex. The function osMutexGetOwner is used to obtain the id of the task
which uses the mutex. The function osMutexDelete is used to delete the mutex.
None of these functions can be used inside an interrupt callback function. We next
provide detailed information about the mentioned mutex functions.

```
osMutexId_t osMutexNew(const osMutexAttr_t *attr)
/*
attr: pointer to the struct containing mutex attributes. Set it
    to NULL in order to use default values.
*/

const char* osMutexGetName(osMutexId_t mutex_id)
/*
id: id of the mutex.
*/

osStatus_t osMutexAcquire(osMutexId_t mutex_id, uint32_t timeout)
/*
id: timeout value.
*/

osStatus_t osMutexRelease(osMutexId_t mutex_id)

osThreadId_t osMutexGetOwner(osMutexId_t mutex_id)

osStatus_t osMutexDelete(osMutexId_t mutex_id)
```

10.5.3.2 Mutex Usage Examples

We next provide an example on the usage of mutex in FreeRTOS. In Listing 10.13, we create an additional task named myTask02 apart from the default task. The priority of this task is set to osPriorityAboveNormal, which is higher than the priority of the default task. We also have the function LEDChange(int condition), which will turn on and off the green LED on the STM32F4 board according to the input condition. If the condition is 0, then the green LED is turned off. Otherwise, the green LED is turned on. The function HAL_Delay is used to halt code execution for 1 s. In the code, the function LEDChange(int condition) is protected by the created mutex.

Listing 10.13 The mutex example in FreeRTOS

```
/* USER CODE BEGIN 0 */
void LEDChange(int condition)
{
osMutexAcquire(myMutex01Handle, osWaitForever);
if (condition == 0)
{
HAL_GPIO_WritePin(GPIOG, GPIO_PIN_13, GPIO_PIN_RESET);
}
else
{
HAL_GPIO_WritePin(GPIOG, GPIO_PIN_13, GPIO_PIN_SET);
}
HAL_Delay(1000);
osMutexRelease(myMutex01Handle);
}
/* USER CODE END 0 */

void StartDefaultTask(void *argument)
{
/* USER CODE BEGIN 5 */
/* Infinite loop */
for (;;)
{
LEDChange(0);
osThreadYield();
}
/* USER CODE END 5 */
}

void StartTask02(void *argument)
{
/* USER CODE BEGIN StartTask02 */
/* Infinite loop */
for (;;)
{
LEDChange(1);
osDelay(500);
}
/* USER CODE END StartTask02 */
}
```

When the code in Listing 10.13 is executed, myTask02 is called first since it has higher priority. This task calls the function LEDChange and the green LED is turned on. After 1 s, myTask2 yields for 0.5 s. Then, the default task is called. Hence,

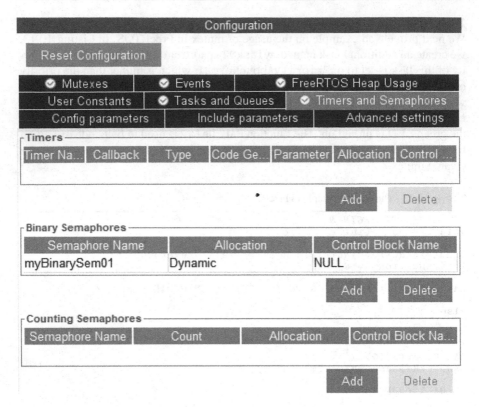

Fig. 10.11 FreeRTOS timers and semaphores tab in STM32CubeMX

the function LEDChange is executed again and the green LED is turned off. After 0.5 s, myTask2 becomes ready and tries to preempt. However, at this time mutex has been locked by the default task. Hence, myTask2 waits for extra 0.5 s even if it has higher priority.   myTask2 is called again when the default task releases (unlocks) the mutex. As a result, green LED toggles every second.

10.5.3.3 Semaphore Setup and Semaphore Functions

We can add a semaphore to our FreeRTOS project under STM32CubeMX by the "Configuration" part of the "FreeRTOS Mode and Configuration" section. There, we should visit the "Timers and Semaphores" tab as in Fig. 10.11. Here, we can choose either a binary or counting semaphore. Without loss of generality, we will continue with binary semaphore usage from this point on.

As we make modifications in STM32CubeMX and save the project files, the following code segments will be added to the two sections of the "main.c" file of the project.

```
/* Definitions for BinarySem1 */
osSemaphoreId_t BinarySem1Handle;
const osSemaphoreAttr_t BinarySem1_attributes = {
```

```
    .name = "BinarySem1"
};
    /* Create the semaphores(s) */
    /* creation of BinarySem1 */
    BinarySem1Handle = osSemaphoreNew(1, 1, &BinarySem1_attributes)
        ;
```

In the first section, the semaphore handle osSemaphoreId_t is created with the name BinarySem1Handle. The semaphore attributes struct osSemaphoreAttr_t is also created with the name BinarySem1. In the second section, the binary semaphore is created with its handle and attributes structure using the function osSemaphoreNew. BinarySem1Handle will be used to call the created binary semaphore afterward.

The functions related to semaphore management in FreeRTOS can be summarized as follows. The function osSemaphoreNew is used to create a new semaphore with initial and maximum number of tokens. The semaphore must be created after the kernel is initialized. The function osSemaphoreGetName is used to obtain the name of the semaphore. The function osSemaphoreAcquire is used to wait until any token becomes available. If a token becomes available, it is taken, and token count is decreased by one. The function osSemaphoreRelease is used to release the token and token count is increased by one. The function osSemaphoreGetCount is used to obtain the number of available tokens. The function osSemaphoreDelete is used to delete the semaphore. The functions osSemaphoreNew, osSemaphoreGetName, and osSemaphoreDelete cannot be used inside an interrupt callback function. The functions osSemaphoreRelease and osSemaphoreGetCount can be used inside interrupt callback functions. The function osSemaphoreAcquire can be used inside an interrupt callback function if its timeout value is set to zero. We next provide detailed information about the mentioned semaphore functions.

```
osSemaphoreId_t osSemaphoreNew(uint32_t max_count, uint32_t
    initial_count, const osSemaphoreAttr_t *attr)
/*
max_count: maximum number of semaphore tokens.
initial_count: initial number of available semaphore tokens.
attr: pointer to the struct containing semaphore attributes. Set
    it to NULL in order to use default values..
*/

const char* osSemaphoreGetName(osSemaphoreId_t semaphore_id)
/*
id: id of the semaphore.
*/

osStatus_t osSemaphoreAcquire(osSemaphoreId_t semaphore_id,
    uint32_t timeout)
/*
timeout: timeout value.
*/

osStatus_t osSemaphoreRelease(osSemaphoreId_t semaphore_id)

uint32_t osSemaphoreGetCount(osSemaphoreId_t semaphore_id)

osStatus_t osSemaphoreDelete(osSemaphoreId_t semaphore_id)
```

10.5.3.4 Semaphore Usage Examples

In this section, we provide two examples on the usage of semaphores in FreeRTOS. In the first example, given in Listing 10.14, we repeat the code in Listing 10.13 now using a binary semaphore. When the code is executed, myTask02 is called first since it has higher priority. This task calls the function LEDChange and the green LED on the STM32F4 board is turned on. After one second, myTask2 yields for 0.5 s. Then, the default task is called. This task calls the function LEDChange and the green LED is turned off. After 0.5 s, myTask2 becomes ready and tries to preempt. However, the binary semaphore token has been acquired by the default task. Hence, myTask2 waits for extra 0.5 s even if it has higher priority. myTask2 is executed again when the default task releases the binary semaphore token. As a result, green LED toggles every second.

Listing 10.14 The first semaphore example in FreeRTOS

```
/* USER CODE BEGIN 0 */
void LEDChange(int condition)
{
osSemaphoreAcquire(myBinarySem01Handle, osWaitForever);
if (condition == 0)
{
HAL_GPIO_WritePin(GPIOG, GPIO_PIN_13, GPIO_PIN_RESET);
}
else
{
HAL_GPIO_WritePin(GPIOG, GPIO_PIN_13, GPIO_PIN_SET);
}
HAL_Delay(1000);
osSemaphoreRelease(myBinarySem01Handle);
}
/* USER CODE END 0 */

void StartDefaultTask(void *argument)
{
/* USER CODE BEGIN 5 */
/* Infinite loop */
for (;;)
{
LEDChange(0);
osThreadYield();
}
/* USER CODE END 5 */
}

void StartTask02(void *argument)
{
/* USER CODE BEGIN StartTask02 */
/* Infinite loop */
for (;;)
{
LEDChange(1);
osDelay(500);
}
/* USER CODE END StartTask02 */
}
```

In the second example, given in Listing 10.15, we benefit from the counting semaphore to control three tasks with different priorities. myTask03 has the highest priority and fills the array arr (with 10 elements) from 0 to 9. myTask02 has the second highest priority and it multiplies each element of the array arr by 5. Finally, the default task has the lowest priority and it prints the elements of array arr. Each task waits for a semaphore token. Therefore, we set the period of Timer 2 to 1 s to generate an interrupt. Three semaphore tokens are released by the Timer 2 callback function. Hence, every time an interrupt is generated by Timer 2, myTask03, myTask02, and the default task are called based on their priority. To note here, the semaphore number is initially reset by the function osSemaphoreNew. Therefore, no operation is done unless the timer interrupt comes. We should modify ITM properties as explained in Sect. 3.4.1.1 to observe printing results in this example.

Listing 10.15 The second semaphore example in FreeRTOS

```
/* USER CODE BEGIN PV */
uint32_t cnt = 0;
uint32_t arr[10];
uint32_t i;
/* USER CODE END PV */

/* USER CODE BEGIN 2 */
HAL_TIM_Base_Start_IT(&htim2);
/* USER CODE END 2 */

void StartDefaultTask(void *argument)
{
/* USER CODE BEGIN 5 */
/* Infinite loop */
for (;;)
{
osSemaphoreAcquire(myCountingSem01Handle, osWaitForever);
for (i = 0; i < 10; i++)
{
cnt = arr[i];
printf("%u \n", cnt);
}
osDelay(100);
}
/* USER CODE END 5 */
}

void StartTask02(void *argument)
{
/* USER CODE BEGIN StartTask02 */
/* Infinite loop */
for (;;)
{
osSemaphoreAcquire(myCountingSem01Handle, osWaitForever);
for (i = 0; i < 10; i++)
{
arr[i] = arr[i] * 5;
}
osDelay(100);
}
/* USER CODE END StartTask02 */
```

```
}
void StartTask03(void *argument)
{
/* USER CODE BEGIN StartTask03 */
/* Infinite loop */
for (;;)
{
osSemaphoreAcquire(myCountingSem01Handle, osWaitForever);
for (i = 0; i < 10; i++)
{
arr[i] = i;
}
osDelay(100);
}
/* USER CODE END StartTask03 */
}

void HAL_TIM_PeriodElapsedCallback(TIM_HandleTypeDef *htim)
{
/* USER CODE BEGIN Callback 0 */

/* USER CODE END Callback 0 */
if (htim->Instance == TIM6)
{
HAL_IncTick();
}
/* USER CODE BEGIN Callback 1 */
else if (htim->Instance == TIM2)
{
osSemaphoreRelease(myCountingSem01Handle);
osSemaphoreRelease(myCountingSem01Handle);
osSemaphoreRelease(myCountingSem01Handle);
}
/* USER CODE END Callback 1 */
}
```

10.5.4 Mutex and Semaphore in Mbed OS

We can handle the mutex and semaphore in Mbed OS in the same way as in FreeRTOS. Hence, we start with mutex followed by semaphore.

10.5.4.1 Mutex Functions

Mbed OS has several functions related to mutex usage. The functions `Mutex` and `Mutex(const char *name)` are used to create and initialize the mutex object with a default or given name, respectively. The function `lock` is used to wait a task until the mutex is available. The function `trylock` is used to try to acquire the mutex and return immediately if it is not available. The function `trylock_for` is used to try to acquire the mutex in a given timeout value and return if it is not available after timeout. The function `trylock_until` is used to try to acquire the mutex until the given absolute time is reached and return if it is not available after this time. The function `unlock` is used to release the mutex. Finally, the function `~Mutex` is used to destruct the mutex object. None of these functions can be used inside an interrupt callback function. We provide the detailed explanation for the mentioned mutex functions next.

```
Mutex()

Mutex(const char *name)
/*
name: name of the mutex.
*/

void lock()

bool trylock()

bool trylock_for(Kernel::Clock::duration_u32 rel_time)
/*
rel_time: timeout value.
*/

bool trylock_until(Kernel::Clock::time_point abs_time)
/*
abs_time: absolute timeout time.
*/

void unlock()

osThreadId_t get_owner()

~Mutex()
```

10.5.4.2 Mutex Usage Examples

We next provide an example on the usage of mutex in Mbed OS. In Listing 10.16, we reconstruct the code in Listing 10.13 now in C++ language. Here, both thread1 and default main threads use the function LEDChange with different inputs. The operation is controlled by the created mutex.

Listing 10.16 The mutex example in Mbed OS

```
#include "mbed.h"

DigitalOut greenLED(LED1);

Thread thread1(osPriorityAboveNormal);
Mutex mutex;

void LEDChange(int condition)
{
mutex.lock();
if (condition == 0)
{
greenLED = 0;
}
else
{
greenLED = 1;
}
wait_us(1000000);
mutex.unlock();
}
```

```
void thread1Fn()
{
while (true)
{
LEDChange(1);
ThisThread::sleep_for(500ms);
}
}

int main()
{
thread1.start(thread1Fn);
while (true)
{
LEDChange(0);
}
}
```

10.5.4.3 Semaphore Functions

Mbed OS has several functions related to semaphore usage. The functions Semaphore(int32_t count=0) and Semaphore(int32_t count, uint16_t max_count) can be used to create and initialize the semaphore object with initial number of tokens or initial and maximum number of tokens, respectively. The function acquire is used to wait until a semaphore token is available. The function try_acquire is used to try to acquire a semaphore token and return immediately if no token is available. The function try_acquire_for is used to try to acquire a semaphore in a given timeout value and return if no token is available after timeout. The function try_acquire_until is used to try to acquire a semaphore until given absolute time is reached and return if no token is available after this time. The function release is used to release the semaphore token. Finally, the function ~Semaphore is used to destruct the semaphore object. The functions Semaphore, ~Semaphore, acquire, and try_acquire_until cannot be used inside an interrupt callback function. The functions release and try_acquire can be used inside the interrupt callback function. The function try_acquire_for can be used inside interrupt callback functions if its timeout value is set to zero. We provide detailed explanation of the mentioned semaphore functions next.

```
Semaphore(int32_t count=0)
/*
count: initial number of semaphore tokens.
*/

Semaphore(int32_t count, uint16_t max_count)
/*
max_count: maximum number of semaphore tokens.
*/

void acquire()

bool try_acquire()

bool try_acquire_for(Kernel::Clock::duration_u32 rel_time)
/*
rel_time: timeout value.
```

```
*/
bool try_acquire_until(Kernel::Clock::time_point abs_time)
/*
abs_time: absolute timeout time.
*/

osStatus release(void)

~Semaphore()
```

10.5.4.4 Semaphore Usage Examples

In this section, we provide two examples on the usage of semaphores in Mbed OS. In the first example, given in Listing 10.17, we reconstruct the C code in Listing 10.14 now in C++ language. Here, both thread1 and the default main threads use the function LEDChange with different inputs. Management of the operation is done by the created binary semaphore.

Listing 10.17 The first semaphore example in Mbed OS

```
#include "mbed.h"

DigitalOut greenLED(LED1);

Thread thread1(osPriorityAboveNormal);
Semaphore oneSlot(1);

void LEDChange(int condition){
oneSlot.acquire();
if(condition == 0){
greenLED = 0;
}
else {
greenLED = 1;
}
wait_us(1000000);
oneSlot.release();
}

void thread1Fn(){
while (true) {
LEDChange(1);
ThisThread::sleep_for(500ms);
}
}

int main()
{
thread1.start(thread1Fn);
while (true){
LEDChange(0);
}
}
```

In the second example, given in Listing 10.18, we reconstruct the C code in Listing 10.15 now in C++ language. Here, we use a semaphore with three tokens. The Ticker1 is used to release these tokens every second. After tokens are released, thread1, thread2, and the default main threads are called according to their priorities.

Listing 10.18 The second semaphore example in Mbed OS

```
#include "mbed.h"

DigitalOut greenLED(LED1);

Thread thread1(osPriorityAboveNormal);
Thread thread2(osPriorityHigh);
Semaphore threeSlot(0, 3);
Ticker ticker1;

uint32_t arr[10];
uint32_t i;

void thread1Fn()
{
while (true)
{
threeSlot.acquire();
for (i = 0; i < 10; i++)
{
arr[i] = arr[i] * 5;
}
ThisThread::sleep_for(100ms);
}
}

void thread2Fn()
{
while (true)
{
threeSlot.acquire();
for (i = 0; i < 10; i++)
{
arr[i] = i;
}
ThisThread::sleep_for(100ms);
}
}

void ticker1Cb()
{
threeSlot.release();
threeSlot.release();
threeSlot.release();
}

int main()
{
thread1.start(thread1Fn);
thread2.start(thread2Fn);
ticker1.attach(&ticker1Cb, 1s);
while (true)
{
threeSlot.acquire();
for (i = 0; i < 10; i++)
{
printf("%u \n", arr[i]);
}
```

```
ThisThread::sleep_for(100ms);
}
}
```

10.6 Queue and Mail

Sometimes we may want to transfer data between tasks. This can be done by either queue or mail usage. To be more specific, we can use queue to transfer data (one variable, pointer to an array or struct) between tasks. On the other hand, mail is the specialized form of queue such that both queue and memory allocation (via memory pool) are used together. Therefore, FreeRTOS does not directly call this operation as mail. Instead, it calls this operation as the joint queue and memory pool usage.

10.6.1 Queue Working Principles

We can transfer data (in the form of single variable) between two tasks by using a queue. Here, one task adds variable to the queue; the other task receives it. The size of the queue can be adjusted. Hence, data with more than one element can be transferred between the tasks in synchronous manner.

We provide general layout of queue usage in Fig. 10.12. As can be seen in this figure, one task adds data to the queue if there is an available slot. Here, the priority is also taken into account such that if more than one task tries to add data to the queue at the same time, the task with higher priority adds the data first. If there is data in the queue, then the task with higher priority gets it. During this process, the data is passed from one task to other in the first in, first out (FIFO) order.

10.6.2 Mail Working Principles

Mail can also be used to transfer data between tasks. Different from the queue, mail can be used to transfer a fixed-sized memory block instead of a single variable. As the transfer ends, this memory block is released. The memory pool is used to create

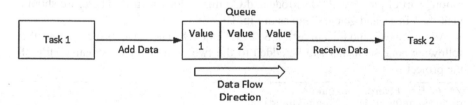

Fig. 10.12 General layout of queue usage

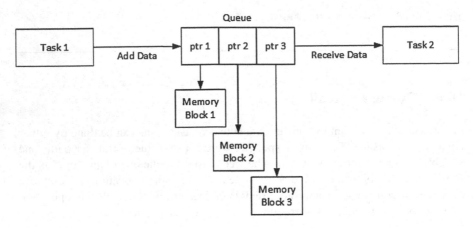

Fig. 10.13 General layout of mail usage

the fixed-sized memory block. Here, only the pointer for the allocated memory block is transferred.

We provide general layout of mail usage in Fig. 10.13. As can be seen in this figure, we have memory blocks in operation. Hence, large or structured data can be transferred between tasks. Here, one task allocates a memory block from the memory pool, adds data to this memory block, and puts the pointer of this memory block into the mail (or queue). Then, another task gets the pointer of the memory block from the mail (or queue), obtains data using this pointer, and releases the allocated memory block.

10.6.3 Queue and Memory Pool in FreeRTOS

In FreeRTOS, queue is used as such. However, mail is formed by jointly using the queue and memory pool. Therefore, some sources call this operation as memory queue. Next, we start with queue operations.

10.6.3.1 Queue Setup and Queue Functions

We can add a queue to our FreeRTOS project in STM32CubeMX by the "Configuration" part of the "FreeRTOS Mode and Configuration" section. There, we should visit the "Tasks and Queues" tab as in Fig. 10.14.

As we make modifications in STM32CubeMX and save project files, the following code segments will be added to the two sections of the "main.c" file of the project.

```
/* Definitions for Queue1 */
osMessageQueueId_t Queue1Handle;
const osMessageQueueAttr_t Queue1_attributes = {
  .name = "Queue1"
};
```

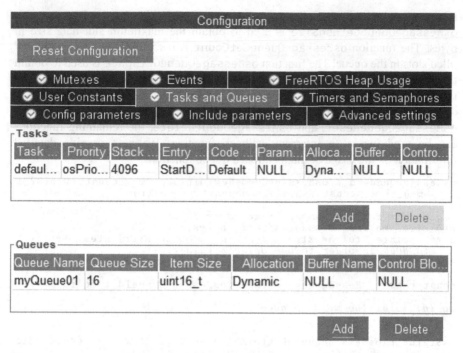

Fig. 10.14 FreeRTOS tasks and queues tab in STM32CubeMX

```
/* Create the queue(s) */
/* creation of Queue1 */
Queue1Handle = osMessageQueueNew (1, sizeof(uint16_t), &
    Queue1_attributes);
```

In the first section, queue handle osMessageQueueId_t is created with the name Queue1Handle. Also, the queue attributes struct osMessageQueueAttr_t is created with the name Queue1. In the second section, the queue is created with its handle and attributes structure using the function osMessageQueueNew. Queue1Handle will be used to call the created queue afterward.

The functions related to queue management in FreeRTOS can be summarized as follows. The function osMessageQueueNew is used to create a new queue with maximum number of slots in the queue and maximum slot data size in bytes. The queue must be created after the kernel is initialized. The function osMessageQueueGetName is used to obtain the name of the queue. The function osMessageQueuePut is used to insert data pointed by given pointer to the queue. Here, priority of the input is used to sort the incoming data on insertion. This function waits until the given timeout value is expired. The function osMessageQueueGet is used to retrieve data from the queue and store it to the location pointed by given pointer. This function waits until the given timeout value is expired. The function osMessageQueueGetCapacity is used to obtain the number of maximum slots in the queue. The function osMessageQueueGetMsgSize is used to obtain the maximum slot data size in bytes. The function osMessageQueueGetCount is used to obtain the number of filled slots in the queue. The function osMessageQueueGetSpace is used to obtain the number of available slots in the queue. The function osMessageQueueReset is used to reset the queue. The function osMessageQueueDelete is used to delete the queue. Except osMessageQueueNew, osMessageQueueGetName, osMessageQueueReset, and osMessageQueueDelete, the remaining functions can be used inside an interrupt callback function. We next provide detailed information on the usage of the mentioned queue functions.

```
osMessageQueueId_t osMessageQueueNew(uint32_t msg_count, uint32_t
    msg_size, const osMessageQueueAttr_t *attr)
/*
msg_count: maximum number of messages in queue.
msg_size: maximum message size in bytes.
attr: pointer to the struct containing queue attributes. Set it
    to NULL in order to use default values.
*/

const char* osMessageQueueGetName(osMessageQueueId_t mq_id)
/*
mq_id: id of the message queue.
*/

osStatus_t osMessageQueuePut(osMessageQueueId_t mq_id, const void
    *msg_ptr, uint8_t msg_prio, uint32_t timeout)
/*
msg_ptr: pointer to the message which will be inserted into queue
    .
msg_prio: message priority.
timeout: timeout value.
*/

osStatus_t osMessageQueueGet(osMessageQueueId_t mq_id, void *
    msg_ptr, uint8_t *msg_prio, uint32_t timeout)

uint32_t osMessageQueueGetCapacity(osMessageQueueId_t mq_id)

uint32_t osMessageQueueGetMsgSize(osMessageQueueId_t mq_id)

uint32_t osMessageQueueGetCount(osMessageQueueId_t mq_id)

uint32_t osMessageQueueGetSpace(osMessageQueueId_t mq_id)

osStatus_t osMessageQueueReset(osMessageQueueId_t mq_id)

osStatus_t osMessageQueueDelete(osMessageQueueId_t mq_id)
```

10.6.3.2 Queue Usage Examples

In this section, we provide two examples on the usage of queue in FreeRTOS. In the first example, given in Listing 10.19, we transfer a variable between two tasks using a queue. Here, we create an additional task named myTask02 apart from the default task. The priority of this task is set to osPriorityNormal, which is same as the priority of the default task. Then, we create a queue named myQueue01 with size one and slot size of `uint16_t`. As the code is executed, the default task increases the variable i by 1 and checks whether the value is odd or even. If the value is odd, the task puts the value 1 to the queue. Otherwise, the task puts the value 0 to the queue. Then, the default task yields for 0.5 s. `myTask02` is kept in the waiting state until this operation ends. Then, it is called and the queue entry is read. If the read value is 0, the green LED on the STM32F4 board is toggled by `myTask02`. If the read value is 1, the red LED on the STM32F4 board is toggled by `myTask02`.

Listing 10.19   The first queue example in FreeRTOS

```
/* Private variables
   ----------------------------------------------------------*/

/* Definitions for defaultTask */
osThreadId_t defaultTaskHandle;
const osThreadAttr_t defaultTask_attributes = {
.name = "defaultTask",
.priority = (osPriority_t)osPriorityNormal,
.stack_size = 128 * 4};

/* Definitions for Task2 */
osThreadId_t myTask02Handle;
const osThreadAttr_t myTask02_attributes = {
.name = "myTask02",
.stack_size = 128 * 4,
.priority = (osPriority_t)osPriorityNormal,
};

/* Definitions for myQueue01 */
osMessageQueueId_t myQueue01Handle;
const osMessageQueueAttr_t myQueue01_attributes = {
.name = "myQueue01"};

void StartDefaultTask(void *argument)
{
/* USER CODE BEGIN 5 */
uint16_t i = 0;
uint16_t message;
/* Infinite loop */
for (;;)
{
i++;
i %= 2;
message = i;
osMessageQueuePut(myQueue01Handle, &message, osPriorityNormal,
    osWaitForever);
osDelay(500);
}
/* USER CODE END 5 */
}

void StartTask02(void *argument)
{
/* USER CODE BEGIN StartTask02 */
uint16_t message;
/* Infinite loop */
for (;;)
{
osMessageQueueGet(myQueue01Handle, &message, NULL, osWaitForever)
    ;
```

```
if (message == 0)
HAL_GPIO_TogglePin(GPIOG, GPIO_PIN_13);
else if (message == 1)
HAL_GPIO_TogglePin(GPIOG, GPIO_PIN_14);
}
/* USER CODE END StartTask02 */
}
```

In the second example, given in Listing 10.20, we show how queue can be used with interrupts and multiple tasks. Here, we configure the timer modules Timer 2 and Timer 3 to trigger interrupts every 1 s and 2 s, respectively. Both timers have their own counter and put a message containing the counter value and timer id to the queue whenever the related interrupt is triggered. At the same time, the default task waits until a message is put in the queue. It then prints the counter value and id of the timer which holds this counter value. In this setup, we do not want to miss any printing. However, priorities of timer interrupts are always higher than task priorities. Hence, we create two more tasks as myTask02 and myTask03 with priorities lower than the default task and associate them with Timer 2 and Timer 3 using an event. Therefore, whenever a timer interrupt is triggered, the associated task is called. The counter is increased and message is put to the queue. The default task prints content of the incoming message. It toggles the green LED on the STM32F4 board if the message originated from Timer 2. It toggles the red LED on the STM32F4 board if the message originated from Timer 3. We should modify ITM properties as explained in Sect. 3.4.1.1 to observe printing results in this example.

Listing 10.20 The second queue example in FreeRTOS

```
/* USER CODE BEGIN PV */
typedef struct
{
uint32_t cnt;
uint8_t id;
} message_t;
/* USER CODE END PV */

/* USER CODE BEGIN 2 */
if (__HAL_TIM_GET_FLAG(&htim2, TIM_FLAG_UPDATE) != RESET)
__HAL_TIM_CLEAR_FLAG(&htim2, TIM_FLAG_UPDATE);
if (__HAL_TIM_GET_FLAG(&htim3, TIM_FLAG_UPDATE) != RESET)
__HAL_TIM_CLEAR_FLAG(&htim3, TIM_FLAG_UPDATE);
HAL_TIM_Base_Start_IT(&htim2);
HAL_TIM_Base_Start_IT(&htim3);
/* USER CODE END 2 */

/* creation of myQueue01 */
myQueue01Handle = osMessageQueueNew (16, sizeof(message_t), &
    myQueue01_attributes);

void StartDefaultTask(void *argument)
{
/* USER CODE BEGIN 5 */
message_t message;
/* Infinite loop */
for (;;)
```

```
{
osMessageQueueGet(myQueue01Handle, &message, NULL, osWaitForever)
   ;
if (message.id == 2)
{
printf("TIM2 cnt = %u \r\n", message.cnt);
HAL_GPIO_TogglePin(GPIOG, GPIO_PIN_13);
}
else if (message.id == 3)
{
printf("TIM3 cnt = %u \r\n", message.cnt);
HAL_GPIO_TogglePin(GPIOG, GPIO_PIN_14);
}
}
/* USER CODE END 5 */
}

void StartTask02(void *argument)
{
/* USER CODE BEGIN StartTask02 */
uint32_t cnt1 = 0;
message_t message1;
/* Infinite loop */
for (;;)
{
osEventFlagsWait(myEvent01Handle, FLAG1, osFlagsWaitAny,
    osWaitForever);
message1.cnt = cnt1;
message1.id = 2;
osMessageQueuePut(myQueue01Handle, &message1, osPriorityNormal,
    osWaitForever);
cnt1++;
osEventFlagsClear(myEvent01Handle, FLAG1);
}
/* USER CODE END StartTask02 */
}

void StartTask03(void *argument)
{
/* USER CODE BEGIN StartTask03 */
uint32_t cnt2 = 0;
message_t message2;
/* Infinite loop */
for (;;)
{
osEventFlagsWait(myEvent01Handle, FLAG2, osFlagsWaitAny,
    osWaitForever);
message2.cnt = cnt2;
message2.id = 3;
osMessageQueuePut(myQueue01Handle, &message2, osPriorityNormal,
    osWaitForever);
cnt2++;
osEventFlagsClear(myEvent01Handle, FLAG2);
}
/* USER CODE END StartTask03 */
}

void HAL_TIM_PeriodElapsedCallback(TIM_HandleTypeDef *htim)
{
/* USER CODE BEGIN Callback 0 */

/* USER CODE END Callback 0 */
if (htim->Instance == TIM6) {
```

```
HAL_IncTick();
}
/* USER CODE BEGIN Callback 1 */
else if (htim->Instance == TIM2)
{
osEventFlagsSet(myEvent01Handle, FLAG1);
}
else if (htim->Instance == TIM3)
{
osEventFlagsSet(myEvent01Handle, FLAG2);
}
/* USER CODE END Callback 1 */
}
```

10.6.3.3 Memory Pool Setup and Memory Pool Functions

There is no memory pool section under STM32CubeMX. Hence, we cannot directly add a memory pool to our FreeRTOS project using STM32CubeMX. Instead, we should add the memory pool via code. Creation of a memory pool handle and its attributes structure can be seen in the first section of the code snippet given below. In the second section, memory pool is created with its handle and attributes structure using the function osMemoryPoolNew. MemoryPool1Handle will be used to call the created memory pool afterward.

```
/* Definitions for Queue1 */
osMemoryPoolId_t MemoryPool1Handle;
const osMemoryPoolAttr_t MemoryPool1_attributes = {
  .name = "MemoryPool1"
};

  /* creation of MemoryPool1 */
  MemoryPool1Handle = osMemoryPoolNew(10, sizeof(uint16_t), &
      MemoryPool1_attributes);
```

Functions related to memory pool management in FreeRTOS are as follows. The function osMemoryPoolNew is used to create new memory pool with maximum number of memory blocks in memory pool and memory block size in bytes. The memory pool must be created after the kernel is initialized. The function osMemoryPoolGetName is used to obtain the name of the memory pool. The function osMemoryPoolAlloc is used to allocate memory block under the memory pool and returns a pointer to the address of the allocated memory block. This function waits for allocation until the given timeout value is expired. The function osMemoryPoolFree is used to free the memory block pointed by given pointer. The function osMemoryPoolGetCapacity is used to obtain the maximum number of memory blocks in the memory pool. The function osMemoryPoolGetBlockSize is used to obtain memory block size in bytes. The function osMemoryPoolGetCount is used to obtain the number of used memory blocks. The function osMemoryPoolGetSpace is used to obtain the number of available memory blocks. The function osMemoryPoolDelete is used to delete the memory pool. Except osMemoryPoolNew, osMemoryPoolGetName, and osMemoryPoolDelete, the remaining functions can be used inside an interrupt callback function. We next provide detailed information on the mentioned memory pool functions.

```
osMemoryPoolId_t osMemoryPoolNew(uint32_t block_count, uint32_t
    block_size, const osMemoryPoolAttr_t *attr)
/*
block_count: maximum number of memory blocks in memory pool.
block_size: memory block size in bytes.
attr: pointer to the struct containing memory pool attributes.
    Set it to NULL in order to use default values.
*/

const char* osMemoryPoolGetName(osMemoryPoolId_t mp_id)
/*
mp_id: id of the memory pool
*/

void* osMemoryPoolAlloc(osMemoryPoolId_t mp_id, uint32_t timeout)
/*
timeout: timeout value
*/

osStatus_t osMemoryPoolFree(osMemoryPoolId_t mp_id, void *block)
/*
block:  address of the allocated memory block to be returned to
    the memory pool.
*/

uint32_t osMemoryPoolGetCapacity(osMemoryPoolId_t mp_id)

uint32_t osMemoryPoolGetBlockSize(osMemoryPoolId_t mp_id)

uint32_t osMemoryPoolGetCount(osMemoryPoolId_t mp_id)

uint32_t osMemoryPoolGetSpace(osMemoryPoolId_t mp_id)

osStatus_t osMemoryPoolDelete(osMemoryPoolId_t mp_id)
```

10.6.3.4 Memory Pool Usage Examples

In this section, we provide two examples on the usage of queue and memory pool in FreeRTOS. In the first example, given in Listing 10.21, we transfer a struct containing uint16_t buffer array with ten elements and id between two tasks using memory pool and queue. Here, we create an additional task named myTask02 apart from the default task. The priority of this task is set to osPriorityHigh, which is higher than the priority of the default task. Then, we create a queue with size one. Data in the queue has the size of constructed structure. The memory pool is initialized at the beginning of the function StartTask02 with ten elements and data size of constructed structure. A variable cnt is increased inside the function StartTask02. Each time the function StartTask02 is called, a memory block is allocated from the memory pool. The value of cnt is written to the first element of the uint16_t buffer array. Then, data is added to the queue. Afterward, myTask02 yields for 0.5 s. The default task will be in the waiting state until data is added to the queue. Then, it resumes its operation and gets data from the queue. If the first element of the received buffer array is even, the green LED on the STM32F4 board is toggled in the function StartDefaultTask. If the first element of the received buffer array is odd, the red LED on the STM32F4 board is toggled in the function `StartDefaultTask`. Then, the allocated memory block is freed and returned to the memory pool.

Listing 10.21 The first memory pool example in FreeRTOS

```
/* Private variables
   ----------------------------------------------------------*/
/* Definitions for defaultTask */
osThreadId_t defaultTaskHandle;
const osThreadAttr_t defaultTask_attributes = {
.name = "defaultTask",
.stack_size = 128 * 4,
.priority = (osPriority_t)osPriorityNormal,
};

/* Definitions for myTask02 */
osThreadId_t myTask02Handle;
const osThreadAttr_t myTask02_attributes = {
.name = "myTask02",
.stack_size = 128 * 4,
.priority = (osPriority_t)osPriorityHigh,
};

/* Definitions for myQueue01 */
osMessageQueueId_t myQueue01Handle;
const osMessageQueueAttr_t myQueue01_attributes = {
.name = "myQueue01"};

/* USER CODE BEGIN PV */
typedef struct
{
uint16_t buffer[10];
uint8_t id;
} message_t;

osMemoryPoolId_t myMemPool01;
/* USER CODE END PV */

myQueue01Handle = osMessageQueueNew(1, sizeof(message_t), &
    myQueue01_attributes);

void StartDefaultTask(void *argument)
{
/* USER CODE BEGIN 5 */
message_t *message;
/* Infinite loop */
for (;;)
{
osMessageQueueGet(myQueue01Handle, &message, NULL, osWaitForever)
    ;
if (((message->buffer[0]) % 2) == 0)
HAL_GPIO_TogglePin(GPIOG, GPIO_PIN_13);
else if (((message->buffer[0]) % 2) == 1)
HAL_GPIO_TogglePin(GPIOG, GPIO_PIN_14);
osMemoryPoolFree(myMemPool01, message);
}
/* USER CODE END 5 */
}

void StartTask02(void *argument)
{
/* USER CODE BEGIN StartTask02 */
```

```
myMemPool01 = osMemoryPoolNew(10, sizeof(message_t), NULL);
message_t *message;
uint16_t cnt = 0;
/* Infinite loop */
for (;;)
{
message = (message_t *)osMemoryPoolAlloc(myMemPool01, 0U);
message->buffer[0] = cnt;
osMessageQueuePut(myQueue01Handle, &message, 0U, 0U);
cnt++;
osDelay(500);
}
/* USER CODE END StartTask02 */
}
```

In the second example, given in Listing 10.22, we show how memory pool and queue can be used with interrupts and multiple tasks. To do so, we use the same structure introduced in the previous example. Now, we create a queue with size two. Data in the queue slots have the size of the constructed structure. The memory pool is initialized at the beginning of the function StartDefaultTask with ten elements and data size of the constructed structure. We configure the timer modules Timer 2 and Timer 3 to trigger interrupts every 1 s and 2 s, respectively. Both timers have their own counter and put data containing the counter value and timer id to the queue whenever related interrupt is triggered. At the same time, the default task waits until data is put to the queue. It then prints the counter value and id of the timer which holds this counter value. As in Listing 10.20, we do not want to miss any printing. However, priorities of timer interrupts are always higher than priority of a task. Hence, we create two more tasks named myTask02 and myTask03 with priorities lower than the default task. We associate them with Timer 2 and Timer 3 respectively using an event. Therefore, whenever a timer interrupt is triggered, the related task is called. Then, the memory block is allocated from the memory pool. The counter value is written to the first element of the buffer array and the id is set according to the timer interrupt source. Afterward, the message is added to the queue. The default task will be waiting for the message to be added to the queue. Afterward, it receives the message. If the id in the incoming message is 2, then the string "TIM2 cnt = " is printed with the cnt value in the message. The green LED on the STM32F4 board is toggled. If the id in the incoming message is 3, then the string "TIM3 cnt = " is printed with the cnt value in the message. The red LED on the STM32F4 board is toggled. Then, the allocated memory block is freed and returned to the memory pool. We should modify ITM properties as explained in Sect. 3.4.1.1 to observe printing results in this example.

Listing 10.22 The second memory pool example in FreeRTOS

```
/* USER CODE BEGIN PV */
typedef struct
{
uint16_t buffer[10];
uint8_t id;
} message_t;

osMemoryPoolId_t myMemPool01;
/* USER CODE END PV */
```

```
/* USER CODE BEGIN 2 */
if (__HAL_TIM_GET_FLAG(&htim2, TIM_FLAG_UPDATE) != RESET)
__HAL_TIM_CLEAR_FLAG(&htim2, TIM_FLAG_UPDATE);
if (__HAL_TIM_GET_FLAG(&htim3, TIM_FLAG_UPDATE) != RESET)
__HAL_TIM_CLEAR_FLAG(&htim3, TIM_FLAG_UPDATE);
HAL_TIM_Base_Start_IT(&htim2);
HAL_TIM_Base_Start_IT(&htim3);
/* USER CODE END 2 */

/* creation of myQueue01 */
myQueue01Handle = osMessageQueueNew (16, sizeof(message_t), &
    myQueue01_attributes);

void StartDefaultTask(void *argument)
{
/* USER CODE BEGIN 5 */
myMemPool01 = osMemoryPoolNew(10, sizeof(message_t), NULL);
message_t *message;
/* Infinite loop */
for (;;)
{
osMessageQueueGet(myQueue01Handle, &message, NULL, osWaitForever)
   ;
if (message->id == 2)
{
        printf("TIM2 cnt = %u \r\n", message->buffer[0]);
        HAL_GPIO_TogglePin(GPIOG, GPIO_PIN_13);
}
else if (message->id == 3)
{
        printf("TIM3 cnt = %u \r\n", message->buffer[0]);
        HAL_GPIO_TogglePin(GPIOG, GPIO_PIN_14);
}
osMemoryPoolFree(myMemPool01, message);
}
/* USER CODE END 5 */
}

void StartTask02(void *argument)
{
/* USER CODE BEGIN StartTask02 */

uint16_t cnt1 = 0;
message_t *message;
/* Infinite loop */
for (;;)
{
osEventFlagsWait(myEvent01Handle, FLAG1, osFlagsWaitAny,
    osWaitForever);
message = (message_t *)osMemoryPoolAlloc(myMemPool01, 0U);
message->buffer[0] = cnt1;
message->id = 2;
osMessageQueuePut(myQueue01Handle, &message, osPriorityNormal,
    osWaitForever);
cnt1++;
osEventFlagsClear(myEvent01Handle, FLAG1);
}
/* USER CODE END StartTask02 */
}

void StartTask03(void *argument)
{
/* USER CODE BEGIN StartTask03 */
```

```
uint16_t cnt2 = 0;
message_t *message;
/* Infinite loop */
for (;;)
{
osEventFlagsWait(myEvent01Handle, FLAG2, osFlagsWaitAny,
    osWaitForever);
message = (message_t *)osMemoryPoolAlloc(myMemPool01, 0U);
message->buffer[0] = cnt2;
message->id = 3;
osMessageQueuePut(myQueue01Handle, &message, osPriorityNormal,
    osWaitForever);
cnt2++;
osEventFlagsClear(myEvent01Handle, FLAG2);
}
/* USER CODE END StartTask03 */
}


void HAL_TIM_PeriodElapsedCallback(TIM_HandleTypeDef *htim)
{
/* USER CODE BEGIN Callback 0 */

/* USER CODE END Callback 0 */
if (htim->Instance == TIM6) {
HAL_IncTick();
}
/* USER CODE BEGIN Callback 1 */
else if (htim->Instance == TIM2)
{
osEventFlagsSet(myEvent01Handle, FLAG1);
}
else if (htim->Instance == TIM3)
{
osEventFlagsSet(myEvent01Handle, FLAG2);
}
/* USER CODE END Callback 1 */
}
```

10.6.4 Queue and Mail in Mbed OS

We can use queue and mail in Mbed OS. We handle them next.

10.6.4.1 Queue Functions

Mbed OS has several functions related to queue usage. The function Queue is used to create and initialize the queue object. The function empty is used to check whether the queue is empty. The function full is used to check whether the queue is full. The function count is used to obtain the number of available slots in the queue. The function try_put is used to put data pointed by the given pointer to the queue. Here, priority of the input is used to sort the incoming data on insertion. This function returns immediately if the queue is full. The function try_put_for is used in the same way as the function try_put. However, it waits until the timeout value is expired. The function try_get is used to retrieve data from the queue and store it to the location pointed by the given pointer. This function returns immediately if the queue is empty. The function `try_get_for` is used in the same way as the function `try_get`. However, it waits until the timeout value is expired. Finally, the function `~Queue` is used to destruct the queue object. Except `Queue` and `~Queue`, the remaining functions can be used inside an interrupt callback function. As a side note, the functions `try_put_for` and `try_get_for` can be used inside interrupt callback functions only if their timeout value is set to zero. We next provide detailed information on the usage of the mentioned queue functions in Mbed OS.

```
Queue()

bool empty()

bool full()

uint32_t count()

bool try_put(T *data, uint8_t prio=0)
/*
data: pointer to the element which will be inserted to the queue.
prio: priority of the operation.
*/

bool try_put_for(Kernel::Clock::duration_u32 rel_time, T *data,
    uint8_t prio=0)
/*
rel_time: timeout value.
*/

bool try_get(T **data_out)
/*
data_out: pointer to memory address which retrieved element will
    be written.
*/

bool try_get_for(Kernel::Clock::duration_u32 rel_time, T **
    data_out)

~Queue()
```

10.6.4.2 Queue Usage Examples

We next provide two examples on the usage of queue in Mbed OS. In the first example, given in Listing 10.23, we reform the C code in Listing 10.19 now in C++ language. Here, we cannot use the macro `osWaitForever` inside the function `try_get_for` to wait indefinitely. Therefore, we use `0xFFFFFFFF`s which is the value for this macro. We add `s` as a suffix to the hexadecimal number to indicate its unit is seconds.

Listing 10.23 The first queue example in Mbed OS

```
#include "mbed.h"

DigitalOut greenLED(LED1);
DigitalOut redLED(LED2);

Thread thread1(osPriorityNormal);
Queue<uint16_t, 1> queue;
```

```
void thread1Fn()
{
uint16_t *message;
while (true)
{
queue.try_get_for(0xFFFFFFFFs, &message);
if (*message == 0)
greenLED = !greenLED;
else if (*message == 1)
redLED = !redLED;
}
}

int main()
{
thread1.start(thread1Fn);
uint16_t i = 0;
uint16_t message;
while (true)
{
i++;
i %= 2;
message = i;
queue.try_put(&message);
ThisThread::sleep_for(500ms);
}
}
```

In the second example, given in Listing 10.24, we reform the C code in Listing 10.20 now in C++ language. The reader can consult the FreeRTOS version of this example to grasp the structure there.

Listing 10.24 The second queue example in Mbed OS

```
#include "mbed.h"

#define FLAG1 0x01
#define FLAG2 0x10

DigitalOut greenLED(LED1);
DigitalOut redLED(LED2);

Thread thread1(osPriorityLow);
Thread thread2(osPriorityLow);
Ticker ticker1;
Ticker ticker2;
EventFlags evt;

typedef struct
{
uint32_t cnt;
uint8_t id;
} message_t;

Queue<message_t, 16> queue;

void thread1Fn()
{
uint32_t cnt1 = 0;
message_t message1;
```

```
while (true)
{
evt.wait_any(FLAG1, osWaitForever);
message1.cnt = cnt1;
message1.id = 2;
queue.try_put(&message1);
cnt1++;
evt.clear(FLAG1);
}
}

void thread2Fn()
{
uint32_t cnt2 = 0;
message_t message2;
while (true)
{
evt.wait_any(FLAG2, osWaitForever);
message2.cnt = cnt2;
message2.id = 3;
queue.try_put(&message2);
cnt2++;
evt.clear(FLAG2);
}
}

void ticker1Cb()
{
evt.set(FLAG1);
}

void ticker2Cb()
{
evt.set(FLAG2);
}

int main()
{
thread1.start(thread1Fn);
thread2.start(thread2Fn);
ticker1.attach(&ticker1Cb, 1s);
ticker2.attach(&ticker2Cb, 2s);
message_t *message;
while (true)
{
queue.try_get_for(0xFFFFFFFFs, &message);
if (message->id == 2)
{
printf("TIM2 cnt = %u \r\n", message->cnt);
greenLED = !greenLED;
}
else if (message->id == 3)
{
printf("TIM3 cnt = %u \r\n", message->cnt);
redLED = !redLED;
}
}
}
```

10.6.4.3 Mail Functions

Mbed OS has several functions related to mail usage. The function Mail is used to create and initialize the mail object. The function empty is used to check whether the mail is empty. The function full is used to check whether the mail is full. The function try_alloc is used to allocate memory block from mail and returns immediately if the mail is full. The function try_alloc_for is used in the same way as the function try_alloc. However, it waits until the timeout value is expired. The function try_alloc_until is used in the same way as the function try_alloc. However, it waits until the given absolute time value is reached. The function try_calloc is used to allocate memory block from the mail and initialize it to zero. This function returns immediately if the mail is full. The function try_calloc_for is used in the same way as the function try_calloc. However, it waits until the timeout value is expired. The function try_calloc_until is used in the same way as the function try_calloc. However, it waits until the given absolute time value is reached. The function put is used to put data to mail. The function try_get is used to get data from mail. The function try_get_for is used to get data from mail until the given timeout is expired. Finally, the function free is used to free a memory block from the mail. The functions Mail, try_alloc_until, and try_calloc_until cannot be used inside an interrupt callback function. The functions empty, full, try_alloc, try_calloc, try_get, put, and free can be used inside interrupt callback functions. The functions try_alloc_for, try_calloc_for, and try_get_for can be used inside an interrupt callback function if their timeout value is set to zero. We next provide detailed information on the mentioned mail functions in Mbed OS.

```
Mail()

bool empty()

bool full()

T* try_alloc()

T* try_alloc_for(Kernel::Clock::duration_u32 rel_time)
/*
rel_time: timeout value.
*/

T* try_alloc_until(Kernel::Clock::time_point abs_time)
/*
abs_time: absolute timeout time.
*/

T* try_calloc()

T* try_calloc_for(Kernel::Clock::duration_u32 rel_time)

T* try_calloc_until(Kernel::Clock::time_point abs_time)

osStatus put(T *mptr)
/*
mptr: pointer to the mail which will be inserted into the queue.
*/
```

```
T* try_get()

T* try_get_for(Kernel::Clock::duration_u32 rel_time)

osStatus free(T *mptr)
```

10.6.4.4 Mail Usage Examples

We next provide three examples on the usage of mail in Mbed OS. In the first example, given in Listing 10.25, we reform the C code in Listing 10.21 now in C++ language. The reader can consult the FreeRTOS version of this example to grasp the structure there.

Listing 10.25 The first mail example in Mbed OS

```cpp
#include "mbed.h"

DigitalOut greenLED(LED1);
DigitalOut redLED(LED2);

Thread thread1;

typedef struct
{
uint16_t buffer[10];
uint8_t id;
} message_t;

MemoryPool<message_t, 10> mempool;
Queue<message_t, 1> queue;

void thread1Fn()
{
uint16_t cnt = 0;
message_t *message;
while (true)
{
message = mempool.try_alloc();
message->buffer[0] = cnt;
queue.try_put(message);
cnt++;
ThisThread::sleep_for(500ms);
}
}

int main()
{
thread1.start(thread1Fn);
message_t *message;
while (true)
{
queue.try_get_for(0xFFFFFFFFs, &message);
if (((message->buffer[0]) % 2) == 0)
{
greenLED = !greenLED;
}
else if (((message->buffer[0]) % 2) == 1)
{
redLED = !redLED;
```

```
}
mempool.free(message);
}
}
```

In the second example, given in Listing 10.26, we reform the C code in Listing 10.21 now in C++ language. Here, we benefit from mail instead of jointly using the queue and memory pool together. The reader can consult the FreeRTOS version of this example to grasp the structure there.

Listing 10.26 The second mail example in Mbed OS

```
#include "mbed.h"

DigitalOut greenLED(LED1);
DigitalOut redLED(LED2);

Thread thread1;

typedef struct
{
uint16_t buffer[10];
uint8_t id;
} mail_t;

Mail<mail_t, 10> myMail;

void thread1Fn()
{
uint16_t cnt = 0;
mail_t *mail;
while (true)
{
mail = myMail.try_alloc();
mail->buffer[0] = cnt;
myMail.put(mail);
cnt++;
ThisThread::sleep_for(500ms);
}
}

int main()
{
thread1.start(thread1Fn);
mail_t *mail;
while (true)
{
mail = myMail.try_get_for(0xFFFFFFFFs);
if (((mail->buffer[0]) % 2) == 0)
{
greenLED = !greenLED;
}
else if (((mail->buffer[0]) % 2) == 1)
{
redLED = !redLED;
}
myMail.free(mail);
}
}
```

In the third example, given in Listing 10.27, we reform the C code in Listing 10.22 now in C++ language. Here, we benefit from mail instead of jointly using the queue and memory pool together. The reader can consult the FreeRTOS version of this example to grasp the structure there.

Listing 10.27 The third mail example in Mbed OS

```
#include "mbed.h"

#define FLAG1 0x01
#define FLAG2 0x10

DigitalOut greenLED(LED1);
DigitalOut redLED(LED2);

Thread thread1(osPriorityLow);
Thread thread2(osPriorityLow);
Ticker ticker1;
Ticker ticker2;
EventFlags evt;

typedef struct
{
uint16_t buffer[10];
uint8_t id;
} mail_t;

Mail<mail_t, 16> myMail;

void thread1Fn()
{
uint16_t cnt1 = 0;
mail_t *mail;
while (true)
{
evt.wait_any(FLAG1, osWaitForever);
mail = myMail.try_alloc();
mail->buffer[0] = cnt1;
mail->id = 2;
myMail.put(mail);
cnt1++;
evt.clear(FLAG1);
}
}

void thread2Fn()
{
uint32_t cnt2 = 0;
mail_t *mail;
while (true)
{
evt.wait_any(FLAG2, osWaitForever);
mail = myMail.try_alloc();
mail->buffer[0] = cnt2;
mail->id = 3;
myMail.put(mail);
cnt2++;
evt.clear(FLAG2);
}
}
```

```
void ticker1Cb()
{
evt.set(FLAG1);
}

void ticker2Cb()
{
evt.set(FLAG2);
}

int main()
{
thread1.start(thread1Fn);
thread2.start(thread2Fn);
ticker1.attach(&ticker1Cb, 1s);
ticker2.attach(&ticker2Cb, 2s);
mail_t *mail;
while (true)
{
mail = myMail.try_get_for(0xFFFFFFFFs);
if (mail->id == 2)
{
printf("TIM2 cnt = %u \r\n", mail->buffer[0]);
greenLED = !greenLED;
}
else if (mail->id == 3)
{
printf("TIM3 cnt = %u \r\n", mail->buffer[0]);
redLED = !redLED;
}
myMail.free(mail);
}
}
```

10.7   Software Timers in FreeRTOS

Tasks in RTOS may depend on time-based operations. The reader can benefit from available timers introduced in Chap. 6 for this purpose. The usage strategy for these timers is generating an interrupt at periodic time instants and jointly using an RTOS component (such as event) to start a task. Although this is an acceptable option, the procedure becomes complex when an interrupt and event are used together.

FreeRTOS provides another solution to generate time-based operations based on its software timers. As the name implies, these timers are not physically available on hardware. They are simple counters based on SysTick. Hence, they cannot perform usual time-based operations such as capture, compare, or PWM generation. However, software timers in FreeRTOS are suitable for periodic operations in which precise timing is not needed. Here, we should emphasize that Systick has 1 ms resolution. Hence, the resolution of the software timers is also 1 ms. On the other hand, time-based task usage can be done fairly easy with software timers.

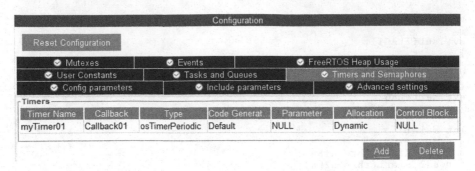

Fig. 10.15 FreeRTOS timers and semaphores tab in STM32CubeMX

10.7.1 Software Timer Setup and Timer Functions

We can add a software timer to our FreeRTOS project in STM32CubeMX by the "Configuration" part of the "FreeRTOS Mode and Configuration" section. There, we should visit the "Timers and Semaphores" tab as in Fig. 10.15. Here, we can choose from periodic or one-time timer.

As we make modifications in STM32CubeMX and save the project files, the following code segments will be added to the three sections of the "main.c" file of the project.

```
/* Definitions for Timer1 */
osTimerId_t Timer1Handle;
const osTimerAttr_t Timer1_attributes = {
.name = "Timer1"
};

/* Private function prototypes
    ---------------------------------------------------*/
...
void Timer_Callback1(void *argument);

/* Create the timer(s) */
/* creation of Timer1 */
Timer1Handle = osTimerNew(Timer_Callback1, osTimerPeriodic, NULL,
    &Timer1_attributes);

/* Timer_Callback1 function */
void Timer_Callback1(void *argument)
{
/* USER CODE BEGIN Timer_Callback1 */

/* USER CODE END Timer_Callback1 */
}
```

In the first section, software timer handle `osTimerId_t` is created with the name `Timer1Handle`. Also, software timer attributes struct `osTimerAttr_t` is created with the name `Timer1`. In the second section, the software timer is created with its handle and attributes structure using the function `osTimerNew`. `Timer1Handle` will be used to call the created software timer afterward. In the third section, the

`Timer_Callback1` function can be used to define what to do under the software timer callback function.

Functions related to software timer management in FreeRTOS can be summarized as follows. The function `osTimerNew` is used to create a new software timer with the given callback function, working behavior, and additional arguments. The software timer must be created after the kernel is initialized. The function `osTimerGetName` is used to obtain the name of the software timer. The function `osTimerStart` is used to start the software timer with given time ticks value. The function `osTimerStop` is used to stop the software timer. The function `osTimerIsRunning` is used to check whether the software timer is running or not. The function `osTimerDelete` is used to delete the software timer. None of these functions can be used inside an interrupt callback function. We next provide detailed information on the mentioned software timer functions.

```
osTimerId_t osTimerNew(osTimerFunc_t func, osTimerType_t type,
    void *argument, const osTimerAttr_t *attr)
/*
func: function pointer to callback function.
type: working behavior of software timer. It can be osTimerOnce
    for one-shot or osTimerPeriodic for periodic behavior.
argument: argument to the timer callback function.
attr: pointer to the struct containing software timer attributes.
    Set it to NULL in order to use default values.
*/
const char* osTimerGetName(osTimerId_t timer_id)
/*
timer_id: id of the software timer
*/

osStatus_t osTimerStart(osTimerId_t timer_id, uint32_t ticks)
/*
ticks:  time ticks value of the software timer.
*/

osStatus_t osTimerStop(osTimerId_t timer_id)

uint32_t osTimerIsRunning(osTimerId_t timer_id)

osStatus_t osTimerDelete(osTimerId_t timer_id)
```

10.7.2 Software Timer Usage Examples

We next provide an example on the usage of the software timer in FreeRTOS. Here, we toggle the onboard green LED on the STM32F4 board by adding the below code snippets to our "main.c" file in addition to the code snippets given at the beginning of this section. As we compile and execute the code, the green LED on the STM32F4 board starts toggling at 0.5 s intervals. We can also add a one-time timer to our project in the same way.

```
/* USER CODE BEGIN RTOS_TIMERS */
/* start timers, add new ones, ... */
osTimerStart(Timer1Handle, 500U);
```

```
/* USER CODE END RTOS_TIMERS */

/* USER CODE BEGIN Timer_Callback1 */

HAL_GPIO_TogglePin(GPIOG,GPIO_PIN_13);

/* USER CODE END Timer_Callback1 */
```

10.8   Memory Management in RTOS

Each created RTOS component needs memory (RAM) space for its operations. As an example, the task needs memory space for its stack and control block. The queue needs memory space for its buffer and control block. Therefore, memory management is an important topic for RTOS operations. There are two ways to perform memory allocation options in RTOS as dynamic and static. Each approach has its own advantages and disadvantages.

In dynamic memory allocation, the memory for an RTOS component is allocated at run time from heap memory. This allocation is handled automatically and unused RTOS components can be deleted at run time to free memory space. However, the deallocation process is very problematic and several problems may occur such as memory fragmentation. Advanced real-time operating systems provide dynamic memory allocation algorithms to prevent these problems.

In static memory allocation, the memory for an RTOS component is allocated during linking the code, and a separate RAM space can be used for this purpose. Hence, there will be no memory fragmentation problem since memory space is allocated beforehand. However, the allocated memory space must be kept when the code is active even if it is not used after some time.

Variables can also be created during RTOS run time. Here, the standard C functions `malloc`, `calloc`, and `free` can be used to create and free memory space for variables. However, these functions are not thread-safe (not suitable to be used under a task) and their execution times are inconsistent. Therefore, real-time operating systems have their own allocation and deallocation functions or make `malloc`, `calloc`, and `free` thread-safe with different approaches. Another way to perform dynamic memory allocation is using memory pool. We explained memory pool in connection with the queue and mail in Sect. 10.6.

10.8.1 Memory Management in FreeRTOS

We can handle memory management in FreeRTOS under STM32CubeMX. To do so, we should first open the "Configuration" part of the "FreeRTOS Mode and Configuration" section. There, we should visit the "Memory management settings" section under the "Config parameters" tab as in Fig. 10.16.

In Fig. 10.16, `TOTAL_HEAP_SIZE` is used to set the desired heap size in bytes. This value is set to 15360 bytes as default. The reader should take different

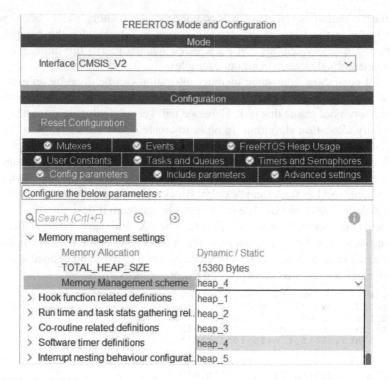

Fig. 10.16 FreeRTOS memory management settings under the Config parameters tab in STM32CubeMX

parameters into account while setting the heap size, such as total RAM and application stack size. In the same section, `Memory Management scheme` is used to select memory allocation algorithm for FreeRTOS. It can be selected between `heap_1` and `heap_5`. In `heap_1`, only memory allocation is permitted. There is no memory-free operation. `heap_2` allows memory freeing operation. However, the freed memory blocks cannot be combined together. `heap_3` permits the usage of standard functions `malloc` and `free` for memory allocation in thread-safe way. In `heap_4`, freed memory blocks can be combined together. In `heap_5`, nonadjacent memory sections can be used as heap. As an example, different parts of SRAM1 and SRAM2 can be used together to form heap memory. Except `heap_3`, in other options the function `pvPortMalloc(size)` can be used for allocating memory, and the function `vPortFree(ptr)` can be used for freeing unused memory (if allowed). The `Memory Management scheme` is set to `heap_4` by default.

In Fig. 10.16, there is also a "FreeRTOS Heap Usage" tab. Through it, the reader can observe the total used and free heap memory. Moreover, memory used for each RTOS component can be observed in this tab.

Each RTOS component can be created statically or dynamically by selecting `Static` or `Dynamic` from the "Allocation" section of its creation window. If the Dynamic option is selected for a component, its memory blocks are created automatically in the heap. If the static option is selected for the component, pointers and memory regions can be selected manually. Till this time, we used dynamic memory allocation in all our examples.

We will next show how static memory allocation can be done by an example. To do so, we will create a task and perform dynamic memory allocation without using memory pool inside this task. Here, we will keep default values for heap size and memory allocation algorithm. In order to create a task statically, go to "Tasks and Queues" tab and click "Add" under "Tasks" to create a new task. After entering the task name, priority, stack size, and task function as before, we will change the allocation selection as "Static." The user can observe that "Buffer Name" (which is the array name used as stack) and "Control Block Name" sections are now available to change their default names. The reader can change them if desired.

As we make modifications in STM32CubeMX and save project files, the following code segment will be added to the "main.c" file of the project. As can be seen here, the task stack and control blocks have been created and defined statically unlike dynamically created tasks given in the previous examples.

```
/* Definitions for myTask04 */
osThreadId_t myTask04Handle;
uint32_t myTask04Buffer[ 128 ];
osStaticThreadDef_t myTask04ControlBlock;
const osThreadAttr_t myTask04_attributes = {
.name = "myTask04",
.cb_mem = &myTask04ControlBlock,
.cb_size = sizeof(myTask04ControlBlock),
.stack_mem = &myTask04Buffer[0],
.stack_size = sizeof(myTask04Buffer),
.priority = (osPriority_t) osPriorityHigh7,
};
```

Next, we should add the following code segments to the specified places in the "main.c" file of the project. Here, memory space for ten uint32_t elements are allocated dynamically in the task function StartTask04. Then, elements of the data array are copied into this new allocated memory space. Afterward, we take square of each element and obtain their average. Finally, allocated memory region is freed at the end of the task function StartTask04.

```
/* USER CODE BEGIN PV */
uint8_t i;
uint16_t data[10] = {1, 2, 3, 4, 5, 6, 7, 8, 9, 10};
uint32_t average;
/* USER CODE END PV */

/* USER CODE END Header_StartTask04 */
void StartTask04(void *argument)
{
/* USER CODE BEGIN StartTask04 */
/* Infinite loop */
for(;;)
{
average = 0;
uint16_t *ptr = NULL;
ptr = pvPortMalloc(10);
for(i = 0; i < 10; i++){
*(ptr+i) = data[i] * data[i];
```

```
}
for(i = 0; i < 10; i++){
average += *(ptr+i);
}
average /= 10;
vPortFree(ptr);
osDelay(1000);
}
/* USER CODE END StartTask04 */
}
```

10.8.2   Memory Management in Mbed OS

In Mbed OS, the ISR/scheduler, idle, timer, and main thread are started by default.
Hence, they are created in static form. The first three of these threads use 2 kB
of RAM and this value cannot be changed. The main thread uses 4 kB of RAM.
This value can be set from the file "mbed_app.json" by changing the parameter
`rtos.main-thread-stack-size` there. Likewise, each formed thread has 4 kB
stack size by default. This value can be changed from the "mbed_app.json" file
by changing the parameter `rtos.thread-stack-size` there. Or the thread stack
size can be changed by setting the `stack_size` parameter of the thread creation
function.

There is no initial "mbed_app.json" file created by default in an Mbed Studio
project. Therefore, we should create an empty "mbed_app.json" file at the project
root folder. Then, we can adjust memory parameters from this file. As an example,
we can add the below code lines to set the main and user thread stack sizes to 2 kB.

```
{
"target_overrides": {
"*": {
"rtos.main-thread-stack-size": 2048,
"rtos.thread-stack-size": 2048
}}}
```

When we check kernel object creation functions in Mbed OS, it can be seen that
there are no predefined initialization parameters for related control blocks. They are
created automatically and placed in the heap which means they cannot be created
statically via Mbed OS functions. We can only place thread stacks inside the static
SRAM region instead of heap region. We will show how this can be achieved by the
example in Listing 10.28.

Listing 10.28   The memory management example in Mbed OS

```
#include "mbed.h"

uint8_t i;
uint16_t data[10] = {1, 2, 3, 4, 5, 6, 7, 8, 9, 10};
uint32_t average;

uint8_t thread1Buffer[1024];
```

```
Thread thread1(osPriorityHigh, 1024, thread1Buffer);

void thread1Fn()
{
while (true)
{
average = 0;
uint16_t *ptr = NULL;
ptr = (uint16_t *)malloc(10);
for (i = 0; i < 10; i++)
{
*(ptr + i) = data[i] * data[i];
}
for (i = 0; i < 10; i++)
{
average += *(ptr + i);
}
average /= 10;
printf("%u \n", average);
free(ptr);
ThisThread::sleep_for(1000ms);
}
}

int main()
{
thread1.start(thread1Fn);

while (true);
}
```

In Listing 10.28, we create a 1 kB buffer for the thread1 stack in static form. In fact, this is the Mbed OS version of the memory management example in the previous section. Here, the functions malloc and free are used to allocate or free memory blocks from the heap. However, these functions are not thread-safe. Hence, they must be used carefully and protected by mutexes when multiple threads use them.

10.9 Application: RTOS-Based Implementation of the Robot Vacuum Cleaner

In this chapter, we migrate the developed software of our robot vacuum cleaner to RTOS. Hence, we benefit from RTOS and show how it can be used to organize complex operations in our robot. We provided the equipment list, circuit diagram, detailed explanation of design specifications, and peripheral unit settings in a separate document in the folder "Chapter10\EoC_applications" of the accompanying supplementary material. In the same folder, we also provided the complete C, C++, and MicroPython codes for the application.

10.10 Summary of the Chapter

RTOS offers a new perspective to realize complex projects. The main difference here compared to bare-metal programming is that the code to be executed is partitioned into tasks. There is also a kernel which coordinates the order of task execution. This setup forms an efficient usage of microcontroller resources. In this chapter, we focused on RTOS components via FreeRTOS and Mbed OS. To do so, we evaluated task, event, mutex, semaphore, queue, and mail in both RTOS types. We provided sample codes for their usage as well. As the end of chapter application, we migrated our robot vacuum cleaner codes to an RTOS-based project. Hence, the reader can observe the fundamental difference between a complex project formed by bare-metal programming and RTOS operations. We believe this will convince the reader that RTOS usage should be considered in forming complex projects.

Problems

10.1. Why has RTOS emerged? What is the main difference between RTOS usage and bare-metal programming?

10.2. What is a kernel in RTOS?

10.3. What is a task?

10.4. Why is kernel tick used in RTOS?

10.5. Why is the code block wrapped by an infinite loop under tasks?

10.6. What should we understand from task states and priorities?

10.7. Is it possible to eliminate the idle task from RTOS completely?

10.8. Form a project with the following specifications. Create two tasks named myTask02 and myTask03 in addition to the default task. Select the priority of these tasks as osPriorityHigh and osPriorityAboveNormal, respectively. The default task will toggle the onboard green LED, keeps CPU busy for 1 second using the appropriate function (HAL_Delay for FreeRTOS and wait_us for Mbed OS), and then yields the process. myTask03 toggles the red LED and then waits 1 second using the appropriate function (HAL_Delay for FreeRTOS and wait_us for Mbed OS). Finally, myTask02 changes the priority of myTask03 between osPriorityAboveNormal and osPriorityLow every 10 s. Form the project using:

(a) FreeRTOS under STM32CubeIDE.
(b) Mbed OS under Mbed.

10.9. Form a project with the following specifications. Create a task named myTask02 in addition to the default task. Select the priority of this task as osPriorityNormal. These two tasks are used to toggle onboard green and red LEDs, respectively. The default task is triggered by the Timer 2 module interrupt via event flag. The time interval for the Timer 2 module is set as 1 second. myTask02 is triggered by the external interrupt of onboard push button via another event flag. Form the project using:

(a) FreeRTOS under STM32CubeIDE.
(b) Mbed OS under Mbed.

10.10. Form a project with the following specifications. Create a task named myTask02 in addition to the default task. Select the priority of this task as osPriorityHigh. These two tasks are used to send 500 consecutive numbers (as number1 and number2) via USART1 module. The default task is used to send number1 and myTask02 is used to send number2. Start of myTask02 will be delayed by 100 ms using an appropriate function. After myTask02 does its job, it is suspended by an appropriate function. The process inside each task must be protected by a mutex. Form the project using:

(a) FreeRTOS under STM32CubeIDE.
(b) Mbed OS under Mbed.

10.11. Repeat Problem 10.10 using a semaphore instead of mutex.

10.12. Form a project with the following specifications. The ADC1 module is configured to obtain 100 consecutive measurements from pin PA5. ADC conversion is triggered by the Timer 3 module with 1 KHz sampling frequency. We also create a structure with a variable for id of the ADC module and array with 100 elements. This structure is filled with obtained measurements and placed in a queue in the ADC interrupt callback function. Default task gets this structure from the queue and takes the average of measurements. The onboard green LED is turned on if the average value is larger than 2.5 V. It is turned off if the average value is less than 2.5 V. Form the project using:

(a) FreeRTOS under STM32CubeIDE.
(b) Mbed OS under Mbed.

10.13. Repeat Problem 10.12 using mail instead of queue.

10.14. Repeat Problem 10.8 in C language under STM32CubeIDE using software timer to change the priority of `myTask03` instead of `myTask02`.

Reference

1. Ünsalan, C., Gürhan, H.D., Yücel, M.E.: Programmable Microcontrollers: Applications on the MSP432 LaunchPad. McGraw-Hill, New York (2018)

LCD, Touch Screen, and Graphical User Interface Formation

11

11.1 LCD

We can group displays into two types as emissive and non-emissive. In the first group, the image is formed directly on the display. Popular devices using this technique are CRT, plasma, LED, and OLED screens. In the second group, light is formed in backside of the display. Then, it is filtered by a panel and the image is formed. The most popular device using this technique is LCD. Since the STM32F4 board has an LCD on it, we will focus on this display type in this chapter.

11.1.1 LCD Structure

There is a layered structure in LCD. These layers are backlight, diffusor, bottom polarization filter, bottom electrode (with TFT, if the LCD is of active matrix type), liquid crystal, top electrode, color filter, top polarization filter, and glass cover from bottom to top. We provide this layered structure in Fig. 11.1.

LCD is a non-emissive display type. Hence, it requires backlight to be generated either by LED or fluorescent lamp. The diffuser layer distributes this light homogeneously on the display. The light is passed through the liquid crystal layer. Bottom and top electrodes generate an electromagnetic field to control crystals in the liquid crystal layer. Through this operation and usage of polarization filters (to control the passing light intensity), we can form an image by stopping, passing, or dimmed

---

Supplementary Information The online version contains supplementary material available at (https://doi.org/10.1007/978-3-030-88439-0_11).

C. Ünsalan et al., *Embedded System Design with Arm Cortex-M Microcontrollers*,
https://doi.org/10.1007/978-3-030-88439-0_11

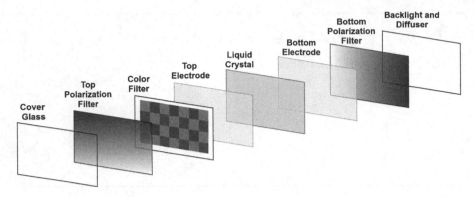

Fig. 11.1 Layered LCD structure

passing of light. The liquid crystal layer does not have color control. Therefore, an extra color filter is used for this purpose. The glass cover protects the display.

11.1.2 LCD Working Principles

We should control the LCD and its smallest elements (called pixels) to display an image on it. This operation is called driving the display. There are four different display driving techniques as direct, multiplex, passive matrix, and active matrix. The direct driving technique works extensively on seven-segment displays. The multiplex technique is an extension of the direct driving technique such that more than one seven-segment module can be driven. Passive and active matrix display driving techniques are used in LCDs. These methods are based on bottom and top electrodes introduced in the previous section. The passive matrix technique is based on the placement and driving of electrodes. There is an additional thin film transistor (TFT) layer at the bottom layer in the active matrix display driving technique.

The LCD on the STM32F4 board has a TFT layer based on the active matrix display driving technique. However, it is not possible to drive the LCD directly. Therefore, there are dedicated display driver modules for this purpose. They take the image to be displayed from a source (such as microcontroller) and feed it to the display with appropriate formatting. We will talk about the specific module on the STM32F4 board in Sect. 11.1.4.

Let's briefly explain working principles of a generic LCD. Beforehand, we should make some definitions. The smallest image element to be displayed on the LCD is called pixel. The pixel definition originates from digital image processing to be introduced in Chap. 14. A row in the display is called line. The visible area in the display is called frame.

The display driver sends pixel information one pixel at a time starting from top left position of the frame. This operation is controlled by the pixel clock signal generated by the display driver module. The pixel clock increases the line pointer.

Hence, the active pixel is moved to right at every pixel clock. As the end of line is reached, the operation restarts from beginning of the next line. This operation is controlled by the HSYNC signal generated by the display driver module. HSYNC resets the display column pointer to the leftmost position while moving to the next line. As the rightmost pixel of the last line in the frame is reached, the operation restarts from the top left position of the frame. This operation is controlled by the VSYNC signal generated by the display driver module. VSYNC resets the display line pointer and sets it to the starting point in the frame. The frequency of the pixel clock has the highest value, with VSYNC and HSYNC having integer division of this value. Frequency of the overall display operation is called refresh or frame rate.

There are five regions in a typical display, four of them non-visible and one of them visible. Non-visible regions are virtual. They are called vertical back porch, vertical front porch, horizontal back porch, and horizontal front porch. They are also called horizontal and vertical blanking. These names originate from CRT displays. They are used in digital displays to adjust the frame buffer keeping the image to be displayed and interrupt generation by the image source (such as microcontroller). The visible region in the display is called active area. We provide the HSYNC and VSYNC signals and non-visible and visible regions in a generic display in Fig. 11.2. The reader can understand the definitions better by observing this figure.

11.1.3 Connecting the LCD to an Image Source

We introduced working principles of the LCD in the previous section. Here, we will focus on how to transmit an image from its source to LCD via display driver module. Since we are using the STM32F4 microcontroller in this book, we will call it as the (image) source from this point on.

There are three digital connection types to be used between the source and LCD. These are MIPI display bus interface (MIPI-DBI), MIPI display parallel interface (MIPI-DPI), and MIPI display serial interface (MIPI-DSI). The Mobile Industry Processor Interface (MIPI) is a group working on camera and LCD connection standards. MIPI-DBI has three types as type A, type B, and type C. These are based on Motorola 6800 bus, Intel 8080 bus, and SPI, respectively.

In MIPI-DBI, the display driver should have a frame buffer. Hence, the image is transferred from the source to this frame buffer. As the image is transferred to the frame buffer, it can be displayed by the display driver continuously. In MIPI-DPI, the display driver does not have a frame buffer. The image to be displayed is kept in the source and directly sent to display. Therefore, the image should be transferred in real time. Otherwise, the image on the display vanishes. In MIPI-DSI, the display driver should have a frame buffer to operate. The image is transferred from the source to this frame buffer by a high-speed differential line. This setup needs an extra PHY circuitry.

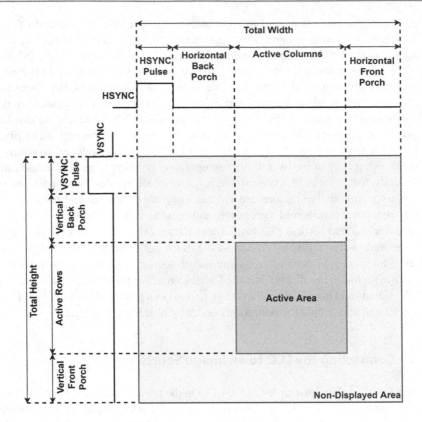

Fig. 11.2 Synchronization signals and regions in a generic display

11.1.4 LCD on the STM32F4 Board

The STM32F4 board has a 2.4 inch color LCD TFT display on it with 320×240 pixel size. There is an integrated display driver module on the LCD called ILI9341 "TFT LCD Single Chip Driver" [3]. This driver has an internal RAM, with size 172800 bytes, used as frame buffer. Hence, each pixel can be represented by 18 bits in this frame buffer. To be more specific, each pixel has red, green, and blue color components each represented by six bits. This color representation format is called RGB666. The total number of color variations in the LCD can be calculated as $2^6 \times 2^6 \times 2^6 = 262144$. The ILI9341 display driver module supports MIPI-DBI type B parallel, MIPI-DBI type C serial (SPI), and MIPI-DPI parallel connection types. The driver has its own internal oscillator. The necessary pixel clock, VSYNC, and HSYNC signals to drive the LCD are generated this way. The reader can check pin connections between the LCD and STM32F4 microcontroller from [2].

11.2   Touch Screen

The user can interact with the GUI running on the LCD via touch screen. The LCD on the STM32F4 board has an integrated touch screen on it. Therefore, we will briefly explore how a touch screen works in this section. Then, we will summarize properties of the touch screen on the LCD of the STM32F4 board.

11.2.1 Touch Screen Working Principles

Touch screen (or panel) is a transparent layer added on top of the display unit. It senses the touch and its location on the display. The touch screen is composed of two modules as sensor and controller. The sensor module is the part detecting touch by the user. It can be of type as resistive or capacitive. In general, the touch sensor is placed on top of the indium tin oxide (ITO) glass on the display. The controller module drives the touch sensor. Moreover, it transfers the sensed touch data to an external device (such as the STM32F4 microcontroller) via I^2C or SPI interface.

The touch screen on the LCD of STM32F4 board is formed by four-wire resistive layer. In it, there is a flexible resistive polyethylene (PET) film layer on top. This layer is conductive to some degree. Below it, there is another hard resistive layer which is also conductive to some degree. There is a free space between these two layers. We can think these two layers as two separate resistors. Each resistor has two wires for connection, hence the name four-wire. We provide this structure in Fig. 11.3. As can be seen in this figure, the resistor wires (electrodes) are placed on the left and right sides of the top layer. They are placed on the top and bottom of the bottom layer. Hence, we can assume that there is one horizontal and one vertical resistor placed on the top and bottom layers, respectively.

We can explain working principles of the four-wire resistive layer based on the setup in Fig. 11.4. The untouched case is given in Fig. 11.4a. Here, the resistors do not touch each other. When the user presses the top layer, it moves downward by the pressure applied to it. This way, the top layer touches the bottom layer. In other words, the resistors touch each other.

Fig. 11.3 Four-wire resistive touch screen layers

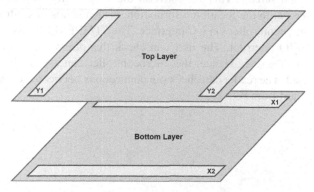

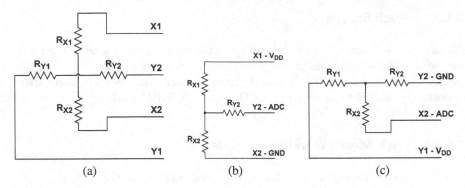

Fig. 11.4 Four-wire resistive touch screen equivalent circuits. (**a**) Untouched case. (**b**) Vertical touch. (**c**) Horizontal touch

We can detect the vertical and horizontal touch location on the screen as follows. We first connect V_{DD} and ground to the bottom layer pins. One pin at the top layer is connected to the ADC module. The touched location forms a voltage divider as in Fig. 11.4b. The middle of the voltage divider is connected to the ADC module through the top layer pin. Hence, the voltage value read there (proportional to the V_{DD} value) gives the vertical touch location with respect to screen height. After this measurement, the top layer pins are connected to V_{DD} and ground, respectively. Then, a bottom layer pin is connected to the ADC module. Now, the touched location forms a voltage divider as in Fig. 11.4c. As with the same reasoning in the previous operation, we can obtain the horizontal touch location with respect to screen width. This way, we can obtain the horizontal and vertical touch locations. This operation is repeated continuously to detect touch locations on the screen.

11.2.2 Touch Screen on the LCD of the STM32F4 Board

The LCD on the STM32F4 board has an integrated four-wire touch screen on it. There is the STMPE811 touch driver IC on the STM32F4 board to control this touch screen. This IC converts the touch information, received from the touch screen, to the location information. It then sends this information to the STM32F4 microcontroller via I^2C interface. The STMPE811 IC also has a pin which generates GPIO interrupt. The user can check this pin when reading the location from the touch screen. Hence, the microcontroller can stay in low power mode when not used. The reader can check pin connections between the touch screen and STM32F4 board from [2].

11.3 Hardware Modules in the STM32F4 Microcontroller for LCD and Touch Screen Control

There are two dedicated modules in the STM32F4 microcontroller for LCD and touch screen control. These are LCD-TFT display controller (LTDC) and DMA2D (or Chrom-ART accelerator). We will introduce these modules in this section. We will also use the SPI and I^2C communication modules, introduced in Chap. 8, besides these. To be more specific, we will benefit from SPI in LCD usage. We will benefit from I^2C in touch screen usage.

11.3.1 LCD-TFT Display Controller

The LCD-TFT display controller (LTDC) is a peripheral unit used for driving LCD-TFT displays in parallel manner (or in MIPI-DPI form). It provides RGB color bits (in red, green, and blue form), synchronization signals, data enable signal, and pixel clock to the LCD in parallel manner. It can support RGB888 format output with 24-bit data. It also supports RGB565 and RGB666 data formats. The LTDC supports maximum 1024×768 pixel resolution. We will explain the mentioned color formats and pixel resolution in detail in Sect. 11.3.2 and Chap. 14.

The LTDC module can be used to program the active area of the display and its size. The module can display two (foreground) images and one background image in layered form as follows. The background image is placed at the bottom. The first image is placed on it. The second image is placed on top of the first image. By using different transparency values, a blended image can be formed. The LTDC module performs all these operations within itself on pixel basis.

The LTDC module supports several input image formats. These are ARGB8888, RGB888, RGB565, ARGB1555, ARGB4444, L8 (8-bit luminance or CLUT), AL44 (4-bit alpha + 4-bit luminance), and AL88 (8-bit alpha + 8-bit luminance). The module converts these formats to the RGB888 format before sending them to the display. It can also apply dithering.

General block diagram of the LTDC module is as in Fig. 11.5. As can be seen in this figure, there are seven block types in the LTDC module as FIFO buffers (layers 0 and 1), pixel format converters, timing generator, registers, blending block, dithering block, and GPIO control block. We will explain working principles of the LTDC module based on these next.

The LTDC module gets the image to be displayed either from DMA2D (to be explained in Sect. 11.3.2) or STM32F4 RAM on a pixel basis. This image is stored in FIFO buffers of the module. If one-layer image will be used, only layer 1 FIFO buffer is used in the LTDC. Afterward, the input image is converted to the suitable format for LCD. If a two-layer image is used, blending is applied. Dithering can also be applied if needed. Then, the image is sent to the LCD via GPIO pins. The timing generator generates all synchronization clock signals.

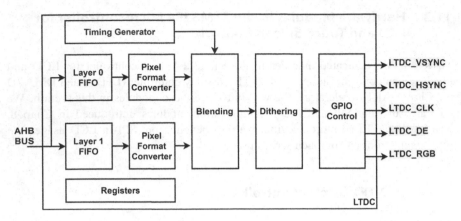

Fig. 11.5 LTDC general block diagram

11.3.2 DMA2D

The DMA2D module works in the same way as with the DMA module. However, it is dedicated to image operations only. The DMA2D module has direct access to memory. It can perform four different tasks, to be explained in the following paragraph, without CPU support. Besides, it can perform these tasks faster than the CPU. STMicroelectronics calls the graphic acceleration operations done via this module as Chrom-ART accelerator. Therefore, we will use both names interchangeably.

The DMA2D module can perform four basic tasks as follows. The first task is filling a rectangular shape in the frame with unique color. This is a register to memory operation. The second task is copying a frame or rectangular part of the frame from one memory location to another. This is a memory-to-memory operation. The third task is converting pixel format of a frame (or rectangular part of a frame) while transferring it from one memory location to another. This is a memory-to-memory operation with pixel conversion. The fourth task is blending two images (with different size and pixel formats) and storing the resulting image in memory. This is a memory-to-memory operation with pixel conversion and blending.

The DMA2D module supports direct and indirect color modes. The direct color mode consists of RGB565, RGB888, and ARGB888 formats. These need two to three bytes per pixel. In the indirect color mode, the color lookup table (CLUT) is used for reference color. CLUT is also called palette. Therefore, only the reference index is kept in the pixel array which has size of the image. As a result, the image size to be displayed decreases. However, this may lead to distortion in color values. This can be minimized by the dithering operation. The DMA2D module has two internal memory locations to keep CLUT values.

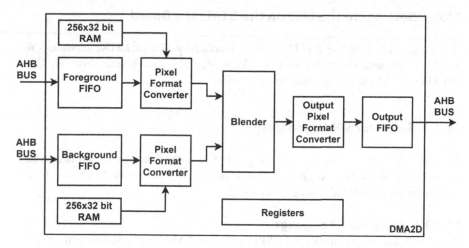

Fig. 11.6 The DMA2D module block diagram

While converting the pixel via the DMA2D module, the input image formats can be one of ARGB8888, RGB888, RGB565, ARGB1555, ARGB4444, L8, AL44, AL88, L4, A8, and A4. The output image format can be one of ARGB8888, RGB888, RGB565, ARGB555, and ARGB4444. Here, ARGB8888, RGB888, RGB565, ARGB555, and ARGB4444 represent the direct color mode formats. L4, L8, A4, A8, AL44, and AL88 represent indexed color mode formats. As an example, L8 indicates that the CLUT is formed by 256 entries, each being either 16-bit RGB565, 24-bit RGB888, or 32-bit ARGB8888 color representation. The pixel array is formed by one byte index data. For more information on direct and indexed color formats, please see [4].

We provide block diagram of the DMA2D module in Fig. 11.6. As can be seen in this figure, the module has 256×32-bit RAM (two blocks), FIFO buffers (one for background and another for foreground), pixel format converters (two blocks), blender, output pixel format converter, output FIFO buffer, and registers. We will explain working principles of the DMA2D module based on these next.

The DMA2D module receives data from the RAM via AHB. Format conversion can be applied to the image on pixel basis if needed. The 256×32-bit RAM block is used for CLUT table if the image will be represented in indexed form such as L4 or L8. The DMA2D module can apply blending if needed. If the blending operation will not be used, only the foreground FIFO buffer is used in operation. The image is converted to the desired output format and sent to RAM via AHB.

11.4 Setting Up the LCD on the STM32F4 Board

We can use LCD on the STM32F4 board either by SPI or LTDC modules. We provide both methods in this section. While doing so, we benefit from C, C++, and MicroPython languages whenever possible.

11.4.1 Setup for SPI-Based Usage

We will consider the setup for LCD usage via SPI as the first method. Although this is a suboptimal solution, we had to follow such a path to use the LCD and digital camera (to be introduced in Chap. 14) together.

11.4.1.1 Setup via C Language
We should start with setting up the SPI module in STM32CubeMX as explained in Chap. 8. More specifically, we should use the SPI5 module in "Half-Duplex Master" mode with the settings in STM32CubeMx as "Frame Format" "Motorola"; "Data Size" "8 Bits"; "First Bit" "MSB First"; "Prescaler (for Baud Rate)" "2"; "Clock Polarity (CPOL)" "Low"; "Clock Phase (CPHA)" "1 Edge"; "CRC Calculation" "Disabled"; NSS Signal Type" "Software." We provide the mentioned settings applied on the "SPI5 Mode and Configuration" tab in Fig. 11.7.

We will be using pins PF7 and PF9 for the SPI module. Both pins should have the "maximum output speed" value set to "very high." This is important; otherwise, the LCD module will not work. Besides, we should set the pins PC2 and PD13 as GPIO output. These pins are used for "command" and "command and data write enable" for the LCD. As a side note, LCD can get both image data and commands to set its own registers. The ILI9341 LCD driver decides on the incoming bits as either being data or command by the pins PC2 and PD13.

We should also set the system clock to 180 MHz. To do so, we should use the clock configuration tab. There, we should set HCLK (MHz) parameter as 180. This way we can drive the LCD at 45 MHz.

11.4.1.2 Setup via C++ Language
We will use the LCD module via SPI in Sect. 11.5.1.2 in C++ language. There, we will also handle the setup via predefined functions. Hence, we do not need any extra adjustments besides these.

11.4.1.3 Setup via MicroPython
MicroPython has no class to perform SPI-based LCD operations. Therefore, we created a lcdSPI class in our custom MicroPython firmware. The reader can find the MicroPython firmware in the book web site. The lcdSPI module can be created and initialized using the function init(rotation). All GPIO and SPI module configurations are done automatically in this function. Initialization sequence for LCD is also realized in this function. Finally, the rotation is used to set screen rotation. It can be set as 0 for vertical, 1 for horizontal, 2 for inverted vertical, and 3

Fig. 11.7 SPI5 module configuration tab for LCD

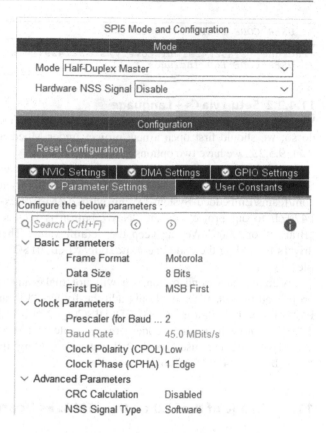

for inverted horizontal direction. After calling this function, SPI-based LCD module will be ready to be used.

11.4.2 Setup for LTDC-Based Usage

We can also use the LCD on the STM32F4 board by the LTDC module. Unfortunately, configuring this module is not easy, and some configurations should be done manually. Therefore, we benefit from the board support package (BSP) drivers provided by STMicroelectronics for this purpose.

11.4.2.1 Setup via C Language
In order to use BSP drivers for the LTDC module, we should first open a new project and press the ".ioc" file to open the "Device Configuration Tool" in STM32CubeMX. Then, we should follow the steps given in Sect. 9.5.2.1. Finally, we should add the below C code script to the appropriate place in the "main.c" file.

```
/* USER CODE BEGIN Includes */
#include "STM32F429I-Discovery/stm32f429i_discovery.h"
#include "STM32F429I-Discovery/stm32f429i_discovery_lcd.h"
/* USER CODE END Includes */
```

11.4.2.2 Setup via C++ Language

We will also benefit from the BSP drivers to use LTDC in C++ language. To do so, we should first open a new project under Mbed Studio. As mentioned in Sect. 9.5.2.2, we have two options at this stage.

In our first method, we will add the BSP library to our project as explained in Sect. 9.5.2.2. Then, we will add the LCD wrapper library located at http://os.mbed.com/teams/Embedded-System-Design-with-ARM-Cortex-M/code/LCD_DISCO_F429ZI/ to our project. As the "Next" button is pressed, the interface asks for "Branch" or "Tag." We can keep the default value. Then, we can use the BSP drivers by adding the code line #include "LCD_DISCO_F429ZI.h" to our main file.

To apply the second option, we will not make any specific modifications in project properties. We should only add the folders "Libraries\Cpp\BSP_DISCO_F429ZI" and "Libraries\Cpp\LCD_DISCO_F429ZI" directly to the project folder. To note here, these codes are available in the accompanying book web site. Then, we can use the BSP drivers by adding the code line #include "LCD_DISCO_F429ZI.h" to our main file.

11.5   Usage of the LCD on the STM32F4 Board

As we set up the LCD on the STM32F4 board, we can use it afterward. To do so, we will benefit from two methods based on SPI and LTDC modules. For both methods, we provide C, C++, and MicroPython language-based codes whenever applicable.

11.5.1   Usage of the LCD via SPI

We can use the LCD via SPI. To do so, we provide methods on C, C++, and MicroPython languages next.

11.5.1.1 Usage via C Language

We formed the "SPI LCD library" in C language to control the LCD on the STM32F4 board via SPI. This library works in connection with the "image processing library" to be introduced in Chap. 14. These libraries are available in the accompanying book web site.

We have six functions in our "SPI LCD library." These are as follows. The function ILI9341_init initializes the LCD to be used by SPI. It also sets the display rotation direction. The function ILI9341_setAddress sets address of the active area of the display. The function ILI9341_setRotation sets the display rotation. The function ILI9341_fillScreen fills the screen by a given color in RGB565 format. If the white color is given, the display is cleared. The function ILI9341_drawPixel plots a pixel with the address in the function ILI9341_setAddress by the given RGB565 color format. If the function ILI9341_drawPixel is successively called, it processes the following pixels in the line. This is done since the LCD increases its internal frame buffer pointer by one at each operation. Here, end of line is also taken into account. When end of frame is reached, the plotting operation restarts from the beginning location of the frame. The function ILI9341_drawImage draws an image on LCD. The image should be of type struct ImageTypeDef as explained in Chap. 14. We provide detailed explanation of the functions next.

```
void ILI9341_init(uint8_t Rotation);
/*
Rotation: Rotation of the screen and can be one of
    SCREEN_VERTICAL, SCREEN_HORIZONTAL, SCREEN_VERTICAL_INV or
    SCREEN_HORIZONTAL_INV.
*/

void ILI9341_setAddress(uint16_t X1, uint16_t Y1, uint16_t X2,
    uint16_t Y2);
/*
X1: x coordinate of the upper left corner of active area
Y1: y coordinate of the upper left corner of active area
X2: x coordinate of the lower right corner of active area
Y2: y coordinate of the lower right corner of active area
*/

void ILI9341_setRotation(uint8_t Rotation);

void ILI9341_fillScreen(uint16_t Color);
/*
Color: color to fill the screen in 16-bit RGB565 format.
*/

void ILI9341_drawPixel(uint16_t Color);
/*
Color: color of the pixel in 16-bit RGB565 format.
*/

void ILI9341_drawImage(ImageTypeDef *Pimg, uint16_t xOffset,
    uint16_t yOffset);
/*
Pimg: pointer to the ImageTypeDef struct
xOffset: image start point offset in horizontal axis in terms of
    pixels
yOffset: image start point offset in vertical axis in terms of
    pixels
*/
```

We next provide two examples on the usage of LCD via SPI. The first example displays an image, given as a header file, on the LCD. We provide the C code for the first example in Listing 11.1.

Listing 11.1 Displaying an image on the LCD via SPI, the C code

```
/* USER CODE BEGIN Includes */
#include "image.h"
#include "ili9341.h"
```

```
#include "mandrill.h"
/* USER CODE END Includes */

/* USER CODE BEGIN 2 */
ILI9341_init(SCREEN_HORIZONTAL);
ILI9341_fillScreen(WHITE);
ILI9341_drawImage(&RGB565_IMG, 0, 0);
/* USER CODE END 2 */
```

To execute the C code in Listing 11.1, we should open a new project under STM32CubeIDE. Then, we should adjust the hardware as explained in Sect. 11.4.1. In order to use the "SPI LCD library," we should add the files "image.c" and "ili9341.c" to the folder "project->Core->Src" and files "image.h" and "ili9341.h" to the folder "project->Core->Inc" in our project. Besides adding the library files, we should also add the "mandrill.h" file provided in the accompanying book web site to the folder "project->Core->Inc" in our project. As we debug and run the code, the Mandrill image will be displayed on the LCD.

In this example, we use the image provided as the header file. However, the reader can also benefit from the Python file "ImageHeaderGenerator.py" provided in the accompanying book web site (under the folder "Libraries\PC_Python"), to transform his or her image file to the corresponding header file. In order to use this file, we should have OpenCV installed in our PC. Please see Chap. 14 for this purpose. Then, the user should enter the path of the input image and output header file name. The user should also enable the related code line in the Python file for the required header type. As the code is executed, the header file will be generated in the same folder as the image.

The second example plots a color bar on the LCD. We provide the C code for this example in Listing 11.2. In this code, the LCD is initialized first. Then, the selected region is filled in the LCD with the functions ILI9341_setAddress and ILI9341_drawPixel. As we run the code, we should see a color bar on the LCD.

Listing 11.2 Plotting a color bar on the LCD via SPI, the C code

```
/* USER CODE BEGIN Includes */
#include "ili9341.h"
/* USER CODE END Includes */

/* USER CODE BEGIN PD */
#define COLORBAR_NUM 18
#define COLORBAR_WIDTH 20
#define SCREEN_WIDTH 320
#define SCREEN_HEIGHT 240
/* USER CODE END PD */

/* USER CODE BEGIN PV */
uint16_t colorBar[18] = {
0x000F, 0x03E0, 0x03EF, 0x7800, 0x780F, 0x7BE0,
0xC618, 0x7BEF, 0x001F, 0x07E0, 0x07FF, 0xF800,
0xF81F, 0xFFE0, 0xFFFF, 0xFD20, 0xAFE5, 0xF81F};

uint32_t i = 0, j = 0;
/* USER CODE END PV */

/* USER CODE BEGIN 2 */
```

```
ILI9341_init(SCREEN_HORIZONTAL);

for (i = 0; i < 18; i++)
{
if (SCREEN_WIDTH > i * COLORBAR_WIDTH)
{
ILI9341_setAddress(i * COLORBAR_WIDTH, 0, (i + 1) *
    COLORBAR_WIDTH - 1, 240);
for (j = 0; j <= SCREEN_HEIGHT * COLORBAR_WIDTH; j++)
{
ILI9341_drawPixel(colorBar[i]);
}
}
}
/* USER CODE END 2 */
```

11.5.1.2 Usage via C++ Language

We formed the "SPI LCD library" in C++ language to control the LCD on the STM32F4 board via SPI. This library works in connection with the "image processing library" to be introduced in Chap. 14. These libraries are available in the accompanying book web site.

We formed a class in the "SPI LCD library" to use the LCD in C++ language. Our class is named ILI9341. It can also work as an initialization function if its constructor has a parameter entered. Within our class, the function init configures the SPI module. Afterward, it sets the ILI9341 LCD driver to SPI mode. Finally, it sets the display rotation property. Our class has functions setAddress, setRotation, fillScreen, drawPixel, and drawImage performing the same operations as their C counterpart. We provide the detailed explanation for the class ILI9341 and its functions next.

```
ILI9341(uint8_t Rotation = SCREEN_HORIZONTAL, uint32_t frequency
    = 45000000)
/*
Rotation: specifies the screen orientation and can be one of
    SCREEN_VERTICAL, SCREEN_HORIZONTAL, SCREEN_VERTICAL_INV or
    SCREEN_HORIZONTAL_INV. The default is SCREEN_HORIZONTAL
frequency: SPI bus clock frequency to be set in Hz. The default
    value is 45000000
*/

void init(uint8_t Rotation = SCREEN_HORIZONTAL, uint32_t
    frequency = 45000000);

void setAddress(uint16_t X1, uint16_t Y1, uint16_t X2, uint16_t
    Y2);
/*
X1: x coordinate of the upper left corner of active area
Y1: y coordinate of the upper left corner of active area
X2: x coordinate of the lower right corner of active area
Y2: y coordinate of the lower right corner of active area
*/

void setRotation(uint8_t Rotation);

void fillScreen(uint16_t Color);
/*
```

```
Color: color to fill screen in 16-bit RGB565 format.
*/

void drawPixel(uint16_t Color);

void drawImage(IMAGE Pimg, uint16_t xOffset, uint16_t yOffset);
/*
Pimg: pointer to the IMAGE object
xOffset: image start point offset in horizontal axis in terms of
    pixels
yOffset: image start point offset in vertical axis in terms of
    pixels
*/
```

We next provide two examples on the usage of our "SPI LCD library" in C++ language. These examples are the same as their C language counterpart. In our first example, given in Listing 11.3, we display an image from a header file as in Listing 11.1.

Listing 11.3 Displaying an image on the LCD via SPI, the C++ code

```cpp
#include "mbed.h"
#include "image.hpp"
#include "ili9341.hpp"
#include "mandrill.hpp"

ILI9341 ili9341;

int main()
{
RGB565_IMG.pData = RGB565_IMG_ARRAY;
RGB565_IMG.width = 160;
RGB565_IMG.height = 120;
RGB565_IMG.size = 160 * 120 * 2;
RGB565_IMG.format = 1;

ili9341.fillScreen(WHITE);
ili9341.drawImage(&RGB565_IMG, 0, 0);

while (true);
}
```

In order to execute the code in Listing 11.3, we should open a new project. We do not need any initialization operations since the `init` function in our library handles all these operations for us. We should add the "SPI LCD library" and "image processing library" files "image.cpp," "ili9341.cpp," "image.hpp," and "ili9341.hpp" to our project folder. We should also add the "mandrill.h" file to our project folder. As we run the code, we should see the Mandrill image on the LCD. As in the previous section, the reader can benefit from the Python file "ImageHeaderGenerator.py" provided in the accompanying book web site, to transform his or her image file to the corresponding header file. The user should also enable the related code line in the Python file for the required header type.

We repeat the operations in Listing 11.2 now in C++ language given in Listing 11.4 as our second example. In this code, the LCD is initialized first. Then, the selected region in the LCD is filled by the functions `setAddress` and `drawPixel`.

Listing 11.4 Plotting a color bar on the LCD via SPI, the C++ code

```cpp
#include "mbed.h"
#include "image.hpp"
#include "ili9341.hpp"
#define COLORBAR_WIDTH 20

ILI9341 ili9341;

uint16_t colorBar[18] = {
0x000F, 0x03E0, 0x03EF, 0x7800, 0x780F, 0x7BE0,
0xC618, 0x7BEF, 0x001F, 0x07E0, 0x07FF, 0xF800,
0xF81F, 0xFFE0, 0xFFFF, 0xFD20, 0xAFE5, 0xF81F};

uint32_t i = 0, j = 0;

int main()
{

for (i = 0; i < 18; i++)
{
if (ili9341.ILI9341_SCREEN_WIDTH > i * COLORBAR_WIDTH)
{
ili9341.setAddress(i * COLORBAR_WIDTH, 0, (i + 1) *
    COLORBAR_WIDTH - 1, 240);
for (j = 0; j <= ili9341.ILI9341_SCREEN_HEIGHT * COLORBAR_WIDTH;
    j++)
{
ili9341.drawPixel(colorBar[i]);
}
}
}

while (true);
}
```

In order to execute the code in Listing 11.3, we should open a new project. We do not need any initialization operations since the `init` function in our library handles all these operations for us. We should add the files "image.cpp," "ili9341.cpp," "image.hpp," and "ili9341.hpp" to our project folder. As we run the code, we should see a color bar on LCD.

11.5.1.3 Usage via MicroPython

The functions `drawPixel(color)` and `fillScreen(color)` can be used to paint one pixel or fill LCD according to `color`, which is 16-bit color data in RGB565 format in the custom MicroPython firmware. The function `setAddress(X1, Y1, X2, Y2)` is used to set the active area of the display. Here, `X1`, `Y1`, `X2`, and `Y2` are used to represent four corner points of the desired active area. The function `drawImage(pData, width, height, size, xOffset, yOffset)` is used to plot the selected image on the LCD. Here, `pData` is the pointer of the image data. `width` is the width of the image in terms of pixels. `height` is the height of the image in terms of pixels. `size` is the total pixel number of the image. `xOffset` is the image starting point offset in horizontal axis in terms of pixels. `yOffset` is the image starting point offset in vertical axis in terms of pixels.

We next provide two examples on the usage of our SPI-based LCD module in MicroPython. These examples are the same as their C language counterpart. In our first example, given in Listing 11.5, we display an image from a ".py" file as in Listing 11.1. Here, we should also add the "mandrill.py" file to the MicroPython file system. As we run the code, we should see the Mandrill image on LCD. As in previous section, the reader can benefit from the Python file "ImageHeaderGenerator.py" provided in the accompanying book web site, to transform his or her image file to the corresponding header file. However, usable free size of MicroPython file system is restricted to nearly 90 kB. Therefore, if the generated file size is larger than this limit, it must be precompiled or frozen into MicroPython firmware as explained in Chap. 15. In this example, width and height of the Mandrill image are selected as 80 and 60, respectively, in order to generate a ".py" file with reasonable size.

Listing 11.5 Displaying an image on the LCD via SPI, the MicroPython code

```
import lcdSPI
import pyb
import mandrill

lcdSPI.init(0)
lcdSPI.fillScreen(0)
lcdSPI.drawImage(mandrill.pData, mandrill.width, mandrill.height,
    mandrill.size, 0, 0)
```

In the second example, given in Listing 11.6, LCD is initialized first. Then, the selected region in the LCD is filled by the functions `setAddress` and `drawPixel` to generate different color bars on the LCD.

Listing 11.6 Plotting a color bar on the LCD via SPI, the MicroPython code

```
import lcdSPI

COLORBAR_NUM = const(16)
COLORBAR_WIDTH = const(20)
SCREEN_WIDTH = const(320)
SCREEN_HEIGHT = const(240)

colorBar = [
    0x000F, 0x03E0, 0x03EF, 0x7800, 0x780F, 0x7BE0,
    0xC618, 0x7BEF, 0x001F, 0x07E0, 0x07FF, 0xF800,
    0xF81F, 0xFFE0, 0xFFFF, 0xFD20, 0xAFE5, 0xF81F]

lcdSPI.init(1)
for i in range(COLORBAR_NUM):
    if SCREEN_WIDTH > i * COLORBAR_WIDTH:
        lcdSPI.setAddress(i * COLORBAR_WIDTH, 0, (i + 1) *
            COLORBAR_WIDTH - 1, 240)
        for j in range(SCREEN_HEIGHT * COLORBAR_WIDTH):
            lcdSPI.drawPixel(colorBar[i])
```

Table 11.1 BSP driver functions for LCD usage via LTDC module: the C language

BSP driver function	Usage
BSP_LCD_Init	Initializes the LCD
BSP_LCD_LayerDefaultInit	Initializes the LCD layers
BSP_LCD_SelectLayer	Selects the LCD layer
BSP_LCD_SetLayerWindow	Sets the display window
BSP_LCD_SetTextColor	Sets the text color
BSP_LCD_SetBackColor	Sets the background color
BSP_LCD_SetFont	Sets the text font
BSP_LCD_DrawPixel	Writes pixel
BSP_LCD_Clear	Clears the LCD
BSP_LCD_FillCircle	Displays a full circle
BSP_LCD_FillRect	Displays a full rectangle
BSP_LCD_DrawRect	Displays a rectangle
BSP_LCD_DrawCircle	Displays a circle
BSP_LCD_DisplayChar	Displays one character
BSP_LCD_DisplayStringAt	Displays a maximum of 60 characters on the LCD

11.5.2 Usage of the LCD via LTDC

We can also use the LCD via LTDC module. We provide methods in C and C++ languages for this purpose next.

11.5.2.1 Usage via C Language

We will benefit from the available BSP driver functions to use the LCD via LTDC module in C language. These functions are tabulated in Table 11.1. For more information on them, please see [1].

While using BSP driver functions for the LTDC, we should always start with the function BSP_LCD_Init first which initializes the LTDC and RAM. Afterward, we should use the function BSP_LCD_LayerDefaultInit to initialize the LTDC layer. This function sets the frame buffer to be used by the layer (within RAM). It also sets the image format to ARGB8888. If two layers will be used, then the function BSP_LCD_SetLayerWindow should be used for selecting the layer.

We can repeat the examples in Sect. 11.5.1 via the LTDC module. To do so, we formed the C code in Listing 11.7 for displaying a given image on LCD. In this code, the LCD is initialized first. Then, the layer is selected and activated. We display the image by the address LCD_FRAME_BUFFER.

Listing 11.7 Displaying an image on the LCD via LTDC, the C code

```
/* USER CODE BEGIN Includes */
#include "STM32F429I-Discovery/stm32f429i_discovery.h"
#include "STM32F429I-Discovery/stm32f429i_discovery_lcd.h"
#include "mandrill.h"
/* USER CODE END Includes */

/* USER CODE BEGIN 2 */
```

```
BSP_LCD_Init();
BSP_LCD_LayerDefaultInit(1, LCD_FRAME_BUFFER);
BSP_LCD_SelectLayer(1);

BSP_LCD_SetLayerWindow(1, 0, 0, 160, 120);
BSP_LCD_SetLayerAddress(1, (uint32_t)RGB8888_IMG_ARRAY);
/* USER CODE END 2 */
```

To execute the C code in Listing 11.7, we should open a new project in STM32CubeIDE. Then, we should apply the settings in Sect. 9.5.2.1. We should also add the BSP driver files. Besides adding them, we should add the "mandrill.h" file to "project->Core->Inc" in our project. As we debug and run the code, the Mandrill image will be displayed on the LCD. The user can benefit from the Python file "ImageHeaderGenerator.py" provided in the accompanying book web site, to transform his or her image file to the corresponding header file. The user should enable the related code line in the Python file for the required header type.

We can repeat the example in Listing 11.2 via the LTDC module. To do so, we formed the C code in Listing 11.8. In this code, the LCD is initialized first. Then, we use the functions BSP_LCD_FillRect and BSP_LCD_SetTextColor. We can debug and run the code as in the previous example. Afterward, we expect to see a color bar on the LCD.

Listing 11.8 Plotting a color bar on the LCD via LTDC, the C code

```
/* USER CODE BEGIN Includes */
#include "STM32F429I-Discovery/stm32f429i_discovery.h"
#include "STM32F429I-Discovery/stm32f429i_discovery_lcd.h"
/* USER CODE END Includes */

/* USER CODE BEGIN PD */
#define COLORBAR_NUM 18
#define COLORBAR_WIDTH 20
#define SCREEN_WIDTH 240
#define SCREEN_HEIGHT 320
/* USER CODE END PD */

/* USER CODE BEGIN PV */
uint32_t colorBar[18] = {
0xFF000078, 0xFF007C00, 0xFF007C78, 0xFF780000, 0xFF780078, 0
    xFF787C00,
0xFFC0C0C0, 0xFF787C78, 0xFF0000F8, 0xFF00FC00, 0xFF00FCF8, 0
    xFFF80000,
0xFFF800F8, 0xFFF8FC00, 0xFFF8FCF8, 0xFFF8A400, 0xFFA8FC28, 0
    xFFF800F8};

uint32_t i = 0, j = 0;
/* USER CODE END PV */

/* USER CODE BEGIN 2 */
BSP_LCD_Init();
BSP_LCD_LayerDefaultInit(1, LCD_FRAME_BUFFER);
BSP_LCD_SelectLayer(1);

for (i = 0; i < 18; i++)
{
if (SCREEN_WIDTH > i * COLORBAR_WIDTH)
{
```

```
BSP_LCD_SetTextColor(colorBar[i]);
BSP_LCD_FillRect(i * COLORBAR_WIDTH, 0, COLORBAR_WIDTH - 1,
    SCREEN_HEIGHT);
}
}
/* USER CODE END 2 */
```

Our third example, given in Listing 11.9, prints characters on the LCD. To do so, we should initialize the LCD first. Then, we should set the font to be used in operation by the function BSP_LCD_SetFont. Afterward, we can print three strings "STM32F429," "LTDC LCD," and "Example" to three lines on the LCD. We can use the functions BSP_LCD_SetTextColor and BSP_LCD_Clear to set the background and text colors, respectively. We should follow the steps as in our second example to execute the code. As the code is run, we should see the strings on the LCD.

Listing 11.9 Printing characters on the LCD via LTDC, the C code

```
/* USER CODE BEGIN Includes */
#include "STM32F429I-Discovery/stm32f429i_discovery.h"
#include "STM32F429I-Discovery/stm32f429i_discovery_lcd.h"
/* USER CODE END Includes */

/* USER CODE BEGIN 2 */
BSP_LCD_Init();
BSP_LCD_LayerDefaultInit(1, LCD_FRAME_BUFFER);

BSP_LCD_SelectLayer(1);
BSP_LCD_SetFont(&Font24);
BSP_LCD_Clear(LCD_COLOR_WHITE);
BSP_LCD_SetTextColor(LCD_COLOR_DARKBLUE);

BSP_LCD_DisplayStringAt(0, 40, (uint8_t *)"STM32F429",
    CENTER_MODE);
BSP_LCD_DisplayStringAt(0, 80, (uint8_t *)"LTDC LCD", CENTER_MODE
    );
BSP_LCD_DisplayStringAt(0, 120, (uint8_t *)"Example", CENTER_MODE
    );
/* USER CODE END 2 */
```

11.5.2.2 Usage via C++ Language

STMicroelectronics officially formed the BSP LCD library for Mbed. This library can be reached from https://os.mbed.com/teams/ST/code/LCD_DISCO_F429ZI/. BSP driver functions in this library are formed as class members. Therefore, all the functions in Table 11.1 can be used by discarding the prefix BSP_LCD_. We forked this library and modified it to be used by the most recent Mbed OS version. The library contains the init function for initialization. The constructor of the class LCD_DISCO_F429ZI performs the initialization operation, selects the layer, clears the screen, and sets the background and font colors. Therefore, the user can proceed with the display operation in C++ language.

We repeat the three examples in the previous section in C++ language. To do so, we formed the C++ codes in Listings 11.10, 11.11, and 11.12. These codes perform the same operations as the ones in Listings 11.7, 11.8, and 11.9, respectively. Therefore, the reader should follow the steps mentioned there. The user can also benefit from the Python file "ImageHeaderGenerator.py" provided in the accompanying book web site, to transform his or her image file to the corresponding header file. The user should enable the related code line in the Python file for the required header type.

Listing 11.10 Displaying an image on the LCD via LTDC, the C++ code

```cpp
#include "mbed.h"
#include "LCD_DISCO_F429ZI.h"
#include "mandrill.h"

LCD_DISCO_F429ZI LCD;

int main()
{
LCD.SetLayerWindow(1, 0, 0, 160, 120);
LCD.SetLayerAddress(1, (uint32_t)RGB8888_IMG_ARRAY);

while (true);
}
```

Listing 11.11 Plotting a color bar on the LCD via LTDC, the C++ code

```cpp
#include "mbed.h"
#include "LCD_DISCO_F429ZI.h"

#define COLORBAR_NUM 18
#define COLORBAR_WIDTH 20
#define SCREEN_WIDTH 240
#define SCREEN_HEIGHT 320

LCD_DISCO_F429ZI LCD;

uint32_t colorBar[18] = {
0xFF000078, 0xFF007C00, 0xFF007C78, 0xFF780000, 0xFF780078, 0
    xFF787C00,
0xFFC0C0C0, 0xFF787C78, 0xFF0000F8, 0xFF00FC00, 0xFF00FCF8, 0
    xFFF80000,
0xFFF800F8, 0xFFF8FC00, 0xFFF8FCF8, 0xFFF8A400, 0xFFA8FC28, 0
    xFFF800F8};

uint32_t i = 0, j = 0;

int main()
{
for (i = 0; i < 18; i++)
{
if (SCREEN_WIDTH > i * COLORBAR_WIDTH)
{
LCD.SetTextColor(colorBar[i]);
LCD.FillRect(i * COLORBAR_WIDTH, 0, COLORBAR_WIDTH, SCREEN_HEIGHT
    );
}
}

while (true);
}
```

Listing 11.12 Printing characters on the LCD via LTDC, the C++ code

```
#include "mbed.h"
#include "LCD_DISCO_F429ZI.h"

LCD_DISCO_F429ZI LCD;

int main()
{
LCD.SetFont(&Font24);
LCD.Clear(LCD_COLOR_WHITE);
LCD.SetTextColor(LCD_COLOR_DARKBLUE);

LCD.DisplayStringAt(0, 40, (uint8_t *)"STM32F429", CENTER_MODE);
LCD.DisplayStringAt(0, 80, (uint8_t *)"LTDC LCD", CENTER_MODE);
LCD.DisplayStringAt(0, 120, (uint8_t *)"Example", CENTER_MODE);

while (true);
}
```

11.6   Setting Up the Touch Screen on the LCD of STM32F4 Board

We should set up the touch screen to use it. We will consider this operation next. As a side note, LCD on the STM32F4 board should have been set beforehand as explained in Sect. 11.4.

11.6.1 Setup via C Language

In order to use BSP drivers for the touch screen, we should first open a new project and enable the LCD via LTDC as explained in Sect. 9.5.2.1. Then, we should add the below C code script to the appropriate place in the "main.c" file.

```
/* USER CODE BEGIN Includes */
#include "STM32F429I-Discovery/stm32f429i_discovery_ts.h"
/* USER CODE END Includes */
```

11.6.2 Setup via C++ Language

We will be using the BSP driver functions available under Mbed provided by STMicroelectronics. In order to use the BSP drivers for the touch screen, we should first open a new project and enable the LCD via LTDC as explained in Sect. 9.5.2.1. There are two options here.

In the first option, we will use the forked library under Mbed with the URL   https://os.mbed.com/teams/Embedded-System-Design-with-ARM-Cortex-M/code/TS_DISCO_F429ZI/. To use it, we should add this library to our project. To do so, right-click on the project and click on the "add library" button. Then, enter the provided URL to the empty space. As we press "Next," it will ask for

"Branch" and "Tag." Keep them in their default state. As these settings are done, we should add the code line #include "TS_DISCO_F429ZI.h" to the main file of our project.

The second option is directly adding the folders under "Libraries\Cpp\ TS_DISCO_F429ZI" to the current project folder. No additional operation is needed for this option. As these settings are done, we should add the code line #include "TS_DISCO_F429ZI.h" to the main file of our project.

11.7 Usage of the Touch Screen on the LCD of STM32F4 Board

As we set up the touch screen, the next step is its usage. We perform this operation in C and C++ languages in the following sections.

11.7.1 Usage via C Language

We will benefit from the available BSP driver functions to use the touch screen in C language. These functions are as follows. The function BSP_TS_Init initializes and configures the touch screen and its hardware resources. The function BSP_TS_GetState returns status and position of the touch on the screen. The struct TS_StateTypeDef stores the touch state structure. For more information on these functions, please see [1]. We can use these functions as follows. We should first initialize the touch screen by the function BSP_TS_Init. As we touch the screen, we can check whether the screen has been touched and the touch location by the function and struct BSP_TS_GetState and TS_StateTypeDef.

We provide an example on the usage of touch screen with the C code in Listing 11.13. Within this code, we draw red circles on the screen. To do so, the code initializes the LCD and then prints the strings to the screen as in Listing 11.9. Then, the touch screen is initialized by the function BSP_TS_Init. The CPU waits for a touch in an infinite loop. If we touch the screen, a red circle will be drawn on that location by the function BSP_LCD_FillCircle.

Listing 11.13 Drawing red circles on the LCD via touching it, the C code

```
/* USER CODE BEGIN Includes */
#include "STM32F429I-Discovery/stm32f429i_discovery.h"
#include "STM32F429I-Discovery/stm32f429i_discovery_ts.h"
#include "STM32F429I-Discovery/stm32f429i_discovery_lcd.h"
/* USER CODE END Includes */

/* USER CODE BEGIN PD */
#define pRadius 5
/* USER CODE END PD */

/* USER CODE BEGIN 1 */
TS_StateTypeDef tsState;
uint16_t tsX, tsY;
/* USER CODE END 1 */
```

```
/* USER CODE BEGIN 2 */
BSP_LCD_Init();
BSP_LCD_LayerDefaultInit(1, LCD_FRAME_BUFFER);

BSP_LCD_SelectLayer(1);
BSP_LCD_SetFont(&Font24);
BSP_LCD_Clear(LCD_COLOR_WHITE);
BSP_LCD_SetTextColor(LCD_COLOR_DARKBLUE);

BSP_LCD_DisplayStringAt(0, 40, (uint8_t *)"STM32F429",
    CENTER_MODE);
BSP_LCD_DisplayStringAt(0, 80, (uint8_t *)"LTDC LCD", CENTER_MODE
    );
BSP_LCD_DisplayStringAt(0, 120, (uint8_t *)"Example", CENTER_MODE
    );

BSP_LCD_SetTextColor(LCD_COLOR_RED);
BSP_TS_Init(240, 320);
/* USER CODE END 2 */

/* Infinite loop */
/* USER CODE BEGIN WHILE */
while (1)
{

BSP_TS_GetState(&tsState);
if (tsState.TouchDetected)
{
tsX = tsState.X;
// tsY = tsState.Y; // uncomment to use with STM32F429I-DISC1
    MB1075C (green board)
tsY = 320 - tsState.Y; // uncomment to use with STM32F429I-DISC1
    MB1075E (blue board)

if (tsX > pRadius && tsY > pRadius && tsX < 240 - pRadius && tsY
    < 320 - pRadius)
{
BSP_LCD_FillCircle(tsX, tsY, pRadius);
}
}
/* USER CODE END WHILE */

/* USER CODE BEGIN 3 */
}
/* USER CODE END 3 */
```

To execute the C code in Listing 11.13, we should open a new project in STM32CubeIDE. Then, we should apply the settings in Sects. 11.4 and 11.6. We should also add the BSP driver files to our project. Please note that there are two different STM32F4 board versions available. Uncomment the corresponding line in the code to get the correct values. As we debug and run the code, we can draw a red circle by touching on the screen.

11.7.2 Usage via C++ Language

We will be using the BSP driver functions available under Mbed for C++ language. STMicroelectronics formed the BSP functions in a library form as class members.

These functions are as follows. The function Init initializes and configures the touch screen functionalities and configures all necessary hardware resources. The function GetState returns status of the touch screen. The struct StateTypeDef stores the touch state structure. For more information on these functions, please see [1]. We can use these functions as follows. We should first initialize the touch screen by the function Init. We can check whether the screen has been touched and its location by the function and struct GetState and TS_StateTypeDef.

We provide the C++ version of the example in Listing 11.13, with the code in Listing 11.14. In order to run this code, we should form a new project. Then, we should add the BSP library to our project as explained in Sect. 11.5.2. Please note that there are two different STM32F4 board versions available. Uncomment the corresponding line in the code to get the correct values. As we run the code, it should work as its C counterpart.

Listing 11.14 Drawing red circles on the LCD via touching it, the C++ code

```cpp
#include "mbed.h"
#include "LCD_DISCO_F429ZI.h"
#include "TS_DISCO_F429ZI.h"

#define pRadius 5

LCD_DISCO_F429ZI LCD;
TS_DISCO_F429ZI TS;

int main()
{
TS_StateTypeDef tsState;
uint16_t tsX, tsY;

LCD.SetFont(&Font24);
LCD.Clear(LCD_COLOR_WHITE);
LCD.SetTextColor(LCD_COLOR_DARKBLUE);

LCD.DisplayStringAt(0, 40, (uint8_t *)"STM32F429", CENTER_MODE);
LCD.DisplayStringAt(0, 80, (uint8_t *)"LTDC LCD", CENTER_MODE);
LCD.DisplayStringAt(0, 120, (uint8_t *)"Example", CENTER_MODE);

LCD.SetTextColor(LCD_COLOR_RED);
while (true)
{
TS.GetState(&tsState);
if (tsState.TouchDetected)
{
tsX = tsState.X;
// tsY = tsState.Y; // uncomment to use with STM32F429I-DISC1
    // MB1075C (green board)
tsY = 320 - tsState.Y; // uncomment to use with STM32F429I-DISC1
    // MB1075E (blue board)
if ((tsX > pRadius) && (tsY > pRadius) && (tsX < (240 - pRadius))
    && (tsY < (320 - pRadius)))
{
LCD.FillCircle(tsY, tsY, pRadius);
}
}
}
}
```

11.8 Graphical User Interface Formation via TouchGFX

Since we have an LCD with integrated touch screen on the STM32F4 board, we can form graphical user interface (GUI) for it. We will benefit from the TouchGFX software provided by STMicroelectronics for this purpose. TouchGFX simplifies GUI formation for the user. We should use it in connection with STM32CubeIDE. To be more precise, TouchGFX uses C language in STM32CubeIDE in connection with the HAL library. It uses C++ language while forming the GUI. This usage will become more clear in the following sections.

11.8.1 Installing TouchGFX

TouchGFX needs a separate installation besides STM32CubeIDE. During this process, two extra programs should be installed as "TouchGFX Generator" and "TouchGFXDesigner." To install TouchGFX Generator, we should open STM32CubeIDE. Within it, we should enter "Help" and "Manage Embedded Software Packages" afterward. In the opening window, we should enter the "STMicroelectronics" tab, expand the "X-CUBE-TOUCHGFX" option, and install the latest version of TouchGFX Generator. We used the version 4.16.0 while writing this book.

We can install the TouchGFXDesigner in two different ways. The first option is via the available "msi" file located at \Users\User_Name\STM32Cube\ Repository\Packs\STMicroelectronics\X-CUBE-TOUCHGFX in our PC. There should be the installation folder for the recent version which is "4.16.0" for us. Within this folder, we should locate the "TouchGFX-4.16.0.msi" file under \4.16.0\Utilities\PC_Software\TouchGFXDesigner. We can install TouchGFXDesigner via double-clicking on this file. The second option is directly installing the TouchGFXDesigner from its web site https://support.touchgfx.com/ docs/introduction/installation. We suggest following the first option in installation for consistency with the TouchGFX Generator.

11.8.2 Setting Up TouchGFX

We should apply necessary settings to use TouchGFX. Therefore, we should first form a new project as explained in Sect. 3.1.3. However, we should select the "Targeted Language" as C++ in the project setup. Afterward, we should set the LTDC. Therefore, we should activate FMC SDRAM, I2C3, SPI5, DMA2D, and LTDC in STM32CubeIDE as explained in Sect. 9.5.2.1. Then, we should adjust parameters under the LTDC configuration menu as in Fig. 11.8.

We should also apply the layer settings in Fig. 11.9. To note here, we should start setting parameters with "Number of Layers: 1 Layer." Then, we should enable the "LTDC global interrupt" under the "NVIC Settings" tab as in Fig. 11.10. Then, we

Fig. 11.8 LTDC configuration menu in STM32CubeMX: parameter settings

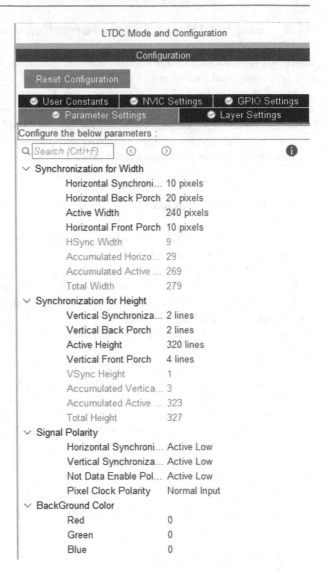

LTDC Mode and Configuration

Configuration

Reset Configuration

✔ User Constants   ✔ NVIC Settings   ✔ GPIO Settings
✔ Parameter Settings   ✔ Layer Settings

Configure the below parameters :

🔍 Search (Crtl+F)   ◁   ▷   ⓘ

⌄ Synchronization for Width
  Horizontal Synchroni... 10 pixels
  Horizontal Back Porch 20 pixels
  Active Width   240 pixels
  Horizontal Front Porch 10 pixels
  HSync Width   9
  Accumulated Horizo...   29
  Accumulated Active ... 269
  Total Width   279
⌄ Synchronization for Height
  Vertical Synchroniza... 2 lines
  Vertical Back Porch   2 lines
  Active Height   320 lines
  Vertical Front Porch   4 lines
  VSync Height   1
  Accumulated Vertica...   3
  Accumulated Active ... 323
  Total Height   327
⌄ Signal Polarity
  Horizontal Synchroni... Active Low
  Vertical Synchroniza... Active Low
  Not Data Enable Pol... Active Low
  Pixel Clock Polarity   Normal Input
⌄ BackGround Color
  Red   0
  Green   0
  Blue   0

should activate the CRC peripheral. To do so, we should only check the "Activated" option under the mode menu which opens up as we select CRC.

Now we are ready to modify the TouchGFX Generator. To do so, we should first open the ".ioc" file and open "Select Components" under "Software Packs" within the "Device Configuration Tool" in STM32CubeMX. Then, we should select TouchGFX Generator. As all these operations are done, a new option comes under the left "Software Packs" menu of device "Configuration Tool" as STMicroelectronics.X-CUBE-TOUCHGFX.4.16.0. To note here, the version of TouchGFX may be different in the future. Hence, the corresponding version will

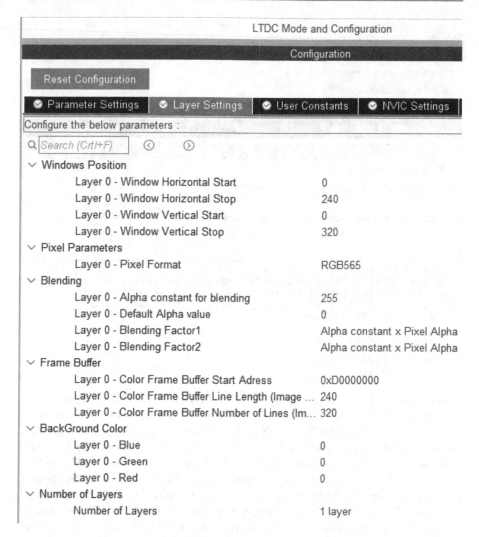

Fig. 11.9 LTDC configuration menu in STM32CubeMX: layer settings

show up there. As we press STMicroelectronics.X-CUBE-TOUCHGFX.4.16.0, we should check the "Graphics Application" checkbox under the mode menu. We should also configure the TouchGFX as in Fig. 11.11. Finally, we should set the system clock (HCLK) to 180 MHz and LCD-TFT clock to 6 MHz.

As we save the ".ioc" file, the "main.c" file will be formed. We should next include the BSP driver files to our project as explained in Sect. 11.5.2. To run the LCD and touch screen, we should add the following lines to the "main.c" file in appropriate places.

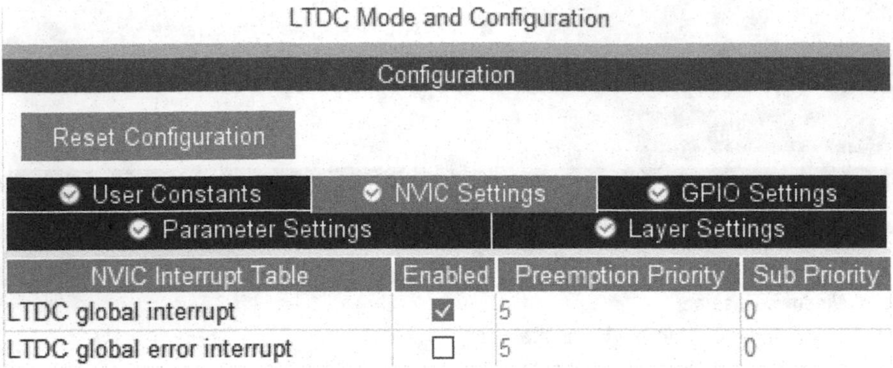

LTDC Mode and Configuration

Configuration			
Reset Configuration			
⊘ User Constants	⊘ NVIC Settings		⊘ GPIO Settings
⊘ Parameter Settings		⊘ Layer Settings	
NVIC Interrupt Table	Enabled	Preemption Priority	Sub Priority
LTDC global interrupt	☑	5	0
LTDC global error interrupt	☐	5	0

Fig. 11.10 LTDC configuration menu: NVIC settings

```
/* USER CODE BEGIN Includes */
#include "STM32F429I-Discovery/stm32f429i_discovery.h"
#include "STM32F429I-Discovery/stm32f429i_discovery_ts.h"
#include "STM32F429I-Discovery/stm32f429i_discovery_lcd.h"
/* USER CODE END Includes */

/* USER CODE BEGIN 2 */
BSP_LCD_Init();
BSP_TS_Init(240, 320);
/* USER CODE END 2 */
```

As we perform project settings in STM32CubeIDE, we should next focus on the design step. To do so, we should open the TouchGFX Designer by double-clicking on the ".part" file under the TouchGFX folder of the project (under project explorer). Then, the reader should click on the "Generate Code" button on the top right part of the screen. Codes of an empty GUI will be formed under the STM32CubeIDE project. To observe them, we should right-click on the project under the project explorer and press "refresh." Afterward, we should open the "application.config" file under TouchGFX folder in "Project Explorer." If the LCD will be used in horizontal mode, we should set "layout_rotation: 90." Otherwise, we should set "layout_rotation: 0."

If there will be interactive areas in the GUI and touch screen needs to be activated, we should add the following code lines to the "/TouchGFX/target/STM32TouchController.cpp" or "/TouchGFX/target/STM32TouchController .hpp" files. These additions can be done in the STM32CubeIDE interface.

```
#include "STM32F429I-Discovery/stm32f429i_discovery.h"
#include "STM32F429I-Discovery/stm32f429i_discovery_ts.h"
#include "STM32F429I-Discovery/stm32f429i_discovery_lcd.h"
```

Besides, the sampleTouch function under the file "/TouchGFX/target/STM32Touch Controller.cpp" should be modified as follows:

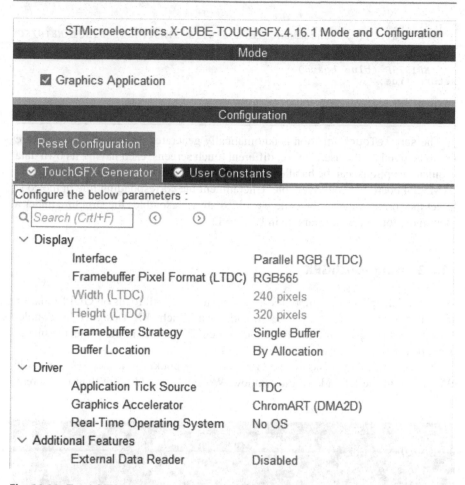

Fig. 11.11 TouchGFX configuration settings in STM32CubeMX

```
bool STM32TouchController::sampleTouch(int32_t& x, int32_t& y)
{
/**
* By default sampleTouch returns false,
* return true if a touch has been detected, otherwise false.
*
* Coordinates are passed to the caller by reference by x and y.
*
* This function is called by the TouchGFX framework.
* By default sampleTouch is called every tick, this can be
     adjusted by HAL::setTouchSampleRate(int8_t);
*
*/
TS_StateTypeDef state;
BSP_TS_GetState(&state);
if (state.TouchDetected)
{
```

```
x = state.X;
// y = state.Y; // uncomment to use with STM32F429I-DISC1 MB1075C
    (green board)
y = 320 - state.Y; // uncomment to use with STM32F429I-DISC1
    MB1075E (blue board)
return true;
}
return false;
}
```

The sampleTouch function is automatically generated. However, its entries are left to be filled by the user. Hence, different touch screens, each having its own data acquisition options, can be handled this way. Please note that there are two different STM32F4 board versions available. Uncomment the corresponding line in the code to get the correct values. The code we added above resembles the structure under a loop in our touch screen example in Listing 11.13.

11.8.3 Using TouchGFX

We formed the project in empty form till this step. We will focus on GUI formation from this point on. To do so, we should open TouchGFX Designer by double-clicking on the ".part" file located under TouchGFX folder in the project explorer. The opened empty screen will be as in Fig. 11.12.

We can add components to the GUI screen by clicking on the "+" sign (Add Widget) from the left side of the window. We can also add more than one screen

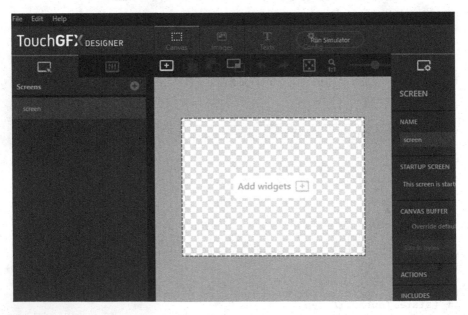

Fig. 11.12 TouchGFX Designer interface

from the screens section if we plan to do so in our application. There are two tabs, as properties and interactions, on the menu located at the right side of the window. The properties tab can be used to adjust properties of the added components on the screen. The interactions tab can be used to set what will be done when the component is touched.

As we design the GUI, we can click on the "Run Simulator" button on the right side of the window. This way we can test the GUI and its properties on PC. We should enter the "Default Image Configuration" menu under the "Config" screen for the simulator to work properly. Here, we should set the "layout_rotation: 90" if we will use the display in horizontal form. If we plan to use the display in vertical form, then we should set it as "layout_rotation: 0."

We next provide a detailed example on the usage of TouchGFX to form a GUI. In our example, we form the background image, add a button, and draw circle to represent an LED. Afterward, we will toggle the LED from the touch screen via GUI usage. We will provide sufficient detail on these steps. If the reader wants to learn all the functions and class structures under TouchGFX, he or she can do so by checking https://support.touchgfx.com/docs/introduction/welcome.

At this step, we assume that a new project has been created as explained in the previous section. We should also add the BSP driver functions during setup. Afterward, we should open the Touch Designer. As the first step, we should select the "screen" from the left side of the Touch Designer menu. Let's set the screen "NAME" as "main" from the right side of the window under "Properties" tab. In order to add a background image, we should open the "+" (Add Widget) menu from left side of the screen and click on the "IMAGE" option. We should click "+" from the pop-up menu and select an image from PC. Afterward, we should set the image "NAME" property as "background_image" in the right side of the screen. The screen should be as in Fig. 11.13.

As we finalize adding the background image to our GUI, we should press the "Generate Code" button on the right side of the window. This way, codes of the GUI and background image will be generated under the STM32CubeIDE project. To observe them, we may need to refresh the Project Explorer. As we build and debug the project in STM32CubeIDE, we should see the image as background on the LCD. We do not need to add any extra code till this time.

The next step in our GUI formation is adding widgets to it. Hence, we switch back to TouchGFX again. We will specifically add buttons and LEDs to our design. To do so, press the "+" add widget menu. Then, we should select "button with label" and add two of them. Afterward, we should select each button from the left menu. We should press the "Properties" tab. In the right-hand side of the opening screen, we should name the buttons as "greenLEDToggle" and "redLEDToggle" in the right "NAME" region, respectively. In the same properties tab, enter strings "Green LED" and "Red LED" in the "TEXT" region, respectively. Afterward, the text on our buttons will change. We should also add two circles from the left "Add Widget" menu by selecting the "+" sign on the left side of the screen. To do so, we should select "Circle." Next, we should select each circle and change their "NAME" property as "greenLEDCircle" and "redLEDCircle," respectively. We should change

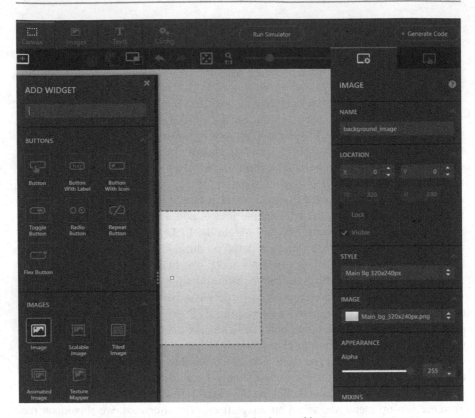

Fig. 11.13 TouchGFX Designer interface: adding background image

the color of these circles from the "IMAGE & COLOR" section as R-0, G-100, and B-0 and R-100, G-0, and B-0, respectively. This way, we colored the circles representing our LEDs as green and red, respectively. The screen should be as in Fig. 11.14. We can press the "Simulator" button on the window to see the final form of our GUI.

Our next step is adding interactions to our GUI in TouchGFX. To do so, we should click on the "Interactions" tab. A new menu appears on the right side of the screen. There is a button named "Add Interaction" there. We should click on it twice to add two separate interactions. We should click on each interaction separately and "Button is clicked" as the trigger for each. Then, we should select "greenLEDToggle" and "redLEDToggle" from the "Choose clicked source" for each, respectively. We should also select "Call new virtual function" as "Action" for both. We should set the function names as "greenLEDToggleFunction" and "redLEDToggleFunction," respectively. As "Interaction," we should call these as "greenLEDToggle" and "redLEDToggle," respectively. The GUI should be as in Fig. 11.15.

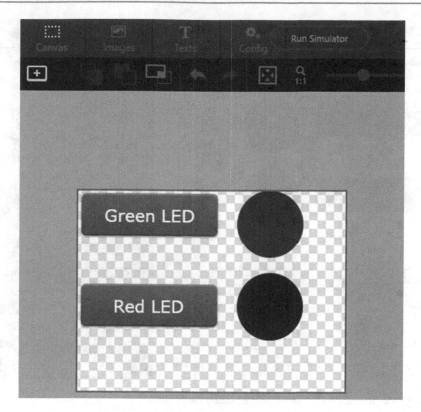

Fig. 11.14 TouchGFX Designer interface: adding buttons and circles

As all the mentioned modifications are done in TouchGFX, we should press the "Generate Code" button on the right side of the screen. Hence, codes of the GUI will be generated under STM32CubeIDE. We may need to refresh the "Project Explorer" to observe the generated codes. From this point on, we will start working in STM32CubeIDE. We should fill in the functions `greenLEDToggleFunction` and `redLEDToggleFunction`. As a reminder, these functions will be called when the associated buttons are pressed. We will perform these operations on the project files located in "/TouchGFX/gui/src" and "/TouchGFX/gui/include/gui."

There will be a folder generated for each screen added to GUI. The folder will be named by the screen name and suffix "_screen." Since our screen is named "main," our folder will be named as "main_screen." Within this folder, we should modify the files "mainView.cpp" and "mainView.hpp." We should add the lines `#include "stm32f4xx_hal.h"` and `#include <touchgfx/Color.hpp>` to the file "mainView.hpp." These lines allow us to reach the HAL library and reach the `Color` class to change colors, respectively. We should also add the functions `virtual void greenLEDToggleFunction()` and `virtual void`

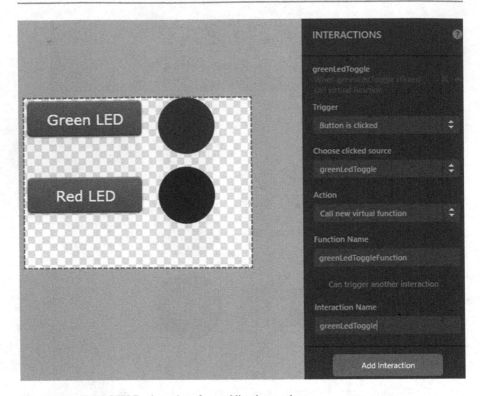

Fig. 11.15 TouchGFX Designer interface: adding interactions

greenLEDToggleFunction() to the class mainView as public method. These are
interaction functions of the buttons added to our GUI.

We should also add private methods to the mainView class as bool
redLEDstatus = false, bool greenLEDstatus = false, uint32_t red
BouncerTime = 0, and uint32_t greenBouncerTime = 0. These will keep
the status of LEDs. They will also keep time to eliminate switch bouncing when the
button is pressed once. Finally, we should add the functions in Listing 11.15
to the "mainView.cpp" file. Here, the objects greenLEDCirclePainter
and redLEDCirclePainter keep the color information of the objects
greenLEDCircle and redLEDCircle, respectively. We can change their color
by the function setColor. Since we set the screen in the RGB565 color format,
we use the same format for the color representation as well. The function
Color::getColorFrom24BitRGB changes the color value from RGB888 to RGB
format of the screen.

Listing 11.15 LED control functions to be added

```
void mainView::redLEDToggleFunction()
{
if ((HAL_GetTick() - redBouncerTime) > 100)
```

```
{
redLEDStatus = !redLEDStatus;
if (redLEDStatus)
{
HAL_GPIO_WritePin(GPIOG, GPIO_PIN_14, GPIO_PIN_SET);
redLEDCirclePainter.setColor(Color::getColorFrom24BitRGB(255, 0,
    0));
}
else
{
HAL_GPIO_WritePin(GPIOG, GPIO_PIN_14, GPIO_PIN_RESET);
redLEDCirclePainter.setColor(Color::getColorFrom24BitRGB(100, 0,
    0));
}
redLEDCircle.invalidate();
}
redBouncerTime = HAL_GetTick();
}


void mainView::greenLEDToggleFunction()
{

if ((HAL_GetTick() - greenBouncerTime) > 100)
{
greenLEDStatus = !greenLEDStatus;
if (greenLEDStatus)
{
HAL_GPIO_WritePin(GPIOG, GPIO_PIN_13, GPIO_PIN_SET);
greenLEDCirclePainter.setColor(Color::getColorFrom24BitRGB(0,
    255, 0));
}
else
{
HAL_GPIO_WritePin(GPIOG, GPIO_PIN_13, GPIO_PIN_RESET);
greenLEDCirclePainter.setColor(Color::getColorFrom24BitRGB(0,
    100, 0));
}
greenLEDCircle.invalidate();
}
greenBouncerTime = HAL_GetTick();
}
```

In Listing 11.15, there is the invalidate function. It should be called when any property of the greenLEDCircle or redLEDCircle object has been changed. Hence, the change can be projected on the screen.

As we debug and run the project in STM32CubeIDE, we should be seeing the background image, circles representing LEDs, and buttons on the LCD. As we press a button on the GUI, the corresponding LED will toggle both on the GUI and STM32F4 board. The reader can benefit from this project to form his or her GUI for another application.

11.9 Application: Improving the Stand-Alone Remote Controller via GUI Formation and Touch Screen Usage

We upgrade the stand-alone remote controller unit for our robot vacuum cleaner in this chapter. Therefore, we represent the physical LEDs and buttons on the second

STM32F4 board via forming a GUI on its LCD and touch screen. We provided the equipment list, circuit diagram, detailed explanation of design specifications, and peripheral unit settings in a separate document in the folder "Chapter11\ EoC_applications" of the accompanying supplementary material. In the same folder, we also provided the complete C, C++, and MicroPython codes for the application.

11.10 Summary of the Chapter

The STM32F4 board has an LCD with integrated touch screen on it. Moreover, STMicroelectronics provides the TouchGFX software to generate professional GUI to run on this hardware. Therefore, we can benefit from them to form an embedded system with professional-looking GUI with its associated hardware components. We considered this topic in this chapter. To do so, we started with the hardware of the LCD, touch screen, and modules available in the STM32F4 microcontroller and STM32F4 board to control them. Afterward, we provided ways of setting up and using the LCD and touch screen modules in C, C++, and MicroPython languages. Then, we used the TouchGFX software to form professional GUI for our embedded systems. We also expanded our previously introduced real-world example by adding LCD and touch screen usage with GUI formation to it as the end of chapter application.

Problems

11.1. Is the LCD on the STM32F4 board of type emissive or non-emissive?

11.2. Briefly explain the LCD structure (with its layers) on the STM32F4 board.

11.3. What are the digital connection types for LCDs?

11.4. How does a four-wire touch screen work?

11.5. Summarize the hardware in the STM32F4 microcontroller and on the STM32F4 board to control the LCD and touch screen on the board.

11.6. Take a photo. Form a project to display it on the LCD via using SPI and LTDC. Form the project using:

(a) C language under STM32CubeIDE
(b) C++ language under Mbed
(c) MicroPython

11.7. Form a project to display two color bars, each with different color, on the LCD via using SPI and LTDC. Form the project using:

(a) C language under STM32CubeIDE
(b) C++ language under Mbed
(c) MicroPython

11.8. Form a project to display your name on the LCD via using LTDC. Form the project using:

(a) C language under STM32CubeIDE
(b) C++ language under Mbed

11.9. Form a project with the following specifications. The C code should count how many times the touch screen has been pressed till that time. Display the result on the LCD. Add power management option to your solution such that the STM32F4 microcontroller waits in an appropriate low power mode when not used. Form the project using:

(a) C language under STM32CubeIDE
(b) C++ language under Mbed

11.10. Form a project with the following specifications. The C code should detect the touched location on the LCD. Display the coordinate of this location on the LCD. Add power management option to your solution such that the STM32F4 microcontroller waits in an appropriate low power mode when not used. Form the project using:

(a) C language under STM32CubeIDE
(b) C++ language under Mbed

11.11. Reconsider Problem 11.6 such that the taken photo is displayed as background image via TouchGFX.

11.12. Form a project via TouchGFX with the following specifications. There will be three screens. Add two buttons to the first screen. Turn on screen two and three when the first and second buttons are pressed, respectively. The second screen will have the same characteristics as the one in Problem 11.9. The third screen will have the same characteristics as the one in Problem 11.10. Screens two and three should also have buttons such that when pressed, the first screen becomes active.

References

1. STMicroelectronics: STM32Cube BSP Drivers Development Guidelines, um2298 edn. (2019)
2. STMicroelectronics: Discovery kit with STM32F429ZI MCU, um1670 edn. (2020)
3. STMicroelectronics: LCD-TFT display controller (LTDC) on STM32 MCUs, an4861 edn. (2020)
4. STMicroelectronics: STM32F405/415, STM32F407/417, STM32F427/437 and STM32F429/439 advanced Arm-based 32-bit MCUs, rm0090 rev 19 edn. (2021)

Introduction to Digital Signal Processing

12

12.1 About Digital Signals

We introduced analog and digital values and signals in Chap. 7. An analog signal can be taken as a function of dependent variable. For example, when we measure room temperature by an analog sensor through time, in fact we form a function. There are infinite number of values in this function. Each value has infinite precision. In Chap. 7, we noted that such values cannot be processed by a digital system. Therefore, we introduced digital values. Let's focus on digital signals formed by such digital values.

12.1.1 Mathematical Definition of the Digital Signal

We know from Chap. 7 that analog values cannot be processed in an embedded system. Hence, they should be sampled and quantized to obtain the corresponding digital values. As a result, we can process them in the embedded system. A digital signal is formed by digital values ordered by an index. If we think about the room temperature as the analog signal, we can apply analog to digital conversion in certain time instants and obtain its quantized samples. As we sort them in time, we can form the corresponding digital signal.

In mathematical terms, the digital signal is represented by an array with index value. Since we sort the samples, we will have the first, second, third, and following elements. In other words, our index should always be an integer value. Otherwise,

---

Supplementary Information The online version contains supplementary material available at (https://doi.org/10.1007/978-3-030-88439-0_12).

© Springer Nature Switzerland AG 2022
C. Ünsalan et al., *Embedded System Design with Arm Cortex-M Microcontrollers*,
https://doi.org/10.1007/978-3-030-88439-0_12

it does not make sense to have the 1.5th sample. As a result, we can represent the digital signal by an array $x[n]$, n being the integer index value.

12.1.2 Representing the Digital Signal in an Embedded System

Array representation of the digital signal can be directly used in the embedded system. To do so, we can benefit from the corresponding array definition in C, C++, and Python languages. Let's consider them next.

We can handle the C and C++ array representations together since they have the same array structure. In C and C++ languages, we can form an array of integer values by the definition `int x[N]`, N being the number of array elements. From the digital signal processing perspective, we will have a digital signal with N integer values. The array index always starts with value zero. We can reach the nth value of our digital signal by `x[n]`. We can also have an array formed by float entries. Hence, we can store signal values with integer and fractional parts. Let's give a simple digital signal example represented in C language. Assume that our digital signal is formed by float entries `float x[4]={0, 0.1, 0.2, 0.3}`. Here, we can reach the first element of the array by `x[1]`.

We can represent a digital signal in Python language as a list. To do so, we should only form the list entries. As an example, we can form the digital signal as `x=[0, 0.1, 0.2, 0.3]`. We can reach the first element of the list by `x[1]`. In Python, list entries start with the zeroth element. We can use the `matplotlib` library to plot the constructed digital signal in Python language.

12.1.3 Forming an Actual Digital Signal from the STM32F4 Board

Digital signals we will consider throughout the book will most of the times originate from a sensor having either analog or digital output. If the sensor output is analog, it can be converted to digital form by the ADC module. If the sensor output is digital, it can be obtained using digital communication modules such as I²C or SPI. We provide one such actual signal in this section. There is a three-axis gyroscope on the STM32F4 board which can be used as a sensor. To be more specific, the gyroscope IC on the STM32F4 boards (having version lower than E) is L3GD20. For the boards with version E or higher, the gyroscope IC is I3G4250D. Both gyroscopes have I²C/SPI digital output interface. We can obtain their values at certain time instants to form a digital signal. To do so, we provide the codes in C, C++, and MicroPython languages next.

12.1.3.1 Data Acquisition via C Language

In order to reach and acquire data from the L3GD20 or I3G4250D three-axis gyroscope in C language, we should first create a project under STM32CubeIDE by initializing all peripherals with their default mode. Hence, all pin assignments for the sensor are realized automatically. We should use board support package (BSP) driver functions to reach the gyroscope. Only driver functions related to the L3GD20 IC are defined in BSP. However, the I3G4250D IC has the same register set as with the L3GD20 IC. Hence, the same driver functions can be used for it with minor modification.

We will benefit from the BSP drivers provided by STMicroelectronics to use the gyroscope as a sensor. To do so, we should first open a new project and press the ".ioc" file to open the "Device Configuration Tool" in STM32CubeMX. Then, we should follow the steps given in Sect. 9.5.2.1. Finally, we should add the below C code script to the appropriate place in the "main.c" file.

```
/* USER CODE BEGIN Includes */
#include "STM32F429I-Discovery/stm32f429i_discovery.h"
#include "STM32F429I -Discovery/stm32f429i_discovery_gyroscope.h"
/* USER CODE END Includes */
```

As the configuration is done, BSP driver functions related to the L3GD20 three-axis gyroscope are ready to be used. We will use the function BSP_GYRO_Init to initialize the sensor. We can read the ID of the gyroscope (212) by the function BSP_GYRO_ReadID. Through it, we can check whether the SPI communication has been established between the STM32F4 microcontroller and gyroscope. This operation also indicates that the gyroscope is ready to be used. Finally, we can use the function BSP_GYRO_GetXYZ(float *xyz) to acquire the gyroscope output as x, y, and z angular rates in mdps (millidegrees per second) and save the result to the xyz buffer.

In order to use the same setup for the I3G4250D three-axis gyroscope, go to the file "l3gd20.h" and change the code lines #define I_AM_L3GD20 ((uint8_t)0xD4) and #define I_AM_L3GD20_TR ((uint8_t)0xD5) with #define I_AM_L3GD20 ((uint8_t)0xD3) and #define I_AM_L3GD20_TR ((uint8_t)0xD4), respectively. Afterward, the same BSP functions can be used for the I3G4250D IC. Only the function BSP_GYRO_ReadID returns 211 instead of 212.

We provide a sample code to acquire a digital signal from the gyroscope in Listing 12.1. Here, the TIM2 timer module is used to obtain sensor output for 1000 samples in 20 s. Hence, the sampling frequency of the acquired digital signal is 50 Hz. There is a function SERIAL_dataSend in the code. This function is used to send the obtained data to PC. Explanation of this function and receiving/sending data at PC side will be handled in Sect. 12.2. We plot the obtained result in Fig. 12.1.

Listing 12.1 Acquiring the gyroscope data, the C code

```
/* USER CODE BEGIN Includes */
#include "STM32F429I-Discovery/stm32f429i_discovery.h"
#include "STM32F429I-Discovery/stm32f429i_discovery_gyroscope.h"
#include "serialData.h"
/* USER CODE END Includes */

/* USER CODE BEGIN PV */
uint8_t readID;
float xyz[3];
float xArr[1000];
```

```
int cnt = 0;
uint8_t sendDataFlag = 0;
/* USER CODE END PV */

/* USER CODE BEGIN 0 */
void HAL_TIM_PeriodElapsedCallback(TIM_HandleTypeDef *htim)
{
if (htim->Instance == TIM2)
{
BSP_GYRO_GetXYZ(xyz);
if (cnt < 1000)
{
xArr[cnt++] = xyz[0] / 1000;
}
else
{
sendDataFlag = 1;
HAL_TIM_Base_Stop_IT(&htim2);
}
}
}
/* USER CODE END 0 */

/* USER CODE BEGIN 2 */
BSP_GYRO_Init();
readID = BSP_GYRO_ReadID();
HAL_TIM_Base_Start_IT(&htim2);
/* USER CODE END 2 */

/* USER CODE BEGIN WHILE */
while (1)
{
if (sendDataFlag == 1)
{
SERIAL_dataSend((uint8_t *)xArr, 4000, 3);
sendDataFlag = 0;
}
/* USER CODE END WHILE */
```

12.1.3.2 Data Acquisition via C++ Language

We can also acquire data from the gyroscope to form a digital signal in C++ language. To do so, we should add the BSP driver functions to our project with the first method described in Sect. 9.5.2.2. Then, we should add the gyro wrapper library located at http://os.mbed.com/teams/Embedded-System-Design-with-ARM-Cortex-M/code/GYRO_DISCO_F429ZI/ to our project. As the "Next" button is pressed, the interface asks for "Branch" or "Tag." We can keep the default value. Then, we can use the BSP driver functions by adding the code line #include "GYRO_DISCO_F429ZI.h" to our main file.

BSP driver functions in the gyro wrapper library are formed as class members. Therefore, all the functions in previous section can be used by discarding the prefix BSP_GYRO_. The reader should adjust the necessary code lines for the specific gyroscope IC as explained in the previous section. We provide the C++ version of the code in Listing 12.1 now in Listing 12.2. There is a function send in the code. This function is used to send the obtained data to PC. Explanation of this function

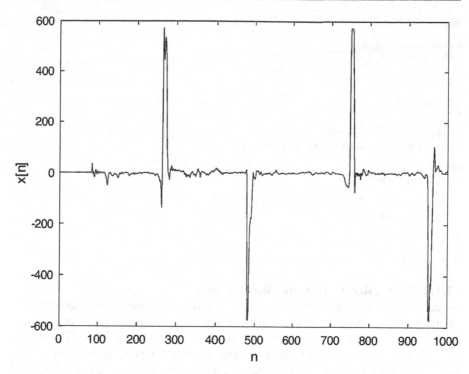

Fig. 12.1 Digital gyroscope signal for 1000 samples in 20 s

and receiving/sending data at PC side will be handled in Sect. 12.2. The reader can run the code to obtain the gyroscope output data.

Listing 12.2 Acquiring the gyroscope data, the C++ code

```cpp
#include "mbed.h"
#include "GYRO_DISCO_F429ZI.h"
#include "serialData.hpp"

uint8_t readID;
Ticker gyroTicker;
float xyz[3];
float xArr[1000];
int cnt = 0;
uint8_t sendDataFlag = 0;

GYRO_DISCO_F429ZI gyro;
SERIALDATA serialData(USBTX, USBRX, 2000000);

void gyro_sample()
{
gyro.GetXYZ(xyz);
if (cnt < 1000)
{
xArr[cnt++] = xyz[0] / 1000;
}
```

```
else
{
gyroTicker.detach();
sendDataFlag = 1;
}
}

int main()
{
gyro.Init();
readID = gyro.ReadID();
gyroTicker.attach(&gyro_sample, 20ms);

while (true){
if (sendDataFlag == 1)
{
serialData.send((uint8_t *)xArr, 4000, 3);
sendDataFlag = 0;
}
}
}
```

12.1.3.3 Data Acquisition via MicroPython

We can also acquire data from the gyroscope to form a digital signal in MicroPython. MicroPython has no class to perform Gyro operations as in the previous section. Therefore, we created a gyro class in our custom MicroPython firmware. The reader can find this custom MicroPython firmware in the accompanying book web site. The gyro module can be created and initialized using the function init(). All gyro module configurations are done automatically in this function. After calling this function, the gyroscope module will be ready to be used. The function read_id() can be used to obtain the ID of the gyroscope. Functions get_x(), get_y(), and get_z() can be used to acquire the gyroscope output as x, y, and z angular rates, respectively. We provide the MicroPython version of the code in Listing 12.1 now in Listing 12.3. There is a function send in the code. This function is used to send the obtained data to PC. Explanation of this function and receiving/sending data at PC side will be handled in Sect. 12.2. The reader can run the code to obtain the gyroscope output data.

Listing 12.3 Acquiring the gyroscope data, the MicroPython code

```
import pyb
import gyro
import serialData

sampleFlag = 0
sendDataFlag = 0

def sampleFn(timer):
    global sampleFlag
    sampleFlag = 1
def main():
    global sampleFlag, sendDataFlag

    xArr = []
    xArr = [0.0 for i in range(1000)]
```

```
cnt = 0
gyro.init()
serialData.init()
readID = gyro.read_id()
print(readID)
timer1 = pyb.Timer(1, prescaler=59, period=55999, mode=pyb.
    Timer.UP)
timer1.callback(sampleFn)
while True:
    if sampleFlag == 1:
        sampleFlag = 0
        if cnt < 1000:
            xArr[cnt] = gyro.get_x()/1000
            cnt = cnt + 1
        else:
            timer1.deinit()
            sendDataFlag = 1
    if sendDataFlag == 1:
        serialData.send(xArr, 4000, 3)
        sendDataFlag = 0

main()
```

12.2   Transferring the Digital Signal Between the PC and STM32F4 Microcontroller

Although it is natural to process the acquired digital signal on an embedded system (such as the STM32F4 microcontroller), it is not easy to observe the signal there. Therefore, it may be beneficiary to transfer the acquired digital signal from the STM32F4 microcontroller to PC. Likewise, we may need to process a predefined signal on the microcontroller. For such cases, we can prepare the digital signal in PC and then transfer it to the STM32F4 microcontroller. We will handle both cases in this section.

12.2.1 Setup in the STM32F4 Microcontroller Side

We can use UART communication to transfer data between the STM32F4 micro-controller and PC. Therefore, we can benefit from the techniques in Chap. 8. We next provide template C, C++, and MicroPython codes for this purpose. These codes can be used for transferring the digital signal between the PC and STM32F4 microcontroller.

12.2.1.1 Setup via C Language
We formed the "data transfer library" to set up and transfer data between the PC and STM32F4 microcontroller in C language. We provide this library in the accompanying book web site. In order to use it, the reader should add the files "serialData.c" and "serialData.h" to the "project->Core->Src" and "project->Core->Inc" folders of the active STM32CubeIDE project, respectively.

Our "data transfer library" has two functions as SERIAL_dataReceive and SERIAL_dataSend. The function SERIAL_dataReceive receives data coming from PC and stores it in RAM of the STM32F4 microcontroller. To do so, this function first sends a request to PC to receive data. Then, the STM32F4 microcontroller starts receiving data from PC. The function SERIAL_dataSend sends data from the STM32F4 microcontroller to PC. To do so, it first sends a request to PC to send data. Then, the STM32F4 microcontroller starts sending data to PC. We provide detailed information on these functions next.

```
void SERIAL_dataReceive(uint8_t *pData, uint16_t Size, uint8_t
     Type)
/*
pData: pointer to the data array to be received
Size: number of bytes to be received
Type: type of the data array. Enter 0 for uint8_t, 1 for uint16_t
    , 2 for uint32_t, and 3 for float data.
*/

void SERIAL_dataSend(uint8_t *pData, uint16_t Size, uint8_t Type
     )
/*
pData: pointer to the data array to be sent
Size: number of bytes to be sent
*/
```

We will use UART communication as explained in Chap. 8 to use the "data transfer library" functions for data transfer. Here, the USART1 module will be used via pins PA9 and PA10. Therefore, they should be configured in UART mode with full-duplex, eight-bit, no parity, one stop bit mode. The baud rate should be set to 2 Mbits/s. Pins PA9 and PA10 should also have "maximum output speed" settings as "very high" under GPIO properties.

12.2.1.2 Setup via C++ Language

We also formed the "data transfer library" to set up and transfer data between the PC and STM32F4 microcontroller in C++ language. This library is available in the accompanying book web site. In order to use this library, we should add the files "serialData.cpp" and "serialData.hpp" to the active project folder. The "data transfer library" has the class SERIALDATA which has three functions as SERIALDATA, receive, and send.

The class SERIALDATA can also be used as the constructor function. UART transmit-receive pin names and desired baud rate will be input for this function. The function receive receives data coming from PC and stores it to the RAM of the STM32F4 microcontroller. To do so, this function first sends a request to PC to receive data. Then, the STM32F4 microcontroller starts receiving data from PC. The function send sends the data from the STM32F4 microcontroller to PC. To do so, this function first sends a request to PC to send data. Then, the STM32F4 microcontroller starts sending data to PC. We provide detailed information on the SERIALDATA class and its functions next.

```
SERIALDATA(PinName p_tx, PinName p_rx, int baudrate);
/*
p_tx: microcontroller pin name for UART transmit pin
p_rx: microcontroller pin name for UART receive pin
baudrate: baudrate, for UART communication.
*/

void receive(uint8_t *pData, uint16_t Size, uint8_t Type);
/*
pData: pointer to the data array to be received
Size: number of bytes to be received
Type: type of the data array. Enter 0 for uint8_t, 1 for uint16_t
    , 2 for uint32_t, and 3 for float data.
*/

void send(uint8_t *pData, uint16_t Size, uint8_t Type);
/*
pData: pointer to the data array to be sent
Size: number of bytes to be sent
*/
```

12.2.1.3 Setup via MicroPython

We also formed the "data transfer library" to set up and transfer data between the PC and STM32F4 microcontroller in MicroPython. This library has been added to the custom MicroPython firmware available in the accompanying book web site. In order to use it, we should add the import serialData line to the "main.py" file and initialize the serialData module using init(). The UART5 module is enabled with full-duplex, eight-bit, no parity, one stop bit mode with this function. The baud rate is set to 921,600 bits/s. Pins PD2 and PC12 are set as Rx and Tx pins, respectively.

The function receive receives data coming from PC and stores it to the RAM of the STM32F4 microcontroller. To do so, this function first sends a request to PC to receive data. Then, the STM32F4 microcontroller starts receiving data from PC. The function send sends data from the STM32F4 microcontroller to PC. To do so, this function first sends a request to PC to send data. Then, the STM32F4 microcontroller starts sending data to PC. We provide detailed information on the serialData functions next.

```
receive(pData, Size, Type);
/*
pData: data array to be received
Size: number of bytes to be received
Type: type of the data array. Enter 0 for uint8_t, 1 for uint16_t
    , 2 for uint32_t, and 3 for float data.
*/

send(pData, Size, Type);
/*
pData: data array to be sent
Size: number of bytes to be sent
*/
```

12.2.2 Setup in the PC Side

Digital signals can be represented and processed easily in Python on PC. Besides, there are several libraries in Python to process the digital signal. Therefore, we pick Python as the digital signal processing medium on PC. Since the STM32F4 microcontroller sends or receives data via UART communication, we should set the UART communication option in Python as well. Therefore, the `pyserial` library should be installed to PC by executing the command `pip install pyserial` at the command window. At this step, we need the COM port in which the STM32F4 board is connected to. Hence, we should check "Device Manager -> Ports (COM & LPT)" under the Windows operating system. There should be a device listed as "STMicroelectronics STLink Virtual COM Port ...." Please note that the same settings should be done for a PC with another operating system such as Linux. Therefore, we kindly ask the reader to check his or her operating system working principles for this purpose.

We formed the "PC Data Transfer Library" in Python to simplify data transfer between the PC and STM32F4 microcontroller. We provide this library as a file in the accompanying book web site. This library contains four functions as follows. The function `init` gets the COM port name as "COMx" and creates the serial communication object. The function `poll_transfer_request` halts execution until the transfer request arrives. This transfer request contains information about transfer direction, transfer size as bytes, and data type. The function `send_data` sends data from PC to the STM32F4 microcontroller according to its size and data type information. The function `read_data` transfers data from the STM32F4 microcontroller to PC according to its size and data type information.

The digital signal obtained from the gyroscope on the STM32F4 board in Sect. 12.1.3 can be transferred to PC and displayed there via our "PC Data Transfer Library." We provide the complete code for this purpose in Listing 12.4. This code first imports `PC_Data_Transfer_Library` as `serialData`. As we run the code on PC, it waits for a request from the STM32F4 microcontroller. As the request comes to receive data, `poll_transfer_request` returns `DATA_WRITE`, and the received data is saved to `receiveArray`.

Listing 12.4 Data transfer code between the PC and STM32F4 microcontroller

```
import PC_Data_Transfer_Library as serialData

dataArray = [1.5 for _i in range(5000)]
serialPort = 'COM8'
serialData.init(serialPort)

while 1:
    reqType = serialData.poll_transfer_request()
    if (reqType == serialData.DATA_WRITE):
        receiveArray = serialData.read_data()
    elif (reqType == serialData.DATA_READ):
        serialData.send_data(dataArray)
```

The code in Listing 12.4 can also be used to send data to the STM32F4 microcontroller. Within the code, we send data in dataArray with size 20,000 (for 5000 float numbers) after a request comes from the STM32F4 microcontroller. As a side note, the code to be executed in the STM32F4 microcontroller can be written in C, C++, or MicroPython language. Since we are using UART communication in all these, the PC side is not affected by the code running in the microcontroller.

12.3 About Digital Systems

Majority of the work in digital signal processing is based on modifying a given input signal. This can be done by digital systems. Hence, we will focus on them in this section.

12.3.1 Mathematical Representation of the Digital System

In mathematical terms, we can define a digital system as an operation applied on the input signal. Output of this operation will be another signal. As an example, our digital system may multiply the input signal and feed it to output. We can represent this system as $y[n] = 3x[n]$ where $x[n]$ and $y[n]$ represent the input and output digital signals, respectively. This example indicates that if we know the input-output relation of the system, we can implement it in code form.

We may want to design a digital system for a specific application. The most popular usage of this option is filtering. Here, the designed filter eliminates or enhances some components in the input signal. We will provide one such example in Sect. 12.5. To note here, we can call the system as filter in digital signal processing. Therefore, we will use system and filter interchangeably.

12.3.2 Linear and Time-Invariant Systems

One digital system type deserves special consideration. This system does not change its characteristics in time. Hence, we call it as time-invariant. The system also has the linearity property. Hence, it will have the same output for the input signal or sum of the signal parts. Besides, multiplying the input signal will result in the multiplied output signal. We call such systems as linear and time-invariant (LTI). For more information on LTI systems, please see [1].

The importance of LTI systems is that their characteristics can be represented by their impulse response. In mathematical terms, this response can be obtained by feeding an impulse signal to the system and obtaining the corresponding output signal. The impulse response of an LTI system is represented as $h[n]$ most of the times. In practice, the impulse response of the system can be obtained by other means such as system identification. For more information on this topic, please see [2].

If the impulse response of the LTI system is available, then we can form its input-output relationship by the convolution sum. Assume that we have a system which does not need future values to calculate the present output value. Then, the convolution sum can be represented as

$$y[n] = \sum_{k=0}^{-\infty} h[k]x[n-k] \tag{12.1}$$

We can represent the convolution sum as an operator between the input signal and the impulse response of the system as $y[n] = x[n] * h[n]$.

As can be seen in Eq. 12.1, the convolution sum only needs multiplication and addition operations. However, there are infinite number of operations to calculate each output value. Hence, it is not possible to directly implement the convolution sum in its present form on an embedded system.

There are two LTI system types based on their impulse response. If the impulse response of the system is of infinite length, then we call it infinite impulse response (IIR) filter. If the response is of finite length, then we call it finite impulse response (FIR) filter.

Based on its definition, input-output relation of the FIR filter becomes

$$y[n] = \sum_{k=0}^{K-1} h[k]x[n-k] \tag{12.2}$$

As can be seen in Eq. 12.2, there are finite number of multiplication and addition operations. Hence, the convolution sum can be implemented on an embedded system for FIR filters.

The convolution sum will consist of infinite number of operations for IIR filters. Therefore, it is not possible to directly implement the filter on an embedded system. Instead, difference equation of the filter can be obtained. The generic difference equation representation between the input and output of an IIR filter is

$$y[n]+a[1]y[n-1]+\cdots+a[L-1]y[n-L+1] = b[0]x[n]+\cdots+b[K-1]x[n-K+1] \tag{12.3}$$

where the difference equation coefficients $a[l]$ and $b[k]$ can be derived from impulse response of the IIR filter. For more information on this topic, please see [1].

12.3.3 Representing the Digital System in an Embedded System

If we know the input-output relation of a digital system, then we can implement it in code form on an embedded system. Moreover, we can implement LTI systems based on their structured input-output relation. We provide the pseudocode to form the input-output relation of FIR filters next.

```
y is the output array with size length N
x is the input array
h is the FIR coefficient array with length K

for(n = 0 to N){
sum = 0
for(k = 0 to K){
if(k <(n+1)){
sum += h[k] * x[n-k]
y[n] = sum
}}}
```

Implementing an IIR filter requires its difference equation representation. Assume that we have this representation. Then, we can use the below pseudocode to implement the IIR filter:

```
y is the output array with size length N
x is the input array
b is the feedforward IIR coefficient array with length K
a is the feedback IIR coefficient array with length L

for(n = 0 to N){
sum = 0
for(k = 0 to K){
if(k <(n+1)){
sum += b[k] * x[n-k]
}}
for(l = 1 to L){
if(l <(n+1)){
sum -= a[l] * y[n-l]
}}
y[n] = sum
}
```

12.4   Digital Signals and LTI Systems in Complex Domain

There are two widely used complex domain representations for digital signals and LTI systems. These are the z-transform and discrete-time Fourier transform (DTFT). Through these, we can analyze characteristics of the signal or system at hand in complex domain.

12.4.1 The z-Transform

We can obtain the z-transform of a signal $x[n]$ by the infinite sum

$$X(z) = \sum_{n=-\infty}^{\infty} x[n]z^{-n} \tag{12.4}$$

where $z = re^{-j\omega}$. Here, r is the radius and ω is the angle term as polar coordinate representation of the complex number.

z-Transform is especially useful for representing digital LTI systems. Here, impulse response of the system is obtained in complex domain as

$$H(z) = \sum_{n=-\infty}^{\infty} h[n]z^{-n} \qquad (12.5)$$

Through $H(z)$, we can obtain valuable information about the system, such as its stability [2]. $H(z)$ can also be used in implementing IIR filters.

Convolution sum in time domain can be represented as multiplication in complex domain. In other words, we have $y[n] = x[n] * h[n] \Leftrightarrow Y(z) = X(z)H(z)$. This allows simplifying the convolution sum operation as multiplication in complex z-domain.

There is also the inverse z-transform to obtain time-domain representation of the signal from its complex domain representation. This way, calculations can be performed in simplified form in z-domain. Then, the result can be transformed back to time domain again. For more information on the inverse z-transform, please see [2].

12.4.2 Discrete-Time Fourier Transform

Frequency domain representation is more important in digital signal processing. Hence, we can fix the radius term as $r = 1$ in Eq. 12.4 and obtain the DTFT. As a result, DTFT of a digital signal $x[n]$ becomes

$$X(e^{j\omega}) = \sum_{n=-\infty}^{\infty} x[n]e^{-j\omega n} \qquad (12.6)$$

As can be seen in Eq. 12.6, the signal $x[n]$ is represented in complex domain with respect to its frequency component ω.

We can represent the convolution sum operation by DTFT as

$$y[n] = x[n] * h[n] \Leftrightarrow Y(e^{j\omega}) = X(e^{j\omega})H(e^{j\omega}) \qquad (12.7)$$

Equation 12.7 is the defining formula for filtering in frequency domain. Hence, we can design a filter with DTFT representation $H(e^{j\omega})$ and filter out the required frequency components in the input signal.

Although frequency domain representation of the signal by DTFT is useful, it cannot be used directly in a digital system since $X(e^{j\omega})$ is a complex and continuous function of the frequency term ω. Therefore, discrete Fourier transform (DFT) has been introduced. We will not implement DFT in this book since it requires heavy mathematical calculations. Instead, we will focus on its modified version called fast Fourier transform (FFT) which is more suitable to be implemented on embedded systems.

We can benefit from the CMSIS DSP library on STM32F4 microcontroller to calculate FFT of a given signal in an STM32CubeIDE project. To do so,

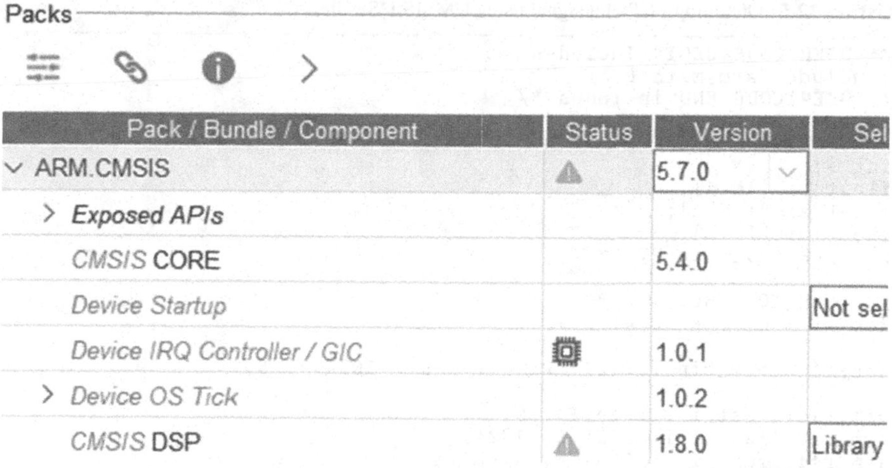

Packs

Pack / Bundle / Component	Status	Version	Sel
∨ ARM.CMSIS	⚠	5.7.0 ∨	
> Exposed APIs			
CMSIS CORE		5.4.0	
Device Startup			Not sel
Device IRQ Controller / GIC	▣	1.0.1	
> Device OS Tick		1.0.2	
CMSIS DSP	⚠	1.8.0	Library

Fig. 12.2 CMSIS configuration in the "Software Packs Component Selector" window

open STM32CubeMX window and press "Alt+O." In the opening "Software Packs Component Selector" window, select "CMSIS" and select "CMSIS DSP" as "Library" as in Fig. 12.2. Then, click "Software Packs" from the left side of "Pinout & Configuration" window. Click "ARM.CMSIS.version" and enable "CMSIS DSP" checkbox from the opening "ARM.CMSIS.version Mode and Configuration" window.

We should add the precompiled CMSIS library files to our project. To do so, create a "Lib" folder under your project folder. Then, go to "STM32Cube/Repository/Packs/ARM/CMSIS/version/CMSIS/DSP/Lib/GCC," and copy the "libarm_cortexM4lf_math.a" lib file to the created "Lib" folder. Finally, right-click on your project, and select "Properties." In the opening window, go to "C/C++ General" and select "Paths and Symbols." In the "Libraries" subwindow, click "Add" and type "arm_cortexM4lf_math." In the "Library Paths" subwindow, click "Add" and type "/project/Lib." Now, CMSIS DSP library functions are ready to be used in your project.

Let's consider the C code given in Listing 12.5. In this code, an 8 Hz sinusoidal signal is sampled with 1024 Hz, and obtained digital signal is stored in the array arr. Here, we use the CMSIS function arm_sin_f32 to create the sinusoidal signal. It can be seen that this signal only contains real data. Hence, we will use the real FFT function of the CMSIS library named arm_rfft_fast_f32. This function uses a special struct arm_rfft_fast_instance_f32 which contains information about the FFT process. We can initialize this structure using the function arm_rfft_fast_init_f32 with the desired FFT size which should be in power of two. Other inputs for the function arm_rfft_fast_f32 are the array containing the input signal, the array to save the FFT result, and the inverse FFT flag. If the inverse FFT flag is set to 0, normal FFT operation is performed. If it is set to 1, inverse FFT operation is performed. Finally, the function arm_cmplx_mag_f32 is used to obtain FFT magnitude is given in Fig. 12.3.

Listing 12.5 Obtaining FFT of a signal using CMSIS DSP library

```
/* USER CODE BEGIN Includes */
#include "arm_math.h"
/* USER CODE END Includes */

/* USER CODE BEGIN PV */
int i;
float arr[1024];
float arr2[1024];
float arr3[512];
/* USER CODE END PV */

/* USER CODE BEGIN 2 */
for (i = 0; i < 1024; i++)
{
arr[i] = arm_sin_f32(2 * PI * i * 8 / 1024) + 5;
}
arm_rfft_fast_instance_f32 S;
arm_rfft_fast_init_f32(&S, 1024);
arm_rfft_fast_f32(&S, arr, arr2, 0);
arm_cmplx_mag_f32(arr2, arr3, 512);
/* USER CODE END 2 */
}
```

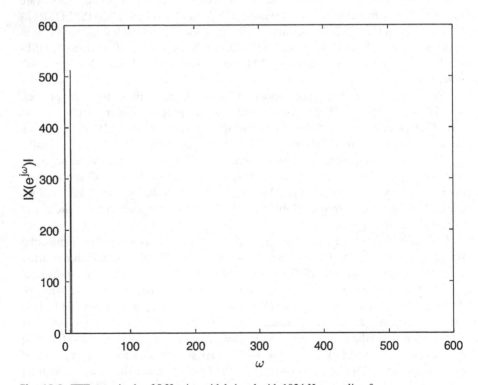

Fig. 12.3 FFT magnitude of 8 Hz sinusoidal signal with 1024 Hz sampling frequency

12.5   Processing Analog Audio Signals on the STM32F4 Microcontroller

We consider an end-to-end digital signal processing system in this section. As a representative example, we pick the equalizer composed of several filters. In implementation, we acquire an analog signal and process it on the STM32F4 microcontroller. Then, we feed the processed signal to output in analog form.

12.5.1 Acquiring the Audio Signal

Our processing of the analog signal on the STM32F4 microcontroller (as the embedded system) starts with its acquisition. Therefore, we should first focus on the microphone circuitry which converts the audio signal to electrical voltage form. We will use the Adafruit SPW2430 microphone breakout board for this purpose. SPW2430 is a small micro-electromechanical system (MEMS) microphone which can operate within 100–10,000 Hz frequency range.

Before explaining how an audio signal is obtained using the SPW2430 microphone and STM32F4 microcontroller, we should provide a brief information about working principles of a generic MEMS microphone. Block diagram of a typical MEMS microphone is as in Fig. 12.4. As can be seen in this figure, the MEMS microphone consists of an application-specific integrated circuit (ASIC), MEMS sensor, and front chamber constructed by a mechanical cover. There is a sound inlet at this mechanical cover which allows sound waves to enter the front chamber.

The MEMS sensor in the microphone is basically a silicon capacitor with two plates. The plate in contact with front chamber is fixed and has acoustic holes in which sound waves move through. The plate in contact with the back chamber is flexible and moves according to the air pressure difference between the front and back chambers. This movement changes the capacitance between plates. The ASIC converts this change to the corresponding electrical voltage. This voltage should be amplified in order to form a useful signal. This amplification process is realized by a separate semiconductor die which works as an audio amplifier. In analog MEMS

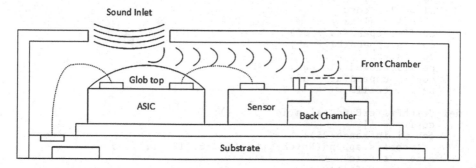

Fig. 12.4 Block diagram of a typical MEMS microphone

microphones, like the SPW2430, this amplified signal is directly fed to output. As we convert the audio signal to electrical form, we can feed it to the ADC module of the STM32F4 microcontroller to obtain its digitized version.

12.5.2 Forming an Equalizer by Digital Filters

We can think of an equalizer as the combination of band-pass filters each passing signals within certain frequency range. We can apply amplification to each filter output and sum the amplified signals. As a result, we will be applying different weights to different frequency components of the signal.

We will first design an IIR notch filter-based equalizer. Numerator and denominator coefficients of the notch filter, as given in Sect. 12.3.3, can be calculated as $b[0] = 1 + \alpha * A$, $b[1] = -2 * \cos(w)$, $b[2] = 1 - \alpha * A$, $a[0] = 1 + \alpha/A$, $a[1] = -2*\cos(w)$, and $a[2] = 1 - \alpha/A$. Here, the constant A, α, and w parameters can be calculated as $A = 10^{g/40}$, $w = 2 * pi * fc/fs$, and $\alpha = \sin(w)/(2 * Q)$ where g is the gain of the filter in dB, fc is the center resonant frequency in Hz, fs is the sampling frequency in Hz, and Q is the quality factor of the filter which is center resonant frequency.

We provide a sample Python code on designing a five-band equalizer using notch filters in Listing 12.6. Here fc for equalizer bands are selected as 100, 300, 1000, 3000, and 10,000 Hz. fs is selected as 48,000 Hz. Q of the filters is selected as 1.

Listing 12.6 Python code for designing a five-band equalizer using notch filters

```
from math import pow, pi, cos, sin

def NotchFilter(fc, Q, gain, fs):
    coeffs = []
    A = pow(10, gain/40)
    w = 2*pi*fc/fs
    sinw = sin(w)
    cosw = cos(w)
    alpha = sinw/(2*Q)
    b0 = 1 + alpha * A
    b1 = -2 * cosw
    b2 = 1 - alpha * A
    a0 = 1 + alpha / A
    a1 = -2 * cosw
    a2 = 1 - alpha / A
    coeffs.append(b0 / a0)
    coeffs.append(b1 / a0)
    coeffs.append(b2 / a0)
    coeffs.append(a1 / a0)
    coeffs.append(a2 / a0)
    return coeffs

def NotchFilterBandDesign(fc, Q, fs):
    coeffs = []
    for _i in range(25):
        coeffs.append(NotchFilter(fc, Q, 12-_i, fs))
    name = str(fc)
    name += "HzFilter.h"
    file = open(name, 'w')
```

```
        file.write("float ")
        file.write("Filter")
        file.write(str(fc))
        file.write("[125] = {")
        file.write("\n")
        for x in range(len(coeffs)):
            for y in range(5):
                file.write(str(coeffs[x][y]))
                file.write(", ")
            file.write("\n")
        file.write("};")
        file.close()
        return coeffs

def main():
    NotchFilterBandDesign(100, 1, 48000)
    NotchFilterBandDesign(300, 1, 48000)
    NotchFilterBandDesign(1000, 1, 48000)
    NotchFilterBandDesign(3000, 1, 48000)
    NotchFilterBandDesign(10000, 1, 48000)

main()
```

When we execute the Python code in Listing 12.6 on PC, it generates filters for each band for different gain values. These gains are 12, 11, ..., −11, and −12 dB. For each filter, there are five filter coefficients as b[0]/a[0], b[1]/a[0], b[2]/a[0], a[1]/a[0], and a[2]/a[0]. Obtained filter coefficients are saved in a header file in a ready-to-use C format.

We can also benefit from the Python SciPy library on PC to design band-pass filters for our equalizer. To do so, we can import the signal library under SciPy. Then, we can use the function iirfilter(N, Wn, rp=None, rs=None, btype='band', analog=False, ftype='butter', output='ba', fs=None) to design the desired band-pass filter. Here, N is the filter order. Wn gives frequency information about critical frequencies. This is a single cutoff frequency for low- and high-pass filters. For band-pass filter, this is a length two array with band cutoff frequencies. rp is the maximum ripple as dB in the pass band for Chebyshev and elliptic filters. rs is the minimum attenuation as dB in the stop band for Chebyshev and elliptic filters. btype is the filter type as 'bandpass', 'lowpass', 'highpass', or 'bandstop'. analog is used to design analog or digital filter by selecting 0 or 1, respectively. ftype is the IIR filter design type as 'butter' for Butterworth, 'cheby1' for Chebyshev type 1, 'cheby2' for Chebyshev type 2, 'ellip' for Cauer/elliptic, or 'bessel' for Bessel/Thomson. output is the filter output formed as 'sos' for second-order sections, 'ba' numerator/denominator, or 'zpk' for pole/zero type. fs is the sampling frequency of the digital filter.

We provide a sample Python code for designing a five-band equalizer using the signal library in Listing 12.7. Here, frequencies for equalizer bands are selected between 20 and 80 Hz for the first band, 80 and 300 Hz for the second band, 300 and 1200 Hz for the third band, 1200 and 5000 Hz for the fourth band, and 5000 and 20,000 Hz for the fifth band. Sampling frequency is selected as 48,000 Hz. Gain of these filters will be adjusted in the microcontroller by multiplying output of the bands by changing variables. As a result, this approach generates additional multiplication for each band. As in the notch filter case, obtained filter coefficients are saved in a header file in a ready-to-use C format.

Listing 12.7 Python code for designing a five-band equalizer using the `signal` library

```
from scipy import signal
import numpy as np
import matplotlib.pyplot as plt

sos1 = signal.iirfilter(2, [20, 80], rs=20, btype='band',
                        analog=False, ftype='cheby2', fs=48000,
                        output='sos')
sos2 = signal.iirfilter(2, [80, 300], rs=20, btype='band',
                        analog=False, ftype='cheby2', fs=48000,
                        output='sos')
sos3 = signal.iirfilter(2, [300, 1200], rs=20, btype='band',
                        analog=False, ftype='cheby2', fs=48000,
                        output='sos')
sos4 = signal.iirfilter(2, [1200, 5000], rs=20, btype='band',
                        analog=False, ftype='cheby2', fs=48000,
                        output='sos')
sos5 = signal.iirfilter(2, [5000, 20000], rs=20, btype='band',
                        analog=False, ftype='cheby2', fs=48000,
                        output='sos')

w1, h1 = signal.sosfreqz(sos1, 48000, fs=48000)
w2, h2 = signal.sosfreqz(sos2, 48000, fs=48000)
w3, h3 = signal.sosfreqz(sos3, 48000, fs=48000)
w4, h4 = signal.sosfreqz(sos4, 48000, fs=48000)
w5, h5 = signal.sosfreqz(sos5, 48000, fs=48000)

b1, a1 = signal.sos2tf(sos1)
b2, a2 = signal.sos2tf(sos2)
b3, a3 = signal.sos2tf(sos3)
b4, a4 = signal.sos2tf(sos4)
b5, a5 = signal.sos2tf(sos5)

file = open("FilterCoeffs.h", 'w')
file.write("float ")
file.write("FilterCoeffs")
file.write("[45] = {")
file.write("\n")
for x in range(5):
    file.write(str(b1[x]))
    file.write(", ")
for x in range(4):
    file.write(str(a1[x+1]))
    file.write(", ")
file.write("\n")
for x in range(5):
    file.write(str(b2[x]))
    file.write(", ")
for x in range(4):
    file.write(str(a2[x+1]))
    file.write(", ")
file.write("\n")
for x in range(5):
    file.write(str(b3[x]))
    file.write(", ")
for x in range(4):
```

```
        file.write(str(a3[x+1]))
        file.write(", ")
file.write("\n")
for x in range(5):
        file.write(str(b4[x]))
        file.write(", ")
for x in range(4):
        file.write(str(a4[x+1]))
        file.write(", ")
file.write("\n")
for x in range(5):
        file.write(str(b5[x]))
        file.write(", ")
for x in range(4):
        file.write(str(a5[x+1]))
        file.write(", ")
file.write("\n")
file.write("};")
file.close()

fig = plt.figure()
ax = fig.add_subplot(1, 1, 1)
ax.semilogx(w1, 20 * np.log10(np.maximum(abs(h1), 1e-5)))
ax.axis((10, 48000, -100, 10))
ax.semilogx(w2, 20 * np.log10(np.maximum(abs(h2), 1e-5)))
ax.semilogx(w3, 20 * np.log10(np.maximum(abs(h3), 1e-5)))
ax.semilogx(w4, 20 * np.log10(np.maximum(abs(h4), 1e-5)))
ax.semilogx(w5, 20 * np.log10(np.maximum(abs(h5), 1e-5)))
plt.show()
```

12.5.3 Feeding the Equalized Digital Signal to Output

The last step in processing the analog audio signal on the STM32F4 microcontroller is feeding the processed digital signal to output in analog form. Afterward, we should feed the analog signal to a speaker to hear it. To do so, we can benefit from the DAC module.

Block diagram of a typical speaker is as in Fig. 12.5. Here, the analog signal is applied to windings of the voice coil, and electrical current flows through them. The voice coil is placed inside a magnetic field of permanent magnet. Hence, the voice coil moves back and forth inside this magnetic field according to the amplitude and direction of the applied current. The voice coil is also attached to the cone. Hence, the cone moves with voice coil. As a result, the cone vibrates surrounding air molecules to create an audio signal.

We should amplify the analog signal before feeding it to the speaker. We will use the Adafruit PAM8302A audio amplifier breakout board for this purpose. PAM8302A is a 2.5 W class D audio amplifier. It is a mono-amplifier which means it can drive only one speaker at a time. It amplifies the analog input signal between 0 and 24 dB which can be adjusted by its onboard potentiometer.

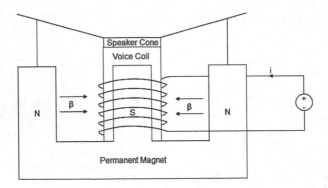

Fig. 12.5 Block diagram of a typical speaker

Fig. 12.6 Circuit layout of
the equalizer system

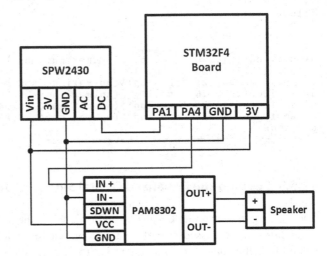

12.5.4 Final Form of the Overall System

We provide circuit layout of the overall five-band equalizer system formed by all mentioned modules in Fig. 12.6. As can be seen in this figure, we connect Vin, GND, and DC pins of the SPW2430 MEMS microphone board to 3 V, GND, and PA1 pins of the STM32F4 board, respectively. 3V and AC pins of SPW2430 MEMS microphone board are left unconnected.

We should connect GND, negative audio input, and positive audio input pins of the PAM8302A board to GND, GND, and PA4 pins of STM32F4 board, respectively. The PAM8302 board can be supplied from 2 to 5 V. Hence, we connect VCC pin of the PAM8302A board to 3 V pin of the STM32F4 board. Shutdown pin of the PAM8302A board is left unconnected. Finally, outputs of the PAM8302A board are connected to the speaker. Here, we use the Adafruit 3 W, 4 Ω mono-enclosed speaker in implementation.

We provide the C code for our five-band equalizer in Listing 12.8. In order to execute it on the STM32F4 microcontroller, open a new project under STM32CubeIDE. In STM32CubeMX, set HCLK frequency to 180000000, and set TIM2 to trigger

ADC module with 48,000 Hz sampling frequency. Pin PA1 should be set as ADC output. Likewise, pin PA4 should be set as DAC output. We will use the filters obtained in the previous section as equalizer band-pass filters. The obtained header file (from either method) should be placed inside the "Core->Inc" folder of the project folder.

Listing 12.8 The C code for the equalizer system

```c
/* USER CODE BEGIN Includes */
#include "math.h"
#include "100HzFilter.h"
#include "300HzFilter.h"
#include "1000HzFilter.h"
#include "3000HzFilter.h"
#include "10000HzFilter.h"
/* USER CODE END Includes */

/* USER CODE BEGIN PD */
#define GAIN100Hz gain12
#define GAIN300Hz gain12
#define GAIN1000Hz gain12
#define GAIN3000Hz gain12
#define GAIN10000Hz gain12

#define NUM_SIZE 3
#define DENUM_SIZE 3
#define INPUT_BUFFER_SIZE 3
#define OUTPUT_BUFFER_SIZE 3
/* USER CODE END PD */

/* USER CODE BEGIN PV */
uint16_t data;
int cnt = 0;
uint16_t DACOut;
float xArr100Hz[INPUT_BUFFER_SIZE] = {0};
float yArr100Hz[OUTPUT_BUFFER_SIZE] = {0};
float xArr300Hz[INPUT_BUFFER_SIZE] = {0};
float yArr300Hz[OUTPUT_BUFFER_SIZE] = {0};
float xArr1000Hz[INPUT_BUFFER_SIZE] = {0};
float yArr1000Hz[OUTPUT_BUFFER_SIZE] = {0};
float xArr3000Hz[INPUT_BUFFER_SIZE] = {0};
float yArr3000Hz[OUTPUT_BUFFER_SIZE] = {0};
float xArr10000Hz[INPUT_BUFFER_SIZE] = {0};
float yArr10000Hz[OUTPUT_BUFFER_SIZE] = {0};
enum gaindB{
gain12, gain11, gain10, gain9, gain8, gain7, gain6, gain5, gain4,
    gain3, gain2, gain1, gain0, gainn1, gainn2, gainn3, gainn4,
    gainn5, gainn6, gainn7, gainn8, gainn9, gainn10, gainn11,
    gainn12
};

/* USER CODE BEGIN 0 */
float IIR_Filter(float *num, float *denum, float *x, float *y,
    int index){
float out = 0;
int i;
for(i = 0; i < NUM_SIZE; i++){
if(index >= i){
out += *(num+i) * *(x+i);
}
}
```

```
for(i = 1; i < DENUM_SIZE; i++){
if(index >= i){
out -= *(denum+i-1) * *(y+i);
}
}
return out;
}

void HAL_ADC_ConvCpltCallback(ADC_HandleTypeDef *hadc)
{
data = HAL_ADC_GetValue(hadc);
xArr100Hz[0] = (data - 870) * 3 / 4095;
xArr300Hz[0] = xArr100Hz[0];
xArr1000Hz[0] = xArr100Hz[0];
xArr3000Hz[0] = xArr100Hz[0];
xArr10000Hz[0] = xArr100Hz[0];
yArr100Hz[0] = IIR_Filter(&Filter100[GAIN100Hz * 5], &Filter100[
    GAIN100Hz * 5 + 3], xArr100Hz, yArr100Hz, cnt);
yArr300Hz[0] = IIR_Filter(&Filter300[GAIN300Hz * 5], &Filter300[
    GAIN300Hz * 5 + 3], xArr300Hz, yArr300Hz, cnt);
yArr1000Hz[0] = IIR_Filter(&Filter1000[GAIN1000Hz * 5], &
    Filter1000[GAIN1000Hz * 5 + 3], xArr1000Hz, yArr1000Hz, cnt);
yArr3000Hz[0] = IIR_Filter(&Filter3000[GAIN3000Hz * 5], &
    Filter3000[GAIN3000Hz * 5 + 3], xArr3000Hz, yArr3000Hz, cnt);
yArr10000Hz[0] = IIR_Filter(&Filter10000[GAIN10000Hz * 5], &
    Filter10000[GAIN10000Hz * 5 + 3], xArr10000Hz, yArr10000Hz,
    cnt);
DACOut = (uint16_t)(((yArr100Hz[0] + yArr300Hz[0] + yArr1000Hz[0]
    + yArr3000Hz[0] + yArr10000Hz[0]) * 4095 / 3)+2048);
HAL_DAC_SetValue(&hdac, DAC_CHANNEL_1, DAC_ALIGN_12B_R, DACOut);
cnt++;
xArr100Hz[2] = xArr100Hz[1];
xArr100Hz[1] = xArr100Hz[0];
yArr100Hz[2] = yArr100Hz[1];
yArr100Hz[1] = yArr100Hz[0];
xArr300Hz[2] = xArr300Hz[1];
xArr300Hz[1] = xArr300Hz[0];
yArr300Hz[2] = yArr300Hz[1];
yArr300Hz[1] = yArr300Hz[0];
xArr1000Hz[2] = xArr1000Hz[1];
xArr1000Hz[1] = xArr1000Hz[0];
yArr1000Hz[2] = yArr1000Hz[1];
yArr1000Hz[1] = yArr1000Hz[0];
xArr3000Hz[2] = xArr3000Hz[1];
xArr3000Hz[1] = xArr3000Hz[0];
yArr3000Hz[2] = yArr3000Hz[1];
yArr3000Hz[1] = yArr3000Hz[0];
xArr10000Hz[2] = xArr10000Hz[1];
xArr10000Hz[1] = xArr10000Hz[0];
yArr10000Hz[2] = yArr10000Hz[1];
yArr10000Hz[1] = yArr10000Hz[0];
}
/* USER CODE END 0 */

/* USER CODE BEGIN 2 */
HAL_TIM_Base_Start(&htim2);
HAL_ADC_Start_IT(&hadc1);
HAL_DAC_Start(&hdac, DAC_CHANNEL_1);
/* USER CODE END 2 */
```

In Listing 12.8, the buffers xarr and yarr are used to keep past input and output values for filtering purposes, respectively. As we obtain an ADC sample as input, it passes through five equalizer bands. Band outputs are summed and obtained data is fed to the DAC module. The user can select desired gain values for each band at the beginning of the code. Here, positive gains are set by using `gainx`, and negative gains are set by using `gainnx` enumerator values. Then, the obtained data is fed to output. The `xarr` and `yarr` buffers are shifted by one element for the next audio sample to be processed.

We provided the C++ and MicroPython codes for our five-band equalizer as supplementary material in the accompanying book web site. The exact name for these files are "Listing12_9.cpp" and "Listing12_10.py." The reader can benefit from them to execute the equalizer in C++ and MicroPython languages, respectively.

12.6 Summary of the Chapter

As microcontrollers have become more powerful, performing digital signal processing operations became feasible on them. This leads to decision-making based on digital signals on the edge. Therefore, we focused on fundamental digital signal processing operations in this chapter. Our approach was more of an introduction to the important concepts than detailed explanation of all topics. Hence, we started with the definition and representation of digital signals on embedded systems. Then, we considered digital systems used for processing digital signals. Here, the main focus was on filtering. Hence, we provided an end-to-end equalizer design and implementation system for analog signals. We also briefly introduced complex domain analysis of digital signals and systems. Throughout the chapter, we performed offline operations on PC via Python language. We provided the C, C++, and MicroPython codes to run on the STM32F4 microcontroller for online operations. We also provided sample codes to transfer data between the PC and STM32F4 microcontroller. Hence, the reader can analyze and observe the acquired signal from the STM32F4 microcontroller. The concepts introduced in this chapter can be used in deep learning-based neural network systems in advanced applications. Hence, we strongly suggest the reader to become familiar to the digital signal processing concepts.

Problems

12.1. What is the difference between FIR and IIR filters? How can they be implemented on an embedded system?

12.2. What is the relationship between the z-transform and DTFT?

12.3. z-Transform and inverse z-transform can be calculated in symbolic form in Python on PC. Let's pick the digital signal as

$$x[n] = \begin{cases} 1 & n = 0, 1, 2 \\ 0 & \text{otherwise} \end{cases} \tag{12.8}$$

Take the z-transform of this signal in Python. Then, take the inverse z-transform of the result. Do you obtain the original signal $x[n]$?

12.4. Find the z-transform and FFT of the equalizer implementation in Sect. 12.5.2 in Python on PC. Plot the magnitude of the FFT result on PC.

12.5. Generate a random signal with 32 elements in the STM32F4 microcontroller. Transfer the generated signal to PC, and plot it in Python. Form the project using

(a) C language under STM32CubeIDE
(b) C++ language under Mbed
(c) MicroPython

12.6. Acquire the gyroscope signal from the STM32F4 board with 32 samples. Transfer the generated signal to PC, and plot it in Python. Form the project using

(a) C language under STM32CubeIDE
(b) C++ language under Mbed
(c) MicroPython

12.7. Median filter is a popular nonlinear filter used in practice. Output of the filter is obtained by the median of the subset of the input signal (called window). Then, this window is moved, and the next output is obtained. This is called the sliding window approach. Implement the median filter on the STM32F4 microcontroller with window size 5. Apply it on the gyroscope signal with 512 elements. Transfer the generated signal to PC, and plot it in Python. Form the project using

(a) C language under STM32CubeIDE
(b) C++ language under Mbed
(c) MicroPython

12.8. Generate a random signal (with 64 elements) from the uniform distribution between 0 and 1 in Python on PC. Transfer the generated signal to the STM32F4 microcontroller. Threshold the transferred signal on the microcontroller such that a value greater than 0.5 is set as 1. A value less than or equal to 0.5 is set as 0. Transfer the thresholded signal to PC and plot it in Python. Form the project using

(a) C language under STM32CubeIDE
(b) C++ language under Mbed
(c) MicroPython

References

1. Ünsalan, C., Yücel, M.E., Gürhan, H.D.: Digital Signal Processing Using Arm Cortex-M Based Microcontrollers: Theory and Practice. Arm Education Media, Cambridge (2018)
2. Ünsalan, C., Barkana, D.E., Gürhan, H.D.: Embedded Digital Control with Microcontrollers: Implementation with C and Python. Wiley - IEEE, Piscataway (2021)

Introduction to Digital Control

13

13.1 About Digital Control

Operations to be performed under digital control are closely related to the digital signal processing concepts introduced in Chap. 12. Although both have similar grounds, they have different aims. Therefore, we will first consider the question what should we understand from the control action? As we provide an answer to this question, we will next focus on how to implement a digital controller on an embedded system (more specifically the STM32F4 microcontroller). Hence, we will make sure that the topics to be considered in the following sections make sense.

13.1.1 The Control Action

We introduced digital systems in Chap. 12. Some of these systems may be used in applications in which their outputs should satisfy predefined criteria. Let's consider an example on this issue. Assume that we pick a DC motor as the system. Our system will have its input signal as voltage. One output that can be obtained from the system is the speed of the motor. Hence, we have a system with associated input and output signals.

Different from digital signal processing, we are interested in the characteristics of the output signal (speed of the motor) obtained from the system (DC motor in our example) for a given input reference signal. In other words, we feed a reference signal to rotate the DC motor at desired speed. During this operation, we may ask

---

Supplementary Information The online version contains supplementary material available at (https://doi.org/10.1007/978-3-030-88439-0_13).

© Springer Nature Switzerland AG 2022
C. Ünsalan et al., *Embedded System Design with Arm Cortex-M Microcontrollers*,
https://doi.org/10.1007/978-3-030-88439-0_13

the system to reach the desired speed within a certain time duration. If the system can satisfy this requirement, then all is fine. If the DC motor cannot satisfy the requirement, then we should add a controller to make sure that the overall system, formed by the controller and DC motor, satisfies the requirement. As a result, the control action, imposed by the controller, aims to modify the overall system characteristics to satisfy the desired requirements. At this step, we should emphasize that the controller is also a system, designed for the given specifications at hand.

There are basically two controller types as transfer function and state-space based. The transfer function-based control takes the input-output relation of the system to be controlled as in Chap. 12. In other words, the controller is represented by its transfer function. Hence, it can be designed according to the transfer function tools. We can also have state-space-based control. Here, we first define states of the system which are the minimum set of parameters that can represent the entire system behavior at any given time. The number of states is equal to the number of first-order difference equations to describe the overall system dynamics. These first-order difference equations are arranged in vector form in state-space-based control. In this book, we will only consider transfer function-based control. For more information on state-space-based control, please see [1].

We can also have open- and closed-loop control action. In the open-loop control, we modify the input signal before feeding it to the system to be controlled. Then, the modified signal is fed to the system to obtain the desired output. In the closed-loop control, we first form an open-loop controller. Then, we observe the output of the system to be controlled by another system and feed the observed signal back to input. Hence, we have a feedback loop. The aim here is to make sure that the output obtained from the overall system follows the desired input reference signal. We will consider the closed-loop control action throughout the chapter. Hence, we provide the general block diagram of a system with closed-loop control action in Fig. 13.1. We will provide detailed explanation on the blocks in this figure in the following sections.

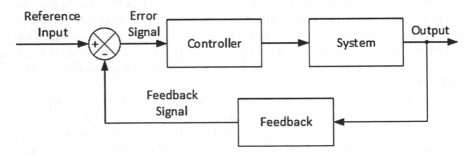

Fig. 13.1 General block diagram of a system with closed-loop control action

13.1.2 Representing the Digital Controller in an Embedded System

We mentioned that the digital controller is also a system. Hence, it can be represented in code form on an embedded system (such as the STM32F4 microcontroller) by the methods introduced in Chap. 12. The STM32F4 microcontroller offers extra advantages besides representing the controller in code form. As an example, sensors to be used to measure output of the system can be used in connection with the microcontroller. Different input signals can be formed by the microcontroller. As a result, we can form a complete system to construct the digital controller and feedback action on the STM32F4 microcontroller (or STM32F4 board to be more precise). Next, we will follow this approach to form such controllers.

13.2   Transfer Function Based Control

As explained in the previous section, transfer function based control can be explained by the concepts introduced in Chap. 12. Therefore, we pick it as the control type in this chapter. Moreover, we will use closed-loop control. Hence, we should also explain what the open-loop control does. As a result, we will provide general definitions on both open- and closed-loop controls in this section. We will also explain the framework in implementing each method on the STM32F4 microcontroller. Finally, we will briefly talk about the transfer function based controller design strategies.

13.2.1 Open-Loop Control

In the open-loop control, the reference input signal is directly applied to the controller. The controller modifies this signal and then feeds it to the system to be controlled. Block diagram of an open-loop control system is given in Fig. 13.2. Here, $G(z)$ is the digital system transfer function. $C(z)$ is the controller system transfer function. $x[n]$ is the reference input signal, and $y[n]$ is the system output. In the open-loop control system, the output can be obtained as $Y(z) = G(z)C(z)X(z)$.

Assume that $G(z)$ and $C(z)$ are represented by IIR transfer functions:

$$G(z) = \frac{Y(z)}{U(z)} = \frac{b_{g0} * z^2 + b_{g1} * z + b_{g2}}{a_{g0} * z^2 + a_{g1} * z + a_{g2}} \tag{13.1}$$

$$C(z) = \frac{U(z)}{X(z)} = \frac{b_{c0} * z + b_{c1}}{a_{c0} * z + a_{c1}} \tag{13.2}$$

Fig. 13.2 General structure of the open-loop control system

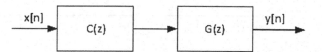

These systems can be represented by their difference equations

$$a_{g0} * y[n] + a_{g1} * y[n-1] + a_{g2} * y[n-2] = b_{g0} * u[n] + b_{g1} * u[n-1]$$
$$+ b_{g2} * u[n-2] \qquad (13.3)$$
$$a_{c0} * u[n] + a_{c1} * u[n-1] = b_{c0} * x[n] + b_{c1} * x[n-1]$$
$$(13.4)$$

We can represent the pseudocode of the overall system as follows:

```
bc is the feedforward coefficient array of the controller with
    length K1
ac is the feedback coefficient array of the controller with
    length L1
bg is the feedforward coefficient array of the system with length
    K2
ag is the feedback coefficient array of the system with length L2
ba is the feedforward coefficient array of overall system with
    length K = K1 + K2 - 1
aa is the feedback coefficient array of overall system with
    length L = L1 + L2 - 1
y is the output array with size length N
x is the input array

for(k1 = 0 to K1){
for(k2 = 0 to K2){
ba[k1+k2] += bc[k1] * bg[k2]
}

for(l1 = 0 to L1){
for(l2 = 0 to L2){
aa[l1+l2] += ac[l1] * ag[l2]
}

for(k3 = 0 to K3)
ba[k3] = ba[k3] / aa[0]

for(l3 = 1 to L3)
aa[l3] = aa[l3] / aa[0]

aa[0] = 1

for(n = 0 to N){
sum = 0
for(k = 0 to K){
if(k <(n+1)){
sum += ba[k] * x[n-k]
}}

for(l = 1 to L){
if(l <(n+1))
sum -= aa[l] * y[n-l]
}

y[n] = sum
}
```

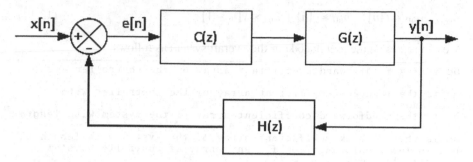

Fig. 13.3 General structure of the closed-loop control system

13.2.2 Closed-Loop Control

In the closed-loop control, the system output is observed, and it affects the controller input through a feedback loop. As a result, the system response depends on its output. Block diagram of a closed-loop control system is given in Fig. 13.3.

As can be seen in Fig. 13.3, the system output, $y[n]$, is measured by a sensor and fed back to input. The transfer function of the feedback is represented by $H(z)$. The difference between the system output and reference input is obtained as error signal, $e[n]$, which is applied to the controller. Based on these definitions, we will have the overall transfer function of the closed-loop system as

$$\frac{Y(z)}{X(z)} = \frac{C(z)G(z)}{1 + H(z)C(z)G(z)} \tag{13.5}$$

Assume that $G(z)$, $C(z)$, and $H(z)$ are represented by IIR transfer functions:

$$G(z) = \frac{Y(z)}{U(z)} = \frac{b_{g0} * z^2 + b_{g1} * z + b_{g2}}{a_{g0} * z^2 + a_{g1} * z + a_{g2}} \tag{13.6}$$

$$C(z) = \frac{U(z)}{E(z)} = \frac{b_{c0} * z + b_{c1}}{a_{c0} * z + a_{c1}} \tag{13.7}$$

$$H(z) = \frac{F(z)}{Y(z)} = \frac{b_{h0} * z + b_{h1}}{a_{h0}} \tag{13.8}$$

These systems can be represented by their difference equations as

$$a_{g0} * y[n] + a_{g1} * y[n-1] + a_{g2} * y[n-2] = b_{g0} * u[n] + b_{g1} * u[n-1]$$
$$+ b_{g2} * u[n-2] \tag{13.9}$$

$$a_{c0} * u[n] + a_{c1} * u[n-1] = b_{c0} * e[n] + b_{c1} * e[n-1] \tag{13.10}$$

$$a_{h0} * f[n] = b_{h0} * y[n] + b_{h1} * y[n-1] \tag{13.11}$$

We can represent the pseudocode of the overall system as follows:

```
bc is the feedforward coefficient array of the controller with
    length K1
ac is the feedback coefficient array of the controller with
    length L1
bg is the feedforward coefficient array of the system with length
    K2
ag is the feedback coefficient array of the system with length L2
bcg is the feedforward coefficient array of controller*system
    with length K3 = K1 + K2 - 1
acg is the feedback coefficient array of overall controller*
    system with length L3 = L1 + L2 - 1

for(k1 = 0 to K1){
for(k2 = 0 to K2){
ba[k1+k2] += bc[k1] * bg[k2]
}

for(l1 = 0 to L1){
for(l2 = 0 to L2){
aa[l1+l2] += ac[l1] * ag[l2]
}

for(k3 = 0 to K3)
bcg[k3] = bcg[k3] / acg[0]

for(l3 = 1 to L3)
acg[l3] = acg[l3] / acg[0]

acg[0] = 1

bh is the feedforward coefficient array of the feedback system
    with length K4
ah is the feedback coefficient array of the feedback system with
    length L4
ba is the feedforward coefficient array of overall system with
    length K = K3 + L4 - 1
aa is the feedback coefficient array of overall overall system
    with length L = max(K3 + K4 - 1, L3 + L4 - 1)
ad1 is the dummy array with length D1 = K3 + K4 - 1
ad2 is the dummy array with length D2 = L3 + L4 - 1
y is the output array with size length N
x is the input array

for(k3 = 0 to K3){
for(l4 = 0 to L4){
ba[k3+l4] += bcg[k3] * ah[l4]
}

for(k3 = 0 to K3){
for(k4 = 0 to K4){
ad1[k3+k4] += bcg[k3] * bh[k4]
}

for(l3 = 0 to L3){
for(l4 = 0 to L4){
ad2[l3+l4] += acg[l3] * ah[l4]
}
if(K3+K4 > L3+L4){
```

```
for(d1 = 0 to D1){
if(d1 < (K3 + K4 - L3 - L4))
aa[d1] = ad1[d1]
else{
aa[d1] = ad1[d1] + ad2[d1 - (K3 + K4 - L3 - L4)]
}}}
else{
for(d2 = 0 to D2){
if(d2 < (L3 + L4 - K3 - K4))
aa[d2] = ad2[d2]
else{
aa[d2] = ad2[d2] + ad1[d2 - (L3 + L4 - K3 - K4)]
}}}}

for(k = 0 to K)
ba[k] = ba[k] / aa[0]

for(l = 1 to L)
aa[l] = aa[l] / aa[0]

aa[0] = 1

for(n = 0 to N){
sum = 0
for(k = 0 to K){
if(k <(n+1)){
sum += ba[k] * x[n-k]
}}

for(l = 1 to L){
if(l <(n+1))
sum -= aa[l] * y[n-l]
}

y[n] = sum
}
```

13.2.3 Designing a Controller

We can design a controller (either in open- or closed-loop form) to satisfy the given performance criteria. There are several well-known design methods developed for this purpose in literature. The reader can pick one of these to form his or her digital controller accordingly. Then, the controller can be implemented on the STM32F4 microcontroller to form the overall digital control system.

The reader can also benefit from available programs such as MATLAB to design the controller. Here, the program can generate a controller to perform the desired control action based on given performance criteria. As a side note, controller design programs (such as MATLAB) are also based on theoretical methods introduced in literature. However, they simplify life for the user by making calculations automatically. We refer the reader to our book for more information on this topic [1]. There is a specific method, called PID controller design, which is extensively used in practical applications. We will consider it next.

13.3   PID Controllers

Proportional integral derivative (PID) is one of the most popular controller structures used in practical applications. The main reason for this popularity is that PID has well-defined design steps. Hence, the user can follow them to construct a fairly good working controller. We will consider the PID controller structure from a general perspective first. Then, we will focus on the design steps. Finally, we will provide ways of implementing the PID controller in C, C++, and MicroPython languages on the STM32F4 microcontroller.

13.3.1   General Structure

The PID controller is initially developed for analog systems. Hence, it has a transfer function representation in s-domain (complex domain representation for continuous-time signals and systems). We can convert this representation to digital form by bilinear transformation [1]. Based on this, we will have the overall transfer function of the PID controller as

$$C(z) = C_p(z) + C_i(z) + C_d(z) \tag{13.12}$$

$$C_p(z) = K_p \tag{13.13}$$

$$C_i(z) = \frac{(K_i * T_s/2) * (z + 1)}{z - 1} \tag{13.14}$$

$$C_d(z) = \frac{(K_d * 2/T_s) * (z - 1)}{z + 1} \tag{13.15}$$

where T_s is the sampling period used in transforming the analog signal to digital form. K_p, K_i, and K_d stand for gain parameters for proportional, integral, and derivative terms, respectively. These should be set to satisfy the design criteria at hand. To note here, one or two of these gains can be set to zero to obtain the PI and P controllers, respectively. We will consider a PID controller design method in Sect. 13.3.2. Beforehand, let's focus on the P, PI, and PID controllers in detail.

13.3.1.1 P Controller
The aim of the proportional (P) controller is to multiply the error signal by a constant in the closed-loop setup and generate the control signal accordingly. Based on this, the closed-loop control setup with the P controller will be as in Fig. 13.4.

13.3.1.2 PI Controller
PI controller has proportional gain the same as in the P controller. It also has an additional integrator which integrates the error until it reaches zero. Hence, it eliminates the steady-state error occurring in the system. The closed-loop control setup with the PI controller will be as in Fig. 13.5.

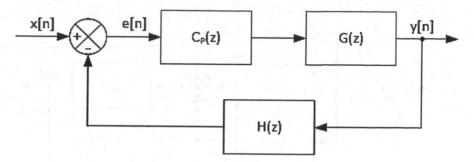

Fig. 13.4 Closed-loop control with the P controller

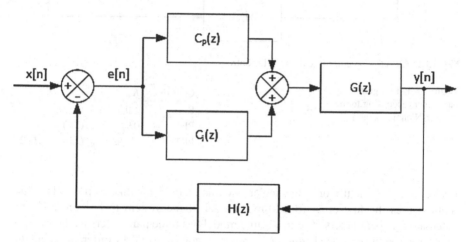

Fig. 13.5 Closed-loop control with the PI controller

13.3.1.3 PID Controller

The PI controller can be extended further by adding a derivative term. This leads to the PID controller which merges both previous input values and future predictions in the control action. The derivative term aims to improve the controller by predicting future input values. However, it may also have a negative effect if the signal to be processed is noisy. For such cases, the derivative term may enhance noise in the signal. As a result, the PID controller may not work as expected or may not work at all in practice. Therefore, this structure should be used with precaution. The closed-loop control setup with the PID controller will be as in Fig. 13.6.

13.3.2 PID Controller Design

As explained in Sect. 13.3.1, the PID controller has three gain parameters K_p, K_i, and K_d. There are several methods in literature to obtain these parameters. In this

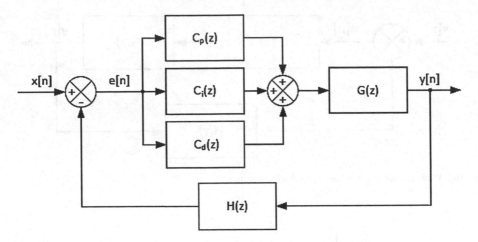

Fig. 13.6 Closed-loop control with the PID controller

Table 13.1 PID gain parameters according to the Ziegler-Nichols method

Controller	K_p	K_i	K_d
P	$1.0/a$	–	–
PI	$0.9/a$	$K_p/(3L)$	–
PID	$1.2/a$	$K_p/(2L)$	$K_p L/2$

book, we will benefit from the Ziegler-Nichols method for this purpose [1]. This method uses the first-order (dead-time) model of the system to be controlled to calculate PID parameters. The dead-time model has three parameters as the system gain (K), the system dead time (L), and time constant (τ). In order to calculate them, step response of the system is generated first. The K parameter is calculated as the ratio of the difference of the final and initial values of the system output to the difference of the final and initial values of the system input. We can estimate τ by multiplying the time it takes the system to reach 28.3% of its step response by 1.5. We can find L by subtracting τ from the time it takes the system to reach 63.2% of its step response. Finally, the a parameter is obtained as $a = K\frac{L}{\tau}$. The K_p, K_i, and K_d can be calculated using K, L, and a parameters as tabulated in Table 13.1. The reader can pick the parameter set from this table for the selected controller type as P, PI, or PID.

13.3.3 Implementing the PID Controller on the STM32F4 Microcontroller

We can implement the PID controller on the STM32F4 microcontroller. To do so, we form a controller via K_p, K_i, and K_d coefficients. Then, we obtain the difference equation of this transfer function to obtain the output for a given input. We provide the C, C++, and MicroPython codes for this purpose next.

13.3.3.1 Implementation in C Language

We formed the PID library to initialize and use PID controller in C language. To use this library, we should add the "PID.c" and "PID.h" files to "project->Core->Src" and "project->Core->Inc" folders for the active STM32CubeIDE project, respectively.

Our PID library has two functions as PID_Init and PID_Output. Both use the struct PID_struct which holds PID controller parameters. We provide the detailed information on them next.

```
typedef struct _pid {
        float Kp, Ki, Kd;
        float Ts;
        float b0, b1, b2;
        float a0, a1, a2;
        float yold1, yold2;
        float xold1, xold2;
} PID_struct;

void PID_Init(PID_struct *pid, float Kp, float Ki, float Kd,
    float Ts)
/*
pid: pointer to the PID_struct struct
Kp: proportional gain
Ki: integral gain
Kd: derivative gain
Ts: sampling time of PID controller in seconds
*/

float PID_Output(float ref, float out)
/*
ref: reference input of the overall system
out: output of the overall system
*/
```

The struct PID_struct holds the general PID parameters Kp, Ki, Kd, and Ts. It also contains PID controller parameters b0, b1, b2, a0, a1, and a2. Finally, the structure keeps the previous controller input and output samples as yold1, yold2, xold1, and xold2. The function PID_Init initializes the PID controller parameters for the given Kp, Ki, Kd, and Ts values. The function PID_Output generates controller output sample for the given reference and overall system output samples.

We provide a sample code in Listing 13.1 on the usage of the given PID functions in C language. First, PID controller is initialized with parameters Kp = 10, Ki = 0, Kd = 0, and Ts = 0.01. Then, the system with parameters bSystem and aSystem is simulated for step input with 250 samples. Obtained output samples are stored to the out array.

Listing 13.1 PID controller applied to a sample system, the C code

```
/* USER CODE BEGIN Includes */
#include "PID.h"
/* USER CODE END Includes */

/* USER CODE BEGIN PV */
PID_struct myPID;
```

```
float bSystem[3] = {0.002, 0.004, 0.002};
float aSystem[3] = {1, -1.82, 0.818};
float xSystem[3] = {0};
float ySystem[3] = {0};
float in[250];
float out[250];
int i;
int k;
/* USER CODE END PV */

/* USER CODE BEGIN 2 */
PID_Init(&myPID, 10, 0, 0, 0.01);
for (i = 0; i < 250; i++)
{
in[i] = 1;
xSystem[0] = PID_Output(&myPID, in[i], ySystem[1]);
ySystem[0] = bSystem[0] * xSystem[0] + bSystem[1] * xSystem[1] +
    bSystem[2] * xSystem[2] - aSystem[1] * ySystem[1] - aSystem
    [2] * ySystem[2];
out[i] = ySystem[0];
for (k = 2; k > 0; k--)
{
ySystem[k] = ySystem[k - 1];
xSystem[k] = xSystem[k - 1];
}
}
/* USER CODE END 2 */
```

13.3.3.2 Implementation in C++ Language

We also formed the PID library to initialize and use PID controller in C++ language. To use this library, we should add the "PID.cpp" and "PID.hpp" files to the active Mbed Studio project. As a side note, all these library files are provided in the accompanying book web site.

We formed a class PID within the PID library. This class has two functions as Init and Output. These functions work in the same way as their C counterparts introduced in the previous section. We provide detailed information on the class and its functions next.

```
PID()

void Init(float Kp, float Ki, float Kd, float Ts)
/*
Kp: proportional gain
Ki: integral gain
Kd: derivative gain
Ts: sampling time of PID controller in seconds
*/

float Output(float ref, float out)
/*
ref: reference input of the overall system
out: output of the overall system
*/
```

The function Init initializes the PID controller parameters for the given Kp, Ki, Kd, and Ts values. The function Output generates controller output sample for the given reference and overall system output samples.

We provide a sample code in Listing 13.2 on the usage of the given PID functions in C++ language. This code works in the same way as its C counterpart given in Listing 13.1.

Listing 13.2 PID controller applied to the sample system, the C++ code

```
#include "mbed.h"
#include "PID.hpp"

float bSystem[3] = {0.002, 0.004, 0.002};
float aSystem[3] = {1, -1.82, 0.818};
float xSystem[3] = {0};
float ySystem[3] = {0};
float in[250];
float out[250];
int i;
int k;

int main()
{
PID myPID(10, 0, 0, 0.01);
for (i = 0; i < 250; i++)
{
in[i] = 1;
xSystem[0] = myPID.Output(in[i], ySystem[1]);
ySystem[0] = bSystem[0] * xSystem[0] + bSystem[1] * xSystem[1] +
    bSystem[2] * xSystem[2] - aSystem[1] * ySystem[1] - aSystem
    [2] * ySystem[2];
out[i] = ySystem[0];
for (k = 2; k > 0; k--)
{
ySystem[k] = ySystem[k - 1];
xSystem[k] = xSystem[k - 1];
}
}

while (true);
}
}
```

13.3.3.3 Implementation in MicroPython

We also formed the PID library to initialize and use the PID controller in MicroPython. To use this library, we should add the "PIDlib.py" file to the active MicroPython project. As a side note, this library file is provided in the accompanying book web site.

We formed a class PID within the PID library. This class has two functions as __init__ and Output. These functions work in the same way as their C counterparts introduced in the previous section. We provide detailed information on the class and its functions next.

```
PID()

__init__(self, Kp, Ki, Kd, Ts)
/*
Kp: proportional gain
Ki: integral gain
Kd: derivative gain
```

```
Ts: sampling time of PID controller in seconds
*/

Output(self, ref, out)
/*
ref: reference input of the overall system
out: output of the overall system
*/
```

The function __init__ is the constructor of the class. It is called when the PID object is created for the given Kp, Ki, Kd, and Ts values. The function Output generates controller output sample for the given reference and overall system output samples.

We provide a sample code in Listing 13.3 on the usage of the given PID functions in MicroPython. This code works in the same way as its C counterpart given in Listing 13.1.

Listing 13.3 PID controller applied to a sample system, the MicroPython code

```
import PIDlib

def main():
    bSystem = [0.002, 0.004, 0.002]
    aSystem = [1, -1.82, 0.818]
    xSystem = [0] * 3
    ySystem = [0] * 3
    inOverall = [0] * 250
    outOverall = [0] * 250
    myPID = PIDlib.PID(10, 0, 0, 0.01)
    for i in range(250):
        inOverall[i] = 1
        xSystem[0] = myPID.Output(inOverall[i], ySystem[1])
        ySystem[0] = bSystem[0] * xSystem[0] + bSystem[1] * \
            xSystem[1] + \
            bSystem[2] * xSystem[2] - aSystem[1] * \
            ySystem[1] - aSystem[2] * ySystem[2]
        outOverall[i] = ySystem[0]
        for k in range(2, 0, -1):
            ySystem[k] = ySystem[k-1]
            xSystem[k] = xSystem[k-1]

main()
```

13.4   PID Control of a DC Motor by the STM32F4 Microcontroller

This section is on speed control of a DC motor. Therefore, we will start with summarizing the DC motor to be used in operation. Then, we will introduce the encoder as the sensor to be used in the feedback loop. Finally, we will provide the PID controller implementation in C, C++, and MicroPython languages for the overall control action.

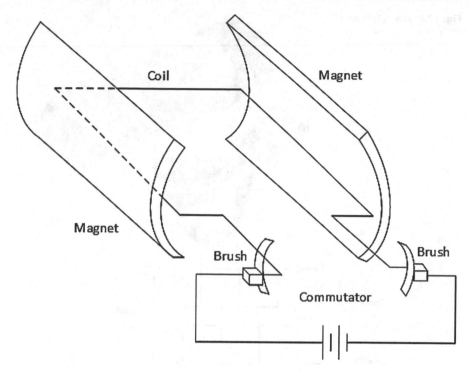

Fig. 13.7 Brushed DC motor

13.4.1 DC Motor as the System to Be Controlled

A brushed DC motor consists of stator magnets, rotor coils, commutators, and brushes as shown in Fig. 13.7. Here, the current flowing through the coil creates an electromagnet. Poles of the electromagnet are attracted to the opposite poles of stator magnets, and it starts rotating. Under normal conditions, rotor coil rotates 180° and stops when opposite poles match. However, there is a gap between commutator parts. Hence, the direction of the current is reversed, and opposite poles are created before rotor coil stops. Hence, a continuous motion is achieved.

We pick the DC motor as in Fig. 13.8. This motor is called Pololu 378:1 Metal Gearmotor 25D×73L mm LP 12 V with 48 CPR encoder. It is a 12 V low-power brushed DC motor with 378:1 gearbox to decrease the rotation speed while increasing torque. The motor has 48 counts per revolution (CPR) quadrature encoder to be explained in detail in the following section.

Fig. 13.8 Pololu brushed
DC motor

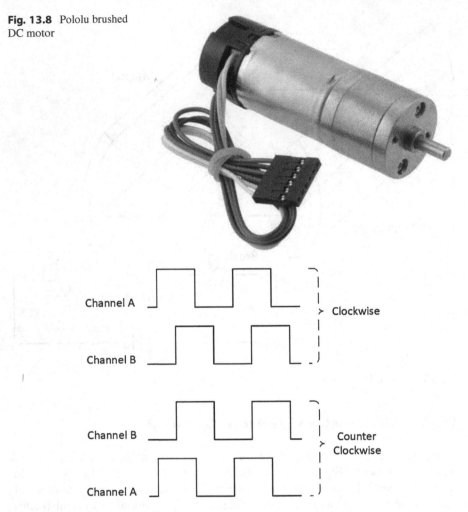

Fig. 13.9 Quadrature encoder outputs

13.4.2 Encoder as the Sensor

As mentioned in the previous section, the selected DC motor has a quadrature encoder with 48 CPR to measure the speed of the DC motor. A quadrature encoder is an incremental encoder with the magnetic two-channel Hall effect sensor. There is a magnetic code disk placed on the motor shaft, and two square waves are created as motor rotates. These waves are called Channel A and B outputs, and they are approximately 90° out of phase as shown in Fig. 13.9.

Direction of the motor can be obtained from the order of the signals in Fig. 13.9. Channel A output leads Channel B output when the motor rotates in clockwise direction. Channel B output leads Channel A output when the motor rotates in counterclockwise direction.

Fig. 13.10 Circuit layout of the DC motor controller system

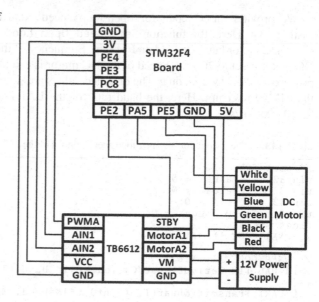

Speed of the motor can be calculated using frequency of the signals in Fig. 13.9 and sensor CPR. If both rising and falling edges of output signals are counted, the sensor provides the resolution of 48 CPR. To find the resolution at the output of the gearbox shaft, the sensor CPR must be multiplied by the gear ratio. Sensor CPR can also be used to measure motor position. We already know that there will be 48 pulses per full rotation. Hence, angular position of the motor with respect to the initial position can be calculated by counting quadrature encoder pulses.

13.4.3 Speed Control of the DC Motor

We provide circuit layout of the overall DC motor control system in Fig. 13.10. As can be seen in this figure, we connect VM and GND pins of the Adafruit TB6612 DC motor driver board to 12 V DC power supply. Here, a 12 V 1 A DC adapter can be used as power supply. Motor output pins MotorA1 and MotorA2 of the TB6612 board should be connected to the red and black cables of the DC motor, respectively. Blue and green cables of the DC motor are VCC and GND pins of the encoder. Hence, they should be connected to the 5V and GND pins of the STM32F4 board, respectively. VCC and GND pins of the TB6612 board should be connected to the 3V and GND pins of the STM32F4 board, respectively. AIN1 and AIN2 pins of TB6612 board should be connected to the pins PE3 and PE4 of the STM32F4 board, respectively. They are used to control the direction of the DC motor. PWMA pin of the TB6612 board should be connected to the pin PC8 of the STM32F4 board. Finally, STBY pin of the TB6612 board should be connected to pin PE2 of the STM32F4 board. This pin should be set to logic level 1 to drive the DC motor.

We provide the C code for open-loop speed control of the DC motor in Listing 13.4. Here, the function Set_Speed_Open_Loop is used to set the speed of the DC motor between −14 and 14 rpm. Frequency of the PWM signal fed to the DC motor is set to 20 kHz. Speed of the DC motor is calculated using input capture property of the TIM2 module. The obtained values are sent to PC every 0.01 s using the TIM10 interrupt. Here, the baud rate for the UART module should be set to 2,000,000 bits/s.

Listing 13.4 The C code for the open-loop speed control of the DC motor

```c
/* USER CODE BEGIN PV */
uint32_t ICValue1 = 0;
uint32_t ICValue2 = 0;
uint32_t ICDiff = 0;
uint8_t CaptureIndex = 0;
float speed = 0;
/* USER CODE END PV */

/* USER CODE BEGIN 0 */
void HAL_TIM_PeriodElapsedCallback(TIM_HandleTypeDef *htim)
{
HAL_UART_Transmit(&huart1, (uint8_t*)&speed, 4, 1000);
}

void HAL_TIM_IC_CaptureCallback(TIM_HandleTypeDef *htim){
if(CaptureIndex == 0)
{
ICValue1 = HAL_TIM_ReadCapturedValue(&htim2, TIM_CHANNEL_1);
CaptureIndex = 1;
}
else
{
ICValue2 = HAL_TIM_ReadCapturedValue(&htim2, TIM_CHANNEL_1);
if (ICValue2 > ICValue1)
{
ICDiff = (ICValue2 - ICValue1);
}
else
{
ICDiff = ((0xFFFFFFFF - ICValue1) + ICValue2) + 1;
}
ICValue1 = ICValue2;
speed = (90000000.0 / (float)ICDiff)* 60.0/ 4535.1975;
if(HAL_GPIO_ReadPin(GPIOE, GPIO_PIN_5) == GPIO_PIN_RESET){
speed = -speed;
}
}
}

void Voltage_to_PWM(float voltage){
uint32_t duty;
if(voltage > 0){
if(voltage > 12)voltage = 12;
HAL_GPIO_WritePin(GPIOE, GPIO_PIN_3, GPIO_PIN_RESET);
HAL_GPIO_WritePin(GPIOE, GPIO_PIN_4, GPIO_PIN_SET);
duty = (uint32_t)(voltage * 1000 / 12);
}
else if(voltage < 0){
if(voltage < -12)voltage = -12;
HAL_GPIO_WritePin(GPIOE, GPIO_PIN_3, GPIO_PIN_SET);
```

```
HAL_GPIO_WritePin(GPIOE, GPIO_PIN_4, GPIO_PIN_RESET);
duty = (uint32_t)(-voltage * 1000 / 12);
}
else if(voltage == 0){
HAL_GPIO_WritePin(GPIOE, GPIO_PIN_3, GPIO_PIN_RESET);
HAL_GPIO_WritePin(GPIOE, GPIO_PIN_4, GPIO_PIN_RESET);
duty = 0;
}

__HAL_TIM_SET_COMPARE (&htim8 , TIM_CHANNEL_3, duty);
}

void Set_Speed_Open_Loop(float speed){
float voltage;
if(speed > 14)speed = 14;
else if(speed < -14)speed = -14;
voltage = 12 * speed / 14;
Voltage_to_PWM(voltage);
}
/* USER CODE END 0 */

/* USER CODE BEGIN 2 */
HAL_TIM_PWM_Start(&htim8 , TIM_CHANNEL_3);
HAL_TIM_IC_Start_IT(&htim2, TIM_CHANNEL_1);
if (__HAL_TIM_GET_FLAG(&htim10, TIM_FLAG_UPDATE) != RESET)
__HAL_TIM_CLEAR_FLAG(&htim10, TIM_FLAG_UPDATE);
HAL_TIM_Base_Start_IT(&htim10);
HAL_GPIO_WritePin(GPIOE, GPIO_PIN_2, GPIO_PIN_SET);
HAL_GPIO_WritePin(GPIOE, GPIO_PIN_3, GPIO_PIN_RESET);
HAL_GPIO_WritePin(GPIOE, GPIO_PIN_4, GPIO_PIN_RESET);
Set_Speed_Open_Loop(6.2);
/* USER CODE END 2 */
```

We next provide the Python code in Listing 13.5 to obtain motor speed data at the PC side. In this code, the function `get_data_from_MC(NoElements, com_port, baud_rate)` is used to obtain the desired number of speed value from the microcontroller in float form. `NoElements` is the number of motor speed samples, and `com_port` is the COM port STM32F4 board is connected to. `baud_rate` is the UART communication speed which should be set to 2,000,000 bits/s.

Listing 13.5 Obtaining motor speed data from the STM32F4 board in Python

```
import serial
from ctypes import *
import matplotlib.pyplot as plt

def get_data_from_MC(NoElements, com_port, baud_rate):
    port = serial.Serial(com_port, baud_rate, timeout=100)
    data1 = [0 for _i in range(NoElements)]
    data1f = [0 for _i in range(NoElements)]
    for x in range(NoElements):
        data = port.read(4)
        data1[x] = data[0] | (data[1] << 8) | (data[2] << 16) | (
            data[3] << 24)
    port.close()
    for x in range(NoElements):
        dummy = hex(data1[x])
        dummy2 = int(dummy, 16)
        cp = pointer(c_int(dummy2))
        fp = cast(cp, POINTER(c_float))
```

```
        data1f[x] = fp.contents.value
    return data1f
def main():
    y = get_data_from_MC(200, 'com9', 2000000)
    print(y)
    t = [0.0 for _i in range(len(y))]
    for n in range(len(t)):
        t[n] = n * 0.001
    plt.plot(t, y)
    plt.show()

main()
```

We provide the C++ and MicroPython codes for the open-loop speed control of
the DC motor in the accompanying book web site. The exact name for these files
are "Listing13_6.cpp" and "Listing13_7.py." The reader can benefit from them to
realize the open-loop controller in C++ and MicroPython languages, respectively.

We provide the C code for the closed-loop PID speed control of the DC motor in
Listing 13.6. As can be seen in the code, the PID control action is performed inside
the TIM10 interrupt callback function. Peripheral configurations for the STM32F4
board are the same as in the open-loop speed control example. Also, the same
Python code used to obtain the speed data at the PC side can be used for this
example. Finally, we should add the "PID.c" and "PID.h" files to "project->Core-
>Src" and "project->Core->Inc" folders of the active STM32CubeIDE project,
respectively. Hence, we can use PID library functions in digital control.

Listing 13.6 The C code for the closed-loop PID speed control of the DC motor

```
/* USER CODE BEGIN Includes */
#include "PID.h"
/* USER CODE END Includes */

/* USER CODE BEGIN PV */
uint32_t ICValue1 = 0;
uint32_t ICValue2 = 0;
uint32_t ICDiff = 0;
uint8_t CaptureIndex = 0;
float speed = 0;

PID_struct myPID;
float xSystem[3] = {0};
float ySystem[3] = {0};
int k;
/* USER CODE END PV */

/* USER CODE BEGIN 0 */
void Voltage_to_PWM(float voltage){
uint32_t duty;
if(voltage > 0){
if(voltage > 12)voltage = 12;
HAL_GPIO_WritePin(GPIOE, GPIO_PIN_3, GPIO_PIN_RESET);
HAL_GPIO_WritePin(GPIOE, GPIO_PIN_4, GPIO_PIN_SET);
duty = (uint32_t)(voltage * 1000 / 12);
}
else if(voltage < 0){
if(voltage < -12)voltage = -12;
```

```c
HAL_GPIO_WritePin(GPIOE, GPIO_PIN_3, GPIO_PIN_SET);
HAL_GPIO_WritePin(GPIOE, GPIO_PIN_4, GPIO_PIN_RESET);
duty = (uint32_t)(-voltage * 1000 / 12);
}
else if(voltage == 0){
HAL_GPIO_WritePin(GPIOE, GPIO_PIN_3, GPIO_PIN_RESET);
HAL_GPIO_WritePin(GPIOE, GPIO_PIN_4, GPIO_PIN_RESET);
duty = 0;
}

__HAL_TIM_SET_COMPARE (&htim8 , TIM_CHANNEL_3, duty);
}

void Set_Speed_Open_Loop(float speed){
float voltage;
if(speed > 14)speed = 14;
else if(speed < -14)speed = -14;
voltage = 12 * speed / 14;
Voltage_to_PWM(voltage);
}

void HAL_TIM_PeriodElapsedCallback(TIM_HandleTypeDef *htim)
{
xSystem[0] = PID_Output(&myPID, 6.2, ySystem[1]);
Voltage_to_PWM(xSystem[0]);
ySystem[0] = speed;
for (k = 2; k > 0; k--)
{
 ySystem[k] = ySystem[k - 1];
 xSystem[k] = xSystem[k - 1];
}
HAL_UART_Transmit(&huart1, (uint8_t*)&speed, 4, 1000);
}

void HAL_TIM_IC_CaptureCallback(TIM_HandleTypeDef *htim){
if(CaptureIndex == 0)
{
ICValue1 = HAL_TIM_ReadCapturedValue(&htim2, TIM_CHANNEL_1);
CaptureIndex = 1;
}
else
{
ICValue2 = HAL_TIM_ReadCapturedValue(&htim2, TIM_CHANNEL_1);
if (ICValue2 > ICValue1)
{
ICDiff = (ICValue2 - ICValue1);
}
else
{
ICDiff = ((0xFFFFFFFF - ICValue1) + ICValue2) + 1;
}
ICValue1 = ICValue2;
speed = (90000000.0 / (float)ICDiff)* 60.0/ 4535.1975;
if(HAL_GPIO_ReadPin(GPIOE, GPIO_PIN_5) == GPIO_PIN_RESET){
speed = -speed;
}
}
}
/* USER CODE END 0 */

/* USER CODE BEGIN 2 */
HAL_TIM_PWM_Start(&htim8 , TIM_CHANNEL_3);
HAL_TIM_IC_Start_IT(&htim2, TIM_CHANNEL_1);
```

```
if (__HAL_TIM_GET_FLAG(&htim10, TIM_FLAG_UPDATE) != RESET)
__HAL_TIM_CLEAR_FLAG(&htim10, TIM_FLAG_UPDATE);
HAL_TIM_Base_Start_IT(&htim10);
HAL_GPIO_WritePin(GPIOE, GPIO_PIN_2, GPIO_PIN_SET);
HAL_GPIO_WritePin(GPIOE, GPIO_PIN_3, GPIO_PIN_RESET);
HAL_GPIO_WritePin(GPIOE, GPIO_PIN_4, GPIO_PIN_RESET);
PID_Init(&myPID, 0.7, 20, 0, 0.01);
/* USER CODE END 2 */
```

We also provide the C++ and MicroPython codes for closed-loop PID speed control of the DC motor in the accompanying book web site. The exact name for these files are "Listing13_9.cpp" and "Listing13_10.py." The reader can benefit from them to realize the closed-loop controller in C++ and MicroPython languages, respectively.

13.5 Summary of the Chapter

Digital control deals with forming a control action to achieve the required performance criteria for the system at hand. Arm® Cortex™-M microcontrollers can be used to realize the control action in software form for this purpose. Therefore, we considered digital control topics in this chapter. To do so, we started with transfer function based control. Next, we focused on implementing PID controllers in software form. Finally, we applied these techniques to control a DC motor. Hence, we provided an actual digital control application via Arm® Cortex™-M microcontrollers. We should mention that the coverage of digital control topics was at the introductory level in this chapter. Therefore, the reader should consult more advanced books on digital control to grasp this important topic better.

Problems

13.1. Give an example on the control action from real life.

13.2. What is the main difference between open- and closed-loop control?

13.3. Give ramp, parabolic, exponential, and sinusoidal input signals to the DC motor. Send the obtained output signals (speed of the DC motor) to PC, and plot them in Python. Form the project using:

(a) C language under STM32CubeIDE
(b) C++ language under Mbed
(c) MicroPython

13.4. What is the effect of each PID controller component (as P, I, and D) on the control action?

13.5. Implement the methods in Sect. 13.4.3 for position control of the DC motor. Form the project using:

(a) C language under STM32CubeIDE
(b) C++ language under Mbed
(c) MicroPython

Reference

1. Ünsalan, C., Barkana, D.E., Gürhan, H.D.: Embedded Digital Control with Microcontrollers: Implementation with C and Python. Wiley - IEEE, Piscataway (2021)

Introduction to Digital Image Processing

14

14.1 About Digital Images

This section summarizes what we should understand from pixel, image representation as a matrix, and grayscale and color images. These definitions will be the building blocks for the following sections. Besides, we will consider how to represent a digital image in an embedded system (such as the STM32F4 microcontroller).

14.1.1 Mathematical Representation of the Digital Image

A digital image is represented in structured form by its smallest element called pixel. In early times, the image was generally called picture. Hence, researchers named the smallest element of the picture as pixel as the shorthand form of "picture element." This naming convention stayed the same since then.

Pixels form the digital image which is mathematically represented as a matrix. Hence, each pixel in the image can be addressed by its two coordinates as horizontal and vertical. As for mathematical form, we will have the image as $I(x, y)$. Here, x and y represent the horizontal and vertical matrix coordinates, respectively.

The pixel value represents light intensity at that image location. This is for grayscale images. For color images, we will have three values as red, green, and blue. Therefore, let's consider color and grayscale images next.

Supplementary Information The online version contains supplementary material available at (https://doi.org/10.1007/978-3-030-88439-0_14).

C. Ünsalan et al., *Embedded System Design with Arm Cortex-M Microcontrollers*, https://doi.org/10.1007/978-3-030-88439-0_14

Fig. 14.1 Grayscale
Mandrill image

14.1.2 Grayscale and Color Images

A grayscale image contains intensity values. Hence, we will have $I(x, y)$ as introduced in the previous section. Entries of this matrix will be in eight-bit unsigned character form most of the times. If the matrix entry has value zero, this means that the intensity there is lowest (black). The maximum value for the eight-bit unsigned integer form is 255. Hence, if the pixel has value 255, this means that the intensity there is highest (white). In a grayscale image, pixel values between 0 and 255 represent the intermediate light intensity (between black and white). We next provide a sample grayscale image, called Mandrill, in Fig. 14.1. This image is formed by 320×240 pixels.

The color image represents not only the intensity value at an image location but the color information there as well. In the following sections, we will see that a digital camera provides different color-formatted output. We can assume that the red, green, and blue color representation is suitable at this point. Hence, each pixel has three color components as red, green, and blue. We may have each color component represented by eight-bit unsigned character form. Hence, the color component value between 0 and 255 will correspond to the lowest and highest values there as in grayscale images. We next provide color version of the Mandrill image in Fig. 14.2. As a side note, this image is in grayscale form in the printed manuscript.

14.1.3 Representing the Digital Image in the STM32F4 Microcontroller

We pick the STM32F4 microcontroller as the embedded system in this chapter. Therefore, we should represent the image such that it can be processed in the microcontroller. We form image representations for this purpose in C, C++, and

Fig. 14.2 Color Mandrill image

Table 14.1 Image size standards for the camera and LCD modules

Format	Name	Size (width × height)
CIF	Common intermediate format	352 × 288
QCIF	Quarter CIF	176 × 144
VGA	Video graphics array	640 × 480
QVGA	Quarter VGA	320 × 240
QQVGA	Quarter QVGA	160 × 120
SVGA	Super VGA	800 × 600
XGA	Extended graphics array	1024 × 768

MicroPython languages. While doing so, we benefit from standard definitions in the camera and LCD modules. Therefore, let's start with the image size.

There are image size standards for both camera and display modules. We tabulate them in Table 14.1. In this table, we only provide common intermediate format (CIF), video graphics array (VGA), and extended graphics array (XGA). CIF has been used in PAL- and NTSC-based TV and video systems. VGA and its derivatives have been used in display modules. Cameras also adopted these standards. Recent display modules have different standards with width and height ratio as 16:9 or 16:10. However, we will not consider them in this book. Since the camera module and LCD on the STM32F4 board support VGA and its derivatives, we will only focus on them in this chapter.

Throughout the book, we will use QVGA and QQVGA image formats in Table 14.1. Therefore, we formed constant declarations for them as IMAGE_RES_QVGA and IMAGE_RES_QQVGA. We will use them in both image- and camera-based operations.

Most camera and display modules have standard grayscale (monochrome) and color image formats. We tabulate them in Table 14.2. We will introduce how a pixel is represented in bitwise form for these formats in Sect. 14.5.4. As a side note, we introduced the ARGB888 format for LCD operations in Sect. 11.3.

Table 14.2 Grayscale and color image formats for the camera and LCD modules

Representation	Bytes per pixel (bpp)
Grayscale	1
RGB444	2
RGB555	2
RGB565	2
RGB888	3
YUV422	2
ARGB888	4

Throughout the book, we will use RGB565, YUV422, and grayscale formats in Table 14.2. Therefore, we formed constant declarations for them as IMAGE_FORMAT_RGB565, IMAGE_FORMAT_YUV422, and IMAGE_FORMAT_GRAYSCALE. We will use them in both image- and camera-based operations.

We have two definitions for representing the digital image in embedded systems. These are frame and line. The frame represents the overall image. The line represents one row of the image. We explained these in connection with the LCD usage in Chap. 11. The same definitions also apply to this chapter.

We can store a digital image either in matrix or array form. Matrix representation is the same as the mathematical form of the image in Sect. 14.1. If we prefer to represent the image in array form, we should perform array operations to reach a specific row and column numbered pixel value. As a side note, both matrix and array representations are kept in the same way in memory. Only how we reach and process them differs in the code.

14.1.3.1 Image Representation in C Language

We can represent an image in structured form in C language. This structure holds all the necessary information about the image. We should also allocate RAM to store the image. Therefore, we provide the image structure and RAM allocation function to be used with it next.

```
typedef struct _image{
uint8_t *pData;     // pointer to the image data array
uint16_t width;     // image width in pixels
uint16_t height;    // image height in pixels
uint32_t size;      // image size in bytes, equal to width*height
    *bpp
uint8_t format;     // image format, can be one of
    IMAGE_FORMAT_RGB565, IMAGE_FORMAT_YUV422 or
    IMAGE_FORMAT_GRAYSCALE
} ImageTypeDef;

void IMAGE_init(Image *Pimg, uint8_t res, uint8_t format);
/*
Pimg: pointer to the ImageTypeDef struct
res: specifies the image resolution to be initialized and can be
    one of  IMAGE_RES_QVGA or IMAGE_RES_QQVGA
format: specifies the image format to be initialized and can be
    one of IMAGE_FORMAT_RGB565, IMAGE_FORMAT_YUV422 or
    IMAGE_FORMAT_GRAYSCALE
*/
```

The constant IMAGE_FORMAT_RGB565 indicates that we are using a color image represented by red, green, and blue components. Moreover, the number of bits assigned to these color bands are five, six, and five, respectively. The same format also applies to other mentioned color constants.

We provide a usage example for the structure and its initialization function for a QVGA color image with RGB565 format next. Here, we first define an image via the struct ImageTypeDef. Then, we set the image properties via the function IMAGE_init.

```
/* USER CODE BEGIN PV */
ImageTypeDef img;
/* USER CODE END PV */

/* USER CODE BEGIN 2 */
IMAGE_init(&img, IMAGE_RES_QVGA, IMAGE_FORMAT_RGB565);
/* USER CODE END 2 */
```

We formed an image processing library in C language. This library has the struct ImageTypeDef and its initialization function. This library is available in the accompanying book web site. The reader can reach this library and use the mentioned structure definition and function in his or her code. To do so, the "image.c" file in the library should be copied to the "project->Core->Src" folder of the active STM32CubeIDE project. Likewise, the "image.h" file in the library should be copied to the "project->Core->Inc" folder of the project.

14.1.3.2 Image Representation in C++ Language

We formed the IMAGE class in C++ language to represent an image. The IMAGE class has all the properties defined in the C struct ImageTypeDef introduced in the previous section. The class has one function called init which allocates RAM space for the constructed image. We provide details of the IMAGE class next.

```
IMAGE(uint8_t res = IMAGE_RES_QVGA, uint8_t format =
      IMAGE_FORMAT_RGB565);
/*
res: specifies the image resolution to be initialized and can be
     one of  IMAGE_RES_QVGA or IMAGE_RES_QQVGA
format: specifies the image format to be initialized and can be
     one of IMAGE_FORMAT_RGB565, IMAGE_FORMAT_YUV422 or
     IMAGE_FORMAT_GRAYSCALE
*/

void init(uint8_t res, uint8_t format);
```

We provide a usage example for the IMAGE class and its initialization function for a QVGA color image with RGB565 format in two different methods next. In the first method, we declare the image by the constructor. Then, we set its parameters by the init function of the class. In the second method, we set image parameters within the constructor and form the image class.

```
//first method
IMAGE img1;
img1.init(IMAGE_RES_QVGA, IMAGE_FORMAT_RGB565);
```

```
//second method
IMAGE img2(IMAGE_RES_QVGA, IMAGE_FORMAT_RGB565);
```

As in C language, we formed an image processing library in C++ language. This library consists of the IMAGE class. This library is available in the accompanying book web site. The reader can reach this library and use the mentioned image class in his or her C++ code. In order to use the class, the files "image.cpp" and "image.hpp" in the library should be copied to the active project folder.

14.1.3.3 Image Representation in MicroPython

We formed the image module in MicroPython to represent, modify, and use an image. In this module, there is a custom struct image_obj_t which holds all the necessary information about the image. The module also has a function called init which allocates RAM space for the constructed image and returns image_obj_t as the MicroPython object. The module also has a function called image which creates the image structure with given image parameters and returns image_obj_t as the MicroPython object. There are also width, height, size, format, resolution, and array functions to reach image structure parameters individually. We provide details of the image module next.

```
typedef struct _image_obj_t {
mp_obj_base_t base;
uint8_t *pData;    // pointer to the image data array
uint16_t width;   // image width in pixels
uint16_t height;  // image height in pixels
uint32_t size;        // image size in bytes, equal to width*height
     *bpp, bpp stands for (bytes per pixel)
uint8_t format;   // image format, can be one of
     IMAGE_FORMAT_RGB565, IMAGE_FORMAT_YUV422 or
     IMAGE_FORMAT_GRAYSCALE
uint8_t resolution; // image resolution, can be one of
     IMAGE_RES_CUSTOM, IMAGE_RES_VGA , IMAGE_RES_QVGA or
     IMAGE_RES_QQVGA
} image_obj_t;

image(pData, width, height, resolution, format)
/*
pData: bytearray which holds the image data as bytes
width: image width in pixels
height: image height in pixels
resolution: specifies the image resolution to be initialized and
     can be one of IMAGE_RES_CUSTOM, IMAGE_RES_VGA,
     IMAGE_RES_QVGA or IMAGE_RES_QQVGA. width and height
     parameters are ignored if resolution is not selected as
     IMAGE_RES_CUSTOM.
format: specifies the image format to be initialized and can be
     one of IMAGE_FORMAT_RGB565, IMAGE_FORMAT_YUV422 or
     IMAGE_FORMAT_GRAYSCALE
*/

init(width, height, resolution, format)

width(img)
 /*
img: created image object based on image_obj_t struct
*/
```

```
height(img)

size(img)

resolution(img)

format(img)
```

We provide a usage example for the image module in two different methods next. In the first method, we create an image object with 2×2 byte array data with GRAYSCALE format. In the second method, we create an empty image object for a QVGA color image with RGB565 format. We also print the width of the first image object and height of the second image object in the code.

```
import image
//first method
imgBytes = bytearray([1, 2, 3, 4])
img1 = image.image(imgBytes, 2, 2, IMAGE_RES_CUSTOM,
    IMAGE_FORMAT_GRAYSCALE)

//second method
img2 = image.init(0, 0, IMAGE_RES_QVGA, IMAGE_FORMAT_RGB565);

print(image.width(img1))
print(image.height(img2))
```

As in C language, we formed an image processing library to be used in MicroPython. The reader can benefit from the custom MicroPython firmware provided in the book web site to use this library. To do so, the code line import image should also be added to the "main.py" file.

14.2 Image Transfer Between the PC and STM32F4 Microcontroller

The first step in digital image processing in the STM32F4 microcontroller is feeding the image to the microcontroller. We have two options here. The first one is transferring the image from PC. We will cover this option in this section. We will also cover the reverse operation such that the processed image in the STM32F4 microcontroller can be sent back to PC for storage or other purposes. The second option for feeding the image to the microcontroller is by acquiring it with a camera module. We will cover this procedure in Sect. 14.5.

14.2.1 Setup in the PC Side

Before feeding the image to the STM32F4 microcontroller, we should read it on PC. The best option for this operation is using the Python OpenCV library. This library can be installed to PC by the command pip install opencv-python. As the library is installed, it allows us to read an image and display it on the PC screen.

As a side note, OpenCV is also available in C++, Java, and JavaScript languages. The reader can also benefit from them for his or her own applications.

We next provide the Python code in Listing 14.1 to read an image and display it on PC screen. In this code, we import the OpenCV library first. Then, we read the color image, display it on the screen, wait for a key press, and close the window. Please make sure that the image to be read is in the same folder as with the executed Python code. Or, the exact address should be given for the image file to be read by OpenCV.

Listing 14.1 Reading and displaying the Mandrill image via OpenCV in Python

```
import cv2

#read the color image
img = cv2.imread('mandrill.tif')

#display the color image
cv2.imshow('Color Image',img)
cv2.waitKey(0)

cv2.destroyAllWindows()

#write the image to a file
#cv2.imwrite('mandrill_store.tif',img)
```

We can save an image in OpenCV. To do so, we should enable the command cv2.imwrite in Listing 14.1. Hence, the reader can save the image to the working directory. This will be helpful when we receive an image from the STM32F4 microcontroller in the following sections.

As we read the image in the PC side, the next operation is transferring it to the STM32F4 microcontroller. Likewise, we may want to transfer an image from the microcontroller to PC. We will benefit from UART communication for both operations. Therefore, we should first install the pyserial library to PC by executing the command pip install pyserial in the command window. Next, we should obtain the COM port the STM32F4 board is connected to. To do so, we should check "Device Manager -> Ports (COM & LPT)" in the Windows operating system. There should be a device listed as "STMicroelectronics STLink Virtual COM Port ...." We will use the COM port defined there in our Python code. Let's assume that the mentioned COM port is "COM5." Please note that the same settings should be done for a PC with another operating system such as Linux. Therefore, we kindly ask the reader to check his or her operating system working principles for this purpose.

In order to simplify data transfer between the PC and STM32F4 microcontroller, we formed the "PC_Image_Transfer_Library." This library contains four functions as follows. The init() function gets the COM port name as input. It opens the port and initializes the image to be transferred. The pollTransferRequest() function starts data transfer. The function sendImage() sends the image from PC to the STM32F4 microcontroller. The function readImage() sends the image from the STM32F4 microcontroller to PC. The user should press the

ESC key while the image window is active to stop the operation. We provide the "PC_Image_Transfer_Library" as a Python file in the accompanying book web site.

We next provide the sample code in Listing 14.2 to transfer the Mandrill image from PC to the STM32F4 microcontroller. This code first installs necessary libraries including "PC_Image_Transfer_Library." As we run the code on PC, it waits for a request from the STM32F4 microcontroller. As the request comes, the code starts transferring data. As a side note, the code to be executed in the STM32F4 microcontroller can be written in C, C++, or MicroPython language since we are using UART communication in all these.

Listing 14.2 Image transfer code between the PC and STM32F4 microcontroller

```python
import PC_Image_Transfer_Library as serialImage

serialPort = 'COM5'
Img = "mandrill.tiff"

serialImage.init(serialPort)

while 1:

    # press ESC to terminate
    reqType = serialImage.pollTransferRequest()

    if (reqType == serialImage.IMAGE_READ):
        serialImage.sendImage(Img)

    elif (reqType == serialImage.IMAGE_WRITE):
        serialImage.readImage()
```

The code in Listing 14.2 works as follows. As the code starts running, it continuously listens to the serial port the STM32F4 board is connected to. If a request comes from the microcontroller, the code checks the first eight bytes consisting of start, request type (image read or write), image size, and format bytes. If the microcontroller has asked for receiving an image, then the predefined image is read in the PC. It is resized and modified for the required image format (such as RGB565 or YUV422). Then, the PC sends the image to the STM32F4 microcontroller. If the microcontroller has asked for sending image data, the Python code in PC receives the image, opens a folder called "capture," and saves the received image to this folder in "jpeg" format. The name of the saved image will contain the actual date and time.

14.2.2 Setup and Display in the STM32F4 Microcontroller Side

As the PC side becomes ready for data transfer, we should next set up the STM32F4 microcontroller. In this section, we provide ways to do this in C, C++, and MicroPython languages. Besides, we also display the received image on the LCD of the STM32F4 board. As a side note, steps for the display operation have been introduced in Chap. 11.

Before explaining the setup functions in C, C++, and MicroPython languages, we should mention some common properties. We will be using the USART1 module for image transfer. We explained how to do this in Chap. 8. In this chapter, we set the pins PA9 and PA10 of the STM32F4 microcontroller for UART communication. We use the USART1 module in UART mode with full-duplex, eight-bit, no parity, one stop bit mode. The baud rate is set to 2 MBits/s. Pins PA9 and PA10 should have "maximum output speed" settings as "very high" under GPIO properties. If all these settings are done correctly, we can transfer a QVGA image in 1.5 s.

14.2.2.1 Setup and Display via C Language

We formed the image transfer library to handle image transfer between the PC and STM32F4 microcontroller in C language. To use this library, we should add the "serialImage.c" and "serialImage.h" files to the folders "project->Core->Src" and "project->Core->Inc" for the active STM32CubeIDE project, respectively. We should also add the image processing library to our project. Then, we should set UART communication parameters on the PC side as explained in the previous section.

Our image transfer library has two functions as SERIAL_imageCapture and SERIAL_imageSend. Both use the struct ImageTypeDef introduced in Sect. 14.1.3. We provide detailed information on them next.

```
void SERIAL_imageCapture(Image *Pimg)
/*
Pimg: pointer to the ImageTypeDef struct
*/

void SERIAL_imageSend(Image *Pimg)
```

The function SERIAL_imageCapture stores the image (sent from PC) to the STM32F4 microcontroller. The function first sends a request to PC to receive image with size and format information. Then, it waits for the response. As the response is received from PC, data transfer begins. The function SERIAL_imageSend sends the image in the STM32F4 microcontroller to PC. To do so, it first sends a request to PC to send data with size and format information. As the code receives the response, the image is sent to PC.

We can also display the received image on the LCD of the STM32F4 board. To do so, we should make necessary adjustments on LCD usage with SPI as explained in Sects. 11.4 and 11.5.1. We provide the C code in Listing 14.3 to show how to perform this operation.

Listing 14.3 Displaying the received image on the LCD of the STM32F4 board, the C code

```
/* USER CODE BEGIN Includes */
#include "image.h"
#include "ili9341.h"
#include "serialImage.h"
/* USER CODE END Includes */

/* USER CODE BEGIN PV */
ImageTypeDef img;
/* USER CODE END PV */
/* USER CODE BEGIN 2 */
```

```
IMAGE_init(&img, IMAGE_RES_QVGA, IMAGE_FORMAT_RGB565);
ILI9341_init(SCREEN_HORIZONTAL);

SERIAL_imageCapture(&img);
ILI9341_drawImage(&img, 0, 0);
/* USER CODE END 2 */
```

In Listing 14.3, an image structure is formed first by ImgeTypeDef. Then, it is initialized. Afterward, the LCD (to be used by SPI) is initialized. Then, the image is received from UART, saved in the structure, and displayed on the LCD. To note here, the Python code in Listing 14.2 should have been running on the PC side at the same time.

14.2.2.2 Setup and Display via C++ Language

We also formed the image transfer library to handle image transfer between the PC and STM32F4 microcontroller in C++ language. To use this library, we should add the files "serialImage.cpp" and "serialImage.hpp" to the active STM32CubeIDE project folder. We should also add the image processing library (for C++ language) to our project. As a side note, all these library files are provided in the accompanying book web site. We formed a class SERIALIMAGE within the image processing library. This class has two functions as capture and send. These work in the same way as their C counterparts introduced in the previous section. We provide detailed information on the class and its functions next.

```
SERIALIMAGE()

void capture(IMAGE Pimg);
/*
Pimg: pointer to the IMAGE object
*/

void send(IMAGE Pimg);
```

The function capture stores the image, initialized by the class object SERIALIMAGE, coming from PC. The function first sends a request to PC to receive image with size and format information. Then, it waits for the response. As the response is received from PC, image transfer begins. The function send sends the image in the STM32F4 microcontroller to PC. To do so, it first sends a request to PC to send an image with size and format information. As the microcontroller receives the response, the image is sent to PC.

We can also display the received image on the LCD of the STM32F4 board. To do so, we should make necessary adjustments for LCD usage with SPI as explained in Sects. 11.4 and 11.5.1. We provide the C++ code in Listing 14.4 to show how to perform this operation.

Listing 14.4 Displaying the received image on the LCD of the STM32F4 board, the C++ code

```
#include "mbed.h"

#include "image.hpp"
#include "ili9341.hpp"
```

```
#include "serialImage.hpp"

IMAGE img(IMAGE_RES_QVGA, IMAGE_FORMAT_RGB565);
ILI9341 ili9341(SCREEN_HORIZONTAL);
SERIALIMAGE serialImage;

int main()
{
serialImage.capture(&img);
ili9341.drawImage(&img, 0, 0);

while (true);
}
```

The C++ code in Listing 14.4 works in the same way as its C counterpart given in Listing 14.3. As the code runs, it asks for receiving a QVGA image in RGB565 format from PC. There is one important issue here. Mbed Studio opens the serial terminal while debugging the code. If the Python code in Listing 14.2 has not been executed on PC before the debug operation, Python gives error indicating that the port is in use. To avoid this error, either the user should run the image transfer code in Listing 14.2 beforehand or, after debugging the code in Mbed Studio, the terminal window there should be closed.

14.2.2.3 Setup and Display via MicroPython

We formed the image transfer module to handle image transfer between the PC and STM32F4 microcontroller in MicroPython. To use this module, the reader should use the custom MicroPython firmware provided in the book web site. The code line `import serialImage` should also be added to the "main.py" file. Then, the reader should initialize the `serialImage` module using `init()` first. UART5 module is enabled with full-duplex, eight-bit, no parity, one stop bit mode with this function. The baud rate is set to 921,600 bits/s. Pins PD2 and PC12 are set as Rx and Tx, respectively. This module also has two functions as `capture` and `send`. These work in the same way as their C counterparts introduced in the previous section. We provide detailed information on the class and its functions next.

```
init();

capture(img);
/*
img: created image object based on image_obj_t struct
*/

send(img);
```

The function `capture` stores the image coming from PC. To do so, it first sends a request to PC to receive the image with size and format information. Then, it waits for the response. As the response is received from PC, image transfer begins. The function `send` sends the image in the STM32F4 microcontroller to PC. To do so, it first sends a request to PC to send the image with size and format information. As the microcontroller receives the response, the image is sent to PC.

We can also display the received image on the LCD of the STM32F4 board. To do so, we should make necessary adjustments for LCD usage with SPI as explained in Sects. 11.4 and 11.5.1. We provide the MicroPython code in Listing 14.5 to show how to perform this operation. This code works in the same way as its C counterpart given in Listing 14.3.

Listing 14.5 Displaying the received image on the LCD of the STM32F4 board, the MicroPython code

```
import image
import serialImage
import lcdSPI

def main():
    img = image.init(0, 0, IMAGE_RES_QVGA, IMAGE_FORMAT_RGB565)
    lcdSPI.init(0)
    serialImage.init()
    serialImage.capture(img)
    lcdSPI.drawImage(img, 0, 0)

main()
```

14.3 Digital Camera as the Image Sensor

Digital camera can be used to acquire images in an embedded system. Therefore, we will introduce working principles of a generic digital camera in this section. Afterward, we will focus on the OV7670 camera module to be used throughout the book and explore its properties. We will also explain how to set up the camera module in this section. These will lead to image acquisition from the OV7670 camera module in Sect. 14.5.

14.3.1 Working Principles of a Digital Camera

We provide a generic digital camera module in Fig. 14.3. As can be seen in this figure, there are five main modules in the camera. These are optics and image sensor, preprocessing, camera interface, timing and synchronization, and control interface and registers. Besides, there are input and output pins for the digital camera as follows. The clock signal is fed to the camera through its XCLK pin. SDA and SCK pins are used for controlling the camera. D[n:0] represents n+1 bit parallel data output pins. PIXCLK is the pixel output clock providing sampling times for data output. Finally, VSYNC and HSYNC are synchronization output signal pins. We provide detailed explanation of digital camera modules as well as related pins next.

14.3.1.1 Optics and Image Sensor

Optics module of the digital camera is composed of the main lens and microlenses. The main lens is located outside the camera module. It focuses light rays coming

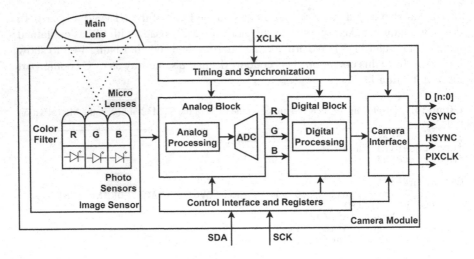

Fig. 14.3 A generic digital camera module

from the outside world to the image sensor. Microlenses distribute the focused light to color filters in the image sensor.

The image sensor is composed of transistors (most of the times being type CMOS) which act as light-sensitive photosensors. Each sensor generates voltage or current proportional to the light falling on it. Hence, an analog signal is generated. We will process this signal and form a digital image by the help of camera modules.

We can associate image sensor elements with pixels in the constructed image. If we are acquiring a grayscale image, we can say that each processed sensor element output corresponds to a pixel in the acquired image. If the acquired image is in color form, then extra operations may be needed to construct a pixel value from neighboring sensor elements.

The image sensor has color filters on it. There is a specific structure in the placement of color filters such as the Bayer pattern. This pattern is implemented for arranging red, green, and blue color filters on the image sensor.

14.3.1.2 Preprocessing

Preprocessing module of the digital camera is composed of analog and digital blocks. The analog block processes the signal coming from image sensors and converts them to digital form. The digital block processes the image formed by digitized analog signals.

We can summarize operations done in the analog block as follows. The first operation is gain control. It adjusts each image sensor element's analog signal response to different light levels. The second operation is black-level compensation. This operation is done to eliminate electronic signal noise in image sensor elements. Hence, we expect to have black regions to be represented as perfect black in the image. The ADC operation converts the analog signal to digital form. Generally, color channels are converted to digital form separately. The ADC module used in the operation generally has eight-bit resolution.

We can summarize the operations done in the digital block as follows. White balance is done to obtain perfect white color in ambient light. This is done by adjusting the gain in color channels. Gamma correction is applied to modify the color intensity obtained from different channels. Color control aims to correct errors caused by non-ideal microlens characteristics. Brightness and contrast control adjusts these values based on the ambient light level. Sharpness adjusts image edges being sharp or smooth. Denoising eliminates noise originating from image sensor elements. Defect pixel correction aims to eliminate constant pixel values (such as always set to white or black) by the help of neighboring pixel values. Lens correction aims to correct nonuniform light distribution in image sensors originating from the lens. Image scale and zoom operations can also be done in the digital block.

14.3.1.3 Camera Interface

Camera interface module feeds the digital image data and synchronization signals from the digital camera. This output can be in serial or parallel form. There are standards defined by MIPI used for this purpose [1]. Most cameras use these standards. For more information on them, please see the mentioned reference. In nonstandard camera outputs, generally UART or SPI communication is used. There is also the low-voltage differential signaling standard (LVDS). For more information on it, please see [1].

14.3.1.4 Timing and Synchronization

The timing and synchronization module in the digital camera is mainly responsible for generating and distributing internal timing and external synchronization signals VSYNC, HSYNC, and PIXCLK. VSYNC indicates that a frame is being transmitted. HSYNC indicates that a line is being transmitted. PIXCLK indicates that a new pixel byte is available. Polarity of these signals can be adjusted via camera registers. Timing diagram of these signals is illustrated in Fig. 14.4a. Output data pin timing diagram can be observed with respect to the PIXCLK signal in Fig. 14.4b. The DCMI module, to be introduced in Sect. 14.4, should be configured according to these signals.

Synchronization can be internal or external. In the external case, we can use VSYNC and HSYNC pins. In the internal case, we send bytecode data at the beginning and end of each frame and line. There are several modes for this operation. The DCMI module in the STM32F4 microcontroller supports two such modes. The first one is called ITU656 which uses four bytecodes. The second one uses one bytecode. Although it is possible to use these modes to acquire image from the camera, we did not pursue this path in this book.

14.3.1.5 Control Interface and Registers

The control interface and registers control all mentioned digital camera modules. Registers are set by serial input from an external device such as microcontroller. The interface is generally based on either I^2C or its derivatives. There are approximately

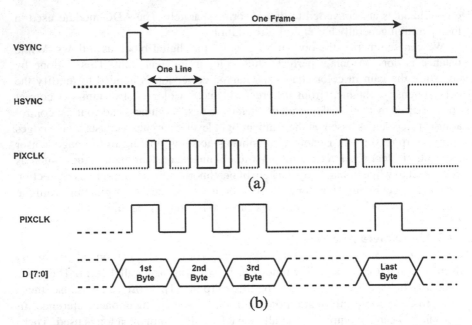

Fig. 14.4 Synchronization signal and data pin timing diagram. (**a**) Synchronization signals. (**b**) Data pin timing with respect to PIXCLK

100 registers in the OV7670 camera. Since these registers depend on the camera at hand, we strongly suggest the reader to check the selected camera datasheet.

14.3.2 Image Data Representation in Digital Cameras

There are different image data representations used in digital cameras. We can group them into three categories as raw, processed, and compressed. Let's explain each category in detail.

14.3.2.1 Raw Form

The first image data representation is called raw. Here, the image sensor output is processed by the analog block in the preprocessing module. Then, the obtained analog signal is converted to digital form by the ADC module. We can either have three separate sensors, with their specific color filters, for three colors as red, green, and blue. This is raw image data. Or, there is a pattern for color image filters on the image sensor. If the color filters on the image sensor are arranged in Bayer pattern, then we call the output as raw Bayer.

We should also explain why the Bayer pattern is used. As mentioned before, each pixel is obtained from three separate color filters in raw form. This means we will need three separate microlenses and color filters for each pixel. If the Bayer pattern is used, each pixel is represented by one sensor, one microlens, and one color filter. Hence, the number of elements used for obtaining the color image with the same image resolution falls to one third in this setup. However, we should apply demosaicing (debayering) to obtain the actual RGB image. To do so, the interpolation operation should be applied to each image pixel and its neighbors. As a result, an approximate RGB representation will be obtained.

14.3.2.2 Processed Form

The second image data representation is called processed. Here, the raw data is fed to the digital block in preprocessing module. The obtained output is called processed. If the raw data is formed by three sensors per pixel, then we can obtain the processed form directly. If the Bayer pattern is used, then we obtain the processed Bayer form.

14.3.2.3 Compressed Form

The third image data representation is called compressed. Some cameras may provide compressed image in jpeg form. This speeds up data transfer between the camera and connected device such as microcontroller. To benefit from the compressed data, the receiving device should have a jpeg decoder. The STM32F4 microcontroller does not have such a decoder in hardware. Instead, this is done by software, more specifically LIBJPEG middleware under STM32CubeMX. The OV7670 camera module does not provide output in jpeg form. Therefore, we did not consider the compressed image data form in this book.

14.3.3 The OV7670 Camera Module

We will use the OV7670 camera module throughout the book. This module is low cost (can be found under $5), is easy to purchase, and can be used with microcontrollers. Front view of the OV7670 camera module is as in Fig. 14.5. In fact, there are two versions of the OV7670 camera module as without and with FIFO buffer. We will use the former version throughout the book.

The OV7670 camera module has CMOS image sensor block produced by OmniVision. This sensor can provide VGA images at 30 frames per second (fps). The camera module has analog and digital blocks. The analog block has exposure control, gain control, white balance, band filter, and black-level calibration properties. The digital block consists of color saturation, hue, gamma, sharpness (edge enhancement), anti-blooming, noise reduction, and defect correction functions. Hence, the camera module can preprocess and send the acquired image to the STM32F4 microcontroller through its output pins. The camera module can be configured through its serial camera control bus (SCCB) interface. SCCB has characteristics close to I^2C. The reader will not deal with this interface since we will handle its usage within our functions.

Pinout of the OV7670 camera module is given in Table 14.3. As can be seen in this table, there are three synchronization signals for the OV7670 camera module

Fig. 14.5 The OV7670
camera module

Table 14.3 Pinout of the
OV7670 camera module

Pin	Description
SIOC	Serial interface clock
SIOD	Serial interface data I/O
VSYNC	Vertical sync output
HREF	Horizontal reference
PCLK	Pixel clock output
D0–D7	Digital data output
PWDN	Power down input
XCLK	System clock input
RESET	Reset input
3V3	Power supply
GND	Ground

as VSYNC, HREF, and PCLK. VSYNC indicates that a frame is being transmitted. HREF indicates that a line is being transmitted. PCLK indicates that a new byte is available. Polarity of these signals can be adjusted via camera registers. Digital data output from the OV7670 camera is obtained from eight pins labeled D0–D7.

The OV7670 image sensor supports several output formats. These can be listed as raw RGB, YUV/YCbCr422, GRB422, RGB565, and RGB555. We will benefit from the RGB565 and YUV/YCbCr422 in the following sections.

14.3.4 Setting Up the OV7670 Camera Module

We should set up the OV7670 camera before acquiring an image from it. Although we will perform these operations by functions in our camera library, we should briefly explain how they work. Therefore, we handled this issue next.

14.3.4.1 Setup via C Language

We formed the camera library to handle OV7670 camera setup functions in C language. To use them, we should add the "ov7670.c" file to "project->Core->Src" folder of the active project. We should also add the "ov7670.h" and "ov7670_regs.h" files to "project->Core->Inc" folder of the project. We should also add the image processing library to our project.

The camera setup functions within our library are as follows. The function OV7670_init initializes the camera to its default settings. It also sets the camera to work in the desired size and format. The function OV7670_init initializes the camera by setting its resolution and image format for image acquisition. The function OV7670_reset resets the camera. Whenever this function is called, the function OV7670_init should be called afterward. We can use the initialization function as OV7670_init(IMAGE_RES_QVGA, IMAGE_FORMAT_RGB565). We provide the detailed explanation of the camera calibration functions next.

```
void OV7670_init(uint8_t res, uint8_t format);
/*
res: specifies the image resolution to be captured and can be
     one of IMAGE_RES_QVGA or IMAGE_RES_QQVGA
format: specifies the image format to be initialized and can be
     one of IMAGE_FORMAT_RGB565 or IMAGE_FORMAT_YUV422
*/

void OV7670_reset();
```

We should also set project properties to use the OV7670 camera module. First and most important of all, the OV7670 camera module needs a clock signal to operate with clock frequency between 10 and 48 MHz. The optimal clock frequency is 24 MHz. We will feed the clock signal to the camera from the STM32F4 microcontroller. We will use MCO2 output of the microcontroller for this purpose. To do so, pin PC9 of the microcontroller should be connected to the XCLK pin of the OV7670 camera. The PC9 pin mode should be set as "RCC_MCO_2." Besides, the GPIO "maximum output speed" setting of the pin should be "very high." Please see Sect. 6.1.3 for these settings. As we perform these adjustments, it becomes possible to acquire images at 30 fps, which is the maximum speed the camera can provide.

We should also set some parameters in the "clock configuration" section of STM32CubeMX. To be more specific, we should set "HCLK" to "180MHz"; "PLLI2S" "N" to "192," "R" to "4"; "MCO2 Source Mux" to "PLLI2SCLK," and the divider at the output of "MCO2 Source Mux" to "4." As all these settings are done, the main system will have 180 MHz clock signal. The MCO2 output, to be connected to the OV7670 camera, will have 24 MHz clock signal.

As clock settings are done, the next step is setting camera registers via SCCB interface. To do so, we will use the I^2C module in blocking mode. We will also discard the ACK signal here. To do so, we should select the "I2C mode for Connectivity - I2C1" in the leftmost part of STM32CubeMX. The communication speed should be set as 100,000 Hz. The remaining parameters can be kept in their default state. We should set pins PB8 and PB9 in the STM32F4 microcontroller as "I2C1_SCL" and "I2C1_SDA" modes, respectively. We should connect the pins

Table 14.4 Connecting the STM32F4 board to the OV7670 camera module

STM32F4 board pin	STM32F4 microcontroller pin function	Camera module pin
VDD	Power supply	3V3
GND	Ground	GND
PB8	I2C1_SCL	SIOC
PB9	I2C1_SDA	SIOD
PG9	DCMI_VSYNC	VSYNC
PA4	DCMI_HSYNC	HREF
PA6	DCMI_PIXCLK	PCLK
PC9	MCO2	XCLK
PE6	DCMI_D7	D7
PE5	DCMI_D6	D6
PD3	DCMI_D5	D5
PE4	DCMI_D4	D4
PG11	DCMI_D3	D3
PG10	DCMI_D2	D2
PC7	DCMI_D1	D1
PC6	DCMI_D0	D0
NRST	Reset	RESET
GND	Ground	PWDN

PB8 and PB9 to the OV7670 camera module pins SIOC and SIOD, respectively. For more information on these settings, please see Sect. 8.4.3.

Finally, we should connect reset pin of the camera to VDD (or NRST) pin of the STM32F4 board. If we connect the reset pin of the camera to NRST pin of the board, then the camera also resets when the STM32F4 microcontroller is reset. We should connect the PWDN pin of the camera to GND pin of the STM32F4 board. We tabulate all the mentioned pin connections between the STM32F4 board and OV7670 camera module in Table 14.4. In this table, we also tabulated the STM32F4 microcontroller pin functions which will be helpful in the following sections.

14.3.4.2 Setup via C++ Language

We can also set up the OV7670 camera module via C++ language. To do so, we formed a camera library. To use it, we should add the "ov7670.cpp," "ov7670.hpp," and "ov7670_regs.h" files to the active project folder. Besides, we should add the image processing library in C++ language to the project folder.

Our camera library has necessary class definitions and associated functions to set up the OV7670 camera. To be more specific, we have a class OV7670. The class function init initializes the camera to its default settings. It also sets the camera to work in the desired image size and format. The function reset resets the camera. Whenever this function is used, the function init should be called afterward. We provide detailed explanation of the OV7670 class and its functions next.

```
OV7670(uint8_t res = IMAGE_RES_QVGA, uint8_t format =
    IMAGE_FORMAT_RGB565);
/*
res: specifies the image resolution to be captured and can be
    one of IMAGE_RES_QVGA or IMAGE_RES_QQVGA. Default value is
    IMAGE_RES_QVGA
format: specifies the image format to be initialized and can be
    one of IMAGE_FORMAT_RGB565 or IMAGE_FORMAT_YUV422.  Default
    value is IMAGE_FORMAT_RGB565
*/

void init(uint8_t res, uint8_t format);

void reset();
```

We provide a usage example of the OV7670 class and its functions next. Here, we form an object called camera from the class OV7670. Then, we reset and initialize the camera object by associated functions.

```
OV7670 camera;

camera.reset();
camera.init(IMAGE_RES_QVGA, IMAGE_FORMAT_RGB565);
```

We do not have to perform any peripheral or clock settings in C++ language. These are done automatically via our camera library functions. Therefore, the user should only connect the OV7670 camera module to the STM32F4 board as tabulated in Table 14.4.

14.3.4.3 Setup via MicroPython

We can also set up the OV7670 camera module via MicroPython. To do so, we formed a camera module named ov7670. To use it, the reader should use the custom MicroPython firmware provided in the book web site. The code line import ov7670 should also be added to the "main.py" file. This module has three functions as init, reset, and read_ID. The function init initializes the camera to its default settings. It also sets the camera to work in the desired image size and format. The function reset resets the camera. Whenever this function is used, the function init should be called afterward. The function read_ID is used to read the predefined ID of the OV7670 camera. We provide the detailed explanation of the OV7670 class and its functions next.

```
init(res, format);
/*
res: specifies the image resolution to be captured and can be
    one of IMAGE_RES_QVGA or IMAGE_RES_QQVGA.
format: specifies the image format to be initialized and can be
    one of IMAGE_FORMAT_RGB565 or IMAGE_FORMAT_YUV422.
*/

reset();

read_ID();
```

We provide a usage example of the ov7670 module and its functions next. Here, we first import the ov7670 module. Then, we reset and initialize the camera module by associated functions. Finally, we read and print the ID of the OV7670 camera module.

```
import ov7670

ov7670.reset()

ov7670.init(IMAGE_RES_QVGA, IMAGE_FORMAT_RGB565)

print(ov7670.read_ID)
```

We do not have to perform any peripheral or clock settings in MicroPython. These are done automatically by our `init` function. Therefore, the user should only connect the OV7670 camera module to the STM32F4 board as tabulated in Table 14.4.

14.4   Digital Camera Interface Module in the STM32F4 Microcontroller

The digital camera interface (DCMI) module in the STM32F4 microcontroller is a peripheral unit which connects the microcontroller to camera modules through its 8-, 10-, 12-, or 14-bit parallel data bus. This way, the DCMI module can receive image data from the camera and send it to a given RAM address through DMA module. The DCMI module can also crop the acquired image. Hence, this operation can be done while acquiring the image.

14.4.1 Working Principles of the DCMI Module

We can explain working principles of the DCMI module based on the definitions in Sect. 8.1 as follows. The DCMI module data can be one of 8, 10, 12, or 14 bits in parallel form. Data format here can be either monochrome or raw Bayer, YCbCr422, RGB565, or compressed jpeg form. The DCMI module uses NRZ data encoding. Voltage levels in operation may change based on the microcontroller and camera module used. Data transfer from the camera to microcontroller is in simplex mode. We can assume that there is a master slave setup in the DCMI module. The reason for this is as follows. The camera initializes the communication. It also generates all data, pixel clock (PIXCLK), and synchronization signals (VSYNC and HSYNC). Hence, the camera module acts as the master device. The microcontroller acts as the slave device. Baud rate for the DCMI module can reach up to 54 MB/s. Rising or falling edge of PIXCLK provides timing of the data to be read from the camera. The DCMI module triggers an interrupt for the following cases: when a new line starts, when a new frame starts, when the frame is grabbed completely, or when an error occurs.

We provide connection diagram between the DCMI and camera module in Fig. 14.6. As can be seen in this figure, there are n+1 data pins in which n can

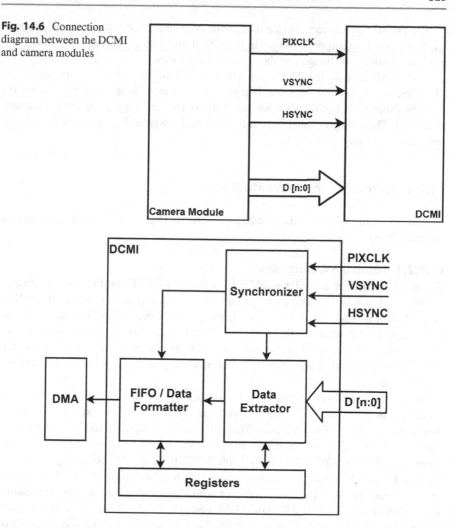

Fig. 14.6 Connection diagram between the DCMI and camera modules

Fig. 14.7 Block diagram of the DCMI module in the STM32F4 microcontroller

be 7, 9, 11, or 13. There are two synchronization pins and one clock pin. All signal directions are from the camera to the DCMI module.

We provide block diagram of the DCMI module in the STM32F4 microcontroller in Fig. 14.7. As can be seen in this figure, there are four main blocks as synchronizer, data extractor, FIFO and data formatter, and register. The synchronizer block captures edges of the clock and synchronization signals to control the data extractor and FIFO blocks. The data extractor block receives data from the pins and transfers it to the FIFO and data formatter block. Data is stored there. Then, it is transferred to a 32-bit data register. Data is transferred to the DMA module afterward. The register block consists of registers to control all mentioned blocks.

The DCMI module can run in either continuous or snapshot mode. In continuous mode, the module acquires images in a continuous manner. In snapshot mode, the module acquires one image and then stops its operation.

The DCMI module has 17 pins of which 3 of them are for synchronization and 14 of them are for data transfer. We provide connection of these pins to GPIO ports in Table 14.4. In this table, we listed the ports which are available via camera connector. There may be other pin configurations here as well. Please see the related references for this purpose [2].

14.4.2 Setting Up the DCMI Module

We should set up the DCMI module before using it. Therefore, we will consider this operation next.

14.4.2.1 Setup via C Language

We will benefit from STM32CubeMX to set up the DCMI module in C language. At this stage, we assume that a project has already been created in STM32CubeIDE as explained in Sect. 3.1.3. First, we should connect the OV7670 camera module pins to the DCMI module. To do so, we should select the mentioned pins and set their property with respect to the associated DCMI function as in Table 14.4. We should also physically connect the camera module pins to the STM32F4 board via this information.

Next, we should set up the DCMI module. To do so, open the left menu on STM32CubeMX, and select the DCMI mode as "Slave 8 bit external Synchro" in the Multimedia - DCMI menu. The following settings should be done in the "DCMI Mode and Configuration" window "Parameter Settings" tab. Set "Pixel clock polarity" to "Active on Rising Edge," "Vertical synchronization polarity" to "Active High," "Horizontal synchronization polarity" to "Active Low," "Frequency of frame capture" to "All frames are captured," and "JPEG mode" to "Disabled." The reader should press "Add" in the "DMA Settings" tab. Then, the "DMA request to the added line" should be set as "DCMI." Afterward, the line should be pressed, and the "DMA mode" should be set as "Circular" in the opening area. In the "DMA Settings" tab, "DCMI global interrupt" should be enabled in the NVIC settings.

14.4.2.2 Setup via C++ Language

The camera library introduced in Sect. 14.3.4 handles all adjustments in C++ language. Therefore, we do not need any extra setup here. However, we should physically connect the OV7670 camera module to the STM32F4 board as tabulated by Table 14.4.

14.4.2.3 Setup via MicroPython

The ov7670 module introduced in Sect. 14.3.4 handles all adjustments in MicroPython. Therefore, we do not need any extra setup here. However, we should physically connect the OV7670 camera module to the STM32F4 board as tabulated by Table 14.4.

14.5 Image Acquisition via Digital Camera

Acquiring an image via digital camera is straightforward once the DCMI and camera modules are configured and initialized. To be more specific, the DCMI module handles the rest of the image acquisition operation as it is initialized. We explore how this is done next. Then, we show how the acquired image can be displayed on LCD of the STM32F4 board. Afterward, we explore how the acquired image can be transferred to PC. Finally, we focus on format conversions applied to the acquired image.

14.5.1 Acquiring the Image

The DCMI module acquires image from the digital camera in pixel basis. Each pixel can be represented by one or more bytes based on the image format. Therefore, we should handle the acquisition operation accordingly. Next, we provide methods in C, C++, and MicroPython languages for this purpose.

14.5.1.1 Image Acquisition via C Language

We have four functions in our camera library in C language for image acquisition as follows. The function OV7670_captureSnapShot acquires one image from the camera in the image structure ImageTypeDef. The function OV7670_startContinuous performs the image acquisition operation in a continuous manner. Each acquired image is stored in the struct ImageTypeDef. The function OV7670_stopContinuous stops the continuous image acquisition operation. The function OV7670_pollForCapture adds a timeout value for the image acquisition operation. This function can be used to eliminate the infinite loop if the acquisition operation becomes unsuccessful in a given timeout value. We will show usage of these functions in Sects. 14.5.2 and 14.5.3 with display on LCD and data transfer to PC, respectively. We next provide detailed explanation of the mentioned functions.

```
void OV7670_captureSnapShot(ImageTypeDef *Pimg);
/*
Pimg: pointer to the ImageTypeDef struct
*/

void OV7670_startContinuous(ImageTypeDef *Pimg);

void OV7670_stopContinuous(void);

void OV7670_pollForCapture(uint32_t timeout);
/*
timeout: timeout duration
*/
```

14.5.1.2 Image Acquisition via C++ Language

We can benefit from our camera library to acquire images in C++ language. The functions in this library share the same name as their C counterpart. These functions are as follows. The function captureSnapShot acquires one image from the camera and stores it in the object created by the IMAGE class. The function startContinuous acquires images continuously from the camera and stores them in the object created by the IMAGE class. The function stopContinuous stops continuous image acquisition operation. The function pollForCapture adds a timeout value for the image acquisition operation. This function can be used to eliminate the infinite loop if the acquisition operation becomes unsuccessful in a given timeout value. To note here, these four functions belong to the OV7670 class. Hence, they should be used accordingly such as OV7670.captureSnapShot(&img). We will show usage of these functions in Sects. 14.5.2 and 14.5.3 with display on LCD and data transfer to PC, respectively. We next provide detailed information on the mentioned functions.

```
void captureSnapShot(IMAGE *Pimg);
/*
Pimg: pointer to the IMAGE object
*/

void startContinuous(IMAGE *Pimg);

void stopContinuous(void);

void pollForCapture(uint32_t timeout);
/*
timeout: timeout duration
*/
```

14.5.1.3 Image Acquisition via MicroPython

The functions in our ov7670 module share the same name as their C counterpart. These functions are as follows. The function captureSnapShot acquires one image from the camera and stores it in the image object. The function startContinuous acquires images continuously from the camera and stores it in the image object. The function stopContinuous stops the continuous image acquisition operation. The function pollForCapture adds a timeout value for the image acquisition operation. This function can be used to eliminate the infinite loop if the acquisition operation becomes unsuccessful in a given timeout value. To note here, these four functions belong to the ov7670 module. Hence, they should be used accordingly such as ov7670.captureSnapShot(img) where img is the created image object. We will show usage of these functions in Sects. 14.5.2 and 14.5.3 with display on LCD and data transfer to PC, respectively. We next provide detailed information on the mentioned functions.

```
captureSnapShot(img);
/*
img: created image object based on image_obj_t struct
*/
```

```
startContinuous(img);

stopContinuous();

pollForCapture(timeout);
/*
timeout: timeout duration in milliseconds
*/
```

14.5.2 Displaying the Acquired Image on LCD

As we acquire an image from the OV7670 camera module, we can display it on LCD of the STM32F4 board. We will benefit from the available functions in our camera and LCD libraries for this purpose. As a side note, the LCD library has been introduced in Chap. 11. We next provide sample codes in C, C++, and MicroPython languages for this purpose.

14.5.2.1 Display via C Language

We provide the C code to display an acquired image from the OV7670 camera module on LCD of the STM32F4 board in Listing 14.6. In this code, we first form the struct `imageTypeDef struct`. Then, it is initialized by the function `IMAGE_init`. Afterward, LCD and camera modules are initialized. Then, the camera runs in continuous mode, and acquired images are displayed on the LCD in an infinite loop.

Listing 14.6 Displaying the acquired image from the camera on the LCD of the STM32F4 board, the C code

```
/* USER CODE BEGIN Includes */
#include "image.h"
#include "ov7670.h"
#include "ili9341.h"
/* USER CODE END Includes */

/* USER CODE BEGIN PV */
ImageTypeDef img;
/* USER CODE END PV */

/* USER CODE BEGIN 2 */
IMAGE_init(&img, IMAGE_RES_QVGA, IMAGE_FORMAT_RGB565);
OV7670_init(IMAGE_RES_QVGA, IMAGE_FORMAT_RGB565);
ILI9341_init(SCREEN_HORIZONTAL);

OV7670_startContinuous(&img);
/* USER CODE END 2 */

/* Infinite loop */
/* USER CODE BEGIN WHILE */
while (1)
{
OV7670_pollForCapture(1000);
ILI9341_drawImage(&img, 0, 0);
/* USER CODE END WHILE */
```

```
/* USER CODE BEGIN 3 */
}
/* USER CODE END 3 */
```

To execute the C code in Listing 14.6, we should create a project first. Then, we should set the OV7670 camera, DCMI, and LCD modules as explained in Sects. 14.3.4, 14.4, and Chap. 11. Please note that the LCD will be set up to be used by SPI. As we debug and run the code, images with the QVGA RGB565 format will be acquired from the camera and displayed on the LCD continuously.

14.5.2.2 Display via C++ Language

We next provide an example to display the acquired image from the OV7670 camera module on the LCD of the STM32F4 board. This example is the same as its C version introduced in the previous section. We provide the C++ code for the example in Listing 14.7.

Listing 14.7 Displaying the acquired image from the camera on LCD of the STM32F4 board, the C++ code

```cpp
#include "mbed.h"

#include "image.hpp"
#include "ov7670.hpp"
#include "ili9341.hpp"

IMAGE img(IMAGE_RES_QVGA, IMAGE_FORMAT_RGB565);
OV7670 ov7670(IMAGE_RES_QVGA, IMAGE_FORMAT_RGB565);
ILI9341 ili9341(SCREEN_HORIZONTAL);

int main()
{
ov7670.startContinuous(&img);
while (true)
{
ov7670.pollForCapture(1000);
ili9341.drawImage(&img, 0, 0);
}
}
```

To execute the C++ code in Listing 14.7, we should create a new project first. Here, we do not need to make any adjustments. The camera library handles all these operations for us. However, we should set the OV7670 camera, DCMI, and LCD modules as explained in Sects. 14.3.4, 14.4, and Chap. 11. Please note that the LCD will be set up to be used by SPI. As we debug and run the code, images with the QVGA RGB565 format will be acquired from the camera and displayed on the LCD continuously.

14.5.2.3 Display via MicroPython

In this example, we display the acquired image from the OV7670 camera module on the LCD of the STM32F4 board as its C version. We provide the MicroPython code for the example in Listing 14.8. As we run the code, images with the QVGA RGB565 format will be acquired from the camera and displayed on the LCD continuously.

Listing 14.8 Displaying the acquired image from the camera on LCD of the STM32F4 board, the MicroPython code

```
import image
import lcdSPI
import ov7670

def main():
    img = image.init(0, 0, IMAGE_RES_QVGA, IMAGE_FORMAT_RGB565)
    lcdSPI.init(0)
    ov7670.reset()
    ov7670.init(IMAGE_RES_QVGA, IMAGE_FORMAT_RGB565)
    ov7670.startContinuous(img)
    while True:
        ov7670.pollForCapture(1000)
        lcdSPI.drawImage(img, 0, 0)

main()
```

14.5.3 Transferring the Acquired Image to PC

As we acquire an image from the camera, we can transfer it to PC for further processing. We will benefit from the available functions in our camera and image transfer libraries for this purpose. Next, we will provide ways of using this in C, C++, and MicroPython languages.

14.5.3.1 Image Transfer via C Language

We provide a sample code in Listing 14.9 to acquire an image from the OV7670 camera module and transfer it to PC. To run the code, we should form a project first. Then, we should set the UART module as explained in Sect. 8.2. Likewise, we should set the OV7670 camera and DCMI modules as explained in Sects. 14.3.4 and 14.4. We should also set the user button on the STM32F4 board as explained in Chap. 4.

Listing 14.9 Transferring the acquired image to PC, the C code

```
/* USER CODE BEGIN Includes */
#include "ov7670.h"
#include "image.h"
#include "serialImage.h"
/* USER CODE END Includes */

/* USER CODE BEGIN PV */
ImageTypeDef img;
/* USER CODE END PV */

/* USER CODE BEGIN 2 */
IMAGE_init(&img, IMAGE_RES_QVGA, IMAGE_FORMAT_RGB565);
OV7670_init(IMAGE_RES_QVGA, IMAGE_FORMAT_RGB565);
/* USER CODE END 2 */

/* Infinite loop */
/* USER CODE BEGIN WHILE */
while (1)
```

```
{
if (HAL_GPIO_ReadPin(GPIOA, GPIO_PIN_0) == GPIO_PIN_SET)
{
OV7670_captureSnapShot(&img);
OV7670_pollForCapture(1000);
SERIAL_imageSend(&img);
}
/* USER CODE END WHILE */

/* USER CODE BEGIN 3 */
}
/* USER CODE END 3 */
```

As the user button on the STM32F4 board is pressed, the OV7670 camera module starts working in snapshot mode and acquires one image. Then, this image is transferred to PC as explained in Sect. 14.2. On the PC side, all necessary settings should have been done beforehand, and the code in Listing 14.2 should have been running. The acquired image will be saved to the "capture" folder in PC.

14.5.3.2 Image Transfer via C++ Language

We next provide a sample code in C++ language to acquire an image from the OV7670 camera module and transfer it to PC. This example is the same as its C version introduced in the previous section. We provide the C++ code in Listing 14.10.

Listing 14.10 Transferring the acquired image to PC, the C++ code

```
#include "mbed.h"

#include "image.hpp"
#include "ov7670.hpp"
#include "serialImage.hpp"

IMAGE img(IMAGE_RES_QVGA, IMAGE_FORMAT_RGB565);
OV7670 ov7670(IMAGE_RES_QVGA, IMAGE_FORMAT_RGB565);
SERIALIMAGE serialImage;
DigitalIn userButton(BUTTON1);

int main()
{
while (true)
{
if (userButton)
{
ov7670.captureSnapShot(&img);
ov7670.pollForCapture(1000);
serialImage.send(&img);
}
}
}
```

To execute the code, we should create a new project. Here, we do not need to make any adjustments. The camera and image transfer libraries handle all these operations for us. As we debug and run the code, it waits for user response. As the user button on the STM32F4 board is pressed, the OV7670 camera module starts working in snapshot mode and acquires one image. Then, this image is transferred to PC as explained in Sect. 14.2. On the PC side, all necessary settings should have been done beforehand, and the code in Listing 14.2 should have been running. The acquired image will be saved to the "capture" folder there.

There is one important issue here. Mbed Studio opens the serial terminal while debugging the code. If the Python code in Listing 14.2 has not been executed on PC before the debug operation, Python gives error indicating that the port is in use. To avoid this error, either the user should run the image transfer code in Listing 14.2 beforehand or the terminal window there should be closed after debugging the code in Mbed Studio.

14.5.3.3 Image Transfer via MicroPython

We next provide a sample code in MicroPython to acquire an image from the OV7670 camera module and transfer it to PC as its C version. We provide the MicroPython code in Listing 14.11.

Listing 14.11 Transferring the acquired image to PC, the MicroPython code

```
import image
import serialImage
import ov7670
import pyb

def main():
    img = image.init(0, 0, IMAGE_RES_QVGA, IMAGE_FORMAT_RGB565)
    ov7670.reset()
    ov7670.init(IMAGE_RES_QVGA, IMAGE_FORMAT_RGB565)
    serialImage.init()
    userButton = pyb.Pin('PA0', mode=pyb.Pin.IN, pull=pyb.Pin.
        PULL_DOWN)
    while True:
        if userButton.value() == 1:
            ov7670.captureSnapShot(img)
            ov7670.pollForCapture(1000)
            serialImage.send(img)

main()
```

As we execute the code, it waits for the user response. As the user button on the STM32F4 board is pressed, the OV7670 camera module starts working in snapshot mode and acquires one image. Then, this image is transferred to PC as explained in Sect. 14.2. On the PC side, all necessary settings should have been done beforehand, and the code in Listing 14.2 should have been running. The acquired image will be saved to the "capture" folder there.

14.5.4 Format Conversions

As mentioned in Sect. 14.1.3, we can acquire images from the digital camera in different color formats. Besides, the LCD on the STM32F4 board has limited options to display images. Therefore, we will consider these formats and conversions between them in this section. To be more specific, we will consider the YUV/YCbCr422 and

RGB565 color formats for the OV7670 camera module. We will only consider the RGB565 color format for the LCD on the STM32F4 board.

In the RGB565 pixel format, one pixel is represented with 16 bits, distributed as 5 bits for red, 6 bits for green, and 5 bits for blue value. The YUV/YCbCr422 color format consists of luminance and two chrominance components as blue and red projections. In this format, two pixels are compressed into four bytes. The Cb/U and Cr/V components are shared between two consecutive pixels.

Since the camera feeds the YUV/YCbCr422 and RGB565 color formats, we did not consider conversion methods between them. Instead, we provided the RGB565 to grayscale and YUV/YCbCr422 to grayscale conversion. As mentioned before, the LCD displays color images in RGB565 format. Hence, we should represent the grayscale image as if it is in RGB565 format to show it on the LCD.

In order to convert an RGB565 color image to grayscale form, we can use the formula $Y = (uint8\_t)((77 * R + 150 * G + 29 * B) \gg 8)$. This is an approximation to the actual conversion $Y = 0.2989 * R + 0.5870 * G + 0.1140 * B$. We picked the former representation since it only uses integer numbers in operation. Hence, the conversion can be done faster. To note here, since the RGB565 color format has limited red, green, and blue values, the corresponding grayscale representation will also be limited in terms of levels. Instead, we can acquire the color image in YUV/YCbCr422 format from the camera. Then, we can use the Y component as the grayscale representation. This way, we do not perform any mathematical operations in obtaining the grayscale image. The implicit formula used here is $Y = 0.59*G + 0.31*R + 0.11*B$ which is almost the same as the one we used. The advantage of this method is that the camera acquires the red, green, and blue values in eight bits within itself. As a result, we can acquire grayscale values between 0 and 255.

We can represent a grayscale image in RGB565 color format to display it on LCD. To do so, we should perform two operations as $RGB565\_H = (Y \& 0xF8) \mid (Y \gg 5)$ and $RGB565\_L = (Y \& 0x1C) \ll 3) \mid (Y \gg 3)$ where Y stands for the grayscale image. Here, RGB565_H and RGB565_L represent the most and least significant eight bits of the RGB565 color data, respectively.

14.5.4.1 Format Conversion in C Language

We next provide functions to perform format conversions mentioned in the previous section in C language. These functions are available in our image processing library. We provide detailed explanation of these functions next.

```
void IMAGE_RGB565toGrayscale(ImageTypeDef *Pimgsrc, ImageTypeDef
    *Pimgdst);
/*
Pimgsrc: pointer to the source ImageTypeDef struct
Pimgdst: pointer to the destination ImageTypeDef struct
*/

void IMAGE_YUV422toGrayscale(ImageTypeDef *Pimgsrc, ImageTypeDef
    *Pimgdst);

void IMAGE_grayscaletoRGB565(ImageTypeDef *Pimgsrc, ImageTypeDef
    *Pimgdst);
```

We provide a sample code, given in Listing 14.12, to show the usage of the mentioned format conversion functions. Within the code, we first form three images with the struct `imageTypeDef`. Then, we initialize them by the function `IMAGE_init`. Afterward, we initialize the OV7670 camera module. Then, we first obtain the color image in YUV422 format, convert it to grayscale form, and send it to PC. Next, we reset the OV7670 camera and initialize it again. Before acquiring the image, we add a 1 s delay. Hence, the camera settings (such as auto white balance and auto exposure) can be done. We then acquire another color image in RGB565 format, convert it to grayscale form, and send it to PC. In order to run the code, we should make necessary settings as explained in Sect. 14.5.3.

Listing 14.12   Format conversion and data transfer to PC, the C code

```
/* USER CODE BEGIN Includes */
#include "ov7670.h"
#include "image.h"
#include "serialImage.h"
/* USER CODE END Includes */

/* USER CODE BEGIN PV */
ImageTypeDef img1;
ImageTypeDef img2;
ImageTypeDef img3;
/* USER CODE END PV */

/* USER CODE BEGIN 2 */
IMAGE_init(&img1, IMAGE_RES_QQVGA, IMAGE_FORMAT_YUV422);
IMAGE_init(&img2, IMAGE_RES_QQVGA, IMAGE_FORMAT_GRAYSCALE);
IMAGE_init(&img3, IMAGE_RES_QQVGA, IMAGE_FORMAT_RGB565);

OV7670_init(IMAGE_RES_QQVGA, IMAGE_FORMAT_YUV422);
HAL_Delay(1000);
OV7670_captureSnapShot(&img1);
OV7670_pollForCapture(1000);
IMAGE_YUV422toGrayscale(&img1, &img2); // convert YUV422 to
    Grayscale
SERIAL_imageSend(&img2);

OV7670_reset();

OV7670_init(IMAGE_RES_QQVGA, IMAGE_FORMAT_RGB565);
HAL_Delay(1000);
OV7670_captureSnapShot(&img3);
OV7670_pollForCapture(1000);
IMAGE_RGB565toGrayscale(&img3, &img2); // convert RGB565 to
    Grayscale
SERIAL_imageSend(&img2);
/* USER CODE END 2 */
```

We provide another example on color image format conversion with the C code given in Listing 14.13. As in the first example, we first form three images with the struct `imageTypeDef`. Then, we initialize them by the function `IMAGE_init`. Afterward, we initialize the OV7670 camera and LCD on the STM32F4 board. We set the camera to work in continuous mode. Therefore, we acquire color images in

YUV/YCbCr422 form in an infinite loop and convert them to grayscale form. Then, we display the acquired images on LCD continuously. In order to run the code, we should make necessary settings as explained in Sect. 14.5.3.

Listing 14.13 Format conversion and image display on LCD, the C code

```c
/* USER CODE BEGIN Includes */
#include "image.h"
#include "ov7670.h"
#include "ili9341.h"
/* USER CODE END Includes */

/* USER CODE BEGIN PV */
ImageTypeDef img3;
ImageTypeDef img1;
ImageTypeDef img2;
/* USER CODE END PV */

/* USER CODE BEGIN 2 */
IMAGE_init(&img1, IMAGE_RES_QQVGA, IMAGE_FORMAT_YUV422);
IMAGE_init(&img2, IMAGE_RES_QQVGA, IMAGE_FORMAT_GRAYSCALE);
IMAGE_init(&img3, IMAGE_RES_QQVGA, IMAGE_FORMAT_RGB565);

ILI9341_init(SCREEN_HORIZONTAL);
OV7670_init(IMAGE_RES_QQVGA, IMAGE_FORMAT_YUV422);
OV7670_startContinuous(&img1);
/* USER CODE END 2 */

/* Infinite loop */
/* USER CODE BEGIN WHILE */
while (1)
{
OV7670_pollForCapture(1000);
IMAGE_YUV422toGrayscale(&img1, &img2); // convert YUV422 to
    Grayscale
IMAGE_grayscaletoRGB565(&img2, &img3); // convert Grayscale to
    RGB565
ILI9341_drawImage(&img3, 0, 0);
/* USER CODE END WHILE */

/* USER CODE BEGIN 3 */
}
/* USER CODE END 3 */
```

14.5.4.2 Format Conversion in C++ Language

We next provide functions to perform format conversions mentioned in the previous section in C++ language. These functions are available in our image processing library. We provide the detailed explanation of these functions next.

```c
void IMAGE_RGB565toGrayscale(IMAGE Pimgsrc, IMAGE Pimgdst);
/*
Pimgsrc: pointer to the source IMAGE object
Pimgdst: pointer to the destination IMAGE object
*/

void IMAGE_YUV422toGrayscale(IMAGE Pimgsrc, IMAGE Pimgdst);

void IMAGE_grayscaletoRGB565(IMAGE Pimgsrc, IMAGE Pimgdst);
```

We provide a sample C++ code, given in Listing 14.14, on format conversion. This example is the same as in its C version introduced in the previous section. We should first open a new project to execute the code. We should make necessary settings as explained in Sect. 14.5.3. As we debug and run the code, two color images will be acquired in QQVGA YUV/YCbCr422 and RGB565 color formats. These are converted to grayscale form separately. Then, they are transferred to PC. On the PC side, all necessary settings should have been done beforehand, and the code in Listing 14.2 should have been running. The converted images will be saved to the "capture" folder in PC.

Listing 14.14 Format conversion and data transfer to PC, the C++ code

```
#include "mbed.h"

#include "image.hpp"
#include "ov7670.hpp"
#include "serialImage.hpp"

IMAGE img1(IMAGE_RES_QQVGA, IMAGE_FORMAT_YUV422);
IMAGE img2(IMAGE_RES_QQVGA, IMAGE_FORMAT_GRAYSCALE);
IMAGE img3(IMAGE_RES_QQVGA, IMAGE_FORMAT_RGB565);

OV7670 ov7670;
SERIALIMAGE serialImage;

int main()
{
ov7670.init(IMAGE_RES_QQVGA, IMAGE_FORMAT_YUV422);
thread_sleep_for(1000);
ov7670.captureSnapShot(&img1);
ov7670.pollForCapture(1000);
IMAGE_YUV422toGrayscale(&img1, &img2);
serialImage.send(&img2);

ov7670.reset();

ov7670.init(IMAGE_RES_QQVGA, IMAGE_FORMAT_RGB565);
thread_sleep_for(1000);
ov7670.captureSnapShot(&img3);
ov7670.pollForCapture(1000);
IMAGE_RGB565toGrayscale(&img3, &img2);
serialImage.send(&img2);

while (true);
}
```

We provide a second example in C++ language in Listing 14.15. This code performs the same operations as its C version in Listing 14.13. In the code, we form three images via the **IMAGE** class. Afterward, we initialize the OV7670 camera and LCD on the STM32F4 board. We set the camera to work in continuous mode. Therefore, we acquire color images in YUV/YCbCr422 format in an infinite loop, convert them to grayscale form, and display them on LCD continuously. We should first create a new project to execute the code. We should make necessary settings as explained in Sect. 14.5.3.

Listing 14.15 Format conversion and image display on LCD, the C++ code

```cpp
#include "mbed.h"

#include "ili9341.hpp"
#include "image.hpp"
#include "ov7670.hpp"

IMAGE img1(IMAGE_RES_QQVGA, IMAGE_FORMAT_YUV422);
IMAGE img2(IMAGE_RES_QQVGA, IMAGE_FORMAT_GRAYSCALE);
IMAGE img3(IMAGE_RES_QQVGA, IMAGE_FORMAT_RGB565);

OV7670 ov7670(IMAGE_RES_QQVGA, IMAGE_FORMAT_YUV422);
ILI9341 ili9341(SCREEN_HORIZONTAL);

int main()
{
ov7670.startContinuous(&img1);

while (true)
{
ov7670.pollForCapture(1000);
IMAGE_YUV422toGrayscale(&img1, &img2);
IMAGE_grayscaletoRGB565(&img2, &img3);
ili9341.drawImage(&img3, 0, 0);
}
}
```

14.5.4.3 Format Conversion in MicroPython

We have functions in our image module to perform format conversions in MicroPython. We provide the detailed explanation of these functions next.

```
IMAGE_RGB565toGrayscale(imgsrc, imgdst);
/*
imgsrc: created source image object based on image_obj_t struct
imgdst: created destination image object based on image_obj_t
    struct
*/

IMAGE_YUV422toGrayscale(imgsrc, imgdst);

IMAGE_grayscaletoRGB565(imgsrc, imgdst);
```

We provide a sample MicroPython code, given in Listing 14.16, on format conversion. This example is the same as its C version. As we run the code, two color images will be acquired in QQVGA YUV/YCbCr422 and RGB565 color formats. These are converted to grayscale form separately. Then, they are transferred to PC. On the PC side, all necessary settings should have been done beforehand, and the code in Listing 14.2 should have been running. The converted images will be saved to the "capture" folder in PC.

Listing 14.16 Format conversion and data transfer to PC, the MicroPython code

```python
import image
import serialImage
import ov7670
```

```
def main():
    img1 = image.init(0, 0, IMAGE_RES_QQVGA, IMAGE_FORMAT_YUV422)
    img2 = image.init(0, 0, IMAGE_RES_QQVGA,
        IMAGE_FORMAT_GRAYSCALE)
    img3 = image.init(0, 0, IMAGE_RES_QQVGA, IMAGE_FORMAT_RGB565)
    serialImage.init()
    ov7670.reset()
    ov7670.init(IMAGE_RES_QQVGA, IMAGE_FORMAT_YUV422)
    ov7670.captureSnapShot(img1)
    ov7670.pollForCapture(1000)
    image.IMAGE_YUV422toGrayscale(img1, img2)
    serialImage.send(img2)

    ov7670.reset()
    ov7670.init(IMAGE_RES_QQVGA, IMAGE_FORMAT_RGB565)
    ov7670.captureSnapShot(Qimg3)
    ov7670.pollForCapture(1000)
    image.IMAGE_RGB565toGrayscale(img3, img2)
    serialImage.send(img2)

main()
```

We provide the second example in MicroPython in Listing 14.17. This code performs the same operations as its C version in Listing 14.13. In the code, we form three images via image module. Afterward, we initialize the OV7670 camera and LCD on the STM32F4 board. We set the camera to work in continuous mode. Therefore, we acquire color images in YUV/YCbCr422 format in an infinite loop, convert them to grayscale form, and display them on the LCD continuously.

Listing 14.17 Format conversion and image display on LCD, the MicroPython code

```
import image
import lcdSPI
import ov7670

def main():
    img1 = image.init(0, 0, IMAGE_RES_QQVGA, IMAGE_FORMAT_YUV422)
    img2 = image.init(0, 0, IMAGE_RES_QQVGA,
        IMAGE_FORMAT_GRAYSCALE)
    img3 = image.init(0, 0, IMAGE_RES_QQVGA, IMAGE_FORMAT_RGB565)

    lcdSPI.init(0)
    ov7670.reset()
    ov7670.init(IMAGE_RES_QQVGA, IMAGE_FORMAT_YUV422)
    ov7670.startContinuous(img1)
    while True:
        ov7670.pollForCapture(1000)
        image.IMAGE_YUV422toGrayscale(img1, img2)
        image.IMAGE_grayscaletoRGB565(img2, img3)
        lcdSPI.drawImage(img3, 0, 0)

main()
```

14.6   Pixel-Based Digital Image Processing Operations

Acquiring an image from the digital camera and displaying it on LCD can be taken as the first phase of digital image processing. The second phase is processing the acquired image based on given requirements. The simplest method to process an image is applying pixel-based operations on it. The idea here is altering the pixel value by a transformation. To do so, we will start with obtaining negative of the image. Then, we will apply intensity transformations to the image. Finally, we will consider thresholding the image.

14.6.1 Obtaining the Negative Image

We can obtain negative of a grayscale image by subtracting each pixel value from 255 (assuming it to be an eight-bit image). To do so, we formed functions in C, C++, and MicroPython languages. We explore them next.

14.6.1.1 C Language

We formed the function IMAGE_grayscaleNegative to take the negative of a given grayscale image in C language. This function is available in our image processing library. Details of this function are given next.

```
void IMAGE_grayscaleNegative(ImageTypeDef *Pimgsrc, ImageTypeDef
    *Pimgdst);
/*
Pimgsrc: pointer to the source ImageTypeDef struct
Pimgdst: pointer to the destination ImageTypeDef struct
*/
```

We provide a sample code in Listing 14.18 on the usage of the function IMAGE_grayscaleNegative. This code works in the same manner as explained in Sect. 14.2 without using the LCD. To check how the code works, we can feed the Mandrill image from PC. As the code is run, we obtain the negated image. It is sent back to PC within the code. As a result, we should have our processed image as in Fig. 14.8.

Listing 14.18 Obtaining the negative image, the C code

```
/* USER CODE BEGIN Includes */
#include "image.h"
#include "serialImage.h"
/* USER CODE END Includes */

/* USER CODE BEGIN PV */
ImageTypeDef img;
/* USER CODE END PV */

/* USER CODE BEGIN 2 */
IMAGE_init(&img, IMAGE_RES_QVGA, IMAGE_FORMAT_GRAYSCALE);

SERIAL_imageCapture(&img);
IMAGE_grayscaleNegative(&img, &img);
```

Fig. 14.8 Negated grayscale Mandrill image

```
SERIAL_imageSend(&img);
/* USER CODE END 2 */
```

In order to run the code, we should set the UART mode only as explained in Sect. 14.2. We should also add the image processing and transfer libraries to our project by adding the files "serialImage.c" and "image.c" to the folder "project->Core->Src" and "serialImage.h" and "image.h" to the folder "project->Core->Inc."

14.6.1.2 C++ Language

We can also obtain negative of a grayscale image in C++ language. To do so, we can use the function IMAGE_grayscaleNegative available in our image processing library. Details of this function are given next.

```
void IMAGE_grayscaleNegative(IMAGE *Pimgsrc, IMAGE *Pimgdst);
/*
Pimgsrc: pointer to the source IMAGE object
Pimgdst: pointer to the destination IMAGE object
*/
```

We provide a sample code in Listing 14.19 to show working principles of the function IMAGE_grayscaleNegative. In order to execute this code, the reader should follow the steps given in Sect. 14.2 without adding the LCD. To check how the code works, we can feed the Mandrill image from PC. As the code is run, the processed image is turned back to PC. As a result, we should have the processed image as in Fig. 14.8.

Listing 14.19 Obtaining the negative image, the C++ code

```
#include "mbed.h"

#include "image.hpp"
#include "serialImage.hpp"

IMAGE img(IMAGE_RES_QVGA, IMAGE_FORMAT_GRAYSCALE);
```

```
SERIALIMAGE serialImage;

int main()
{
serialImage.capture(&img);
IMAGE_grayscaleNegative(&img, &img);
serialImage.send(&img);

while (true);
}
```

In order to run the code, we do not need any adjustments. The C++ image processing and transfer libraries handle all settings for us. We should only include them by adding the files "serialImage.cpp," "serialImage.hpp," "image.cpp," and "image.hpp" to the active project folder.

14.6.1.3 MicroPython

We can also obtain negative of a grayscale image in MicroPython. To do so, we can use the function IMAGE_grayscaleNegative available in our image module. Details of this function are given next.

```
IMAGE_grayscaleNegative(imgsrc, imgdst);
/*
imgsrc: created source image object based on image_obj_t struct
imgdst: created destination image object based on image_obj_t
    struct
*/
```

We provide a sample code in Listing 14.20 to show working principles of the function IMAGE_grayscaleNegative. In order to execute this code, the reader should follow the steps given in Sect. 14.2 without adding the LCD. To check how the code works, we can feed the Mandrill image from PC. As the code is run, the processed image is turned back to PC. As a result, we should have the processed image as in Fig. 14.8.

Listing 14.20 Obtaining the negative image, the MicroPython code

```
import image
import serialImage

def main():
    img = image.init(0, 0, IMAGE_RES_QVGA, IMAGE_FORMAT_GRAYSCALE
        )
    serialImage.init()
    serialImage.capture(img)
    image.IMAGE_grayscaleNegative(img, img)
    serialImage.send(img)

main()
```

14.6.2 Intensity Transformation Applied to the Image

Intensity transformation is generally used to increase or decrease brightness of an image. One way of doing this is by obtaining power of a pixel value. Therefore, this specific operation is also called power transformation. As we apply the transformation, we should normalize the transformed image to the range 0 and 255. Next, we provide functions to perform power transformations in C, C++, and MicroPython languages.

14.6.2.1 C Language

We formed the function IMAGE_grayscalePowerTransform to obtain the power transformation of a given image in C language. This function is available in our image processing library. Details of the function are given next.

```
void IMAGE_grayscalePowerTransform(ImageTypeDef *Pimgsrc,
    ImageTypeDef *Pimgdst, float_t pw);
/*
Pimgsrc: pointer to the source ImageTypeDef struct
Pimgdst: pointer to the destination ImageTypeDef struct
pw: power exponent
*/
```

The function IMAGE_grayscalePowerTransform gets the source image, takes the power of each pixel by the pw value, and returns the transformed image. The power function works slow in C language. Therefore, we initially formed a lookup table and used it in obtaining the power transformation in our function.

We provide a sample code in Listing 14.21 on the usage of the function IMAGE_grayscalePowerTransform. This code works in the same manner as explained in Sect. 14.6.1.1. In order to execute the code, we should feed the Mandrill image from PC. As the code is run, we obtain the power transformed image turned back to PC. The result should be as in Fig. 14.9.

Listing 14.21 Obtaining power transformation of the image, the C code

```
/* USER CODE BEGIN Includes */
#include "image.h"
#include "serialImage.h"
/* USER CODE END Includes */

/* USER CODE BEGIN PV */
ImageTypeDef img;
/* USER CODE END PV */

/* USER CODE BEGIN 2 */
IMAGE_init(&img, IMAGE_RES_QVGA, IMAGE_FORMAT_GRAYSCALE);

SERIAL_imageCapture(&img);
IMAGE_grayscalePowerTransform(&img, &img, 0.5);
SERIAL_imageSend(&img);
/* USER CODE END 2 */
```

Fig. 14.9 Power transformation applied to the grayscale Mandrill image

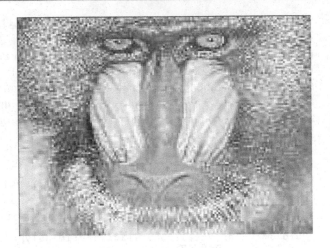

14.6.2.2 C++ Language

We can also obtain power transform of a grayscale image in C++ language. To do so, we formed the function IMAGE_grayscalePowerTransform in our image processing library. Details of this function are given next.

```
void IMAGE_grayscalePowerTransform(IMAGE *Pimgsrc, IMAGE *Pimgdst
    , float_t pw);
/*
Pimgsrc: pointer to the source IMAGE object
Pimgdst: pointer to the destination IMAGE object
pw: power exponent
*/
```

The function IMAGE_grayscalePowerTransform works similar to its C counterpart. We provide a sample code in Listing 14.22 to show how the power transformation works. To execute this code, the reader should follow the steps given in Sect. 14.6.1.2. To check how the code works, we can feed the Mandrill image from PC. As the code is run, we should obtain the transformed image sent back to PC. As a result, we should have the resultant image as in Fig. 14.9.

Listing 14.22 Obtaining power transformation of the image, the C++ code

```
#include "mbed.h"

#include "image.hpp"
#include "serialImage.hpp"

IMAGE img(IMAGE_RES_QVGA, IMAGE_FORMAT_GRAYSCALE);

SERIALIMAGE serialImage;

int main()
{
serialImage.capture(&img);
IMAGE_grayscalePowerTransform(&img, &img, 0.5);
serialImage.send(&img);
```

```
while (true);
}
```

14.6.2.3 MicroPython

We can also obtain the power transform of a grayscale image in MicroPython. To do so, we formed the function IMAGE_grayscalePowerTransform in our image processing library. Details of this function are given next.

```
IMAGE_grayscalePowerTransform(imgsrc, imgdst, pw);
/*
imgsrc: created source image object based on image_obj_t struct
imgdst: created destination image object based on image_obj_t
    struct
pw: power exponent
*/
```

The function IMAGE_grayscalePowerTransform works similar to its C counterpart. We provide a sample code in Listing 14.23 to show how the power transformation works. To execute this code, the reader should follow the steps given in Sect. 14.6.1.3. To check how the code works, we can feed the Mandrill image from PC. As the code is run, we should obtain the transformed image sent back to PC. As a result, we should have the image as in Fig. 14.9.

Listing 14.23 Obtaining power transformation of the image, the MicroPython code

```
import image
import serialImage

def main():
    img = image.init(0, 0, IMAGE_RES_QVGA, IMAGE_FORMAT_GRAYSCALE
        )
    serialImage.init()
    serialImage.capture(img)
    image.IMAGE_grayscalePowerTransform(img, img, 0.5)
    serialImage.send(img)

main()
```

14.6.3 Thresholding the Image

Thresholding corresponds to transforming a grayscale image to binary form. This operation may be used to emphasize an object of interest in the image. Next, we provide functions to threshold a given grayscale image in C, C++, and MicroPython languages.

14.6.3.1 C Language

We formed the function IMAGE_grayscaleThreshold to threshold a given grayscale image in C language. This function is available in our image processing library. Details of this function are given next.

Fig. 14.10 Thresholding applied to the grayscale Mandrill image

```
void IMAGE_grayscaleThreshold(ImageTypeDef *Pimgsrc,
    ImageTypeDef *Pimgdst, uint8_t threshold);
/*
Pimgsrc: pointer to the source ImageTypeDef struct
Pimgdst: pointer to the destination ImageTypeDef struct
threshold: threshold level
*/
```

The function `IMAGE_grayscaleThreshold` takes the source image, thresholds it, and returns the thresholded image. Here, the thresholded image is represented by two values as 0 (black) and 255 (white).

We provide a sample code in Listing 14.24 on the usage of the function `IMAGE_grayscaleThreshold`. This code works in the same manner as explained in Sect. 14.6.1.1. In order to execute the code, we can provide the Mandrill image from PC. As the code is run, it turns back the thresholded image to PC. We should obtain the thresholded image as in Fig. 14.10.

Listing 14.24 Obtaining the thresholded image, the C code

```
/* USER CODE BEGIN Includes */
#include "image.h"
#include "serialImage.h"
/* USER CODE END Includes */

/* USER CODE BEGIN PV */
ImageTypeDef img;
/* USER CODE END PV */

/* USER CODE BEGIN 2 */
IMAGE_init(&img, IMAGE_RES_QVGA, IMAGE_FORMAT_GRAYSCALE);

SERIAL_imageCapture(&img);
IMAGE_grayscaleThreshold(&img, &img, 128);
SERIAL_imageSend(&img);
/* USER CODE END 2 */
```

14.6.3.2 C++ Language

We can also obtain thresholded version of a grayscale image in C++ language. To do so, we formed the function IMAGE_grayscaleThreshold in our image processing library. Details of this function are given next.

```
void IMAGE_grayscaleThreshold(IMAGE *Pimgsrc, IMAGE *Pimgdst,
    uint8_t threshold);
/*
Pimgsrc: pointer to the source IMAGE object
Pimgdst: pointer to the destination IMAGE object
threshold: threshold level
*/
```

The function IMAGE_grayscaleThreshold works similar to its C counterpart. We provide a sample code in Listing 14.25. To execute this code, the reader should follow the steps given in Sect. 14.6.1.2. To check how the code works, we can feed the Mandrill image from PC. As the code is run, we should have the resultant image as in Fig. 14.10.

Listing 14.25 Obtaining the thresholded image, the C++ code

```
#include "mbed.h"

#include "image.hpp"
#include "serialImage.hpp"

IMAGE img(IMAGE_RES_QVGA, IMAGE_FORMAT_GRAYSCALE);

SERIALIMAGE serialImage;

int main()
{
serialImage.capture(&img);
IMAGE_grayscaleThreshold(&img, &img, 128);
serialImage.send(&img);

while (true);
}
```

14.6.3.3 MicroPython

We can also obtain thresholded version of a grayscale image in MicroPython. To do so, we formed the function IMAGE_grayscaleThreshold in our image processing library. Details of this function are given next.

```
IMAGE_grayscaleThreshold(imgsrc, imgdst, threshold);
/*
imgsrc: created source image object based on image_obj_t struct
imgdst: created destination image object based on image_obj_t
    struct
threshold: threshold level
*/
```

The function IMAGE_grayscaleThreshold works similar to its C counterpart. We provide a sample code in Listing 14.26. To execute this code, the reader should follow the steps given in Sect. 14.6.1.3. To check how the code works, we can feed

the Mandrill image from PC. As the code is run, we should have the resultant image
as in Fig. 14.10.

Listing 14.26 Obtaining the thresholded image, the MicroPython code

```
import image
import serialImage

def main():
    img = image.init(0, 0, IMAGE_RES_QVGA, IMAGE_FORMAT_GRAYSCALE
        )
    serialImage.init()
    serialImage.capture(img)
    image.IMAGE_grayscaleThreshold(img, img, 128)
    serialImage.send(img)

main()
```

14.7   Summary of the Chapter

Embedded digital image processing systems are becoming popular in our daily life.
Besides, digital image processing methods are used in the related computer vision
and machine learning systems (especially deep learning-based ones). In other words,
an embedded computer vision or machine learning system requires implementing
digital image processing methods. These observations led us to add a chapter on this
topic to the book. Therefore, we started with the fundamental definitions in digital
image representation on Arm® Cortex™-M microcontrollers. Then, we explained
image transfer between the PC and STM32F4 microcontroller. We next focused on
digital cameras and their usage. Then, we considered the digital camera interface
module available in the STM32F4 microcontroller. In the final part of the chapter,
we introduced pixel-based digital image processing methods. As in the previous
chapters on digital signal processing and digital control, the reader should take this
chapter as a brief introduction to this important topic. Therefore, we suggest the
reader to consult more advanced books on digital image processing to grasp this
important topic better.

Problems

14.1. Transfer an image of your choice from PC to the STM32F4 microcontroller.
Display the transferred image on the LCD of the STM32F4 board. Form the project
using:

(a) C language under STM32CubeIDE
(b) C++ language under Mbed
(c) MicroPython

14.2. Summarize working principles of a digital camera starting from optics to feeding image data to output.

14.3. What should we understand from the image sensor?

14.4. Why is the preprocessing step needed in digital camera working steps?

14.5. How is the camera interface module used in operation?

14.6. Explain the timing and synchronization step as well as the clock signals available in a generic digital camera.

14.7. In how many different ways image data can be represented in a digital camera?

14.8. Summarize properties of the OV7670 camera module based on the definitions in Sect. 14.3.1.

14.9. Explain working principles of the DCMI module in the STM32F4 microcontroller.

14.10. Set up the OV7670 camera and DCMI modules and take your selfie. Display the image on the LCD of the STM32F4 board. Transfer the image to PC and display it there via Python code. Form the project using:

(a) C language under STM32CubeIDE
(b) C++ language under Mbed
(c) MicroPython

14.11. Form a project to apply format conversion to the image in Problem 14.10. Hence, its grayscale version can be displayed on the LCD of the STM32F4 board. Form the project using:

(a) C language under STM32CubeIDE
(b) C++ language under Mbed
(c) MicroPython

14.12. Form a project to obtain negative version of your grayscale selfie image. Display the negated image on the LCD of the STM32F4 board. Form the project using:

(a) C language under STM32CubeIDE
(b) C++ language under Mbed
(c) MicroPython

14.13. Form a project to apply intensity transformation to your grayscale selfie image such that the transformed image becomes brighter. Display the transformed image on the LCD of the STM32F4 board. Form the project using:

(a) C language under STM32CubeIDE
(b) C++ language under Mbed
(c) MicroPython

14.14. Form a project to threshold the grayscale version of your selfie image. Use Otsu's method for finding the threshold value. Display the thresholded image on the LCD of the STM32F4 board. Form the project using:

(a) C language under STM32CubeIDE
(b) C++ language under Mbed
(c) MicroPython

References

1. MIPI Alliance: https://www.mipi.org/current-specifications. Accessed 4 June 2021
2. STMicroelectronics: Digital Camera Interface (DCMI) for STM32 MCUs, an5020 edn. (2017)

Advanced Topics

15

15.1 Assembly Language Programming

Although assembly language is an important topic, we postponed it till this chapter for two main reasons. First, this book focuses more on high-level programming languages (compared to assembly language). Second, the Arm® Cortex™-M4 instruction set is fairly complex and requires detailed treatment to handle it appropriately. On the other hand, an embedded programmer should at least get a feeling on assembly language programming. Therefore, we will start with the requirements to form a complete assembly code in this section. Afterward, we will provide a method to form an STM32CubeIDE project to execute the code. Finally, we will consider inline assembly usage in C, C++, and MicroPython languages.

15.1.1 Forming a Complete Assembly Code

The instruction set is necessary but not sufficient to form an assembly code to be executed on the microcontroller. The reason is as follows. The compiler (introduced in Chap. 3) is responsible for all operations related to forming a complete binary file to be embedded on the microcontroller flash memory when the code is written in C or C++ language. The programmer is responsible for these tasks while forming the code in assembly language. Assembler directives are used for this purpose. The list of basic directives, with their brief description, to be used in this book is as follows:

Supplementary Information Information The online version contains supplementary material available at (https://doi.org/10.1007/978-3-030-88439-0_15).

© Springer Nature Switzerland AG 2022
C. Ünsalan et al., *Embedded System Design with Arm Cortex-M Microcontrollers*,
https://doi.org/10.1007/978-3-030-88439-0_15

- **.text** assembles the following code into text (executable code) section.
- **.align** defines the length of instructions to be processed in terms of bytes.
- **.global** defines one or more global symbols.
- **.end** identifies end of the code.

We provide a template assembly code using the mentioned directives in Listing 15.1. To note here, this structure is specific to STM32CubeIDE. There may be extra directives or a different format for other development environments. Therefore, the reader should check them before preparing his or her assembly code in other IDEs.

Listing 15.1 Template assembly code

```
.text
.align 2
.global main

main:

// Assembly Code Goes Here

END: B END
.end
```

Let's consider the code in Listing 15.1. The first directive `.text` indicates that the following lines correspond to assembly code. Hence, they should be placed in the code section of the flash memory. The second directive `.align` defines the length of instructions to be processed in terms of bytes. If the instructions to be executed have 16-bit length, then we should use `.align 2` as in our case. The third directive `.global` defines the global variable `main` which indicates the starting address of our code. Finally, the directive `.end` indicates the end of the assembly code.

15.1.2 Creating an Assembly Project in STM32CubeIDE

There is no direct method to create an assembly project in STM32CubeIDE. Therefore, we should first follow the steps to form an empty C project as explained in Sect. 3.1.3. To create the debug file (with extension ".debug"), we should debug our (empty) C code for the first time. As the mentioned file is created, we should delete the "main.c" file from the "Src" folder of the project. In fact, we can delete all files with extension ".c" from the project's "Src" folder. Then, we should add a new file with extension ".s" to this folder. To do so, right-click on the "Src" folder, and press "New" and "File." In the opening window, enter the "File Name" such as "main.s" and click "Finish." An empty file will be created under the "Src" folder. We can add our assembly code to this file. Here, we will use the sample code given in Listing 15.2.

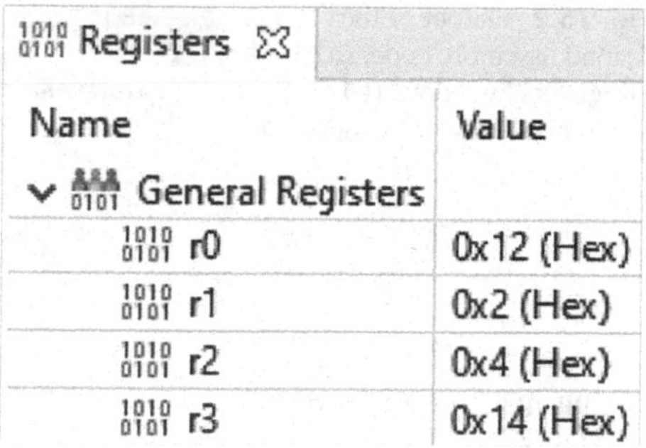

Fig. 15.1 The "Registers" window after executing the first assembly code

1010 0101 Registers ⊠	
Name	**Value**
∨ 🔢 General Registers	
1010 0101 r0	0x12 (Hex)
1010 0101 r1	0x2 (Hex)
1010 0101 r2	0x4 (Hex)
1010 0101 r3	0x14 (Hex)

Listing 15.2 The first assembly code for the STM32F4 microcontroller

```
.text
.align 2
.global main

main:

 MOVS R0, #0x12
 MOVS R1, #0x2
 MOVS R2, #0x4
 ADD R3, R0, R1

END: B END
 .end
```

The assembly code in Listing 15.2 modifies the register content and applies the addition operation on them. As we compile the code, we can execute it either by the "step into" option in STM32CubeIDE or directly by the "Resume" command. We suggest using the former option to observe all execution steps. We can observe output of the code by inspecting the "Registers" window as in Fig. 15.1. As can be seen in this figure, the registers have been modified according to our assembly code.

We can apply operations on the memory of the STM32F4 microcontroller by assembly language. We provide one such example in Listing 15.3. In this code, we modify and reach content of the memory address 0x20000020 by the LDR and STR instructions. As we compile and run the code, we can observe its output in the "Registers" and "Memory Browser" windows as in Fig. 15.2.

Listing 15.3 The second assembly code for the STM32F4 microcontroller

```
.text
.align 2
.global main

Mem_Addr=0x20000020

main:

 MOVS R0, #0x09
 MOVS R1, #0x11
 MOVS R2, #10
```

Fig. 15.2 Output of the second assembly code. (**a**) "Registers" window. (**b**) "Memory Browser" window

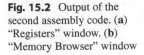
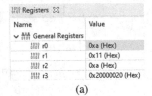
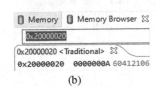

(a)

(b)

```
    LDR R3, =Mem_Addr
    STR R2, [R3]
    LDR R0, [R3]

END: B END
    .end
```

15.1.3 Inline Assembly in C Language

Assembly language commands can also be added to the C code. This is called inline assembly. If the user wants to execute single line of assembly code, then this can be done by the command __asm(). The same structure can also be used to execute more than one line of assembly code. Here, each line should be separated by the new line command. We provide usage examples for both cases in Listing 15.4. This code modifies the register content. Hence, the reader can observe its output from the "Registers" window in STM32CubeIDE.

Listing 15.4 The first inline assembly usage for the STM32F4 microcontroller, the C code

```
int main(void)
{
// Single line of code
//       __asm(" MOVS R0, #0x12");

// Multi code lines
__asm("MOVS R0, #0x12 \n"
      "MOVS R1, #0x2 \n"
      "MOVS R2, #0x4 \n"
      "ADD R3, R0, R1");

while(1);
}
```

We can effectively use inline assembly and directly reach memory content in C language. We provide one such example in Listing 15.5. Here, we store two numbers to two successive memory locations via directly reaching memory content in C language. Then, we add the memory content and store it to another memory address in inline assembly. Finally, we assign this memory content to the variable a. The reader can observe working steps of this code from the "Memory Browser" and "Expressions" window in STM32CubeIDE.

Listing 15.5 The second inline assembly usage for the STM32F4 microcontroller, the C code

```c
#define memLoc1 (*(unsigned long *)0x20000020)
#define memLoc2 (*(unsigned long *)0x20000024)
#define memLoc3 (*(unsigned long *)0x20000028)

unsigned int a;

int main(void)
{
memLoc1 = 0x3;
memLoc2 = 0x4;

__asm(" LDR R7, =0x20000020 \n"
" LDR R0, [R7] \n"
" ADD R7, R7, 0x4 \n"
" LDR R1, [R7] \n"
" ADD R2, R1, R0 \n"
" ADD R7, R7, 0x4 \n"
" STR R2, [R7]");

a = memLoc3;

for (;;);
}
```

15.1.4 Inline Assembly in C++ Language

We can also benefit from inline assembly in C++ language via Mbed Studio. We modify the C code in Listing 15.5 for this purpose and provide the corresponding C++ code in Listing 15.6. This code performs the same operations as its C counterpart. The reader can observe its output from the "Output" window in Mbed Studio.

Listing 15.6 Inline assembly usage for the STM32F4 microcontroller, the C++ code

```cpp
#include "mbed.h"

#define WAIT_TIME_MS 500

#define memLoc1 (*(unsigned long *)0x20000020)
#define memLoc2 (*(unsigned long *)0x20000024)

unsigned int *memLoc;

unsigned int a = 0;

int main()
{

memLoc1 = 0x3;
memLoc2 = 0x4;

memLoc = (unsigned int *)0x20000020;

printf("memory location value is %X\n", memLoc);
printf("content of the memory location is %X\n", *memLoc);
```

```
printf("\n");

memLoc++;

printf("memory location value is %X\n", memLoc);
printf("content of the memory location is %X\n", *memLoc);
printf("\n");

__asm("LDR R7, =0x20000020 \n"
"LDR R0, [R7] \n"
"ADD R7, R7, 0x4 \n"
"LDR R1, [R7] \n"
"ADD R2, R1, R0 \n"
"ADD R7, R7, 0x4 \n"
"STR R2, [R7]");
memLoc++;

printf("memory location value is %X\n", memLoc);
printf("content of the memory location is %X\n", *memLoc);

a = *memLoc;

printf("a value is %X\n", a);

while (1)
{
thread_sleep_for(WAIT_TIME_MS);
}
}
```

15.1.5 Inline Assembly in MicroPython

We can use assembly language commands under MicroPython. To do so, we should add the line @micropython.asm_thumb to our code. Then, we can define an inline assembly function below this line. Each defined function should be composed of assembly language instructions. To note here, these are not the exact Arm® assembly language instructions. For more information on them, please see [1, 2].

We next provide an example on the usage of inline assembly commands in MicroPython on the STM32F4 microcontroller. We provide the Python code in Listing 15.7 for this purpose. As can be seen in this code, the red LED on the STM32F4 board is toggled every second via assembly commands.

Listing 15.7 Toggling the red LED on the STM32F4 board every second via inline assembly, the MicroPython code

```
import pyb

@micropython.asm_thumb
def led_on():
    movwt(r0, stm.GPIOG)
    movw(r1, 1)
    lsl(r1, r1, 13)
    strh(r1, [r0, stm.GPIO_BSRRL])

@micropython.asm_thumb
def led_off():
```

```
        movwt(r0, stm.GPIOG)
        movw(r1, 1)
        lsl(r1, r1, 13)
        strh(r1, [r0, stm.GPIO_BSRRH])
def main():
    while True:
        led_on()
        pyb_delay(2000)
        led_off()
        pyb_delay(2000)

main()
```

15.2   Customizing the MicroPython Firmware

As explained in Sect. 9.3, some applications may require modifying the MicroPython firmware to effectively use RAM or speed up code execution. We will consider the methods used for this purpose in this section. To do so, we will start with the necessary settings. Then, we will consider precompiling scripts, frozen bytecode formation, and adding C functions to the MicroPython firmware.

15.2.1 Necessary Settings to Modify the MicroPython Firmware

In this section, we will explain the steps needed to modify the MicroPython firmware under Ubuntu. Therefore, the reader needs either a PC with Ubuntu on it or a Windows OS with Ubuntu virtual machine running on it. As the Ubuntu environment is ready, we should run the commands given below sequentially to compile the MicroPython firmware properly:

- "sudo apt-get install build-essential libreadline-dev libffi-dev git pkg-config gcc-arm-none-eabi libnewlib-arm-none-eabi"—to install necessary packages for compiling the MicroPython firmware.
- "sudo apt-get install git"—to install git.
- "git clone –recurse-submodules https://github.com/micropython/micropython.git"—to clone the MicroPython repository to the computer.
- "cd micropython"—to go to the MicroPython folder.
- "make -C mpy-cross"—to compile the code via mpy-cross compiler. After the compilation is done, mpy-cross executable can be found in the "mpy-cross" folder.
- "cd ports/stm"—to go to the "stm32" folder.
- "make BOARD=STM32F429DISC"—to compile the MicroPython firmware for the STM32F4 board.

As all these commands are executed, the MicroPython firmware for the STM32F4 board will be ready and can be found under the folder "ports/stm32/build-STM32F429DISC" as "firmware.dfu," "firmware.elf," and "firmware.hex" files. The reader should follow the steps in Chap. 3 to download the firmware to the STM32F4 board.

15.2.2 Precompiling Scripts Usage

The reader can use the MicroPython code by creating his or her own ".py" files and uploading them to the microcontroller having MicroPython firmware on it. The uploaded files are converted to ".mpy" files (consisting of bytecode data) and kept in the microcontroller RAM. We can decrease the RAM requirement in this method by precompiling the MicroPython code and loading it directly to the microcontroller RAM.

We will explain how to precompile a custom module into MicroPython firmware next. Therefore, create a file named "mymodule.py," and add the code lines given below to this file:

```
def add(a,b):
    return a + b
```

Then, go to the folder "mpy-cross" with the command cd "micropython/mpy-cross." Afterward, run the command "./mpy-cross FileDirectory/mymodule.py." As a result, the file "mymodule.mpy" will be created under the same folder. Upload this file to the MicroPython file system. When the commands "import mymodule" and "mymodule.add(1,2)" are run in the REPL window, the reader can observe the result as 3.

15.2.3 Frozen Bytecode Usage

The reader can also use the MicroPython code by compiling it along with the MicroPython firmware. Here, the user modules are kept in flash memory besides the firmware. Hence, RAM will not be used during the operation. In order to apply this process, we should first create a folder to keep the files to be frozen under the folder "micropython/drivers." Therefore, we should run the command "mkdir /micropython/drivers/mymodules." We should next copy the "mymodule.py" file (introduced in the previous section) to the created folder by running the command "cp mymodule.py /micropython/drivers/mymodules." Finally, we should go to the folder "micropython/ports/stm32/boards/" and open the read-only file "manifest.py" with the command "gedit manifest.py." This is the file that shows the files to be frozen into the MicroPython firmware. Add the line "freeze("$(MPY_DIR)/drivers/mymodules", "mymodule.py") at the bottom of this file. Then, save and close it.

Now, we are ready to compile the MicroPython firmware with the module we have added. Therefore, go to the folder "ports/stm32" with the command "cd ports/stm32," and run the command "make BOARD=STM32F429DISC" there. If the MicroPython firmware has been compiled before, only the made change will be added to the firmware. Otherwise, the overall compilation process will be done. Afterward, the MicroPython firmware for the STM32F4 board will be ready. The reader should follow the steps in Chap. 3 to download it to the STM32F4 board. When the commands "import mymodule" and "mymodule.add(1,2)" are run in the REPL window, the reader can observe the result as 3.

15.2.4 Adding C Functions to MicroPython

We can also add custom C functions to the MicroPython firmware. This process has been explained in detail in https://docs.micropython.org/en/latest/develop/cmodules.html. We kindly ask the reader to visit this web site for the necessary steps to be followed.

15.3 Mbed Simulator

Arm® provided an online tool to simulate Mbed. Hence, Mbed can be used without an actual board. The web site for the Mbed simulator is at https://simulator.mbed.com/. There are several available demonstration codes in this web site. Hence, the reader can test his or her ideas by taking one of these codes and modifying it accordingly. We also provided extra codes to work on the Mbed simulator on the topics explained in this book. The reader can reach them in a separate folder under the accompanying book web site.

Mbed simulator can also be used offline. To do so, the reader should construct the simulator on PC. This can be done by following the steps explained in https://os.mbed.com/blog/entry/introducing-mbed-simulator/. We strongly suggest the reader to benefit from Mbed simulator either in online or offline form.

References

1. http://docs.micropython.org/en/v1.9.3/pyboard/pyboard/tutorial/assembler.html. Accessed 4 June 2021
2. https://docs.micropython.org/en/latest/reference/asm_thumb2_index.html. Accessed 4 June 2021

Index

© Springer Nature Switzerland AG 2022
C. Ünsalan et al., *Embedded System Design with Arm Cortex-M Microcontrollers*,
https://doi.org/10.1007/978-3-030-88439-0

Printed in the United States
by Baker & Taylor Publisher Services